Machine Learning for Speaker Recognition

Understand fundamental and advanced statistical models and deep learning models for robust speaker recognition and domain adaptation. This useful toolkit enables you to apply machine learning techniques to address practical issues, such as robustness under adverse acoustic environments and domain mismatch, when deploying speaker recognition systems. Presenting state-of-the-art machine learning techniques for speaker recognition and featuring a range of probabilistic models, learning algorithms, case studies, and new trends and directions for speaker recognition based on modern machine learning and deep learning, this is the perfect resource for graduates, researchers, practitioners, and engineers in electrical engineering, computer science, and applied mathematics.

Man-Wai Mak is an associate professor in the Department of Electronic and Information Engineering at The Hong Kong Polytechnic University.

Jen-Tzung Chien is the chair professor at the Department of Electrical and Computer Engineering at the National Chiao Tung University. He has published extensively, including the book *Bayesian Speech and Language Processing* (Cambridge University Press). He is currently serving as an elected member of the IEEE Machine Learning for Signal Processing (MLSP) Technical Committee.

Machine Learning for Speaker Recognition

MAN-WAI MAK
The Hong Kong Polytechnic University

JEN-TZUNG CHIEN
National Chiao Tung University

CAMBRIDGE
UNIVERSITY PRESS

CAMBRIDGE
UNIVERSITY PRESS

University Printing House, Cambridge CB2 8BS, United Kingdom

One Liberty Plaza, 20th Floor, New York, NY 10006, USA

477 Williamstown Road, Port Melbourne, VIC 3207, Australia

314–321, 3rd Floor, Plot 3, Splendor Forum, Jasola District Centre, New Delhi – 110025, India

79 Anson Road, #06–04/06, Singapore 079906

Cambridge University Press is part of the University of Cambridge.

It furthers the University's mission by disseminating knowledge in the pursuit of education, learning, and research at the highest international levels of excellence.

www.cambridge.org
Information on this title: www.cambridge.org/9781108428125
DOI: 10.1017/9781108552332

First published 2021

Printed in the United Kingdom by TJ Books Limited, Padstow Cornwall

A catalogue record for this publication is available from the British Library.

Library of Congress Cataloging-in-Publication Data
Names: Mak, M. W., author. | Chien, Jen-Tzung, author.
Title: Machine learning for speaker recognition / Man-Wai Mak, Hong Kong Polytechnic University, Jen-Tzung Chien, National Chiao Tung University.
Description: New York : Cambridge University Press, 2021. | Includes bibliographical references and index.
Identifiers: LCCN 2019051262 (print) | LCCN 2019051263 (ebook) | ISBN 9781108428125 (hardback) | ISBN 9781108552332 (epub)
Subjects: LCSH: Automatic speech recognition. | Biometric identification. | Machine learning.
Classification: LCC TK7882.S65 M35 2021 (print) | LCC TK7882.S65 (ebook) | DDC 006.4/54–dc23
LC record available at https://lccn.loc.gov/2019051262
LC ebook record available at https://lccn.loc.gov/2019051263

ISBN 978-1-108-42812-5 Hardback

Additional resources for this publication at www.cambridge.org/makchien.

Contents

Preface

In the last 10 years, many methods have been developed and deployed for real-world biometric applications and multimedia information systems. Machine learning has been playing a crucial role in these applications where the model parameters could be learned and the system performance could be optimized. As for speaker recognition, researchers and engineers have been attempting to tackle the most difficult challenges: noise robustness and domain mismatch. These efforts have now been fruitful, leading to commercial products starting to emerge, e.g., voice authentication for e-banking and speaker identification in smart speakers.

Research in speaker recognition has traditionally been focused on signal processing (for extracting the most relevant and robust features) and machine learning (for classifying the features). Recently, we have witnessed the shift in the focus from signal processing to machine learning. In particular, many studies have shown that model adaptation can address both robustness and domain mismatch. As for robust feature extraction, recent studies also demonstrate that deep learning and feature learning can be a great alternative to traditional signal processing algorithms.

This book has two perspectives: machine learning and speaker recognition. The machine learning perspective gives readers insights on what makes state-of-the-art systems perform so well. The speaker recognition perspective enables readers to apply machine learning techniques to address practical issues (e.g., robustness under adverse acoustic environments and domain mismatch) when deploying speaker recognition systems. The theories and practices of speaker recognition are tightly connected in the book.

This book covers different components in speaker recognition including front-end feature extraction, back-end modeling, and scoring. A range of learning models are detailed, from Gaussian mixture models, support vector machines, joint factor analysis, and probabilistic linear discriminant analysis (PLDA) to deep neural networks (DNN). The book also covers various learning algorithms, from Bayesian learning, unsupervised learning, discriminative learning, transfer learning, manifold learning, and adversarial learning to deep learning. A series of case studies and modern models based on PLDA and DNN are addressed. In particular, different variants of deep models and their solutions to different problems in speaker recognition are presented. In addition, the book highlights some of the new trends and directions for speaker recognition based on deep learning and adversarial learning. However, due to space constraints, the book has overlooked many promising machine learning topics and models, such as reinforcement learning, recurrent neural networks, etc. To those numerous contributors, who deserve many more credits than are given here, the authors wish to express their most sincere apologies.

The book is divided into two parts: fundamental theories and advanced studies.

1 **Fundamental theories**: This part explains different components and challenges in the construction of a statistical speaker recognition system. We organize and survey speaker recognition methods according to two categories: learning algorithms and learning models. In learning algorithms, we systematically present the inference procedures from maximum likelihood to approximate Bayesian for probabilistic models and error backpropagation algorithm for DNN. In learning models, we address a number of linear models and nonlinear models based on different types of latent variables, which capture the underlying speaker and channel characteristics.
2 **Advanced studies**: This part presents a number of deep models and case studies, which are recently published for speaker recognition. We address a range of deep models ranging from DNN and deep belief networks to variational auto-encoders and generative adversarial networks, which provide the vehicle to learning representation of a true speaker model. In case studies, we highlight some advanced PLDA models and i-vector extractors that accommodate multiple mixtures, deep structures, and sparsity treatment. Finally, a number of directions and outlooks are pointed out for future trend from the perspectives of deep machine learning and challenging tasks for speaker recognition.

In the Appendix, we provide exam-style questions covering various topics in machine learning and speaker recognition.

In closing, *Machine Learning for Speaker Recognition* is intended for one-semester graduate-school courses in machine learning, neural networks, and speaker recognition. It is also intended for professional engineers, scientists, and system integrators who want to know what state-of-the-art speaker recognition technologies can provide. The prerequisite courses for this book are calculus, linear algebra, probabilities, and statistics. Some explanations in the book may require basic knowledge in speaker recognition, which can be found in other textbooks.

Acknowledgments

This book is the result of a number of years of research and teaching on the subject of neural networks, machine learning, speech and speaker recognition, and human–computer interaction. The authors are very much grateful to their students for their questions on and contribution to many examples and exercises. Some parts of the book are derived from the dissertations of several postgraduate students and their joint papers with the authors. We wish to thank all of them, in particular Dr. Eddy Zhili Tan, Dr. Ellen Wei Rao, Dr. Na Li, Mr. Wei-Wei Lin, Mr. Youzhi Tu, Miss Xiaomin Pang, Mr. Qi Yao,

Miss Ching-Huai Chen, Mr. Cheng-Wei Hsu, Mr. Kang-Ting Peng, and Mr. Chun-Lin Kuo. We also thank Youzhi Tu for proofreading the earlier version of the manuscript.

We have benefited greatly from the enlightening exchanges and collaboration with colleagues, particularly Professor Helen Meng, Professor Brian Mak, Professor Tan Lee, Professor Koichi Shinoda, Professor Hsin-min Wang, Professor Sadaoki Furui, Professor Lin-shan Lee, Professor Sun-Yuan Kung, and Professor Pak-Chung Ching. We have been very fortunate to have worked with Ms. Sarah Strange, Ms. Julia Ford, and Mr. David Liu at Cambridge University Press, who have provided the highest professional assistance throughout this project. We are grateful to the Department of Electronic and Information Engineering at The Hong Kong Polytechnic University and the Department of Electrical and Computer Engineering at the National Chiao Tung University for making available such a scholarly environment for both teaching and research.

We are pleased to acknowledge that the work presented in this book was in part supported by the Research Grants Council, Hong Kong Special Administrative Region (Grant Nos. PolyU 152117/14E, PolyU 152068/15E, PolyU 152518/16E, and PolyU 152137/17E); and The Ministry of Science and Technology, Taiwan (Grant Nos. MOST 107-2634-F-009-003, MOST 108-2634-F-009-003 and MOST 109-2634-F-009-024).

We would like to thank the researchers who have contributed to the field of neural networks, machine learning, and speaker recognition. The foundation of this book is based on their work. We sincerely apologize for the inevitable overlooking of many important topics and references because of time and space constraints.

Finally, the authors wish to acknowledge the kind support of their families. Without their full understanding throughout the long writing process, this project would not have been completed so smoothly.

Abbreviations

AA-PLDA	adversarial augmentation PLDA
AAE	adversarial autoencoder
AC-GAN	auxiliary classifier GAN
AEVB	autoencoding variational Bayes
AfV	audio from video
AM-PLDA	adversarial manifold PLDA
CD	contrastive divergence
CNN	convolutional neural network
CTS	conversational telephone speech
DA	domain adaptation
DAE	denoising autoencoder
DBM	deep Boltzmann machine
DBN	deep belief network
DCF	decision cost function
DET	detection error tradeoff
DICN	dataset-invariant covariance normalization
DNN	deep neural network
EER	equal error rate
ELBO	evidence lower bound
EM	expectation-maximization
FA	factor analysis
FAR	false acceptance rate
FFT	fast Fourier transform
FRR	false rejection rate
GAN	generative adversarial network
GMM	Gaussian mixture model
HMM	hidden Markov model
IDVC	inter-dataset variability compensation
JFA	joint factor analysis
JS	Jensen–Shannon
KL	Kullback–Leibler
LSTM	long short-term memory
MAP	maximum *a posteriori*

MCMC	Markov chain Monte Carlo
MFCC	mel-frequency cepstral coefficient
ML	maximum likelihood
MLP	multilayer perceptron
MMD	maximum mean discrepancy
NAP	nuisance attribute projection
NDA	nonparametric discriminant analysis
NIST	National Institute of Standards and Technology
PCA	principal component analysis
PLDA	probabilistic linear discriminant analysis
RBF	radial basis function
RBM	restricted Boltzmann machine
RKHS	reproducing kernel Hilbert space
ReLU	rectified linear unit
SD-mPLDA	SNR-dependent mixture of PLDA
SDI-PLDA	SNR- and duration-invariant PLDA
SI-mPLDA	SNR-independent mixture of PLDA
SGD	stochastic gradient descent
SGVB	stochastic gradient variational Bayes
SNE	stochastic neighbor embedding
SRE	speaker recognition evaluation
SVDA	support vector discriminant analysis
SVM	support vector machine
t-SNE	t-distributed stochastic neighbor embedding
UBM	universal background model
VDANN	variational domain adversarial neural network
VAE	variational autoencoder
VB	variational Bayesian
VB-EM	variational Bayesian expectation-maximization
VM-PLDA	variational manifold PLDA
WCC	within-class covariance correction
WCCN	within-class covariance normalization

Notations

$\mathbf{o}$	Acoustic vectors (observations)
$\mathcal{O}$	Set of acoustic vectors
π_c	Prior probability of the cth mixture component
$\boldsymbol{\mu}_c$	The mean vector of the cth mixture component
$\boldsymbol{\Sigma}_c$	The covariance matrix of the cth mixture component
C	The number of mixture components in GMMs
ℓ_c	Indicator variable for the cth mixture in a GMM
$\gamma(\ell_c)$ and $\gamma_c(\cdot)$	Posterior probability of mixture c
$\mathbf{x}$	I-vector
$\mathcal{X}$	A set of i-vectors
$\mathbf{V}$	Speaker loading matrix of PLDA and JFA
$\mathbf{V}^{\mathsf{T}}$	Transpose of matrix $\boldsymbol{V}$
$\mathbf{U}$	SNR loading matrix in SNR-invariant PLDA and channel loading matrix in JFA
$\mathbf{G}$	Channel loading matrix in PLDA model
ϵ	Residue term of the PLDA model or the factor analysis model in i-vector systems
$\boldsymbol{\Sigma}$	Covariance matrix of ϵ in PLDA
$\mathbf{z}$	Latent factor in PLDA
$\mathcal{Z}$	Set of latent factors $\mathbf{z}$
$\mathbf{m}$	Global mean of i-vectors
$\mathbb{E}\{\mathbf{x}\}$	Expectation of $\mathbf{x}$
$\langle\mathbf{x}\rangle$	Expectation of $\mathbf{x}$
$\boldsymbol{\Lambda}$	A set of model parameters
$\mathbf{0}$	Vector with all elements equal to 0
$\mathbf{1}$	Vector with all elements equal to 1
$\mathbf{I}$	Identity matrix
$\langle\mathbf{z}_i\|\mathbf{x}_i\rangle$	Conditional expectation of $\mathbf{z}_i$ given $\mathbf{x}_i$
$\mathcal{Q}(\cdot)$	Auxiliary function of EM algorithms
$\mathbf{T}$	Total variability matrix in i-vector systems
$\mathbf{w}_i$	Latent factor of the factor analysis (FA) model in i-vector systems
$\boldsymbol{\mu}^{(b)}$	Supervector of the UBM in the FA model of i-vector systems
$\boldsymbol{\mu}_i$	Utterance-dependent supervector in i-vector systems

$\mathbf{y}$	Indicator vector in PLDA mixture models
$\mathcal{Y}$	Set of indicator vectors, complete data or target labels
$y_{\cdot,\cdot,c}$	indicator variables for the cth mixture in i-vector FA model
$\mathcal{H}_{\cdot,c}$	Set comprising the frame indexes whose acoustic vectors are aligned to mixture c
$\mathbf{N}$	Matrix comprising the zeroth order sufficient statistics in its diagonal
n_c	Zeroth order sufficient statistics of mixture c
$\tilde{\mathbf{f}}$	Vector comprising the first order sufficient statistics
ξ_i	Slack variables in SVM
α_i	Lagrange multipliers in SVM
$\phi(\cdot)$	Feature map in SVM
$K(\cdot,\cdot)$	Kernel function in SVM
b	Bias
$\mathcal{L}(\cdot)$	Lower bound of a log likelihood function
$\mathbb{R}^D$	Real numbers in D-dimensional space
$\mathbf{v}$	Visible units $\{v_i\}$ in RBM
$\mathbf{h}$	Hidden units $\{h_j\}$ in RBM
$E(\mathbf{v},\mathbf{h})$	Energy function of visible units $\mathbf{v}$ and hidden units $\mathbf{h}$ in RBM
$E(\mathbf{w})$	Error function with DNN parameters $\mathbf{w}$
$\mathbf{X}_n$	Minibatch data with length T_n
E_n	Error function using minibatch data $\mathbf{X}_n$
$\mathbf{y}(\mathbf{x}_t,\mathbf{w})$	Regression outputs corresponding to inputs $\mathbf{x}_t$ in DNN
$\mathbf{r}_t$	Regression targets in DNN
$\mathbf{z}_t$	Hidden units in DNN
a_{tk}	Activation of unit k
$\mathbb{H}[\cdot]$	Entropy function
$q(\mathbf{h}\|\mathbf{v})$	Variational distribution of hidden units $\mathbf{h}$ given visible units $\mathbf{v}$ in RBM
$\widetilde{\mathbf{x}}$	Corrupted version of an original sample $\mathbf{x}$ in DAE
$\widehat{\mathbf{x}}$	Reconstructed data in DAE
$\mathbf{h}$	Deterministic latent code in DAE
θ	Model parameter in VAE
ϕ	Variational parameter in VAE
L	Total number of samples
$\mathbf{z}^{(l)}$	The lth latent variable sample
$\mathcal{D}_{\mathrm{KL}}(q\,\\|\,p)$	Kullback–Leibler divergence between distributions q and p
$\mathcal{D}_{\mathrm{JS}}(q\,\\|\,p)$	Jensen–Shannon divergence between distributions q and p
$p_{\mathrm{data}}(\mathbf{x})$	Data distribution
$p_{\mathrm{model}}(\mathbf{x})$	Model distribution
$p_g(\mathbf{x})$	Distribution of generator (or equivalently model distribution $p_{\mathrm{model}}(\mathbf{x})$)
G	Generator with distribution $p_g(\mathbf{x})$ in GAN

D	Discriminator in GAN
$\boldsymbol{\theta}_g$	Parameter of generator in GAN
$\boldsymbol{\theta}_d$	Parameter of discriminator in GAN
$\boldsymbol{\theta}_e$	Parameter of encoder in an AAE
$\boldsymbol{\theta}_{\text{enc}}$	Parameter of encoder in manifold or adversarial learning
$\boldsymbol{\theta}_{\text{dec}}$	Parameter of decoder in manifold or adversarial learning
$\boldsymbol{\theta}_{\text{dis}}$	Parameter of discriminator in manifold or adversarial learning
$\boldsymbol{\theta}_{\text{gen}}$	Parameter of generator in manifold or adversarial learning
t_n	Target value of an i-vector $\mathbf{x}_n$
t_{nm}	Target value for indication if $\mathbf{x}_n$ and $\mathbf{x}_m$ belong to the same class
$\mathcal{T}$	A set of target values
θ_g	Parameter of decoder in VDANN
θ_e	Parameter of encoder in VDANN
θ_c	Parameter of classifier in VDANN
θ_d	Parameter of discriminator in VDANN
$\boldsymbol{\mu}_\phi$	Mean vector of encoder's output in a VAE
$\boldsymbol{\sigma}_\phi$	Standard deviation vector of encoder's output in a VAE

Part I

Fundamental Theories

1 Introduction

This chapter defines the term "speaker recognition" and looks at this technology from a high-level perspective. Then, it introduces the fundamental concepts of speaker recognition, including feature extraction, speaker modeling, scoring, and performance measures.

1.1 Fundamentals of Speaker Recognition

From time to time we hear that the information of millions of customers of remote services has been compromised. These security leaks cause concerns about the security of the remote services that everyone uses on a daily basis. While these remote services bring convenience and benefit to users, they are also gold mines for criminals to carry out fraudulent acts. The conventional approach to user authentication, such as usernames and passwords, is no longer adequate for securing these services. A number of companies have now introduced voice biometrics as a complement to the conventional username–password approach. With this new authentication method, it is much harder for the criminals to imitate the legitimate users. Voice biometrics can also reduce the risk of leaking customers' information caused by social engineering fraudulence. Central to voice biometrics authentication is speaker recognition.

Another application domain of voice biometrics is to address the privacy issues of smartphones, home assistants, and smart speakers. With the increasing intelligence capabilities of these devices, we can interact with them as if they were human. Because these devices are typically used solely by their owners or their family members and speech is the primary means of interaction, it is natural to use the voice of the owners for authentication, i.e., a device can only be used by its owner.

Speaker recognition is a technique to recognize the identity of a speaker from a speech utterance. As shown in Figure 1.1, in terms of recognition tasks, speaker recognition can be categorized into speaker identification, speaker verification, and speaker diarization. In all of these tasks, the number of speakers involved can be fixed (closed set) or varied (open set).

Speaker identification is to determine whether the voice of an unknown speaker matches one of the N speakers in a dataset, where N could be very large (thousands). It is a one-to-many mapping and it is often assumed that the unknown voice

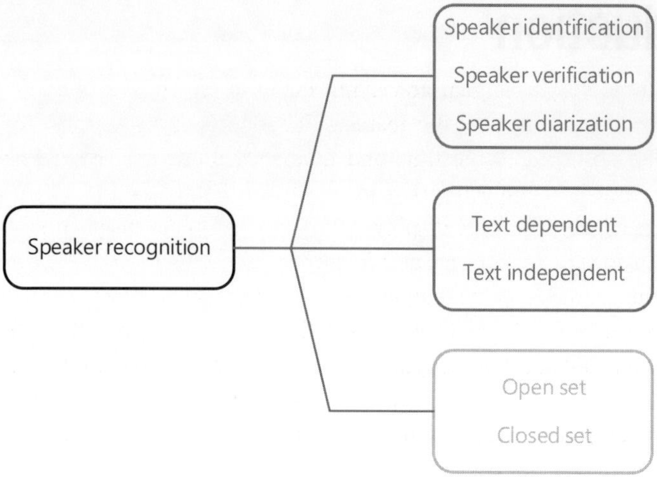

Figure 1.1 Categorization of speaker recognition in terms of tasks (top), text restriction (middle), and datasets (bottom).

must come from a set of known speakers – referred to as closed-set identification. Adding a "none of the above" option to closed-set identification gives us open-set identification.

Speaker verification is to determine whether the voice of an unknown speaker matches a *specific* speaker. It is a one-to-one mapping. In closed-set verification, the population of clients is fixed, whereas in open-set verification, new clients can be added without having to redesign the system.

Speaker diarization [5] is to determine when speaker changes have occurred in speech signals, which is an analogy to the speech segmentation task in speech recognition. When it is also needed to group together speech segments corresponding to the same speaker, we have speaker clustering. In both cases, prior speaker information may or may not be available.

There are two input modes in speaker recognition systems: text dependent and text independent. Text-dependent recognition systems know the texts that will be spoken by the speakers or expect legitimate users to speak some fixed or prompted phrases. The phrases are typically very short. Because the phrases are known, speech recognition can be used for checking spoken text to improve system performance. Text-dependent systems are mainly used for applications with strong control over user input, e.g., biometric authentication. On the contrary, in text-independent systems, there is no restriction on the spoken text. Typically, conversational speech is used as the input. So, the sentences are much longer than those in text-dependent systems. While the spoken text is unknown, speech recognition can still be used for extracting high-level features to boost performance. Text-independent systems are mainly used in applications with less control over user input, e.g., forensic speaker ID. Compared with text-dependent systems, text-independent systems are more flexible but recognition is more difficult.

1.2 Feature Extraction

Speech is a time-varying signal conveying multiple layers of information, including words, speaker identities, acoustic features, languages, and emotions. Information in speech can be observed in the time and frequency domains. The most widely used visualization tool is the spectrogram in which the frequency spectra of consecutive short-term speech segments are displayed as an image. In the image, the horizontal and vertical dimensions represent time and frequency, respectively, and the intensity of each point in the image indicates the magnitude of a particular frequency at a particular time.

While spectrograms are great for visualization of speech signals, they are not appropriate for speech and speaker recognition. There are two reasons for this. First, the frequency dimension is still too high. For 1024-point fast Fourier transform (FFT), the frequency dimension is 512, which is far too large for statistical modeling. Second, the frequency components after FFT are highly correlated with each other, which do not facilitate the use of diagonal covariance matrices to model the variability in the feature vectors. To obtain a more compact representation of speech signals and to de-correlate the feature components, cepstral representation of speech is often used. The most widely used representation is the Mel-frequency cepstral coefficients (MFCCs) [6]. Figure 1.2 shows the process of extracting MFCCs from a frame of speech. In the figure, $s(n)$ represents a frame of speech, $X(m)$ is the logarithm of the spectrum at frequencies defined by the mth filter in the filter bank, and

$$o_i = \sum_{m=1}^{M} \cos\left[i\left(m - \frac{1}{2}\right)\frac{\pi}{M}\right] X(m), \quad i = 1, \ldots, P \tag{1.1}$$

are its MFCCs. In Figure 1.2, the symbols Δ and $\Delta\Delta$ represent velocity and acceleration of MFCCs, respectively. Denoting e as the log-energy, an acoustic vector corresponding to $s(n)$ is given by

$$\mathbf{o} = [e, o_1, \ldots, o_P, \Delta e, \Delta o_1 \ldots, \Delta o_P, \Delta\Delta e, \Delta\Delta o_1, \ldots, \Delta\Delta o_P]^{\mathsf{T}}. \tag{1.2}$$

In most speaker recognition systems, $P = 19$, giving $\mathbf{o} \in \Re^{60}$.

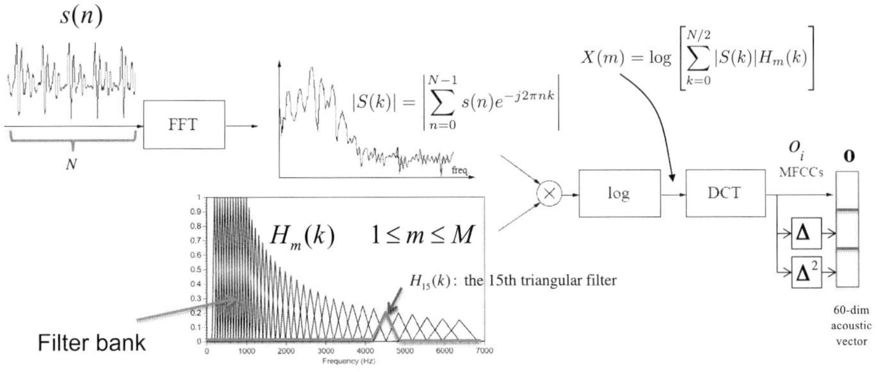

Figure 1.2 Process of extracting MFCCs from a frame of speech. Refer to Eq. 1.1 and Eq. 1.2 for o_i and $\mathbf{o}$, respectively.

1.3 Speaker Modeling and Scoring

For text-independent speaker recognition, we assume that the acoustic vectors are independent. Therefore, speaker modeling amounts to modeling a batch of independent acoustic vectors derived from speech waveform as shown in Figure 1.3. The process can be considered as mapping the consecutive frames independently to an acoustic space (Figure 1.4(a)). In the most traditional approach, the distribution of these vectors is represented by a Gaussian mixture model (GMM) as shown in Figure 1.4(b).

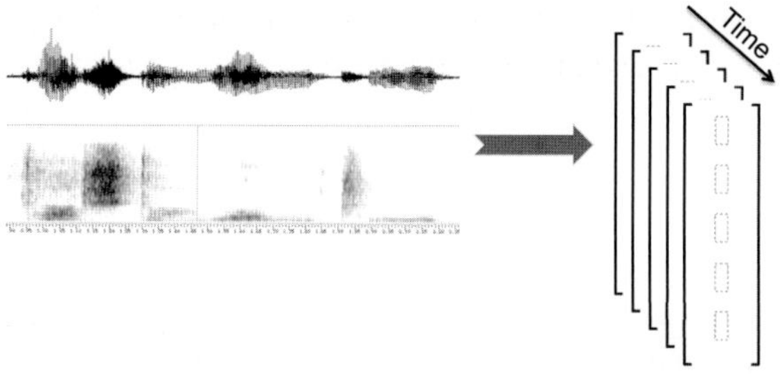

Figure 1.3 From waveform to a sequence of acoustic vectors.

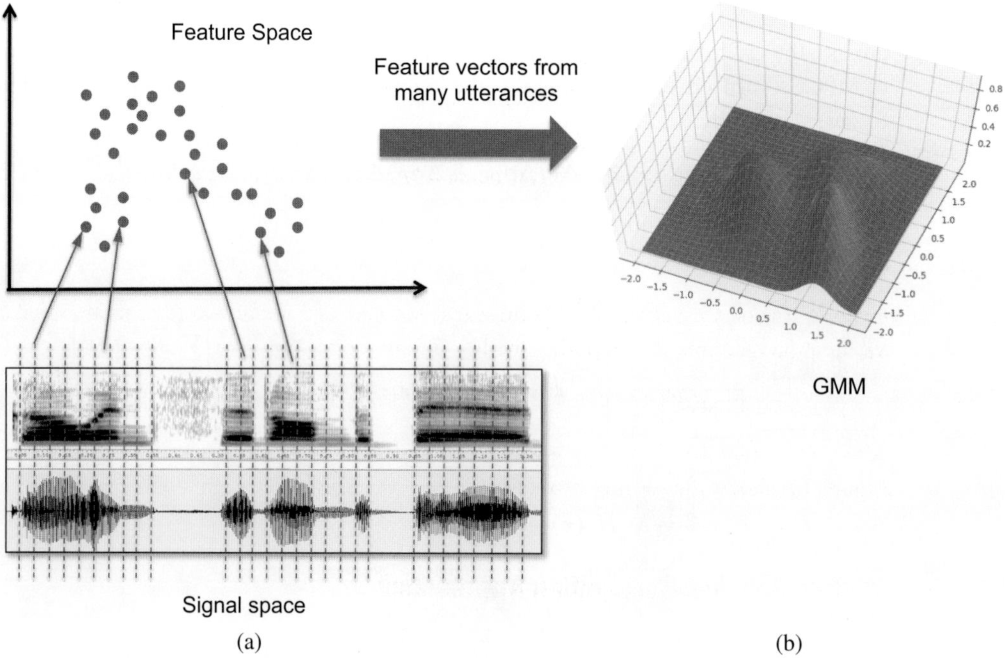

Figure 1.4 Modeling the statistical properties of acoustic vectors by a Gaussian mixture model.

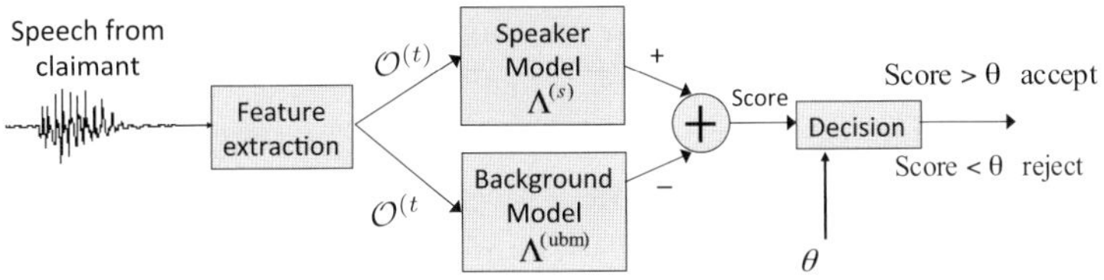

Figure 1.5 GMM–UBM scoring process.

1.3.1 Speaker Modeling

In 2000, Reynolds [7] proposed using the speech of a large number of speakers to train a GMM to model the acoustic characteristics of a general population. The resulting model is called the universal background model (UBM). The density of acoustic vectors $\mathbf{o}$'s is given by

$$p(\mathbf{o}|\text{UBM}) = p(\mathbf{o}|\Lambda^{\text{ubm}}) = \sum_{c=1}^{C} \pi_c^{\text{ubm}} \mathcal{N}(\mathbf{o}|\boldsymbol{\mu}_c^{\text{ubm}}, \boldsymbol{\Sigma}_c^{\text{ubm}}). \tag{1.3}$$

The UBM parameters $\Lambda^{\text{ubm}} = \{\pi_c^{\text{ubm}}, \boldsymbol{\mu}_c^{\text{ubm}}, \boldsymbol{\Sigma}_c^{\text{ubm}}\}_{c=1}^{C}$ are estimated by the expectation-maximization (EM) algorithm [8] using the speech of many speakers. See Section 3.1.1 for the detail of the EM algorithm.

While the UBM represents the general population, individual speakers are modeled by speaker-dependent Gaussian mixture models. Specifically, for a target-speaker s, his/her GMM is given by

$$p(\mathbf{o}|\text{Spk } s) = p(\mathbf{o}|\Lambda^{(s)}) = \sum_{c=1}^{C} \pi_c^{(s)} \mathcal{N}(\mathbf{o}|\boldsymbol{\mu}_c^{(s)}, \boldsymbol{\Sigma}_c^{(s)})$$

where $\Lambda^{(s)} = \{\pi_c^{(s)}, \boldsymbol{\mu}_c^{(s)}, \boldsymbol{\Sigma}_c^{(s)}\}_{c=1}^{C}$ are learned by using a maximum *a posteriori* (MAP) adaptation [7]. See Section 3.1.3 for the details of a MAP adaptation.

1.3.2 Speaker Scoring

Given the acoustic vectors $O^{(t)}$ from a test speaker and a claimed identity s, speaker verification amounts to computing the log-likelihood ratio:

$$S_{\text{GMM–UBM}}(O^{(t)}|\Lambda^{(s)}, \Lambda^{\text{ubm}}) = \log p(O^{(t)}|\Lambda^{(s)}) - \log p(O^{(t)}|\Lambda^{\text{ubm}}), \tag{1.4}$$

where $\log p(O^{(t)}|\Lambda^{(s)})$ is the log-likelihood of $O^{(t)}$ given the speaker model $\Lambda^{(s)}$. Figure 1.5 shows the scoring process.

1.4 Modern Speaker Recognition Approaches

GMM–UBM is a frame-based approach in that the speaker models (GMMs) describe the distribution of acoustic frames. Since its introduction in 2000, it has been the state-of-the-art method for speaker verification for a number of years. However, it has its own limitations. One major drawback is that the training of the GMMs and the UBM is disjointed, meaning that contrastive information between target speakers and impostors cannot be explicitly incorporated into the training process. Another drawback is that suppressing nonspeaker information (e.g., channel and noise) is difficult. Although attempts have been made to suppress channel and noise variabilities in the feature [9, 10], model [11, 12], and score domains [13], they are not as effective as the modern approaches outlined below and explained in detail in this book.

In 2006, researchers started to look at the speaker verification problem from another view. Instead of accumulating the frame-based log-likelihood scores of an utterance, researchers derived methods to map the acoustic characteristics of the entire utterance to a high-dimensional vector. These utterance-based vectors live on a high-dimensional space parameterized by GMMs. Because the dimension is the same regardless of the utterance duration, standard machine learning methods such as support vector machines and factor analysis can be applied on this space. The three influential methods based on this idea are GMM–SVM, joint factor analysis, and i-vectors.

In GMM–SVM [14] (see Section 3.2), supervectors are constructed from MAP-adapted target-speaker GMMs. For each target speaker, a speaker-dependent SVM is then trained to discriminate his/her supervectors from those of the impostors. To reduce channel mismatch, the directions corresponding to nonspeaker variability are projected out. Scoring amounts to computing the SVM scores of the test utterances and decisions are made by comparing the scores with a decision threshold.

In joint factor analysis [15] (see Section 3.7), speaker and session variabilities are represented by latent variables (speaker factors and channel factors) in a factor analysis model. During scoring, session variabilities are accounted for by integrating over the latent variables, e.g., the channel factors.

In an i-vector system [16] (see Section 3.6), utterances are represented by the posterior means of latent factors, called the i-vectors. I-vectors capture both speaker and channel information. During scoring, the unwanted channel variability is removed by linear discriminant analysis (LDA) or by integrating out the latent factors in a probabilistic LDA model [17].

1.5 Performance Measures

For closed-set speaker identification, recognition rate (accuracy) is the usual performance measure:

$$\text{Recognition rate} = \frac{\text{No. of correct recognitions}}{\text{Total no. of trials}}.$$

Speaker verification, on the other hand, has a rich set of performance measures. Although different datasets have slightly different measures, their principles remain the same. The common measures include false rejection rate (FRR), false acceptance rate (FAR), equal error rate (EER), minimum decision cost function (minDCF), and actual decision cost function (DCF).

1.5.1 FAR, FRR, and DET

The definition of FAR and FRR are as follows:

$$\text{False rejection rate (FRR)} = \text{Miss probability} = \frac{\text{No. of true-speakers rejected}}{\text{Total no. of true-speaker trials}};$$

$$\text{False acceptance rate (FAR)} = \text{False alarm probability} = \frac{\text{No. of impostors accepted}}{\text{Total no. of impostor attempts}}.$$

Equal error rate (EER) corresponds to the operating point at which FAR = FRR. The concept of FAR, FRR, and EER can be explained by using the distributions of speaker scores and impostor scores in two speaker verification systems (System A and System B) shown in Figure 1.6. When the decision threshold θ is swept from low to high, FAR drops from 100 percent gradually to 0 percent but FRR gradually increases from 0 percent to 100 percent. When θ is very large, we have a secure but user-unfriendly system. On the other hand, when θ is very small, we have a user-friendly but nonsecure system. A system developer chooses a decision threshold such that the FAR and FRR meet the requirements of the application.

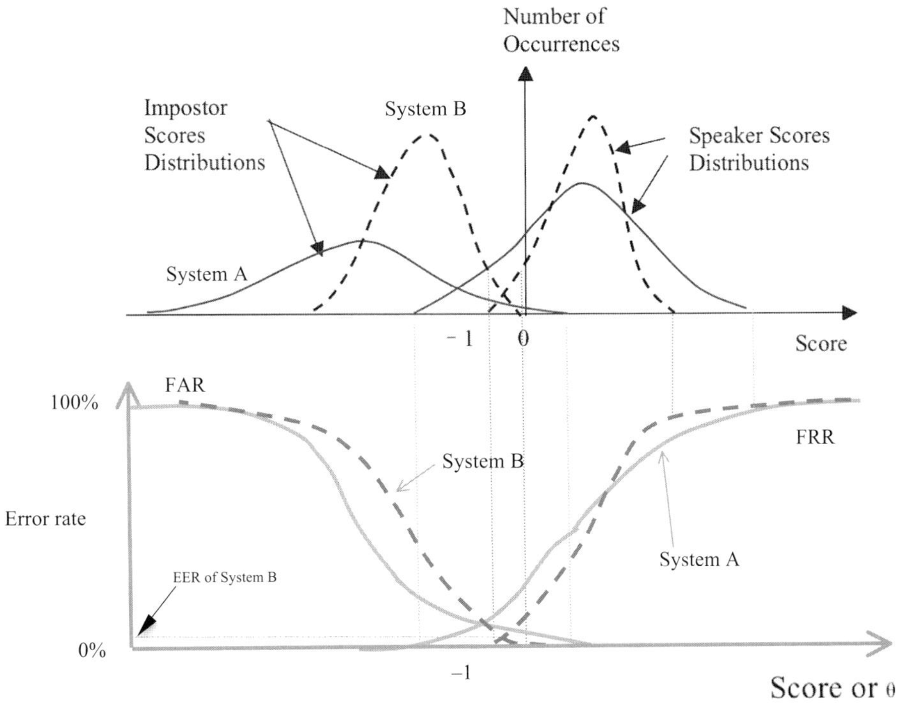

Figure 1.6 *Top:* Distributions of true-speaker scores and impostor scores of two speaker verification systems. *Bottom*: The FAR, FRR, and EER of the two systems.

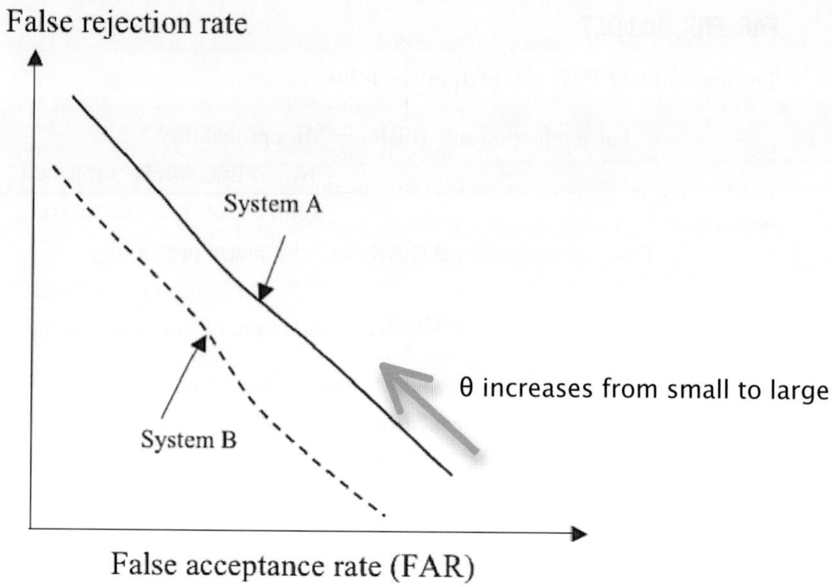

Figure 1.7 The DET curves correspond to system A and system B in Figure 1.6. System B performs better because its DET curve is closer to the origin.

Detection error tradeoff (DET) curves [18] are similar to receiver operating characteristic curves but with nonlinear x- and y-axis. The advantage of the nonlinear axes is that the DET curves of systems with verification scores following Gaussian distributions will be displayed as straight lines, which facilitate comparison of systems with similar performance. A DET curve is produced by sweeping the decision threshold of a system from low to high. Figure 1.7 shows the DET curves of System A and System B in Figure 1.6.

1.5.2 Decision Cost Function

The decision cost function (DCF) [18] is a weighted sum of the FRR ($P_{\text{Miss}|\text{Target}}$) and FAR ($P_{\text{FalseAlarm}|\text{Nontarget}}$):

$$\begin{aligned}C_{\text{Det}}(\theta) = C_{\text{Miss}} \times P_{\text{Miss}|\text{Target}}(\theta) \times P_{\text{Target}} + \\ C_{\text{FalseAlarm}} \times P_{\text{FalseAlarm}|\text{Nontarget}}(\theta) \times (1 - P_{\text{Target}}),\end{aligned}$$

where θ is a decision threshold, C_{Miss} is the cost of false rejection, $C_{\text{FalseAlarm}}$ is the cost of false acceptance, and P_{Target} is the prior probability of target speakers. With the DCF, a normalized cost can be defined:

$$C_{\text{Norm}}(\theta) = C_{\text{Det}}(\theta)/C_{\text{Default}},$$

where

$$C_{\text{Default}} = \min \begin{cases} C_{\text{Miss}} \times P_{\text{Target}} \\ C_{\text{FalseAlarm}} \times (1 - P_{\text{Target}}). \end{cases} \tag{1.5}$$

The normalized DCF and the measures derived from it are the primary performance index of the NIST speaker recognition evaluations (SRE).[1] The parameters in Eq. 1.5 are set differently for different years of evaluations:

- SRE08 and earlier:

$$C_{\text{Miss}} = 10; \quad C_{\text{FalseAlarm}} = 1; \quad P_{\text{Target}} = 0.01.$$

- SRE10:

$$C_{\text{Miss}} = 1; \quad C_{\text{FalseAlarm}} = 1; \quad P_{\text{Target}} = 0.001.$$

In SRE12, the decision cost function has been changed to:

$$\begin{aligned} C_{\text{Det}}(\theta) = C_{\text{Miss}} \times P_{\text{Miss|Target}}(\theta) \times P_{\text{Target}} + C_{\text{FalseAlarm}} \times (1 - P_{\text{Target}}) \\ \times \left[P_{\text{FalseAlarm|KnownNontarget}}(\theta) \times P_{\text{Known}} \right. \\ \left. + P_{\text{FalseAlarm|UnKnownNontarget}} \times (1 - P_{\text{Known}}) \right], \end{aligned} \tag{1.6}$$

where “KnownNontarget” and “UnknownNontarget” mean that the impostors are known and unknown to the evaluator, respectively. Similar to previous years' SRE, the DCF is normalized, giving

$$C_{\text{Norm}}(\theta) = C_{\text{Det}}(\theta)/(C_{\text{Miss}} \times P_{\text{Target}}). \tag{1.7}$$

The parameters for core test conditions are set to

$$C_{\text{Miss}} = 1; \; C_{\text{FalseAlarm}} = 1; \; P_{\text{Target1}} = 0.01; \; P_{\text{Target2}} = 0.001; \; P_{\text{Known}} = 0.5.$$

Substituting the above P_{Target1} and P_{Target2} into Eq. 1.6 and Eq. 1.7, we obtain $C_{\text{Norm1}}(\theta_1)$ and $C_{\text{Norm2}}(\theta_2)$, respectively. Then, the primary cost of NIST 2012 SRE can be obtained:

$$C_{\text{Primary}}(\theta_1, \theta_2) = \frac{C_{\text{Norm1}}(\theta_1) + C_{\text{Norm2}}(\theta_2)}{2}. \tag{1.8}$$

In SRE16, the decision cost was changed to

$$\begin{aligned} C_{\text{Det}}(\theta) = C_{\text{Miss}} \times P_{\text{Miss|Target}}(\theta) \times P_{\text{Target}} + \\ C_{\text{FalseAlarm}} \times P_{\text{FalseAlarm|Nontarget}}(\theta) \times (1 - P_{\text{Target}}). \end{aligned} \tag{1.9}$$

The normalized DCF remains the same as Eq. 1.7, and the parameters for the core test conditions are:

$$C_{\text{Miss}} = 1; \; C_{\text{FalseAlarm}} = 1; \; P_{\text{Target1}} = 0.01; \; P_{\text{Target2}} = 0.001.$$

The primary cost is computed as in Eq. 1.8.

[1] www.nist.gov/itl/iad/mig/speaker-recognition

Table 1.1 Cost parameters and prior of target-speakers in NIST 2018 SRE.

Speech Type	Parameter ID	C_{Miss}	$C_{\text{FalseAlarm}}$	P_{Target}
CTS	1	1	1	0.01
CTS	2	1	1	0.005
AfV	3	1	1	0.05

Note that in Eq. 1.8, the decision thresholds θ_1 and θ_2 are assumed unknown and can be set by system developers. When θ_1 and θ_2 are optimized to achieve the lowest C_{Norm1} and C_{Norm2}, respectively, then we have the minimum primary cost. Sometimes, researchers simply call it minDCF.

In practical applications of speaker verification, application-independent decision thresholds [19] are more appropriate. The goal is to minimize not only the EER and minDCF but also the actual DCF (actDCF) or C_{primary} at specific thresholds. For each SRE, NIST defined specific thresholds to evaluate how well systems were calibrated. For example in SRE16, θ_1 and θ_2 were set to 4.5951 and 5.2933, respectively. The primary cost using these two thresholds are called actual DCF or actual primary cost. Specifically, we have

$$C_{\text{actPrimary}} = \frac{C_{\text{Norm1}}(4.5951) + C_{\text{Norm2}}(5.2933)}{2}.$$

Any systems with scores that deviate significantly from these two thresholds will get a high $C_{\text{actPrimary}}$, which could be larger than 1.0.

The decision cost function in SRE18 is identical to that of SRE16 in Eq. 1.9. However, there are two types of speech: conversational telephone speech (CTS) and audio from video (AfV). The prior of target-speakers are different for these two types of speech, as shown in Table 1.1. With the three sets of parameters, the primary cost in Eq. 1.8 is extended to

$$C_{\text{Primary}}(\theta_1, \theta_2, \theta_3) = \frac{1}{2}\left[\frac{C_{\text{Norm1}}(\theta_1) + C_{\text{Norm2}}(\theta_2)}{2} + C_{\text{Norm3}}(\theta_3)\right]. \tag{1.10}$$

Using the P_{Target}'s in Table 1.1, the actual DCF in SRE18 is

$$C_{\text{Primary}} = \frac{1}{2}\left[\frac{C_{\text{Norm1}}(4.5951) + C_{\text{Norm2}}(5.2933)}{2} + C_{\text{Norm3}}(2.9444)\right].$$

2 Learning Algorithms

This chapter addresses the background knowledge about machine learning and deep learning, which is used as a foundation for statistical speaker recognition. The paradigms based on statistical inference and neural networks are presented. In particular, we address the basics of expectation-maximization (EM) algorithms, variational EM algorithms, and general approximate inference algorithms. Bayesian learning is presented to deal with regularization in model-based speaker recognition.

2.1 Fundamentals of Statistical Learning

In general, the speech signal in speaker recognition is collected as a set of observation vectors. The underlying factors including channels, noises, and environments, etc., are unknown and treated as latent variables when training the speaker recognition model. Hierarchical latent variables or features can be explored to characterize the structural information in latent variable or feature space. In machine learning community, the latent space models are constructed in different ways with different interpretations. Machine learning provides a wide range of model-based approaches for speaker recognition. A model-based approach aims to incorporate the physical phenomena, measurements, uncertainties, and noises in the form of mathematical models. Such an approach is developed in a unified manner through different algorithms, examples, applications, and case studies. Mainstream methods are based on the statistical models. There are two popular categories of statistical methods in modern machine learning literature. One is the probabilistic models and the other is the neural network models that have been developing widely in various speaker recognition systems. A brief introduction to these two broad categories is addressed in what follows.

2.1.1 Probabilistic Models

The first category of statistical learning methods is grouped into probabilistic models or latent variable models that aim to relate a set of observable variables $\mathcal{X}$ to a set of latent variables $\mathcal{Z}$ based on a number of probability distributions [1]. Under a specification of model structure, the model parameters $\mathbf{\Lambda}$ of probability functions are estimated by maximizing the likelihood function with respect to model parameters $\mathbf{\Lambda}$. Basically, probabilistic model is configured as a *top-down* structure or representation. Starting

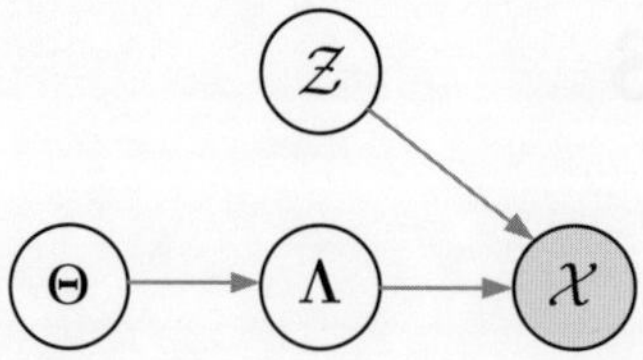

Figure 2.1 A hierarchical Bayesian representation for a latent variable model with hyperparameters $\mathbf{\Theta}$, parameters $\mathbf{\Lambda}$, latent variables $\mathcal{Z}$, and observed variables $\mathcal{X}$. Shaded node means observations. Unshaded nodes means latent variables.

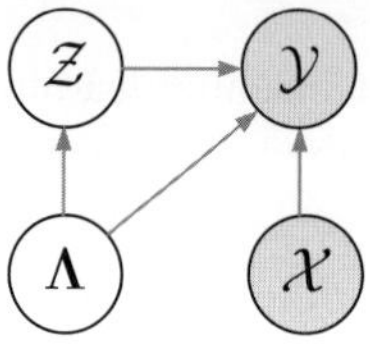

Figure 2.2 A supervised probabilistic model with parameters $\mathbf{\Lambda}$, latent variables $\mathcal{Z}$, and observed samples $\mathcal{X}$ and observed labels $\mathcal{Y}$.

from the hyperparameters $\mathbf{\Theta}$, model parameters are drawn or represented by a prior density $p(\mathbf{\Lambda}|\mathbf{\Theta})$. Having the model parameters $\mathbf{\Lambda}$, the speech training samples $\mathcal{X}$ are generated by a likelihood function $p(\mathcal{X}|\mathbf{\Lambda})$, which is marginalized over discrete latent variables by

$$p(\mathcal{X}|\mathbf{\Lambda}) = \sum_{\mathcal{Z}} p(\mathcal{X}, \mathcal{Z}|\mathbf{\Lambda}) \tag{2.1}$$

or over continuous latent variable by

$$p(\mathcal{X}|\mathbf{\Lambda}) = \int p(\mathcal{X}, \mathcal{Z}|\mathbf{\Lambda}) d\mathcal{Z}. \tag{2.2}$$

Probabilistic representation of data generation is intuitive with the graphical representation as shown in Figure 2.1. It is straightforward to build a supervised regression or classification model by directly maximizing the *conditional likelihood* $p(\mathcal{Y}|\mathcal{X}, \mathbf{\Lambda})$ where the training samples consist of observation samples $\mathcal{X}$ as well as their regression or class labels $\mathcal{Y}$. Figure 2.2 illustrates a latent variable model for supervised training. The conditional likelihood with discrete latent variables is expressed by

$$\begin{aligned} p(\mathcal{Y}|\mathcal{X}, \mathbf{\Lambda}) &= \sum_{\mathcal{Z}} p(\mathcal{Y}, \mathcal{Z}|\mathcal{X}, \mathbf{\Lambda}) \\ &= \sum_{\mathcal{Z}} p(\mathcal{Y}|\mathcal{X}, \mathbf{\Lambda}).p(\mathcal{Z}|\mathbf{\Lambda}) \end{aligned} \tag{2.3}$$

The supervised, unsupervised, and semi-supervised models can be flexibly performed and constructed by optimizing the corresponding likelihood functions in individual or hybrid style. The uncertainties of observations, parameters, and hyperparameters are easily incorporated in the resulting solution based on the probability functions. Domain knowledge can be represented as constraints in the likelihood function, which essentially turns the maximum-likelihood problem into a constrained optimization problem.

For real-world applications, the probabilistic models are complicated with different latent variables in various model structures. There are a number of approximate inference algorithms that are available for solving the optimization problems in probabilistic models. However, the approximate inference or decoding algorithm is usually difficult to derive. Direct optimization over a likelihood function is prone to be intractable. Alternatively, indirect optimization over a lower bound of the likelihood function (or called the evidence lower bound [ELBO]) is implemented as an analytical solution. There are many latent variable models in various information systems. The latent variable models in speaker recognition, addressed in this book, include the Gaussian mixture model (GMM), factor analysis (FA), probabilistic linear discriminant analysis (PLDA), joint factor analysis (JFA), and mixture of PLDA, which will be addressed in Sections 3.1, 3.3, 3.5, 3.7, and 5.4.1, respectively.

2.1.2 Neural Networks

Deep neural networks (DNNs) have successfully boosted the performance of numerous information processing systems in many domains. The key to the success of DNNs is originated from the deep structured and hierarchical learning, which mimic the information processing functions of human's neural system. Multiple layers of nonlinear processing units are introduced to represent very complicated mapping between input samples and output targets. High-level abstraction is extracted to judge the target values corresponding to raw input samples. For example, the deeper layers in Figure 2.3 are capable of identifying the high-level semantic meaning of the input image and use the meaning for predicting the type of action in the image.

Table 2.1 illustrates the conceptual differences between the paradigms of probabilistic models and neural network models. Different from probabilistic models, neural networks are constructed as a bottom-up structure. The observed samples are received and propagated in a layer-wise model that is known as the multilayer perceptron (MLP) [20]. MLP is established as a distributed computer. The computation style in neural networks is consistent in different layers. In general, a neural network is seen as a black box and is basically hard to analyze and interpret. Neural networks can be considered

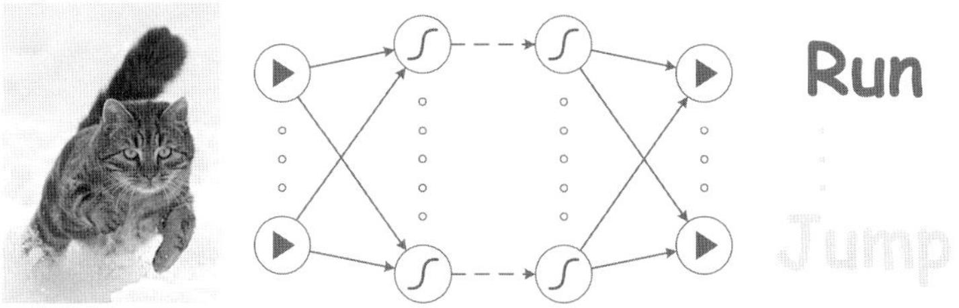

Figure 2.3 A deep neural network is constructed to retrieve the semantic meaning from an input image.

Table 2.1 Probabilistic models versus neural networks from different machine learning perspectives.

	Probabilistic model	Neural network
Structure	Top-down	Bottom-up
Representation	Intuitive	Distributed
Interpretation	Easier	Harder
Supervised/unsupervised learning	Easier	Harder
Incorporation of domain knowledge	Easy	Hard
Incorporation of constraint	Easy	Hard
Incorporation of uncertainty	Easy	Hard
Learning algorithm	Many algorithms	Back-propagation
Inference/decode process	Harder	Easier
Evaluation on	ELBO	End performance

as deterministic learning machines where the uncertainty of model parameters is disregarded. A fixed style of computation model is implemented so that it is difficult to incorporate domain knowledge and constraint in the optimization procedure. However, neural networks are easy to implement and infer. The learning objective is based on the error backpropagation algorithm, which optimizes the end performance by propagating the gradients from the output layer back to the input layer. Only the computations of gradients and their propagations are required. Computation procedure is simple and intuitive.

A desirable machine learning algorithm for speaker recognition should preserve good properties from both paradigms. Our goal is to pursue modern machine learning models and optimization procedures that are explainable and intuitive. The details of deep neural networks will be discussed in Chapter 4. In what follows, we introduce the fundamentals of model inference that will be used in the construction of probabilistic models for speaker recognition.

2.2 Expectation-Maximization Algorithm

This subsection addresses the general principle of a maximum likelihood (ML) estimation of probabilistic models and the detailed procedure of how the Jensen's inequality is used to derive the expectation-maximization (EM) algorithm [8]. The EM algorithm provides an efficient approach to tackling the incomplete data problem in a ML estimation; it also provides a general solution to the estimation of model parameters based on the ML principle.

2.2.1 Maximum Likelihood

We assume that the spectral matrix of a speech signal $\mathcal{X}$ is observed, which is generated by a distribution $p(\mathcal{X}|\mathbf{\Lambda})$ parameterized by $\mathbf{\Lambda}$ under a specialized model. The optimal

ML parameters $\mathbf{\Lambda}_{\text{ML}}$ of a speaker recognition model are estimated by maximizing the accumulated likelihood function from T frames of signal spectra $\mathcal{X} = \{\mathbf{x}_t\}_{t=1}^{T}$, i.e.,

$$\mathbf{\Lambda}_{\text{ML}} = \underset{\mathbf{\Lambda}}{\operatorname{argmax}}\, p(\mathcal{X}|\mathbf{\Lambda}). \tag{2.4}$$

Basically, the missing data problem happens in the estimation in Eq. 2.4 owing to the absence of latent variables $\mathcal{Z}$ in the representation. We could not directly work on such an optimization. To deal with this problem in model inference, the complete data $\{\mathcal{X}, \mathcal{Z}\}$, consisting of observation data $\mathcal{X}$ and latent variables $\mathcal{Z}$, are introduced in expression of a ML estimation in a form of

$$\mathbf{\Lambda}_{\text{ML}} = \underset{\mathbf{\Lambda}}{\operatorname{argmax}} \sum_{\mathcal{Z}} p(\mathcal{X}, \mathcal{Z}|\mathbf{\Lambda}). \tag{2.5}$$

This optimization considers discrete latent variables $\mathcal{Z}$, which are integrated by summation.

2.2.2 Iterative Procedure

In general, the exact solution to global optimum in optimization problem of Eq. 2.4 is intractable because of an incomplete data problem. To cope with this problem, the expectation-maximization (EM) algorithm [8] was developed to find an analytical solution to ML estimation. This algorithm is implemented by alternatively running the expectation step (also called E-step) and the maximization step (M-step) and iteratively finding the convergence to local optimum for ML estimate of model parameters $\mathbf{\Lambda}_{\text{ML}}$. In this ML estimation, the first step (E-step) is performed by building an auxiliary function $Q(\mathbf{\Lambda}|\mathbf{\Lambda}^{(\tau)})$ or calculating an expectation function of the log likelihood function with new parameters $\mathbf{\Lambda}$ conditional on the parameters $\mathbf{\Lambda}^{(\tau)}$ at previous iteration τ

$$\begin{aligned} Q(\mathbf{\Lambda}|\mathbf{\Lambda}^{(\tau)}) &= \mathbb{E}_{\mathcal{Z}}[\log p(\mathcal{X}, \mathcal{Z}|\mathbf{\Lambda})|\mathcal{X}, \mathbf{\Lambda}^{(\tau)}] \\ &= \sum_{\mathcal{Z}} p(\mathcal{Z}|\mathcal{X}, \mathbf{\Lambda}^{(\tau)}) \log p(\mathcal{X}, \mathcal{Z}|\mathbf{\Lambda}). \end{aligned} \tag{2.6}$$

The second step (M-step) is to estimate the updated parameters $\mathbf{\Lambda}^{(\tau+1)}$ at iteration $\tau + 1$ by maximizing the auxiliary function based on

$$\mathbf{\Lambda}^{(\tau+1)} = \underset{\mathbf{\Lambda}}{\operatorname{argmax}}\, Q(\mathbf{\Lambda}|\mathbf{\Lambda}^{(\tau)}). \tag{2.7}$$

In this iterative procedure, new parameters $\mathbf{\Lambda}^{(\tau+1)}$ are then viewed as the old parameters when moving to next EM iteration $\tau + 2$ for estimating the newly updated parameters $\mathbf{\Lambda}^{(\tau+2)}$. Such an iterative estimation only assures a local optimum while the global optimum solution is not guaranteed. Instead of directly maximizing the likelihood in Eq. 2.5, the EM algorithm indirectly maximizes the auxiliary function and theoretically proves a nondecreasing auxiliary function through a number of EM iterations [8]. This property is attractive and useful in order to run debugging over the likelihood values in successive EM iterations in the implementation procedure. Nevertheless, the increase of auxiliary function may be rapidly saturated in the EM algorithm. But, a well-tuned setting of initialization of model parameters $\mathbf{\Lambda}^{(0)}$ is helpful to globally seek the optimal parameters.

It is important to investigate how the iterative optimization over auxiliary function $Q(\mathbf{\Lambda}|\mathbf{\Lambda}^{(\tau)})$ is performed to continuously increase the log likelihood $\log p(\mathcal{X}|\mathbf{\Lambda})$. Here, $\mathcal{Y} \triangleq \{\mathcal{X}, \mathcal{Z}\}$ denotes the complete data. Log likelihood function is denoted by $L(\mathbf{\Lambda}) \triangleq \log p(\mathcal{X}|\mathbf{\Lambda})$. We would like to investigate the increase of likelihood function $p(\mathcal{X}|\mathbf{\Lambda})$ by means of evaluating the increase of log likelihood function $L(\mathbf{\Lambda})$ since logarithm log is a monotonically increasing function. To do so, we first have the equation

$$p(\mathcal{Y}|\mathcal{X},\mathbf{\Lambda}) = \frac{p(\mathcal{Y},\mathcal{X}|\mathbf{\Lambda})}{p(\mathcal{X}|\mathbf{\Lambda})} = \frac{p(\mathcal{X}|\mathcal{Y},\mathbf{\Lambda})p(\mathcal{Y}|\mathbf{\Lambda})}{p(\mathcal{X}|\mathbf{\Lambda})} = \frac{p(\mathcal{Y}|\mathbf{\Lambda})}{p(\mathcal{X}|\mathbf{\Lambda})}, \tag{2.8}$$

which is obtained by using $p(\mathcal{X}|\mathcal{Y},\mathbf{\Lambda}) = p(\mathcal{X}|\mathcal{X},\mathcal{Z},\mathbf{\Lambda}) = 1$. Having this equation, we manipulate it by taking logarithm and then taking expectation over both sides of Eq. 2.8 with respect to $p(\mathcal{Y}|\mathcal{X},\mathbf{\Lambda}^{(\tau)})$ to yield

$$\begin{aligned}
L(\mathbf{\Lambda}) &= \log p(\mathcal{Y}|\mathbf{\Lambda}) - \log p(\mathcal{Y}|\mathcal{X},\mathbf{\Lambda}) && \text{(2.9a)}\\
&= \mathbb{E}_{\mathcal{Y}}[\log p(\mathcal{Y}|\mathbf{\Lambda})|\mathcal{X},\mathbf{\Lambda}^{(\tau)}] - \mathbb{E}_{\mathcal{Y}}[\log p(\mathcal{Y}|\mathcal{X},\mathbf{\Lambda})|\mathcal{X},\mathbf{\Lambda}^{(\tau)}] && \text{(2.9b)}\\
&= Q(\mathbf{\Lambda}|\mathbf{\Lambda}^{(\tau)}) - H(\mathbf{\Lambda}|\mathbf{\Lambda}^{(\tau)}) && \text{(2.9c)}
\end{aligned}$$

where $H(\mathbf{\Lambda}|\mathbf{\Lambda}^{(\tau)})$ is defined by

$$H(\mathbf{\Lambda}|\mathbf{\Lambda}^{(\tau)}) \triangleq \mathbb{E}_{\mathcal{Y}}[\log p(\mathcal{Y}|\mathcal{X},\mathbf{\Lambda})|\mathcal{X},\mathbf{\Lambda}^{(\tau)}]. \tag{2.10}$$

Here, the RHS of Eq. 2.9c is equal to $L(\mathbf{\Lambda})$ because

$$\mathbb{E}_{\mathcal{Y}}[L(\mathbf{\Lambda})] = \int p(\mathcal{Y}|\mathcal{X},\mathbf{\Lambda}^{(\tau)})L(\mathbf{\Lambda})d\mathcal{Y} = L(\mathbf{\Lambda}). \tag{2.11}$$

By considering the negative logarithm as a convex function, the Jensen's inequality can be applied to derive

$$\begin{aligned}
&H(\mathbf{\Lambda}^{(\tau)}|\mathbf{\Lambda}^{(\tau)}) - H(\mathbf{\Lambda}|\mathbf{\Lambda}^{(\tau)})\\
&= \mathbb{E}_{\mathcal{Y}}\left[\log \frac{p(\mathcal{Y}|\mathcal{X},\mathbf{\Lambda}^{(\tau)})}{p(\mathcal{Y}|\mathcal{X},\mathbf{\Lambda})}\,\middle|\,\mathcal{X},\mathbf{\Lambda}^{(\tau)}\right]\\
&= \int p(\mathcal{Y}|\mathcal{X},\mathbf{\Lambda}^{(\tau)})\left(-\log\frac{p(\mathcal{Y}|\mathcal{X},\mathbf{\Lambda})}{p(\mathcal{Y}|\mathcal{X},\mathbf{\Lambda}^{(\tau)})}\right)d\mathcal{Y}\\
&= \mathcal{D}_{\text{KL}}(p(\mathcal{Y}|\mathcal{X},\mathbf{\Lambda}^{(\tau)})\|p(\mathcal{Y}|\mathcal{X},\mathbf{\Lambda}))\\
&\geq -\log\left(\int p(\mathcal{Y}|\mathcal{X},\mathbf{\Lambda}^{(\tau)})\frac{p(\mathcal{Y}|\mathcal{X},\mathbf{\Lambda})}{p(\mathcal{Y}|\mathcal{X},\mathbf{\Lambda}^{(\tau)})}d\mathcal{Y}\right) = 0
\end{aligned} \tag{2.12}$$

where $\mathcal{D}_{\text{KL}}(\cdot\|\cdot)$ means the Kullback–Leibler divergence [21]. Substituting Eq. 2.12 into Eq. 2.9c yields

$$
\begin{aligned}
L(\mathbf{\Lambda}) &= Q(\mathbf{\Lambda}|\mathbf{\Lambda}^{(\tau)}) - H(\mathbf{\Lambda}|\mathbf{\Lambda}^{(\tau)}) \\
&\geq Q(\mathbf{\Lambda}^{(\tau)}|\mathbf{\Lambda}^{(\tau)}) - H(\mathbf{\Lambda}^{(\tau)}|\mathbf{\Lambda}^{(\tau)}) = L(\mathbf{\Lambda}^{(\tau)}),
\end{aligned}
\tag{2.13}
$$

which is derived by adopting the useful inequality

$$
\begin{aligned}
& Q(\mathbf{\Lambda}|\mathbf{\Lambda}^{(\tau)}) \geq Q(\mathbf{\Lambda}^{(\tau)}|\mathbf{\Lambda}^{(\tau)}) \\
& \Rightarrow L(\mathbf{\Lambda}) \geq L(\mathbf{\Lambda}^{(\tau)}) \\
& \Rightarrow p(\mathcal{X}|\mathbf{\Lambda}) \geq p(\mathcal{X}|\mathbf{\Lambda}^{(\tau)}).
\end{aligned}
\tag{2.14}
$$

In summary, we increase the auxiliary function $Q(\mathbf{\Lambda}|\mathbf{\Lambda}^{(\tau)})$, which equivalently increase the log likelihood function $L(\mathbf{\Lambda})$ or the likelihood function $p(\mathcal{X}|\mathbf{\Lambda})$. Auxiliary function is maximized to increase the likelihood function given by the updated parameters $\mathbf{\Lambda}^{(\tau)} \rightarrow \mathbf{\Lambda}^{(\tau+1)}$. However, we don't directly optimize the original likelihood function. Indirectly optimizing the auxiliary function will result in a local optimum in estimation of ML parameters.

2.2.3 Alternative Perspective

An alternative view of the EM algorithm is addressed by introducing a variational distribution $q(\mathcal{Z})$ for latent variables $\mathcal{Z}$. No matter what distribution $q(\mathcal{Z})$ is chosen, we can derive a general formula for decomposition of log likelihood in a form of

$$
\log p(\mathcal{X}|\mathbf{\Lambda}) = \mathcal{D}_{\text{KL}}(q\|p) + \mathcal{L}(q, \mathbf{\Lambda}).
\tag{2.15}
$$

This formula is obtained because

$$
\begin{aligned}
\mathcal{D}_{\text{KL}}(q\|p) &= -\sum_{\mathcal{Z}} q(\mathcal{Z}) \log\left\{\frac{p(\mathcal{Z}|\mathcal{X},\mathbf{\Lambda})}{q(\mathcal{Z})}\right\} \\
&= -\mathbb{E}_q[\log p(\mathcal{Z}|\mathcal{X},\mathbf{\Lambda})] - \mathbb{H}_q[\mathcal{Z}]
\end{aligned}
\tag{2.16}
$$

and

$$
\begin{aligned}
\mathcal{L}(q,\mathbf{\Lambda}) &= \sum_{\mathcal{Z}} q(\mathcal{Z}) \log\left\{\frac{p(\mathcal{X},\mathcal{Z}|\mathbf{\Lambda})}{q(\mathcal{Z})}\right\} \\
&= \mathbb{E}_q[\log p(\mathcal{X},\mathcal{Z}|\mathbf{\Lambda})] + \mathbb{H}_q[\mathcal{Z}],
\end{aligned}
\tag{2.17}
$$

where the entropy of $q(\mathcal{Z})$ is defined by

$$
\mathbb{H}_q[\mathcal{Z}] \triangleq -\sum_{\mathcal{Z}} q(\mathcal{Z}) \log q(\mathcal{Z}).
\tag{2.18}
$$

We have the following interpretation. The log likelihood $\log p(\mathcal{X}|\mathbf{\Lambda})$ is decomposed as a KL divergence between $q(\mathcal{Z})$ and posterior distribution $p(\mathcal{Z}|\mathcal{X},\mathbf{\Lambda})$ and a lower bound $\mathcal{L}(q,\mathbf{\Lambda})$ of the log likelihood function $\log p(\mathcal{X}|\mathbf{\Lambda})$. We call $\mathcal{L}(q,\mathbf{\Lambda})$ as the lower bound since KL term in Eq. 2.15 meets the property $\mathcal{D}_{\text{KL}}(q\|p) \geq 0$. This property was also shown in Eq. 2.12. Figure 2.4(a) illustrates the relation among log likelihood function, KL divergence, and lower bound. ML estimation of model parameters $\mathbf{\Lambda}$

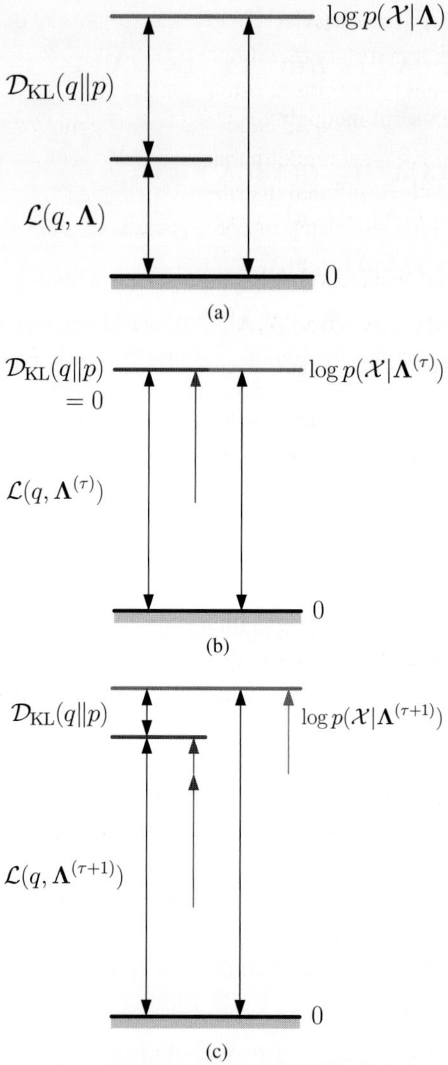

Figure 2.4 Illustration for (a) decomposing a log likelihood $\log p(\mathcal{X}|\mathbf{\Lambda})$ into a KL divergence $\mathcal{D}_{\mathrm{KL}}(q\|q)$ and a lower bound $\mathcal{L}(q,\mathbf{\Lambda})$, (b) updating the lower bound by setting KL divergence to zero with new distribution $q(\mathcal{Z}) = p(\mathcal{Z}|\mathcal{X},\mathbf{\Lambda}^{(\tau)})$, and (c) updating lower bound again by using new parameters $\mathbf{\Lambda}^{(\tau+1)}$ [1]. [Based on *Pattern Recognition and Machine Learning (Figure. 9.11–Figure. 9.13), by C. M. Bishop, 2006, Springer.*]

turns out as an indirect optimization via finding a variational distribution $q(\mathcal{Z})$. This indirect optimization is carried out by alternatively executing two steps to estimate the distribution of latent variables $q(\mathcal{Z})$ and the model parameters $\mathbf{\Lambda}$. The first step is to use the current estimate $\mathbf{\Lambda}^{(\tau)}$ to estimate the approximate distribution $q(\mathcal{Z})$ by

$$q(\mathcal{Z}) = p(\mathcal{Z}|\mathcal{X},\mathbf{\Lambda}^{(\tau)}), \tag{2.19}$$

which is obtained by imposing $\mathcal{D}_{\text{KL}}(q\|p) = 0$. Namely, the variational distribution is derived as the posterior distribution given by the old parameters $\mathbf{\Lambda}^{(\tau)}$ at iteration τ. As seen in Figure 2.4(b), the variational lower bound is improved accordingly. Then, in the second step, shown in Figure 2.4(c), we fix $q(\mathcal{Z})$ and maximize again the lower bound $\mathcal{L}(q, \mathbf{\Lambda})$ with respect to $\mathbf{\Lambda}$ to estimate new model parameters $\mathbf{\Lambda}^{(\tau+1)}$ at iteration $\tau + 1$. It is noted that this lower bound with variational distribution $q(\mathcal{Z})$ in Eq. 2.19 is exactly the same as the auxiliary function for the E-step in EM algorithm as shown below[1]

$$\begin{aligned}\mathcal{L}(q, \mathbf{\Lambda}) &= \mathbb{E}_q[\log p(\mathcal{X}|\mathcal{Z}, \mathbf{\Lambda}) + \log p(\mathcal{Z}|\mathbf{\Lambda})] + \text{const} \\ &= \mathbb{E}_{\mathcal{Z}}[\log p(\mathcal{X}, \mathcal{Z}|\mathbf{\Lambda})|\mathcal{X}, \mathbf{\Lambda}^{(\tau)}] + \text{const} \\ &= \mathcal{Q}(\mathbf{\Lambda}|\mathbf{\Lambda}^{(\tau)}) + \text{const.}\end{aligned} \tag{2.20}$$

Here, the terms independent of $\mathbf{\Lambda}$ are merged in a constant. Given the parameters $\mathbf{\Lambda}^{(\tau+1)}$, the distribution of latent variables is further updated at the next EM iteration as $\mathbf{\Lambda}^{(\tau+2)} \leftarrow \mathbf{\Lambda}^{(\tau+1)}$. A series of parameter updating are obtained by

$$\mathbf{\Lambda}^{(0)} \to \mathbf{\Lambda}^{(1)} \to \mathbf{\Lambda}^{(2)} \to \cdots \tag{2.21}$$

This updating assures that the lower bound $\mathcal{L}(q, \mathbf{\Lambda})$ is sequentially elevated and the log likelihood function $\log p(\mathcal{X}|\mathbf{\Lambda})$ is accordingly improved. Therefore, finding $q(\mathcal{Z})$ in Eq. 2.19 is equivalent to calculating the auxiliary function in E-step. Maximizing the lower bound $\mathcal{L}(q, \mathbf{\Lambda})$ with respect to model parameters $\mathbf{\Lambda}$ corresponds to fulfill the M-step in EM iterative updating. Originally, the log likelihood $\log p(\mathcal{X}|\mathbf{\Lambda})$ under latent variable model with $\mathcal{Z}$ is seen a non-convex or non-concave function. However, the inference based on the lower bound or expectation function in Eqs. 2.6 and 2.20 under Gaussian observation distribution becomes a concave optimization as illustrated in Figure 2.5. In this figure, the continuously maximizing the lower bound or the iterative updating of model parameters from $\mathbf{\Lambda}^{(\tau)}$ to $\mathbf{\Lambda}^{(\tau+1)}$ via EM steps assures the optimum in a ML estimate $\mathbf{\Lambda}_{\text{ML}}$. Owing to a nondecreasing lower bound or auxiliary function, we guarantee the convergence of model parameters within a few EM steps.

2.2.4 Maximum *A Posteriori*

From a Bayesian perspective, it is meaningful to extend the optimization over the posterior distribution $p(\mathbf{\Lambda}|\mathcal{X})$ and address how to seek the MAP estimate $\mathbf{\Lambda}_{\text{MAP}}$ by maximizing the posterior distribution of model parameters $\mathbf{\Lambda}$ by giving observations $\mathcal{X}$. The discrete latent variables $\mathcal{Z}$ are introduced in the following MAP estimation to estimate the mode $\mathbf{\Lambda}_{\text{MAP}}$ of posterior distribution from the collection of training data $\mathcal{X}$

$$\begin{aligned}\mathbf{\Lambda}_{\text{MAP}} &= \underset{\mathbf{\Lambda}}{\operatorname{argmax}}\, p(\mathbf{\Lambda}|\mathcal{X}) \\ &= \underset{\mathbf{\Lambda}}{\operatorname{argmax}} \sum_{\mathcal{Z}} p(\mathcal{X}, \mathcal{Z}|\mathbf{\Lambda}) p(\mathbf{\Lambda}).\end{aligned} \tag{2.22}$$

[1] Note that in Eq. 2.9c, we take the expectation with respect to the posterior $p(\mathcal{Y}|\mathcal{X}, \mathbf{\Lambda}^{(r)})$, which is equivalent to taking expectation with respect to $p(\mathcal{Z}|\mathcal{X}, \mathbf{\Lambda}^{(r)})$ as $\mathcal{Y} = \{\mathcal{X}, \mathcal{Z}\}$.

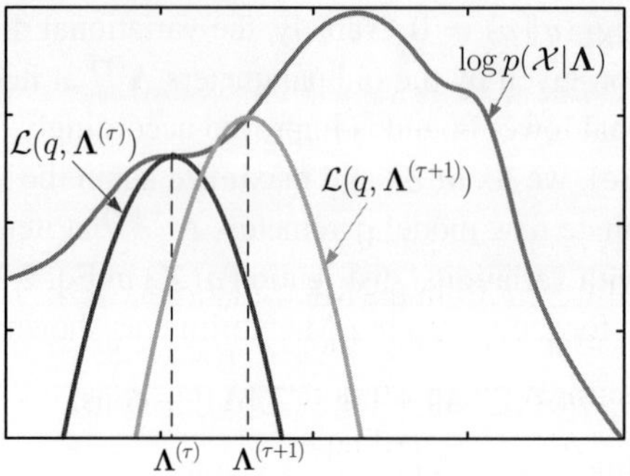

Figure 2.5 Illustration of updating the lower bound of log likelihood function $\mathcal{L}(q, \Lambda)$ in EM iteration from τ to $\tau + 1$. Lower bound is seen as an iterative log Gaussian approximation to the original log likelihood function $\log p(\mathcal{X}|\Lambda)$. [Based on *Pattern Recognition and Machine Learning (Figure. 9.14)*, by C. M. Bishop, 2006, Springer.]

$$\underbrace{p(\Lambda|\mathcal{X})}_{\text{Posterior}} = \frac{\overbrace{p(\mathcal{X}|\Lambda)}^{\text{Likelihood}} \overbrace{p(\Lambda)}^{\text{Prior}}}{\underbrace{\int p(\mathcal{X}|\Lambda)p(\Lambda)d\Lambda}_{\text{Evidence}}}$$

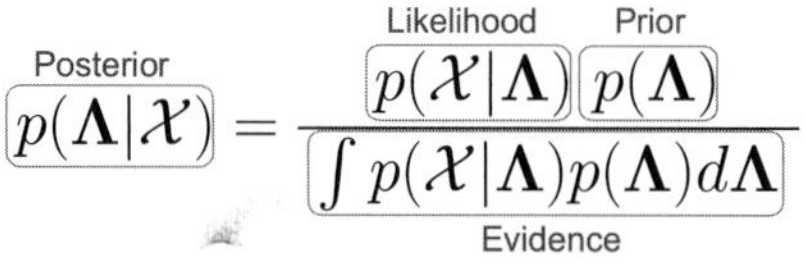

Figure 2.6 Bayes rule for illustrating the relation among the posterior distribution, likelihood function, prior distribution, and evidence function.

In Eq. 2.22, the summation is taken due to the discrete latent variables $\mathcal{Z}$. Figure 2.6 shows the relation among posterior distribution, likelihood function, prior distribution, and evidence function, which is obtained by the Bayes theorem. We don't really need to compute the integral in evidence function for the MAP estimation because the evidence function $p(\mathcal{X}) = \int p(\mathcal{X}|\Lambda)p(\Lambda)d\Lambda$ is independent of Λ. The MAP estimation is an extension of the ML estimation by additionally considering the prior distribution that acted as a regularization term to reshape the estimation of model parameters. It is obvious that MAP provides advantage over ML for a richer parameter estimation. Importantly, if a conjugate prior distribution is chosen and combined with the likelihood function in a form of exponential distribution, a MAP estimation comparably searches the mode of the corresponding posterior distribution, which is formulated as the same distribution but with different parameters as the prior distribution. Conjugate prior is crucial in a MAP estimation. If we don't adopt the conjugate prior, the analytical posterior distribution does not exist for finding the mode or optimum of posterior distribution.

Here, we address an EM algorithm for the MAP estimation, which is an extension from EM based on a ML estimation. The incomplete data problem in a MAP estimation is tackled via EM steps to derive an analytical solution to local optimum Λ_{MAP}. The resulting MAP-EM algorithm is formulated as E-step and M-step in an iterative learning procedure that updates new estimate Λ from the old parameters $\Lambda^{(\tau)}$ at iteration τ. In the first step (E-step), we calculate the posterior auxiliary function, which is an expectation over posterior distribution

$$
\begin{aligned}
R(\mathbf{\Lambda}|\mathbf{\Lambda}^{(\tau)}) &= \mathbb{E}_{\mathcal{Z}}[\log p(\mathcal{X},\mathcal{Z},\mathbf{\Lambda})|\mathcal{X},\mathbf{\Lambda}^{(\tau)}] \\
&= \underbrace{\mathbb{E}_{\mathcal{Z}}[\log p(\mathcal{X},\mathcal{Z}|\mathbf{\Lambda})|\mathcal{X},\mathbf{\Lambda}^{(\tau)}]}_{Q(\mathbf{\Lambda}|\mathbf{\Lambda}^{(\tau)})} + \log p(\mathbf{\Lambda}).
\end{aligned} \tag{2.23}
$$

In Eq. 2.23, the prior term $\log p(\mathbf{\Lambda})$ is independent of $\mathcal{Z}$ and is accordingly outside the expectation function. Obviously, different from the ML estimation, there is an additional factor $\log p(\mathbf{\Lambda})$ in the MAP estimation due to the prior density of model parameters. In the second step (M-step), we maximize the posterior expectation function to estimate the new model parameters $\mathbf{\Lambda}^{(\tau)} \to \mathbf{\Lambda}^{(\tau+1)}$ according to

$$
\mathbf{\Lambda}^{(\tau+1)} = \underset{\mathbf{\Lambda}}{\operatorname{argmax}}\, R(\mathbf{\Lambda}|\mathbf{\Lambda}^{(\tau)}). \tag{2.24}
$$

The parameters are updated by $\mathbf{\Lambda}^{(\tau)} \to \mathbf{\Lambda}^{(\tau+1)}$ at iteration $\tau + 1$ and then treated as the old parameters for estimation of new parameters at iteration $\tau + 2$ in the next EM iteration.

We may further investigate the optimization of expectation function $R(\mathbf{\Lambda}|\mathbf{\Lambda}^{(\tau)})$, which can result in local optimum of $p(\mathbf{\Lambda}|\mathcal{X})$ or $p(\mathcal{X}|\mathbf{\Lambda})p(\mathbf{\Lambda})$. To do so, we first have the definition of logarithm of joint distribution $L_{\text{MAP}}(\mathbf{\Lambda}) \triangleq \log(p(\mathcal{X}|\mathbf{\Lambda})p(\mathbf{\Lambda}))$ and use the relation

$$
p(\mathcal{X}|\mathbf{\Lambda}) = \frac{p(\mathcal{X},\mathcal{Z}|\mathbf{\Lambda})}{p(\mathcal{Z}|\mathcal{X},\mathbf{\Lambda})}, \tag{2.25}
$$

to rewrite the expression as

$$
L_{\text{MAP}}(\mathbf{\Lambda}) = \log p(\mathcal{X},\mathcal{Z}|\mathbf{\Lambda}) - \log p(\mathcal{Z}|\mathcal{X},\mathbf{\Lambda}) + \log p(\mathbf{\Lambda}). \tag{2.26}
$$

By referring Eq. 2.9c and taking expectation of both sides of Eq. 2.26 with respect to posterior distribution $p(\mathcal{Z}|\mathcal{X},\mathbf{\Lambda}^{(\tau)})$, we yield

$$
\begin{aligned}
L_{\text{MAP}}(\mathbf{\Lambda}) &= \mathbb{E}_{\mathcal{Z}}[\log p(\mathcal{X},\mathcal{Z}|\mathbf{\Lambda})|\mathcal{X},\mathbf{\Lambda}^{(\tau)}] \\
&\quad - \mathbb{E}_{\mathcal{Z}}[\log p(\mathcal{Z}|\mathcal{X},\mathbf{\Lambda})|\mathcal{X},\mathbf{\Lambda}^{(\tau)}] + \log p(\mathbf{\Lambda}) \\
&= \underbrace{\mathbb{E}_{\mathcal{Z}}[\log p(\mathcal{X},\mathcal{Z}|\mathbf{\Lambda})|\mathcal{X},\mathbf{\Lambda}^{(\tau)}] + \log p(\mathbf{\Lambda})}_{=R(\mathbf{\Lambda}|\mathbf{\Lambda}^{(\tau)})} \\
&\quad - \underbrace{\mathbb{E}_{\mathcal{Z}}[\log p(\mathcal{Z}|\mathcal{X},\mathbf{\Lambda})|\mathcal{X},\mathbf{\Lambda}^{(\tau)}]}_{\triangleq H(\mathbf{\Lambda}|\mathbf{\Lambda}^{(\tau)})}.
\end{aligned} \tag{2.27}
$$

The terms $L_{\text{MAP}}(\mathbf{\Lambda})$ and $\log p(\mathbf{\Lambda})$ are independent of $\mathcal{Z}$ and are unchanged in Eq. 2.27. The posterior auxiliary function is decomposed as

$$
R(\mathbf{\Lambda}|\mathbf{\Lambda}^{(\tau)}) = L_{\text{MAP}}(\mathbf{\Lambda}) + H(\mathbf{\Lambda}|\mathbf{\Lambda}^{(\tau)}). \tag{2.28}
$$

Because it has been shown in Eq. 2.12 that $H(\mathbf{\Lambda}|\mathbf{\Lambda}^{(\tau)})$ is continuously decreasing, i.e.,

$$
H(\mathbf{\Lambda}|\mathbf{\Lambda}^{(\tau)}) \le H(\mathbf{\Lambda}^{(\tau)}|\mathbf{\Lambda}^{(\tau)}), \tag{2.29}
$$

we can derive

$$\begin{aligned} & R(\boldsymbol{\Lambda}|\boldsymbol{\Lambda}^{(\tau)}) \geq R(\boldsymbol{\Lambda}^{(\tau)}|\boldsymbol{\Lambda}^{(\tau)}) \\ & \Rightarrow L_{\text{MAP}}(\boldsymbol{\Lambda}) \geq L_{\text{MAP}}(\boldsymbol{\Lambda}^{(\tau)}) \\ & \Rightarrow p(\boldsymbol{\Lambda}|\mathcal{X}) \geq p(\boldsymbol{\Lambda}^{(\tau)}|\mathcal{X}). \end{aligned} \tag{2.30}$$

Namely, the increasing posterior auxiliary function $R(\boldsymbol{\Lambda}|\boldsymbol{\Lambda}^{(\tau)})$ will lead to an increasing logarithm of posterior distribution $L(\boldsymbol{\Lambda})$ or equivalently an increasing posterior distribution $p(\boldsymbol{\Lambda}|\mathcal{X})$. An EM algorithm is completed for a MAP estimation, which attains a converged local optimum for MAP parameters.

2.3 Approximate Inference

Variational Bayesian (VB) inference has been originated in machine learning literature since the late 1990s [22, 23]. VB theory has become a popular tool for approximate Bayesian inference in the presence of a latent variable model by implementing the EM-like algorithm. This section would like to construct a general latent variable model from a set of training data $\mathcal{X} = \{\mathbf{x}_t\}_{t=1}^T$ based on the set of variables including latent variables $\mathcal{Z}$, parameters $\boldsymbol{\Lambda}$, and hyperparameters $\boldsymbol{\Psi}$. All such variables are used to build an assumed topology $\mathcal{T}$ in learning representation. In this learning process, a fundamental issue in VB algorithm is to infer the joint posterior distribution

$$p(\mathcal{Z}|\mathcal{X}) \qquad \text{where } \mathcal{Z} = \{Z_j\}_{j=1}^J, \tag{2.31}$$

where all variables are included in the inference problem. Given the posterior distribution $p(\mathcal{Z}|\mathcal{X})$, the auxiliary functions in Eq. 2.23 are calculated to fulfill the E-step of an EM algorithm for both a ML and a MAP estimation procedures, respectively. Nevertheless, a number of latent variables do exist in probabilistic speaker recognition systems. These variables are coupled each other in calculation of posterior distribution. For instance, in a joint factor analysis model shown in Eq. 3.165, the speaker factor $\mathbf{y}(s)$ and the channel factor $\mathbf{x}_h(s)$ are correlated in their posterior distributions.

However, the analytical solution to exact calculation of posterior distribution $p(\mathcal{Z}|\mathcal{X})$ does not exist. The variational method is applied to carry out an approximate inference to handle this issue. Using the variational inference, an approximate or variational posterior distribution $q(\mathcal{Z})$ is factorized and viewed as a proxy to the original true distribution $p(\mathcal{Z}|\mathcal{X})$. Figure 2.7 illustrates the idea of approximate inference that analytically uses the variational distribution $q(\mathcal{Z})$ to approximate true posterior $p(\mathcal{Z}|\mathcal{X})$.

In the sequel, the underlying concept of approximating true posterior $p(\mathcal{Z}|\mathcal{X})$ using variational distribution $q(\mathcal{Z})$ is addressed. We formulate the variational Bayesian expectation-maximization VB-EM algorithm to implement a ML or MAP estimation where multiple latent variables are coupled in original posterior distribution. For notation simplicity, we formulate the approximation in the presence of continuous latent variables. The marginalization of probabilistic distribution is performed by integrating latent variables $\int d\mathcal{Z}$ instead of summing latent variables $\sum_{\mathcal{Z}}$.

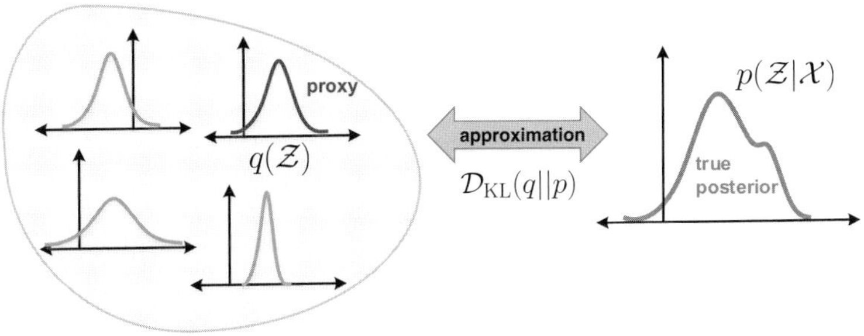

Figure 2.7 Illustration of finding an approximate Gaussian distribution $q(\mathcal{Z})$ that is factorizable and optimally close to the true posterior $p(\mathcal{Z}|\mathcal{X})$.

2.3.1 Variational Distribution

Approximate inference needs to approximate the true posterior distribution $p(\mathcal{Z}|\mathcal{X})$ by using a variational distribution $q(\mathcal{Z})$ based on a distance metric for two distributions. In Section 2.2.3, a Kullback–Leibler (KL) divergence [21] was adopted as a measure of distance between two distance in a ML–EM algorithm. In this section, a KL divergence is used to estimate the variational distribution. A KL divergence between $q(\mathcal{Z})$ and $p(\mathcal{Z}|\mathcal{X})$ is defined and decomposed by

$$\begin{aligned}\mathcal{D}_{\mathrm{KL}}(q(\mathcal{Z})\|p(\mathcal{Z}|\mathcal{X})) &\triangleq \int q(\mathcal{Z})\log\frac{q(\mathcal{Z})}{p(\mathcal{Z}|\mathcal{X})}d\mathcal{Z}\\ &= -\mathbb{E}_q[\log p(\mathcal{Z}|\mathcal{X})] - \mathbb{H}_q[\mathcal{Z}]\\ &= \int q(\mathcal{Z})\log\frac{q(\mathcal{Z})}{\frac{p(\mathcal{X},\mathcal{Z})}{p(\mathcal{X})}}d\mathcal{Z}\\ &= \log p(\mathcal{X}) - \underbrace{\int q(\mathcal{Z})\log\frac{p(\mathcal{X},\mathcal{Z})}{q(\mathcal{Z})}d\mathcal{Z}}_{\triangleq \mathcal{L}(q(\mathcal{Z}))},\end{aligned} \tag{2.32}$$

where

$$\begin{aligned}\mathcal{L}(q(\mathcal{Z})) &\triangleq \int q(\mathcal{Z})\log\frac{p(\mathcal{X},\mathcal{Z})}{q(\mathcal{Z})}d\mathcal{Z}\\ &= \mathbb{E}_q[\log p(\mathcal{X},\mathcal{Z})] + \mathbb{H}_q[\mathcal{Z}].\end{aligned} \tag{2.33}$$

Here, $\mathcal{L}(q(\mathcal{Z}))$ is seen as the the variational lower bound of $\log p(\mathcal{X})$, which is also known as the evidence lower bound (ELBO) $\mathcal{L}(q(\mathcal{Z}))$. It is because KL divergence is nonnegative so as to obtain $\log p(\mathcal{X}) \geq \mathcal{L}(q(\mathcal{Z}))$. This result is similar to what we have addressed in Eqs. 2.15–2.17. The variational distribution $q(\mathcal{Z})$, which is closest to the original posterior distribution $p(\mathcal{Z}|\mathcal{X})$, is obtained by minimizing the KL divergence between them or equivalently maximizing the variational lower bound $\mathcal{L}(q(Z|\mathcal{X}))$. This is because the evidence function $\log p(\mathcal{X})$ is independent of $\mathcal{Z}$. Therefore, we have

$$\widehat{q}(\mathcal{Z}) = \underset{q(\mathcal{Z})}{\operatorname{argmax}}\, \mathcal{L}(q(\mathcal{Z})). \tag{2.34}$$

In the implementation, an analytical solution to Eq. 2.34 is derived under the constraint that the variational posterior distribution $q(\mathcal{Z})$ is factorizable. We therefore derive the variational inference procedure where a surrogate of original posterior $p(\mathcal{Z}|\mathcal{X})$, here an alternative posterior based on the variational distribution $q(\mathcal{Z})$, is estimated to find a ML $\mathbf{\Lambda}_{\text{ML}}$ or a MAP model parameters $\mathbf{\Lambda}_{\text{MAP}}$. The resulting variational Bayesian (VB) inference is also known as the factorized variational inference, since the approximate inference is realized by the constraint of factorization.

2.3.2 Factorized Distribution

According to the factorized variational inference, the variational distribution is assumed to be conditionally independent over J latent variables

$$q(\mathcal{Z}) = \prod_{j=1}^{J} q(Z_j). \tag{2.35}$$

Notably, the factorization is *not* imposed in true posterior $p(\mathcal{Z}|\mathcal{X})$. We preserve the original posterior and arrange the posterior of a target latent variable $p(Z_i|\mathcal{X})$ in a form of marginal distribution of $p(\mathcal{Z}|\mathcal{X})$ over all Z_j except Z_i

$$\begin{aligned} p(Z_i|\mathcal{X}) &= \int \cdots \int p(\mathcal{Z}|\mathcal{X}) \prod_{j \neq i}^{J} dZ_j \\ &\triangleq \int p(\mathcal{Z}|\mathcal{X}) dZ_{\setminus i}. \end{aligned} \tag{2.36}$$

Here, $Z_{\setminus i}$ is defined as the complementary set of Z_i. Eq. 2.36 is then used to calculate the following KL divergence between $q(Z_i)$ and $p(Z_i|\mathcal{X})$

$$\begin{aligned} &\mathcal{D}_{\text{KL}}(q(Z_i) \| p(Z_i|\mathcal{X})) = \int q(Z_i) \log \frac{q(Z_i)}{\int p(\mathcal{Z}|\mathcal{X}) dZ_{\setminus i}} dZ_i \\ &= \int q(Z_i) \log \frac{q(Z_i)}{\int \frac{p(\mathcal{X},\mathcal{Z})}{p(\mathcal{X})} dZ_{\setminus i}} dZ_i \\ &= \log p(\mathcal{X}) - \int q(Z_i) \log \frac{\int p(\mathcal{X},\mathcal{Z}) dZ_{\setminus i}}{q(Z_i)} dZ_i. \end{aligned} \tag{2.37}$$

By considering the negative logarithm in the second term of Eq. 2.37 as a convex function, similar to Eq. 2.12, the Jensen's inequality is applied to yield

$$\begin{aligned} &\int q(Z_i) \log \left(\frac{\int p(\mathcal{X},\mathcal{Z}) dZ_{\setminus i}}{q(Z_i)} \right) dZ_i \\ &= \int q(Z_i) \log \left(\int q(Z_{\setminus i}) \frac{p(\mathcal{X},\mathcal{Z})}{q(\mathcal{Z})} dZ_{\setminus i} \right) dZ_i \\ &\geq \int q(Z_i) \int q(Z_{\setminus i}) \log \left(\frac{p(\mathcal{X},\mathcal{Z})}{q(\mathcal{Z})} \right) dZ_{\setminus i} dZ_i \\ &= \int q(\mathcal{Z}) \log \left(\frac{p(\mathcal{X},\mathcal{Z})}{q(\mathcal{Z})} \right) d\mathcal{Z} = \mathcal{L}(q(\mathcal{Z})). \end{aligned} \tag{2.38}$$

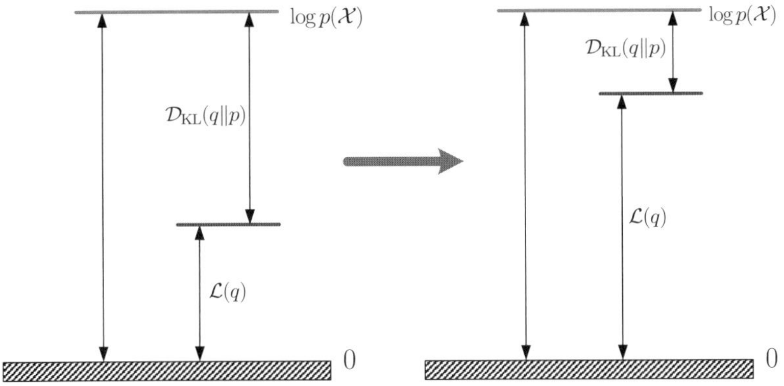

Figure 2.8 Illustration of minimizing KL divergence $\mathcal{D}_{\text{KL}}(q\|p)$ via maximization of variational lower bound $\mathcal{L}(q)$ where $\log p(\mathcal{X})$ is fixed.

We therefore obtain the inequality

$$\mathcal{D}_{\text{KL}}(q(Z_i)\|p(Z_i|\mathcal{X})) \leq \log p(\mathcal{X}) - \mathcal{L}(q(\mathcal{Z})). \tag{2.39}$$

Correspondingly, we minimize the KL divergence $\mathcal{D}_{\text{KL}}(q(Z_i)\|p(Z_i|\mathcal{X}))$ or maximize the variational lower bound to derive the variational distribution for individual latent variable $q(Z_i)$ as follows:

$$\widehat{q}(Z_i) = \underset{q(Z_i)}{\operatorname{argmax}}\, \mathcal{L}(q(\mathcal{Z})). \tag{2.40}$$

The joint posterior distribution $\widehat{q}(\mathcal{Z})$ is then obtained by Eq. 2.35 by using J variational posterior distributions $\{\widehat{q}(Z_i)\}_{i=1}^{J}$ found by Eq. 2.40. An illustration of maximizing the variational lower bound $\mathcal{L}(q(\mathcal{Z}))$ or minimizing the KL divergence $\mathcal{D}_{\text{KL}}(q(\mathcal{Z})\|p(\mathcal{Z}|\mathcal{X}))$ for estimation of variational distribution $q(\mathcal{X})$ is shown in Figure 2.8.

We can formulate the optimization problem in a form of

$$\begin{aligned} &\max_{q(Z_i)} \;\mathbb{E}_q[\log p(\mathcal{X},\mathcal{Z})] + \mathbb{H}_q[\mathcal{Z}] \\ &\text{subject to} \int_{\mathcal{Z}} q(d\mathcal{Z}) = 1. \end{aligned} \tag{2.41}$$

In this optimization, we may arrange the variational lower bound in Eq. 2.33 as

$$\mathcal{L}(q(\mathcal{Z})) = \int \prod_j q_j \left\{\log p(\mathcal{X},\mathcal{Z}) - \sum\nolimits_j \log q_j\right\} d\mathcal{Z} \tag{2.42a}$$

$$= \int q_i \underbrace{\int \log p(\mathcal{X},\mathcal{Z}) \prod_{j\neq i} q_j dZ_j}_{\triangleq \log \widetilde{p}(\mathcal{X},Z_i)} dZ_i - \int q_i \log q_i dZ_i + \text{const} \tag{2.42b}$$

$$= -\mathcal{D}_{\text{KL}}(q(Z_i)\|\widetilde{p}(\mathcal{X},Z_i)) + \text{const}. \tag{2.42c}$$

In Eq. 2.42, we simply denote $q(Z_i)$ as q_i and define the term

$$\log \widetilde{p}(\mathcal{X}, Z_i) = \mathbb{E}_{\mathcal{Z}_{\setminus i}}[\log p(\mathcal{X}, \mathcal{Z})] + \text{const.} \tag{2.43}$$

The second term of Eq. 2.42(b) is obtained by noting that

$$\begin{aligned}
&\int \prod_j q_j \Big(\sum_j \log q_j\Big) d\mathcal{Z} \\
&= \int \Big(\prod_j q_j\Big) \Big(\log q_i + \sum_{j \neq i} \log q_j\Big) d\mathcal{Z} \\
&= \int \Big(\prod_j q_j\Big) \log q_i d\mathcal{Z} + \int \Big(\prod_j q_j\Big) \Big(\sum_{j \neq i} \log q_j\Big) d\mathcal{Z} \\
&= \Big(\prod_{j \neq i} \int q_j dZ_j\Big) \int q_i \log q_i dZ_i + \int \Big(\prod_{j \neq i} q_j\Big) q_i \Big(\sum_{j \neq i} \log q_j\Big) d\mathcal{Z} \\
&= \int q_i \log q_i dZ_i + \int \Big(\prod_{j \neq i} q_j\Big) \Big(\sum_{j \neq i} \log q_j\Big) d\mathcal{Z}_{\setminus i} \int q_i dZ_i \\
&= \int q_i \log q_i dZ_i + \text{const independent of } q_i,
\end{aligned} \tag{2.44}$$

where we have used the property of distribution $\int q_i dZ_i = 1$. The factorized variational distribution is finally obtained by

$$\widehat{q}(Z_i) \propto \exp\left(\mathbb{E}_{\mathcal{Z}_{\setminus i}}\left[\log p(\mathcal{X}, \mathcal{Z})\right]\right). \tag{2.45}$$

More specifically, the exact solution to general form of the factorized distribution is expressed by

$$\widehat{q}(Z_i) = \frac{\exp\left(\mathbb{E}_{\mathcal{Z}_{\setminus i}}\left[\log p(\mathcal{X}, \mathcal{Z})\right]\right)}{\int \exp\left(\mathbb{E}_{\mathcal{Z}_{\setminus i}}\left[\log p(\mathcal{X}, \mathcal{Z})\right]\right) dZ_i} \tag{2.46}$$

where the normalization constant is included. From Eq. 2.45, we can see that the approximate posterior distribution of a target latent variable Z_i is inferred through calculating the joint distribution of training data $\mathcal{X}$ and all latent variables $\mathcal{Z}$. The logarithm of the derived variational distribution of a target variable Z_i is obtained by calculating the logarithm of the joint distribution over all visible and hidden variables and taking the expectation over all the other hidden variables $\mathcal{Z}_{\setminus i}$. Although the variational distribution is assumed to be factorizable, the learned target posterior distribution $\widehat{q}(Z_i)$ and the other posterior distributions $q(\mathcal{Z}_{\setminus i}) = \prod_{j \neq i}^{J} q(Z_j)$ are correlated each other. This finding can be seen from Eq. 2.45. As a result, starting from the initial variational distributions, we iteratively perform optimization for factorized distributions of all $q(Z_i)$.

2.3.3 EM versus VB-EM Algorithms

In general, we may follow the alternative perspective of the EM algorithm, as addressed in Section 2.2.3, to carry out the estimation of factorized posterior distributions $\{\widehat{q}(Z_j)\}_{j=1}^{J}$. This is comparable to fulfill an expectation step (E-step) similar to Eq. 2.19 in the standard EM algorithm. The key difference of VB-EM algorithm compared to EM algorithm is an additional factorization step due to the coupling of various latent variables in true posterior distribution $p(\mathcal{Z}|\mathcal{X}, \mathbf{\Lambda}^{(\tau)})$. Using the VB-EM algorithm, the factorized distribution $q(Z_i) = \widetilde{p}(\mathcal{X}, Z_i)$ does not lead to the result $\mathcal{D}_{\text{KL}}(q(\mathcal{Z})\|p(\mathcal{Z}|\mathcal{X})) = 0$ in Eq. 2.32. But, using EM algorithm, the variational distribution $q(\mathcal{Z}) = p(\mathcal{Z}|\mathcal{X}, \mathbf{\Lambda}^{(\tau)})$ is imposed to force $\mathcal{D}_{\text{KL}}(q(\mathcal{Z})\|p(\mathcal{Z}|\mathcal{X})) = 0$ in Eq. 2.15. Figures 2.8 and 2.4(a)–(b) also show the difference between the EM and VB-EM algorithms. Such a difference is mainly because of the factorization of variational distribution that is required in VB-EM algorithm in the presence of various latent variables.

In addition to inferring the factorized variational distribution, it is crucial to address the general formula of VB-EM algorithm for a ML or MAP estimation. In this case, the model parameters in a latent variable model are merged in the variational lower bound $\widetilde{\mathcal{L}}(q(\mathcal{Z}), \mathbf{\Lambda})$ which is similar to Eq. 2.17 in the EM algorithm. Obviously, the variational lower bound of a VB-EM algorithm $\widetilde{\mathcal{L}}(q, \mathbf{\Lambda})$ is different from that of an EM algorithm $\mathcal{L}(q, \mathbf{\Lambda})$. To tell the difference, we denote the variational lower bound of a VB-EM inference as $\widetilde{\mathcal{L}}(\cdot)$ while the bound of an EM inference is denoted by $\mathcal{L}(\cdot)$. Similar to an EM algorithm, there are also two steps in a VB-EM algorithm. The first step (called the VB-E step) is to estimate the variational distribution of each individual latent variable Z_i at new iteration $\tau + 1$ by maximizing the variational lower bound

$$q^{(\tau+1)}(Z_i) = \underset{q(Z_i)}{\operatorname{argmax}}\, \widetilde{\mathcal{L}}(q(\mathcal{Z}), \mathbf{\Lambda}^{(\tau)}), \tag{2.47}$$

where $\{Z_i\}_{i=1}^J$. Given the updated variational distribution

$$q^{(\tau+1)}(\mathcal{Z}) = \prod_{j=1}^{J} q^{(\tau+1)}(Z_j), \tag{2.48}$$

the second step (also called the VB-M step) is to estimate the ML or MAP parameters by maximizing the new lower bound

$$\mathbf{\Lambda}^{(\tau+1)} = \underset{\mathbf{\Lambda}}{\operatorname{argmax}}\, \widetilde{\mathcal{L}}(q^{(\tau+1)}(\mathcal{Z}), \mathbf{\Lambda}). \tag{2.49}$$

The updating of the ML or MAP parameters $\mathbf{\Lambda}^{(0)} \to \mathbf{\Lambda}^{(1)} \to \mathbf{\Lambda}^{(2)} \cdots$ based on the VB-EM steps does improve the variational lower bound $\widetilde{\mathcal{L}}(q(\mathcal{Z}), \mathbf{\Lambda})$ or increase the corresponding likelihood function $p(\mathcal{X}|\mathbf{\Lambda})$ or posterior distribution $p(\mathbf{\Lambda}|\mathcal{X})$. Without loss of generality, the learning process using the VB-EM algorithm is here named as the variational Bayesian learning.

2.4 Sampling Methods

Exact inference in most latent variable models is intractable. As a result, it is necessary to approximate the posterior distribution of the latent variables in these models. In addition to variational Bayesian inference, Markov chain Monte Carlo [24, 25] is an alternative implementation of the full Bayesian treatment for practical solutions.

MCMC is popular as a stochastic approximation, which is different from the deterministic approximation based on variational Bayesian (VB) as addressed in Section 2.3. It is interesting to make a comparison between VB inference and MCMC inference. In general, the approximate inference using VB theory aims to approximate the true posterior distribution based on the factorized variational distribution over a set of latent variables. VB inference is easy to scale up for large systems. However, on the other hand, MCMC inference involves the numerical computation for sampling process, which is computationally demanding. This is different from finding analytical solutions to solving the integrals and expectations in VB inference. Moreover, MCMC can freely use any distributions in mind, which provides an avenue to highly flexible models for a wide range of applications. The advanced speaker recognition systems are seen as the complicated probabilistic models with a number of latent variables.

A basic issue in MCMC inference is to calculate the expectation for a latent variable function $f(\mathcal{Z})$ by taking into account the probability distribution $p(\mathcal{Z})$. The individual latent variables in $\mathcal{Z}$ may be discrete or continuous and may contain some factors or parameters that are estimated for the construction of a probabilistic speaker recognition model. In case of continuous latent variables in $\mathcal{Z}$, the expectation function is evaluated by taking an integral

$$\mathbb{E}_{\mathcal{Z}}[f(\mathcal{Z})] = \int f(\mathcal{Z})p(\mathcal{Z})d\mathcal{Z}. \tag{2.50}$$

If $\mathcal{Z}$'s are discrete, the integral is replaced by a summation. The reason of resorting to sampling methods is that such expectations are too complicated to be evaluated analytically. That is, the sampling methods avoid the requirement of an analytical solution for Eq. 2.50. Instead, what we need is to obtain a set of samples $\{\mathcal{Z}^{(l)}, l = 1, \ldots, L\}$ that are drawn independently from the distribution $p(\mathcal{Z})$. The integral is then approximated by finding an ensemble mean of function $f(\mathbf{\Lambda})$ given by these samples $\mathcal{Z}^{(l)}$

$$\widehat{f} = \frac{1}{L}\sum_{l=1}^{L} f(\mathcal{Z}^{(l)}). \tag{2.51}$$

Because the samples $\mathcal{Z}^{(l)}$ are collected from the distribution $p(\mathcal{Z})$, the estimator $\widehat{f}$ acts as a good approximation to the true expectation in Eq. 2.50. Roughly speaking, 10–20 independent samples are sufficient to approximate the expectation function. However, in real implementation, we may not draw the samples $\{\mathcal{Z}^{(l)}\}_{l=1}^{L}$ independently. The number of effective samples may be much less than the number of total samples L. This implies that a larger sample size $L' > L$ is needed to ensure a good approximation of the true expectation.

To apply Eq. 2.51, we need to sample from $p(\mathcal{Z})$. If $p(\mathcal{Z})$ is not a simple standard distribution, drawing samples from such a distribution is usually difficult. Therefore, we may use the technique of *importance sampling* in which the samples are drawn from a *proposal distribution* $q(\mathcal{Z})$ instead of original distribution $p(\mathcal{Z})$. The idea is that sampling from proposal distribution $q(\mathcal{Z})$ is much easier than sampling from original distribution $p(\mathcal{Z})$. The expectation in Eq. 2.50 is accordingly calculated by means of a finite sum over samples $\{\mathcal{Z}^{(l)}\}_{l=1}^{L}$ drawn from $q(\mathcal{Z})$

$$\begin{aligned}\mathbb{E}_{\mathcal{Z}}[f(\mathcal{Z})] &= \int f(\mathcal{Z})\frac{p(\mathcal{Z})}{q(\mathcal{Z})}q(\mathcal{Z})d\mathcal{Z} \\ &\simeq \frac{1}{L}\sum_{l=1}^{L}\frac{p(\mathcal{Z}^{(l)})}{q(\mathcal{Z}^{(l)})}f(\mathcal{Z}^{(l)}) \\ &= \frac{1}{L}\sum_{l=1}^{L} r^{(l)} f(\mathcal{Z}^{(l)}),\end{aligned} \tag{2.52}$$

where $r^{(l)} = p(\mathcal{Z}^{(l)})/q(\mathcal{Z}^{(l)})$ is the so-called importance weight. This importance weight is seen as the correction for the bias due to sampling based on the approximate proposal distribution rather than the original true distribution.

2.4.1 Markov Chain Monte Carlo

The strategy of importance sampling suffers from the severe limitation of high dimensional space when evaluating the expectation function. In this subsection, we discuss a sampling framework called the Markov chain Monte Carlo (MCMC) that not only is able to sample from a large class of distributions but also has the desirable property of scaling well with the dimension of the sample space for large applications.

In the context of latent variable models, a first-order Markov chain is employed to express the probability function in the presence of a series of latent variables $\mathcal{Z}^{(1)}, \ldots, \mathcal{Z}^{(\tau)}$ in different states, iterations, or learning epochs τ such that the *conditional independence*

$$p(\mathcal{Z}^{(\tau+1)}|\mathcal{Z}^{(1)}, \ldots, \mathcal{Z}^{(\tau)}) = p(\mathcal{Z}^{(\tau+1)}|\mathcal{Z}^{(\tau)}) \tag{2.53}$$

holds for different states $\tau \in \{1, \ldots, T-1\}$. Starting from the initial sample or state $p(\mathcal{Z}^{(0)})$, the Markov chain is applied and operated with the transition probability

$$T(\mathcal{Z}^{(\tau)}, \mathcal{Z}^{(\tau+1)}) \triangleq p(\mathcal{Z}^{(\tau+1)}|\mathcal{Z}^{(\tau)}). \tag{2.54}$$

If the transition probabilities $T(\mathcal{Z}^{(\tau)}, \mathcal{Z}^{(\tau+1)})$ are unchanged for all τ, a *homogeneous* Markov chain is implemented. This means that the evolution of Markov chain depends on the current state and a fixed transition matrix. For a homogeneous Markov chain with transition probability $T(\mathcal{Z}', \mathcal{Z})$, a sufficient condition for the distribution $p^\star(\mathcal{Z})$ to be invariant is that the following reversibility condition is met:

$$p^\star(\mathcal{Z}')T(\mathcal{Z}', \mathcal{Z}) = p^\star(\mathcal{Z})T(\mathcal{Z}, \mathcal{Z}').$$

Summing over all $\mathcal{Z}'$ on both size of this equation, we have

$$\begin{aligned}\sum_{\mathcal{Z}'} p^\star(\mathcal{Z}')p(\mathcal{Z}|\mathcal{Z}') &= \sum_{\mathcal{Z}'} p^\star(\mathcal{Z})p(\mathcal{Z}'|\mathcal{Z}) \\ &= p^\star(\mathcal{Z})\sum_{\mathcal{Z}'} p(\mathcal{Z}'|\mathcal{Z}) \\ &= p^\star(\mathcal{Z}),\end{aligned}$$

which enables us to write the marginal probability of $\mathcal{Z}^{(\tau+1)}$ at epoch $\tau+1$ as follows:

$$p(\mathcal{Z}^{(\tau+1)}) = \sum_{\mathcal{Z}^{(\tau)}} p(\mathcal{Z}^{(\tau+1)}|\mathcal{Z}^{(\tau)})p(\mathcal{Z}^{(\tau)}). \tag{2.55}$$

The idea of MCMC is to construct a Markov chain so that samples obtained through the chain mimic the samples drawn from the target distribution $p(\mathcal{Z})$. The chain is constructed such that it spends more time on the important regions in the sample space and that the desired distribution $p^\star(\mathcal{Z})$ is invariant. It is also necessary for the distribution $p(\mathcal{Z}^{(\tau)})$ to converge to the required invariant distribution $p^\star(\mathcal{Z})$ when $\tau \to \infty$, irrespective of the choice of initial distribution $p(\mathcal{Z}^{(0)})$. Such an invariant distribution is also called the *equilibrium* distribution.

In Section 2.2.2, we outline the general EM algorithm without stating how to evaluate the posterior distribution $p(\mathcal{Z}|\mathcal{X}, \mathbf{\Lambda})$. In case the posterior cannot be written in an analytical form, we may use MCMC to draw samples through a Markov chain and replace $\log p(\mathcal{Z}|\mathcal{X}, \mathbf{\Lambda})$ with the average log-probability obtained from the drawn samples [26, Figure. 13].

2.4.2 Gibbs Sampling

Gibbs sampling [25, 27] is a realization of MCMC algorithm that has been widely applied for generating a sequence of observations that follow a specified distribution. This realization is seen as a special variant of Metropolis–Hastings algorithm [28]. Suppose we are given the distribution of a set of latent variables $p(\mathcal{Z}) = p(Z_1, \ldots, Z_J)$ and the distribution of initial state of a Markov chain $p(\mathcal{Z}^{(0)})$. In each step of sampling procedure, an arbitrary latent variable, say Z_i, has a value drawn from $p(Z_i|\mathcal{Z}_{\backslash i})$, where $\mathcal{Z}_{\backslash i} = \{Z_1, \ldots, Z_{i-1}, Z_{i+1}, \ldots, Z_J\}$. We repeat the sampling procedure where different variables in Z are cycled in random or in a specific order. The distribution $p(\mathcal{Z})$ is sampled in a Gibbs sampling procedure. This distribution is invariant at each learning epoch or in the whole Markov chain. Correspondingly, the marginal distribution $p(\mathcal{Z}_{\backslash i})$ is invariant and the conditional distribution $p(Z_i|\mathcal{Z}_{\backslash i})$ is corrected at each sampling step i. The Gibbs sampling for a latent variable model with J variables running by T steps is performed according to the following algorithm

- Initialize $\{Z_i^{(1)} : i = 1, \ldots, J\}$
- For $\tau = 1, \ldots, T$:
  - Sample $Z_1^{(\tau+1)} \sim p(Z_1|Z_2^{(\tau)}, Z_3^{(\tau)}, \ldots, Z_J^{(\tau)})$.

– Sample $Z_2^{(\tau+1)} \sim p(Z_2|Z_1^{(\tau+1)}, Z_3^{(\tau)}, \ldots, Z_J^{(\tau)})$.
$\vdots$
– Sample $Z_i^{(\tau+1)} \sim p(Z_i|Z_1^{(\tau+1)}, \ldots, Z_{i-1}^{(\tau+1)}, Z_{i+1}^{(\tau)}, \ldots, Z_J^{(\tau)})$.
$\vdots$
– Sample $Z_J^{(\tau+1)} \sim p(Z_J|Z_1^{(\tau+1)}, Z_2^{(\tau+1)}, \ldots, Z_{J-1}^{(\tau+1)})$.

We can adopt the Metropolis–Hastings algorithm to sample a target variable Z_i by fixing the rest of variables $\mathcal{Z}_{\backslash i}$ and adopting the transition probability from $\mathcal{Z}$ to $\mathcal{Z}^\star$ in Gibbs sampling, which is obtained by

$$q_i(\mathcal{Z}^\star|\mathcal{Z}) = p(Z_i^\star|\mathcal{Z}_{\backslash i}). \tag{2.56}$$

It is because that the rest of variables is unchanged by a sampling step, then $(\mathcal{Z}^\star)_{\backslash i} = \mathcal{Z}_{\backslash i}$. According to the property $p(\mathcal{Z}) = p(Z_i|\mathcal{Z}_{\backslash i})p(\mathcal{Z}_{\backslash i})$, we can determine the acceptance probability in Metropolis–Hastings algorithm obtained by

$$\begin{aligned} A_i(\mathcal{Z}^\star, \mathcal{Z}) &= \frac{p(\mathcal{Z}^\star)q_i(\mathcal{Z}|\mathcal{Z}^\star)}{p(\mathcal{Z})q_i(\mathcal{Z}^\star|\mathcal{Z})} \\ &= \frac{p(Z_i^\star|(\mathcal{Z}^\star)_{\backslash i})p((\mathcal{Z}^\star)_{\backslash i})p(Z_i|(\mathcal{Z}^\star)_{\backslash i})}{p(Z_i|\mathcal{Z}_{\backslash i})p(\mathcal{Z}_{\backslash i})p(Z_i^\star|\mathcal{Z}_{\backslash i})} = 1, \end{aligned} \tag{2.57}$$

which means that we can always accept the steps of the Gibbs sampling in the Metropolis–Hastings algorithm. Eq. 2.57 is obtained by applying $(\mathcal{Z}^\star)_{\backslash i} = \mathcal{Z}_{\backslash i}$.

2.5 Bayesian Learning

A challenge in machine learning and big data analytics is that data are collected from heterogeneous environments or under adverse conditions. The collected data are usually noisy, non-labeled, non-aligned, mismatched, and ill-posed. Bayesian inference and statistics play an essential role in dealing with uncertainty modeling and probabilistic representation in machine learning, including classification, regression, supervised learning, and unsupervised learning, to name a few. However, probabilistic models may be improperly assumed, overestimated, or underestimated. It is important to address the issue of model regularization in construction of speaker recognition system. Bayesian solutions to model regularization in different information systems have been popular in the last decade. Sections 2.3 and 2.4 have surveyed the approximate Bayesian inference and learning based on variational inference and sampling methods, respectively. This section started with a general description of model regularization (Section 2.5.1) and then presented how Bayesian learning is developed to tackle the regularization issue in speaker recognition (Section 2.5.2). In general, model regularization is known as a process that incorporates additional information to reflect model uncertainty or randomness that can deal with the ill-posed or overfitting problem. The uncertainty or randomness can be characterized by a prior distribution $p(\mathbf{\Lambda})$ or even a prior process for finding a drawing process for prior distributions [29]. The prior process can be Dirichlet process, beta process, Gaussian process, etc., which is the distribution over distribution functions [30].

2.5.1 Model Regularization

Speaker recognition models based on GMM, PLDA, SVM, and DNN are basically learned or adapted from a collection of training utterances or an online test utterance with learning strategy either in supervised mode or in unsupervised or semi-supervised mode. The goal of building a learning machine for speaker recognition is to learn a regularized model that is robust and can sufficiently generalize to recognize an unheard speaker utterance. Our ultimate goal is to achieve a desirable prediction performance under a learning machine. It is inevitable that the overfitting or underfitting problem may happen because the assumed model may not faithfully reflect the conditions in speaker utterances. The number of training utterances may be too large or too small. It is important to characterize the stochastic property in speaker utterances $\mathcal{X}$, in model assumption, and in parameter estimation during the construction of speaker model.

We therefore need a tool or framework to sufficiently represent or factorize speech signal, and faithfully characterize the mapping between collected data and assumed model. This tool should be flexible to probe speech signals and understand the production of real-world speaker utterances in the presence of various scenarios and environments. Using this tool, a scalable model topology can be learned or adapted with different size of training utterances. The learned representation can be robust under the mismatched or ill-posed conditions in training and test utterances. The noise interference and channel mismatch can be accommodated in the model to implement a noise-robust and channel-aware speaker recognition system. We need a general tool or theory to analyze, represent, and understand the speaker utterances in adverse environments. This tool can handle the issue of model selection and provide an efficient calculation or selection for optimal hyperparameter or regularization parameters for an assumed model without requirement of additional validation data. The optimization does not only consider the likelihood function or the goodness-of-fit measure but also the regularization term for model penalty. Accordingly, the modeling tool based on Bayesian learning is adopted. Occam's razor [31] is taken into account for model construction and hyperparameter selection. Bayesian theory provides a complete solution, platform, tool, or infrastructure to cope with various issues in model regularization. In what follows, we summarize how Bayesian learning can work for speaker recognition [30]

- tackle the overfitting or underfitting problem
- characterize the randomness for probabilistic model construction
- achieve the robust model under mismatched and ill-posed environments
- choose the model topology and hyperparameters
- be insensitive to noise contamination and microphone variations
- be scalable for large systems or applications
- be adaptive, flexible, and autonomous

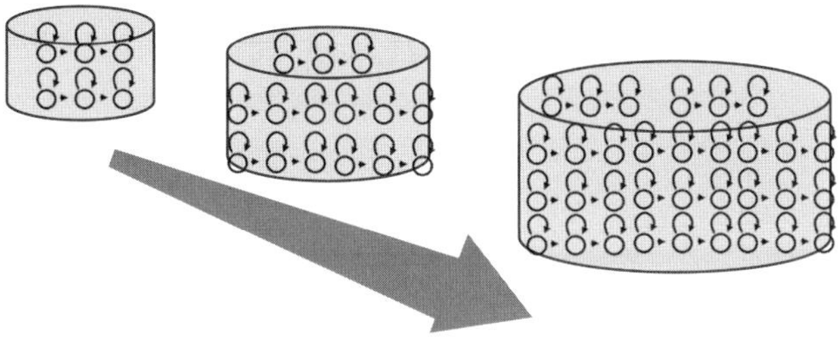

Figure 2.9 Illustration of a Bayesian speaker model where the model complexity is scalable by the amount of training utterances.

Figure 2.9 shows the concept that the Bayesian speaker model is flexible and scalable in the presence of changing amounts of speaker utterances.

2.5.2 Bayesian Speaker Recognition

The real-world speaker recognition encounters a number of challenging issues that could be tackled by Bayesian approaches. First, the learning processes for speaker recognition are usually operated in an unsupervised fashion. The target information is basically none or very little during the training phase. The prediction performance under unsupervised learning for unknown test utterances is sometimes unreliable and difficult. Second, the number of factors or Gaussian components is unknown in factor analysis models or Gaussian mixture models. Determining the model size is an unsupervised learning problem, and the ability to determine the model size from heterogeneous data is crucial because a suitable model structure provides meaningful and reliable predictions for future data. Too complicated models will be unstable or over-expressive with weak generalization. Third, how to build a balanced solution or how to tune a tradeoff between overdetermined and under-determined systems is a critical issue in speaker recognition. Fourth, the enrollment utterances in speaker recognition system are usually very short. It is challenging to verify the right speaker using limited data. In addition, speaker recognition performance is substantially affected by the inter- and intra-speaker variabilities. The robustness to these variabilities is essential to achieve desirable performance. Finally, the variabilities in channels and noise are also severe in real-world speaker recognition. How to build a representative model for these variabilities is important.

To deal with the abovementioned issues, Bayesian perspectives and Bayesian methods are introduced and implemented. From this perspective, we need to explore the latent space containing latent variables to characterize speaker identities. Bayesian learning plays a natural role to carry out latent variable models for speaker recognition. Based on Bayesian approaches, the model regularization issue is handled and the uncertainty modeling is performed. Approximate Bayesian inference methods are feasible to realize these ideas. Using these ideas, the prediction performance on future data will be improved. The robustness of the resulting model will be enhanced.

3 Machine Learning Models

This chapter provides detailed descriptions and derivations of various machine learning models for speaker verification. Instead of discussing these models in chronological order, we will start from the fundamental models, such as Gaussian mixture models and support vector machines and progressively move to the more complicated ones that are built on these fundamental models. We will leave the recent development in DNN-based models to the next two chapters.

3.1 Gaussian Mixture Models

A Gaussian mixture model (GMM) is a weighted sum of Gaussian distributions. In the context of speech and speaker recognition, a GMM can be used to represent the distribution of acoustic vectors $\mathbf{o}$'s:[1]

$$p(\mathbf{o}|\Lambda) = \sum_{c=1}^{C} \pi_c \mathcal{N}(\mathbf{o}|\boldsymbol{\mu}_c, \boldsymbol{\Sigma}_c), \tag{3.1}$$

where $\Lambda = \{\pi_c, \boldsymbol{\mu}_c, \boldsymbol{\Sigma}_c\}_{c=1}^{C}$ contains the parameters of the GMM. In Eq. 3.1, π_c, $\boldsymbol{\mu}_c$ and $\boldsymbol{\Sigma}_c$ are the weight, mean vector, and covariance matrix of the cth mixture whose density has the form

$$\mathcal{N}(\mathbf{o}|\boldsymbol{\mu}_c, \boldsymbol{\Sigma}_c) = \frac{1}{(2\pi)^{\frac{D}{2}} |\boldsymbol{\Sigma}_c|^{\frac{1}{2}}} \exp\left\{-\frac{1}{2}(\mathbf{o} - \boldsymbol{\mu}_c)^{\mathsf{T}} \boldsymbol{\Sigma}_c^{-1} (\mathbf{o} - \boldsymbol{\mu}_c)\right\}.$$

The mixture weight π_c's can be considered as the priors of the mixtures and they should sum to 1.0, i.e., $\sum_{c=1}^{C} \pi_c = 1$. Figure 3.1 shows an example of a one-dimensional GMM with three mixtures.

To estimate Λ, we express the log-likelihood of a set of training data in terms of the parameters and find the parameters in Λ that lead to the maximum-likelihood of the training data. Specifically, given a set of T independent and identically distributed (iid) acoustic vectors $\mathcal{O} = \{\mathbf{o}_t; t = 1, \ldots, T\}$, the log-likelihood function (function of Λ) is given by

[1] In this book, we use lower case boldface letters to denote random vectors. Depending on the context, lower-case boldface letters can also represent the realization (i.e., observations) of the corresponding random vectors. Note that this is different from the usual probabilistic convention that random variables are represented by capital letters (e.g., X) and their realization (x), as in $P(X < x)$.

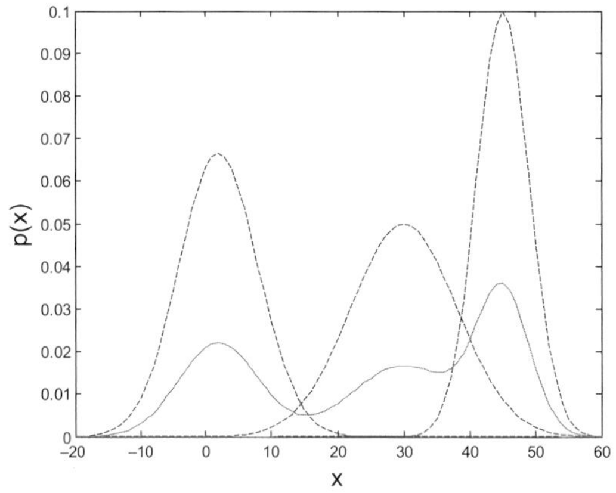

Figure 3.1 A one-D GMM with three mixture components. [Adapted from *Bayesian Speech and Language Processing (Figure 3.2), by S. Watanabe and J.T. Chien, 2015, Cambridge University Press.*]

$$\log p(\mathcal{O}|\Lambda) = \log \left\{ \prod_{t=1}^{T} \sum_{c=1}^{C} \pi_c \mathcal{N}(\mathbf{o}_t|\boldsymbol{\mu}_c, \boldsymbol{\Sigma}_c) \right\} = \sum_{t=1}^{T} \log \left\{ \sum_{c=1}^{C} \pi_c \mathcal{N}(\mathbf{o}_t|\boldsymbol{\mu}_c, \boldsymbol{\Sigma}_c) \right\}.$$

To find Λ that maximizes $\log p(\mathcal{O}|\Lambda)$, we may set $\frac{\partial \log p(\mathcal{O})}{\partial \Lambda} = 0$ and solve for Λ. But this method will not give a closed-form solution for Λ. The trouble is that the summation appears inside the logarithm. The problem, however, can be readily solved by using the EM algorithm to be explained next.

3.1.1 The EM Algorithm

The EM algorithm (see Section 2.2.2) is a popular iterative method for estimating the parameters of statistical models. It is also an elegant method for finding maximum-likelihood solutions for models with latent variables, such as GMMs. This is achieved by maximizing a log-likelihood function of the model parameters through two iterative steps: **E**xpectation and **M**aximization.

Denote the acoustic vectors from a large population as $\mathcal{O} = \{\mathbf{o}_t; t = 1, \ldots, T\}$. In GMM, for each data point $\mathbf{o}_t$, we do not know which Gaussian generates it. Therefore, the latent information is the Gaussian identity for each $\mathbf{o}_t$. Define

$$\mathcal{L} = \{\ell_{tc};\ t = 1, \ldots, T \text{ and } c = 1, \ldots, C\}$$

as the set of latent variables, where $\ell_{tc} = 1$ if $\mathbf{o}_t$ is generated by the cth Gaussian; otherwise $\ell_{tc} = 0$. Figure 3.2 shows a graphical representation of a GMM. This graphical model suggests that $\boldsymbol{o}_t$ depends on $\boldsymbol{\ell}_t = [\ell_{t1} \ \ldots \ \ell_{tC}]^{\mathsf{T}}$ so that the marginal likelihood $p(\mathbf{o}_t)$ can be obtained by marginalizing out the latent variable $\boldsymbol{\ell}_t$:

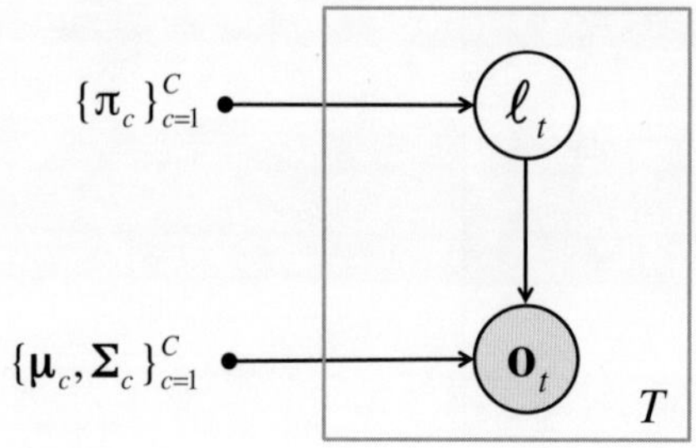

Figure 3.2 Graphical representation of a GMM. $\boldsymbol{\ell}_t = [\ell_{t1} \ \ldots \ \ell_{tC}]^\mathsf{T}$ and $\mathbf{o}_t$, with $t = 1, \ldots, T$, are the latent variables and observed vectors, respectively.

$$
\begin{aligned}
p(\mathbf{o}_t) &= \sum_{\boldsymbol{\ell}_t} p(\boldsymbol{\ell}_t) p(\mathbf{o}_t|\boldsymbol{\ell}_t) \\
&= \sum_{c=1}^{C} P(\ell_{tc} = 1) p(\mathbf{o}_t|\ell_{tc} = 1) \\
&= \sum_{c=1}^{C} \pi_c \mathcal{N}(\mathbf{o}_t|\boldsymbol{\mu}_c, \boldsymbol{\Sigma}_c),
\end{aligned} \tag{3.2}
$$

which has the same form as Eq. 3.1.

In the EM literatures, $\{\mathcal{O}, \mathcal{L}\}$ is called the *complete* data set, and $\mathcal{O}$ is the *incomplete* data set. In EM, we maximize $\log p(\mathcal{O}, \mathcal{L}|\Lambda)$ with respect to Λ instead of maximizing $\log p(\mathcal{O}|\Lambda)$. The important point is that maximizing the former is much more straightforward and will lead to closed-form solutions for each EM iteration, as will be shown below.

However, we actually do not know $\mathcal{L}$. So, we could not compute $\log p(\mathcal{O}, \mathcal{L}|\Lambda)$. Fortunately, we know its posterior distribution, i.e., $P(\mathcal{L}|\mathcal{O}, \Lambda)$, through the Bayes theorem. Specifically, for each $\mathbf{o}_t$, we compute the posterior probability:[2]

$$
\begin{aligned}
\gamma(\ell_{tc}) &\equiv P(\ell_{tc} = 1|\mathbf{o}_t, \Lambda) \\
&= \frac{P(\ell_{tc} = 1|\Lambda) p(\mathbf{o}_t|\ell_{tc} = 1, \Lambda)}{p(\mathbf{o}_t|\Lambda)} \\
&= \frac{\pi_c \mathcal{N}(\mathbf{o}_t|\boldsymbol{\mu}_c, \boldsymbol{\Sigma}_c)}{\sum_{j=1}^{C} \pi_j \mathcal{N}(\mathbf{o}_t|\boldsymbol{\mu}_j, \boldsymbol{\Sigma}_j)}.
\end{aligned} \tag{3.3}
$$

In the speech and speaker recognition literatures, computing the posterior probabilities of mixture components is called *alignment*. Its aim is to determine how close a vector $\mathbf{o}_t$ is to the individual Gaussians, accounting for both the priors, means and covariances of the mixture components. Figure 3.3 illustrates the alignment process.

With the posteriors $\gamma(\ell_{tc})$, given the current estimate of the model parameters Λ^{old}, we can find its new estimate Λ by computing the expected value of $\log p(\mathcal{O}, \mathcal{L}|\Lambda)$ under

[2] We denote probabilities and probability mass functions of discrete random variables using capital letter P, and we denote the likelihoods and probability density functions of continuous random variables using lower case letter p.

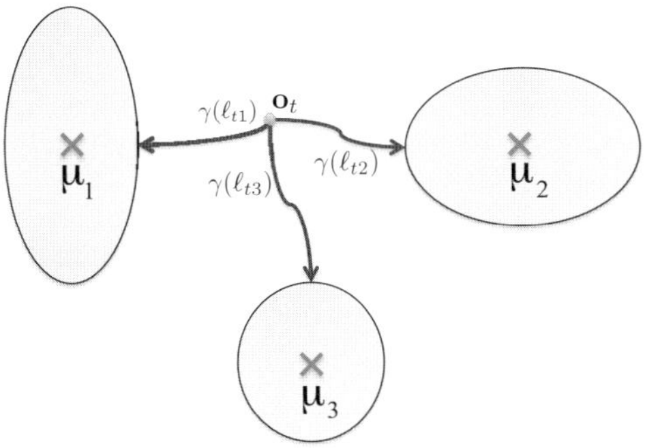

Figure 3.3 Aligning an acoustic vector $\mathbf{o}_t$ to a GMM with three mixture components.

the posterior distribution of $\mathcal{L}$:

$$\begin{aligned} Q(\Lambda|\Lambda^{\text{old}}) &= \mathbb{E}_{\mathcal{L}}\{\log p(\mathcal{O},\mathcal{L}|\Lambda)|\mathcal{O},\Lambda^{\text{old}}\} \\ &= \sum_{t=1}^{T}\sum_{c=1}^{C} P(\ell_{tc}=1|\mathbf{o}_t,\Lambda^{\text{old}})\log p(\mathbf{o}_t,\ell_{tc}=1|\Lambda) \\ &= \sum_{t=1}^{T}\sum_{c=1}^{C} \gamma(\ell_{tc})\log p(\mathbf{o}_t,\ell_{tc}=1|\Lambda) \\ &= \sum_{t=1}^{T}\sum_{c=1}^{C} \gamma(\ell_{tc})\log p(\mathbf{o}_t|\ell_{tc}=1,\Lambda)P(\ell_{tc}=1|\Lambda) \\ &= \sum_{t=1}^{T}\sum_{c=1}^{C} \gamma(\ell_{tc})\log\left[\mathcal{N}(\mathbf{o}_t|\boldsymbol{\mu}_c,\boldsymbol{\Sigma}_c)\pi_c\right], \end{aligned} \tag{3.4}$$

where $Q(\Lambda|\Lambda^{\text{old}})$ is called the auxiliary function or simply Q-function (see Eq. 2.6). The E-step consists in computing $\gamma(\ell_{tc})$ for all training samples so that $Q(\Lambda|\Lambda^{\text{old}})$ can be expressed as a function of $\boldsymbol{\mu}_c$, $\boldsymbol{\Sigma}_c$, and π_c for $c = 1,\ldots,C$.

Then, in the M-step, we maximize $Q(\Lambda|\Lambda^{\text{old}})$ with respect to Λ by setting $\frac{\partial Q(\Lambda|\Lambda^{\text{old}})}{\partial\Lambda} = 0$ to obtain (see [32, Ch. 3]):

$$\boldsymbol{\mu}_c = \frac{\sum_{t=1}^{T}\gamma(\ell_{tc})\mathbf{o}_t}{\sum_{t=1}^{T}\gamma(\ell_{tc})} \tag{3.5a}$$

$$\boldsymbol{\Sigma}_c = \frac{\sum_{t=1}^{T}\gamma(\ell_{tc})(\mathbf{o}_t-\boldsymbol{\mu}_c)(\mathbf{o}_t-\boldsymbol{\mu}_c)^{\mathsf{T}}}{\sum_{t=1}^{T}\gamma(\ell_{tc})} \tag{3.5b}$$

$$\pi_c = \frac{1}{T}\sum_{t=1}^{T}\gamma(\ell_{tc}), \tag{3.5c}$$

where $c = 1,\ldots,C$. Eq. 3.5a–Eq. 3.5c constitute the M-step of the EM algorithm. In practical implementation of the M-step, we compute the sufficient statistics:

$$n_c = \sum_{t=1}^{T} \gamma(\ell_{tc}) \tag{3.6a}$$

$$\mathbf{f}_c = \sum_{t=1}^{T} \gamma(\ell_{tc})\mathbf{o}_t \tag{3.6b}$$

$$\mathbf{S}_c = \sum_{t=1}^{T} \gamma(\ell_{tc})\mathbf{o}_t\mathbf{o}_t^{\mathsf{T}}, \tag{3.6c}$$

where $c = 1,\ldots,C$. Then, Eq. 3.5a–Eq. 3.5c become

$$\boldsymbol{\mu}_c = \frac{1}{n_c}\mathbf{f}_c \tag{3.7a}$$

$$\boldsymbol{\Sigma}_c = \frac{1}{n_c}\mathbf{S}_c - \boldsymbol{\mu}_c\boldsymbol{\mu}_c^{\mathsf{T}} \tag{3.7b}$$

$$\pi_c = \frac{1}{T}n_c. \tag{3.7c}$$

In summary, the EM algorithm iteratively performs the E- and M-steps until $Q(\Lambda|\Lambda^{\text{old}})$ no longer increases.

- **Initialization**: Randomly select C samples from $\mathcal{O}$ and assign them to $\{\boldsymbol{\mu}_c\}_{c=1}^{C}$; Set $\pi_c = \frac{1}{C}$ and $\boldsymbol{\Sigma}_c = \mathbf{I}$, where $c = 1,\ldots,C$.
- **E-Step**: Find the distribution of the latent (unobserved) variables, given the observed data and the current estimate of the parameters;
- **M-Step**: Re-estimate the parameters to maximize the likelihood of the observed data, under the assumption that the distribution found in the E-step is correct.

The iterative process guarantees to increases the true likelihood or leaves it unchanged (if a local maximum has already been reached).

3.1.2 Universal Background Models

If we use the speech of a large number of speakers to train a GMM using Eq. 3.3 and Eqs. 3.5a–3.5c, we obtain a universal background model (UBM), Λ^{ubm}, with density function

$$p(\mathbf{o}|\Lambda^{\text{ubm}}) = \sum_{c=1}^{C} \pi_c^{\text{ubm}}\mathcal{N}(\mathbf{o}|\boldsymbol{\mu}_c^{\text{ubm}}, \boldsymbol{\Sigma}_c^{\text{ubm}}). \tag{3.8}$$

Specifically, the UBM is obtained by iteratively carried out the E- and M-steps as follows.

- E-step: Compute the conditional distribution of mixture components:

$$\gamma(\ell_{tc}) \equiv \Pr(\text{Mixture} = c|\mathbf{o}_t) = \frac{\pi_c^{\text{ubm}} \mathcal{N}(\mathbf{o}_t|\boldsymbol{\mu}_c^{\text{ubm}}, \boldsymbol{\Sigma}_c^{\text{ubm}})}{\sum_{c=1}^{C} \pi_c^{\text{ubm}} \mathcal{N}(\mathbf{o}_t|\boldsymbol{\mu}_c^{\text{ubm}}, \boldsymbol{\Sigma}_c^{\text{ubm}})} \tag{3.9}$$

  where $c = 1, \ldots, C$.
- M-step: Update the model parameters:
  - Mixture weights: $\pi_c^{\text{ubm}} = \frac{1}{T} \sum_{t=1}^{T} \gamma(\ell_{tc})$
  - Mean vectors: $\boldsymbol{\mu}_c^{\text{ubm}} = \frac{\sum_{t=1}^{T} \gamma(\ell_{tc})\mathbf{o}_t}{\sum_{t=1}^{T} \gamma(\ell_{tc})}$
  - Covariance matrices: $\boldsymbol{\Sigma}_c^{\text{ubm}} = \frac{\sum_{t=1}^{T} \gamma(\ell_{tc})\mathbf{o}_t\mathbf{o}_t^T}{\sum_{t=1}^{T} \gamma(\ell_{tc})} - \boldsymbol{\mu}_c^{\text{ubm}}(\boldsymbol{\mu}_c^{\text{ubm}})^{\mathsf{T}}$

  where $c = 1, \ldots, C$.

In a GMM–UBM system, each of the genuine speakers (also called target speakers) has his/her own GMM, and the UBM serves as a reference for comparing likelihood ratios. More precisely, if the likelihood of a test utterance with respect to a genuine-speaker's GMM is larger than its likelihood with respect to the UBM, there is a high chance that the test utterance is spoken by the genuine speaker.

3.1.3 MAP Adaptation

If each target speaker has a large number of utterances for training his/her GMM, the EM algorithm in Section 3.1 can be directly applied. However, in practice, the amount of speech for each speaker is usually small. As a result, directly applying the EM algorithm will easily cause overfitting. A better solution is to apply the maximum *a posteriori* (MAP) adaptation in which target-speaker models are adapted from the UBM.

The MAP algorithm finds the parameters of target-speaker's GMM given UBM parameters $\Lambda^{\text{ubm}} = \{\pi_c^{\text{ubm}}, \boldsymbol{\mu}_c^{\text{ubm}}, \boldsymbol{\Sigma}_c^{\text{ubm}}\}_{c=1}^{C}$ using the EM algorithm. The objective is to estimate the mode of the posterior of the model parameters:

$$\begin{aligned}
\Lambda^{\text{map}} &= \underset{\Lambda}{\text{argmax}}\; p(\Lambda|\mathcal{O}) \\
&= \underset{\Lambda}{\text{argmax}}\; p(\mathcal{O}|\Lambda)p(\Lambda) \\
&= \underset{\Lambda}{\text{argmax}} \prod_{t=1}^{T} p(\mathbf{o}_t|\Lambda)p(\Lambda).
\end{aligned} \tag{3.10}$$

Direct optimization of Eq. 3.10 is difficult. However, we may iteratively find the optimal solution via the EM algorithm. Similar to the EM algorithm for GMM in Section 3.1, instead of maximizing Eq. 3.10, we maximize the auxiliary function:

$$\begin{aligned}
Q(\Lambda|\Lambda^{\text{old}}) &= \mathbb{E}_{\mathcal{L}}\left\{\log\left[p(\mathcal{O},\mathcal{L}|\Lambda)p(\Lambda)\right]|\mathcal{O},\Lambda^{\text{old}}\right\}\\
&= \mathbb{E}_{\mathcal{L}}\left\{\log p(\mathcal{O},\mathcal{L}|\Lambda)|\mathcal{O},\Lambda^{\text{old}}\right\} + \mathbb{E}_{\mathcal{L}}\left\{\log p(\Lambda)|\mathcal{O},\Lambda^{\text{old}}\right\}\\
&= \sum_{t=1}^{T}\sum_{c=1}^{C} P(\ell_{tc}=1|\mathbf{o}_t,\Lambda^{\text{old}})\log p(\mathbf{o}_t,\ell_{tc}=1|\Lambda) + \log p(\Lambda)\\
&= \sum_{t=1}^{T}\sum_{c=1}^{C} \gamma(\ell_{tc})\log p(\mathbf{o}_t,\ell_{tc}=1|\Lambda) + \log p(\Lambda)\\
&= \sum_{t=1}^{T}\sum_{c=1}^{C} \gamma(\ell_{tc})\left[\log p(\mathbf{o}_t|\ell_{tc}=1,\Lambda) + \log P(\ell_{tc}=1|\Lambda)\right] + \log p(\Lambda)\\
&= \sum_{t=1}^{T}\sum_{c=1}^{C} \gamma(\ell_{tc})\left[\log \mathcal{N}(\mathbf{o}_t|\boldsymbol{\mu}_c,\boldsymbol{\Sigma}_c) + \log \pi_c\right] + \log p(\Lambda), \qquad (3.11)
\end{aligned}$$

If only the mean vectors are adapted, we have $\pi_c = \pi_c^{\text{ubm}}$ and $\boldsymbol{\Sigma}_c = \boldsymbol{\Sigma}_c^{\text{ubm}}$. Also, we only need to assume a prior over $\boldsymbol{\mu}_c$:

$$p(\boldsymbol{\mu}_c) = \mathcal{N}(\boldsymbol{\mu}_c|\boldsymbol{\mu}_c^{\text{ubm}}, r^{-1}\boldsymbol{\Sigma}_c^{\text{ubm}}), \qquad (3.12)$$

where r is called the relevant factor [7]. Figure 3.4 shows the graphical model of the Bayesian GMM when only the mean vectors $\{\boldsymbol{\mu}_c\}_{c=1}^{C}$ of the GMM are assumed random with prior distribution given by Eq. 3.12.

Dropping terms independent of $\boldsymbol{\mu}_c$, Eq. 3.11 can be written as:

$$Q(\boldsymbol{\mu}|\boldsymbol{\mu}^{\text{ubm}}) = \sum_{t=1}^{T}\sum_{c=1}^{C} \gamma(\ell_{tc})\log \mathcal{N}(\mathbf{o}_t|\boldsymbol{\mu}_c,\boldsymbol{\Sigma}_c^{\text{ubm}}) + \log \mathcal{N}(\boldsymbol{\mu}_c|\boldsymbol{\mu}_c^{\text{ubm}}, r^{-1}\boldsymbol{\Sigma}_c^{\text{ubm}}). \qquad (3.13)$$

where $\boldsymbol{\mu} = [\boldsymbol{\mu}_1^{\mathsf{T}} \ \ldots \ \boldsymbol{\mu}_C^{\mathsf{T}}]^{\mathsf{T}}$.

The E-step is similar to the E-step in training GMMs. Specifically, given T_s acoustic vectors $\mathcal{O}^{(s)} = \{\mathbf{o}_1,\ldots,\mathbf{o}_{T_s}\}$ from speaker s, we compute the sufficient statistics:

$$n_c = \sum_{t=1}^{T_s} \gamma(\ell_{tc}) \quad \text{and} \quad E_c(\mathcal{O}^{(s)}) = \frac{1}{n_c}\sum_{t=1}^{T_s} \gamma(\ell_{tc})\mathbf{o}_t, \qquad (3.14)$$

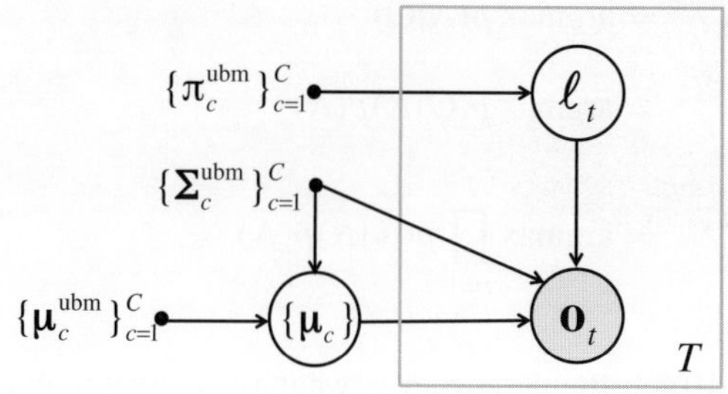

Figure 3.4 Graphical model of Bayesian GMMs with prior over the GMM's mean vectors given by Eq. 3.12.

where $\gamma(\ell_{tc})$ is computed as in Eq. 3.3. In the M-step, we differentiate $Q(\boldsymbol{\mu}|\boldsymbol{\mu}^{\text{old}})$ in Eq. 3.13 with respect to $\boldsymbol{\mu}_c$, which gives

$$\frac{\partial Q(\boldsymbol{\mu}|\boldsymbol{\mu}^{\text{ubm}})}{\partial \boldsymbol{\mu}_c} = \sum_{t=1}^{T_s} \gamma(\ell_{tc})(\boldsymbol{\Sigma}_c^{\text{ubm}})^{-1}(\mathbf{o}_t - \boldsymbol{\mu}_c) - r(\boldsymbol{\Sigma}_c^{\text{ubm}})^{-1}(\boldsymbol{\mu}_c - \boldsymbol{\mu}_c^{\text{ubm}}). \quad (3.15)$$

By setting Eq. 3.15 to $\mathbf{0}$, we obtain the adapted mean

$$\begin{aligned} \boldsymbol{\mu}_c &= \frac{\sum_t \gamma(\ell_{tc})\mathbf{o}_t}{\sum_t \gamma(\ell_{tc}) + r} + \frac{r\boldsymbol{\mu}_c^{\text{ubm}}}{\sum_t \gamma(\ell_{tc}) + r} \\ &= \alpha_c E_c(\mathcal{O}^{(s)}) + (1 - \alpha_c)\boldsymbol{\mu}_c^{\text{ubm}}, \end{aligned} \quad (3.16)$$

where

$$\alpha_c = \frac{n_c}{n_c + r}. \quad (3.17)$$

The relevance factor r is typically set to 16. Figure 3.5 shows the MAP adaptation process and Figure 3.6 illustrates a two-dimensional examples of the adaptation process. Note that in practice only the mean vectors will be adapted.

According to Eq. 3.14 and Eq. 3.17, $\alpha_c \to 1$ when $\mathcal{O}^{(s)}$ comprises lots of vectors (long utterances) and $\alpha_c \to 0$ otherwise. This means that $\boldsymbol{\mu}_c^{(s)}$ will be closed to the observed vectors from Speaker s when the utterance is long and will be similar to the cth Gaussian of the UBM when not many frames are aligned to the cth mixture. This property agrees with the Bayesian philosophy.

In Eq. 3.17, r determines the minimum number of frames aligning to mixture c to have the adaptation effect on $\boldsymbol{\mu}_c^{(s)}$. Specifically, when r is very small, say $r = 1$, a very small number of frames aligning to mixture c will be enough to cause the mean vector $\boldsymbol{\mu}_c^{(s)}$ to be adapted to $\mathcal{O}^{(s)}$. On the other hand, when r is large, say $r = 30$, a lot more frames aligning to mixture c are required to have any adaptation effect.

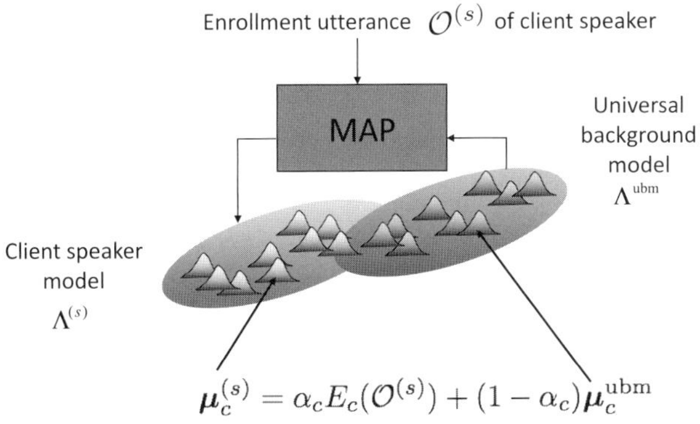

Figure 3.5 The MAP adaptation process.

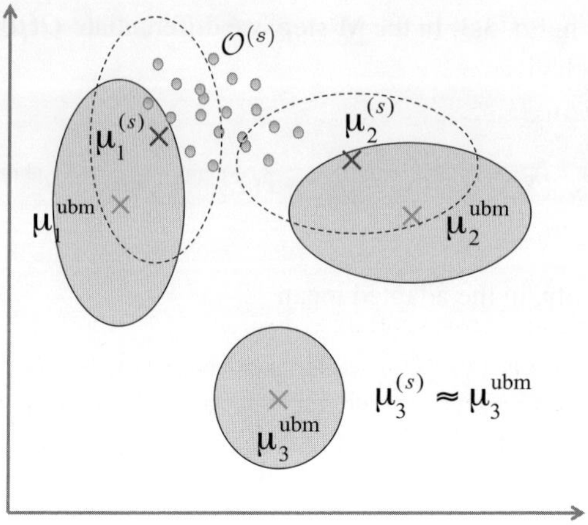

Figure 3.6 A two-dimensional example illustrating the adaptation of mixture components in a UBM. μ_1^{ubm} and μ_2^{ubm} will move toward the speaker-dependent samples $\mathcal{O}^{(s)}$ as the samples are close enough to these two Gaussians, whereas μ_3^{ubm} will remain unchanged because the samples are too far away from it.

3.1.4 GMM–UBM Scoring

Given the acoustic vectors $\mathcal{O}^{(t)}$ from a test speaker and a claimed identity s, speaker verification can be formulated as a two-class hypothesis problem:

- H_0: $\mathcal{O}^{(t)}$ comes from the true speaker s
- H_1: $\mathcal{O}^{(t)}$ comes from an impostor

Verification score is a log-likelihood ratio:

$$S_{\text{GMM–UBM}}(\mathcal{O}^{(t)}|\Lambda^{(s)}, \Lambda^{\text{ubm}}) = \log p(\mathcal{O}^{(t)}|\Lambda^{(s)}) - \log p(\mathcal{O}^{(t)}|\Lambda^{\text{ubm}}), \tag{3.18}$$

where $\log p(\mathcal{O}^{(t)}|\Lambda^{(s)})$ is the log-likelihood of $\mathcal{O}^{(t)}$ given the speaker model $\Lambda^{(s)}$, which is given by

$$\log p(\mathcal{O}^{(t)}|\Lambda^{(s)}) = \sum_{\mathbf{o}\in\mathcal{O}^{(t)}} \log \sum_{c=1}^{C} \pi_c^{\text{ubm}} \mathcal{N}(\mathbf{o}|\mu_c^{(s)}, \mathbf{\Sigma}_c^{\text{ubm}}). \tag{3.19}$$

Note that only the mean vectors in the speaker model are speaker-dependent, i.e., $\Lambda^{(s)} = \{\pi_c^{\text{ubm}}, \mu_c^{(s)}, \mathbf{\Sigma}_c^{\text{ubm}}\}_{c=1}^{C}$. This is because only the means are adapted in practice. Figure 3.5 shows the scoring process.

A side benefit of MAP adaptation is that it keeps the correspondence between the mixture components of the target-speaker model $\Lambda^{(s)}$ and the UBM. This property allows for fast scoring when the GMM and the UBM cover a large region of the feature space so that only a few Gaussians contribute to the likelihood value for each acoustic vector $\mathbf{o}$. Figure 3.6 illustrates such situation in which the contribute of the third Gaussian (with mean μ_3^{ubm}) can be ignored for any test vectors far away from it. Therefore, we may express the log-likelihood ratio of a test utterance with acoustic vectors $\mathcal{O}^{(t)}$ as

$$
\begin{aligned}
S_{\text{GMM-UBM}}(\mathcal{O}^{(t)}|\Lambda^{(s)}, \Lambda^{\text{ubm}}) &= \log p(\mathcal{O}^{(t)}|\Lambda^{(s)}) - \log p(\mathcal{O}^{(t)}|\Lambda^{\text{ubm}}) \\
&= \sum_{\mathbf{o}\in\mathcal{O}^{(t)}} \left[\log \sum_{c=1}^{C} \pi_c^{\text{ubm}} \mathcal{N}(\mathbf{o}|\boldsymbol{\mu}_c^{(s)}, \boldsymbol{\Sigma}_c^{\text{ubm}}) - \log \sum_{c=1}^{C} \pi_c^{\text{ubm}} \mathcal{N}(\mathbf{o}|\boldsymbol{\mu}_c^{\text{ubm}}, \boldsymbol{\Sigma}_c^{\text{ubm}}) \right] \\
&\approx \sum_{\mathbf{o}\in\mathcal{O}^{(t)}} \left[\log \sum_{c\in\Omega} \pi_c^{\text{ubm}} \mathcal{N}(\mathbf{o}|\boldsymbol{\mu}_c^{(s)}, \boldsymbol{\Sigma}_c^{\text{ubm}}) - \log \sum_{c\in\Omega} \pi_c^{\text{ubm}} \mathcal{N}(\mathbf{o}|\boldsymbol{\mu}_c^{\text{ubm}}, \boldsymbol{\Sigma}_c^{\text{ubm}}) \right],
\end{aligned}
\tag{3.20}
$$

where Ω is a set of indexes corresponding to the C' largest likelihoods in $\{\pi_c^{\text{ubm}} \mathcal{N}(\mathbf{o}|\boldsymbol{\mu}_c^{\text{ubm}}, \boldsymbol{\Sigma}_c^{\text{ubm}})\}_{c=1}^{C}$. Typically, $C' = 5$. The third equation in Eq. 3.20 requires $C + C'$ Gaussian evaluations for each $\mathbf{o}$, whereas the second equation requires $2C$ Gaussian evaluations. When $C = 1024$ and $C' = 5$, substantial computation saving can be achieved.

3.2 Gaussian Mixture Model–Support Vector Machines

A drawback of GMM–UBM systems is that the GMMs and UBM are trained separately, which means that information that discriminates the target speakers from the background speakers cannot be fully utilized. In 2006, Campbell [14] proposed to turn the GMMs into vectors so that target-speakers' GMMs and background-speakers' GMMs can be treated as positive- and negative-class vectors for training discriminative classifiers, one for each target speaker. Because of the high dimensionality of the resulting vectors, linear support vector machines are a natural choice.

3.2.1 Support Vector Machines

To understand the concepts of support vector machines (SVMs), we need to ask ourself "what is a good decision boundary?" Consider a two-class linearly separable classification problem shown in Figure 3.7. There are plenty of methods to find a boundary that can separate the two classes. A naive way is to find a line that is perpendicular to the line joining the two centers and is of equal distance to the centers, as shown in Figure 3.7. Setting the boundary position based on the centers of the two classes means that all samples are considered equally important. But is the boundary in Figure 3.7 good? Obviously it is not.

In 1995, Vapnik [33] advocated finding a subset of vectors that are more relevant to the classification task and using these vectors to define the decision boundary. Moreover, the decision boundary should be as far away from the data of both classes as possible (of course, not infinitely far). These two criteria can be fulfilled by maximizing the margin shown in Figure 3.8. In the figure, relevant vectors highlighted by the big circles are called support vectors, which give rise to the name support vector machines.

Linearly Separable Problems

In Figure 3.8, the decision boundary is given by

$$\mathbf{w}^{\mathsf{T}}\mathbf{x} + b = 0.$$

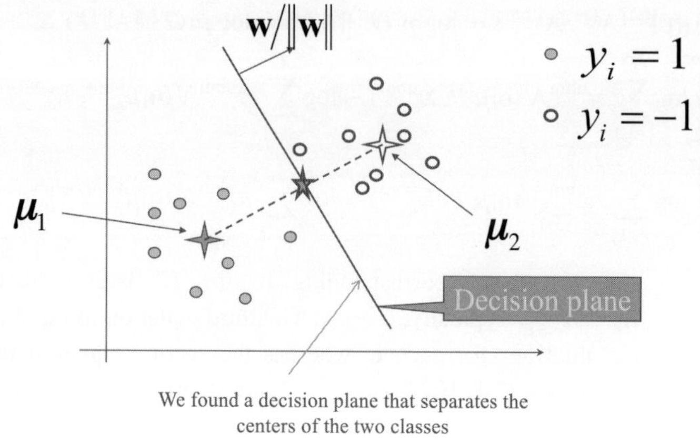

Figure 3.7 A naive way of finding a linear decision boundary that separates two classes. For two-dimensional problems, the decision plane is of equal distance to the two class-centers and is perpendicular to the line joining the centers. [Based on *Biometric Authentication: A Machine Learning Approach (Figure 4.1), by S.Y. Kung, M.W. Mak and S.H. Lin, 2005, Prentice Hall.*]

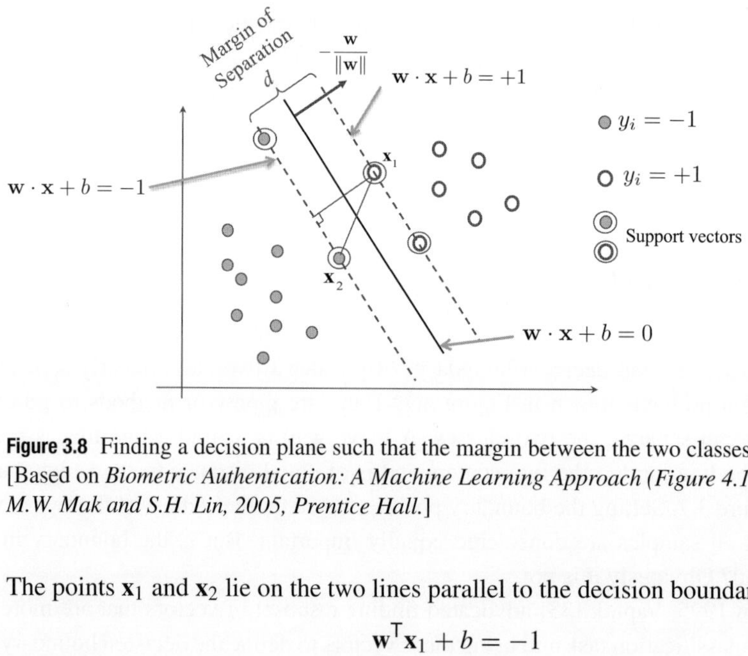

Figure 3.8 Finding a decision plane such that the margin between the two classes is the largest. [Based on *Biometric Authentication: A Machine Learning Approach (Figure 4.1), by S.Y. Kung, M.W. Mak and S.H. Lin, 2005, Prentice Hall.*]

The points $\mathbf{x}_1$ and $\mathbf{x}_2$ lie on the two lines parallel to the decision boundary, i.e.,

$$\begin{aligned} \mathbf{w}^{\mathsf{T}}\mathbf{x}_1 + b &= -1 \\ \mathbf{w}^{\mathsf{T}}\mathbf{x}_2 + b &= 1. \end{aligned} \tag{3.21}$$

The margin of separation d is the projection of $(\mathbf{x}_2 - \mathbf{x}_1)$ onto the direction perpendicular to the decision boundary, i.e.,

$$\begin{aligned} d &= \frac{\mathbf{w}^{\mathsf{T}}}{\|\mathbf{w}\|}(\mathbf{x}_2 - \mathbf{x}_1) \\ &= \frac{2}{\|\mathbf{w}\|} \qquad \text{(Using Eq. 3.21).} \end{aligned}$$

Therefore, maximizing d is equivalent to minimizing $\|\mathbf{w}\|^2$ subject to the constraint that no data points fall inside the margin of separation. This can be solved by using constrained optimization. Specifically, given a set of training data points $\mathcal{X} = \{\mathbf{x}_1, \ldots, \mathbf{x}_N\}$ with class labels $\mathcal{Y} = \{y_1, \ldots, y_N\}$ where $y_i \in \{+1, -1\}$, the weight vector $\mathbf{w}$ can be obtained by solving the following *primal* problem:

$$\begin{aligned}&\text{Minimize } \frac{1}{2}\|\mathbf{w}\|^2\\&\text{subject to } y_i(\mathbf{w}^\mathsf{T}\mathbf{x}_i + b) \geq 1, \quad \text{for } i = 1, \ldots, N.\end{aligned} \tag{3.22}$$

To solve this constrained optimization problem, we introduce Lagrange multipliers $\alpha_i \geq 0$ to form the Lagrangian function

$$L(\mathbf{w}, b, \alpha) = \frac{1}{2}\|\mathbf{w}\|^2 - \sum_{i=1}^{N} \alpha_i \left(y_i(\mathbf{w}^\mathsf{T}\mathbf{x}_i + b) - 1 \right), \tag{3.23}$$

where $\alpha = \{\alpha_1, \ldots, \alpha_N\}$. Setting the gradient of $L(\mathbf{w}, b, \alpha)$ with respect to $\mathbf{w}$ and b to zero, we have

$$\mathbf{w} = \sum_{i=1}^{N} \alpha_i y_i \mathbf{x}_i \text{ and } \sum_{i=1}^{N} \alpha_i y_i = 0. \tag{3.24}$$

Substituting Eq. 3.24 into Eq. 3.23, we have [32, Ch. 4]

$$L(\alpha) = \sum_{i=1}^{N} \alpha_i - \frac{1}{2}\sum_{i=1}^{N}\sum_{j=1}^{N} \alpha_i \alpha_j y_i y_j \mathbf{x}_i^\mathsf{T}\mathbf{x}_j,$$

which is a function of α_i's instead of $\mathbf{w}$. As Eq. 3.24 suggests that if we know α_i's, we will know $\mathbf{w}$. Therefore, we may solve the primal problem by solving the following *dual* problem:[3]

$$\begin{aligned}&\text{maximize } L(\alpha) = \sum_{i=1}^{N} \alpha_i - \frac{1}{2}\sum_{i=1}^{N}\sum_{j=1}^{N} \alpha_i \alpha_j y_i y_j \mathbf{x}_i^\mathsf{T}\mathbf{x}_j\\&\text{subject to } \alpha_i \geq 0 \text{ and } \sum_{i=1}^{N} \alpha_i y_i = 0.\end{aligned} \tag{3.25}$$

Note that because $L(\alpha)$ is quadratic in α_i, a global maximum of α_i can be found.

The solution of Eq. 3.25 comprises two kinds of Lagrange multipliers:

- $\alpha_i = 0$: The corresponding $\mathbf{x}_i$ are irrelevant to the classification task,
- $\alpha_i > 0$: The corresponding $\mathbf{x}_i$ are critical to the classification task,

where $\mathbf{x}_i$ for which $\alpha_i > 0$ are called *support vectors*. The parameter b can be computed by using the Karush-Kuhn-Tucker (KKT) condition [34], i.e., for any k such that $y_k = 1$ and $\alpha_k > 0$, we have

[3] Because Eq. 3.23 is convex in $\mathbf{w}$, the duality gap is zero. As a result, the primal and dual problems have the same solution.

$$\alpha_k[y_k(\mathbf{w}^\mathsf{T}\mathbf{x}_k + b) - 1] = 0$$
$$\Longrightarrow b = 1 - \mathbf{w}^\mathsf{T}\mathbf{x}_k.$$

The SVM output is given by

$$f(\mathbf{x}) = \mathbf{w}^\mathsf{T}\mathbf{x} + b$$
$$= \sum_{i\in\mathcal{S}} \alpha_i y_i \mathbf{x}_i^\mathsf{T}\mathbf{x} + b, \tag{3.26}$$

where $\mathcal{S}$ is the set of indexes for which $\alpha_k > 0$.

Linearly Non-Separable Problems

If the data patterns are not separable by a linear hyperplane (see Figure 3.9), a set of *slack* variables $\{\xi = \xi_1, \ldots, \xi_N\}$ is introduced with $\xi_i \geq 0$ such that the inequality constraints in Eq. 3.22 become

$$y_i(\mathbf{w}^\mathsf{T}\mathbf{x}_i + b) \geq 1 - \xi_i \quad \forall i = 1, \ldots, N. \tag{3.27}$$

The slack variables $\{\xi_i\}_{i=1}^N$ allow some data to violate the constraints in Eq. 3.22. For example, in Figure 3.9, $\mathbf{x}_1$ and $\mathbf{x}_3$ violate the constraints in Eq. 3.22 and their slack variables ξ_1 and ξ_3 are nonzero. However, they satisfy the relaxed constraints in Eq. 3.27. The value of ξ_i indicates the degree of violation. For example, in Figure 3.9, $\mathbf{x}_3$ has a higher degree of violation than $\mathbf{x}_1$, and $\mathbf{x}_2$ does not violate the constraint in Eq. 3.22.

With the slack variables, the minimization problem becomes

$$\min \frac{1}{2}\|\mathbf{w}\|^2 + C\sum_{i=1}^{N} \xi_i, \quad \text{subject to} \quad y_i(\mathbf{w}^\mathsf{T}\mathbf{x}_i + b) \geq 1 - \xi_i, \tag{3.28}$$

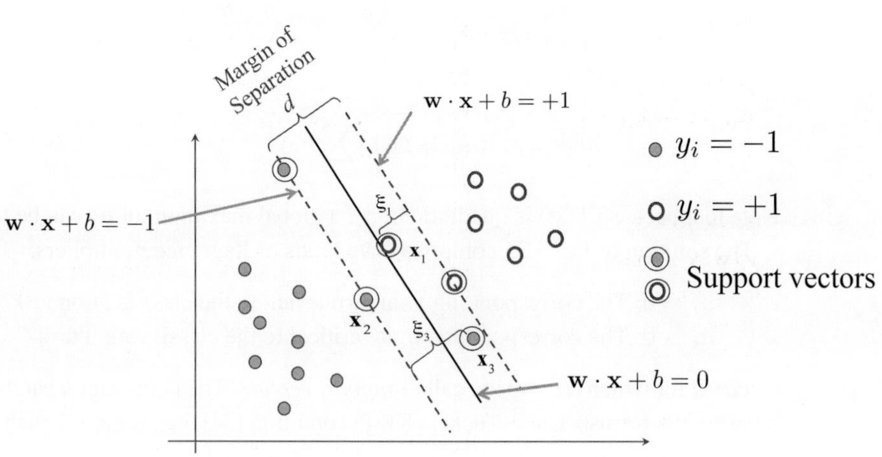

Figure 3.9 An example classification problem that cannot be perfectly separated by a linear hyperplane. The nonzero slack variables ξ_1 and ξ_3 allow $\mathbf{x}_1$ and $\mathbf{x}_3$ to violate the primal constraints in Eq. 3.22 so that the problem can still be solved by a linear SVM.

where C is a user-defined penalty parameter to penalize any violation of the safety margin for all training data. The new Lagrangian is

$$L(\mathbf{w}, b, \alpha) = \frac{1}{2}\|\mathbf{w}\|^2 + C\sum_{i}^{N} \xi_i - \sum_{i=1}^{N} \alpha_i (y_i(\mathbf{x}_i \cdot \mathbf{w} + b) - 1 + \xi_i) - \sum_{i=1}^{N} \beta_i \xi_i, \tag{3.29}$$

where $\alpha_i \geq 0$ and $\beta_i \geq 0$ are, respectively, the Lagrange multipliers to ensure that $y_i(\mathbf{x}_i \cdot \mathbf{w} + b) \geq 1 - \xi_i$ and that $\xi_i \geq 0$. Differentiating $L(\mathbf{w}, b, \alpha)$ w.r.t. $\mathbf{w}$, b, and ξ_i and set the results to zero, we obtain Eq. 3.24 and $C = \alpha_i + \beta_i$. Substituting them into Eq. 3.29, we obtain the Wolfe dual:

$$\begin{aligned} &\text{maximize } L(\alpha) = \sum_{i=1}^{N} \alpha_i - \frac{1}{2}\sum_{i=1}^{N}\sum_{j=1}^{N} \alpha_i \alpha_j y_i y_j \mathbf{x}_i^{\mathsf{T}} \mathbf{x}_j \\ &\text{subject to } 0 \leq \alpha_i \leq C, i = 1, \ldots, N, \text{ and } \sum_{i=1}^{N} \alpha_i y_i = 0. \end{aligned} \tag{3.30}$$

The solution of Eq. 3.30 comprises three types of support vectors shown in Figure 3.10: (1) on the margin hyperplanes, (2) inside the margin of separation, and (3) outside the margin of separation but on the wrong side of the decision boundary. Note that for the nonsupport vectors and support vectors on the margin hyperplanes, their slack variables ξ_i are 0.

Nonlinear SVM

The earlier discussions cover large-margin classifiers with a linear decision boundary only. However, linear SVMs are not rich enough to solve complex problems. For exam-

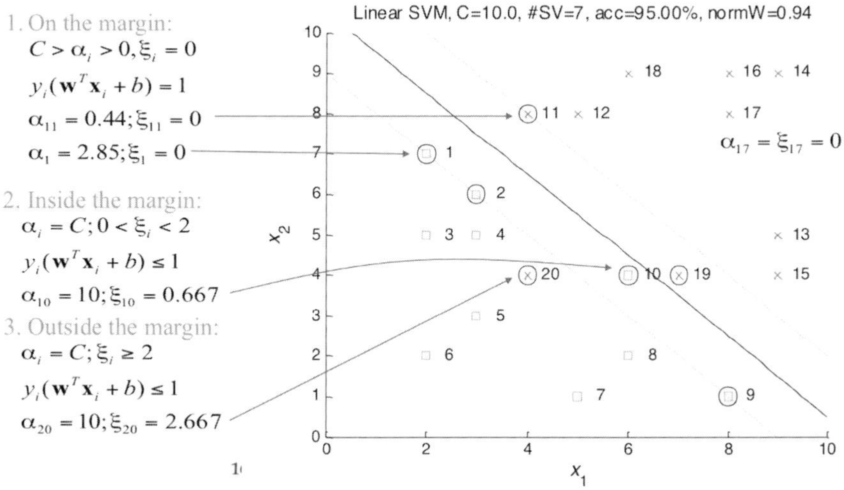

Figure 3.10 Three types of support vectors in a linear SVM with slack variables in its optimization constraints.

ple, in Figure 3.10, no straight line can *perfectly* separate the two classes. Figure 3.11(a) shows another example in which two decision boundaries (thresholds) are required to separate the two classes. Again, a one-D linear SVM could not solve this problem because it can only provide one decision threshold.

In case the training data $\mathbf{x} \in \mathcal{X}$ are not linearly separable, we may use a nonlinear function $\phi(\mathbf{x})$ to map the data from the input space to a new high-dimensional space (called feature space) where data become linearly separable. For example, in Figure 3.11, the nonlinear function is to perform the mapping:

$$\phi : x \rightarrow [x \;\; x^2]^\mathsf{T}.$$

The decision boundary in Figure 3.11(b) is a straight line that can perfectly separate the two classes. Specifically, it can be written as

$$x^2 - c = [0 \;\; 1] \begin{bmatrix} x \\ x^2 \end{bmatrix} - c = 0$$

Or equivalently,

$$\mathbf{w}^\mathsf{T} \phi(x) + b = 0, \qquad (3.31)$$

where $\mathbf{w} = [0 \;\; 1]^\mathsf{T}$, $\phi(x) = [x \;\; x^2]^\mathsf{T}$, and $b = -c$. Note that Eq. 3.31 is linear in ϕ, which means that by mapping x to $[x \;\; x^2]^\mathsf{T}$, a nonlinearly separable problem becomes linearly separable.

Figure 3.12(a) shows a two-dimensional example in which linear SVMs will not be able to perfectly separate the two classes. However, if we transform the input vectors $\mathbf{x} = [x_1 \;\; x_2]^\mathsf{T}$ by a nonlinear map:

$$\phi : \mathbf{x} \rightarrow [x_1^2 \;\; \sqrt{2}x_1x_2 \;\; x_2^2]^\mathsf{T}, \qquad (3.32)$$

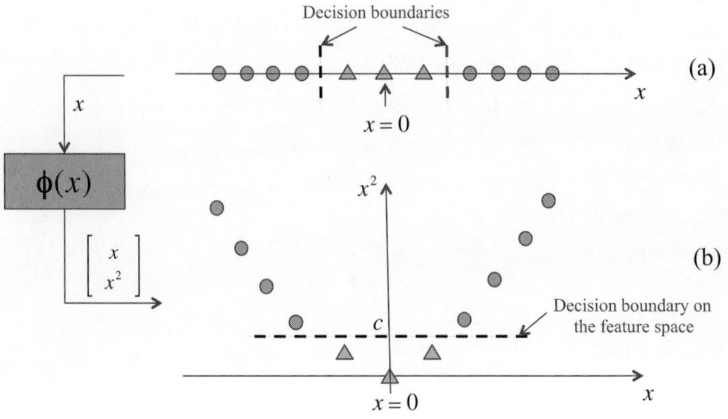

Figure 3.11 (a) One-dimensional example that cannot be solved by linear SVMs. (b) The same problem on a two-dimensional feature space in which linear SVMs can easily solve the problem.

we will be able to use a linear SVM to separate the two classes in three-dimensional space, as shown in Figure 3.12(b). The linear SVM has the form

$$\begin{aligned} f(\mathbf{x}) &= \sum_{i \in \mathcal{S}} \alpha_i y_i \boldsymbol{\phi}(\mathbf{x}_i)^{\mathsf{T}} \boldsymbol{\phi}(\mathbf{x}) + b \\ &= \mathbf{w}^{\mathsf{T}} \boldsymbol{\phi}(\mathbf{x}) + b, \end{aligned}$$

where $\mathcal{S}$ is the set of support vector indexes and $\mathbf{w} = \sum_{i \in \mathcal{S}} \alpha_i y_i \boldsymbol{\phi}(\mathbf{x}_i)$. Note that in this simple problem, the dot products $\boldsymbol{\phi}(\mathbf{x}_i)^{\mathsf{T}} \boldsymbol{\phi}(\mathbf{x}_j)$ for any $\mathbf{x}_i$ and $\mathbf{x}_j$ in the input space can be easily evaluated

$$\boldsymbol{\phi}(\mathbf{x}_i)^{\mathsf{T}} \boldsymbol{\phi}(\mathbf{x}_j) = x_{i1}^2 x_{j1}^2 + 2x_{i1}x_{i2}x_{j1}x_{j2} + x_{i2}^2 x_{j2}^2 = (\mathbf{x}_i^{\mathsf{T}} \mathbf{x}_j)^2. \tag{3.33}$$

While the mapping in Eq. 3.32 can transform the nonlinearly separable problem in Figure 3.12 into a linearly separable one, its capability is rather limited, as demonstrated in the two-spiral problem in Figure 3.13(a) and 3.13(b). The failure of this mapping in the two-spiral problem suggests that it is necessary to increase the dimension of the feature space. To achieve this without increasing the degree of the polynomial, we may change Eq. 3.33 to

$$\boldsymbol{\phi}(\mathbf{x}_i)^{\mathsf{T}} \boldsymbol{\phi}(\mathbf{x}_j) = (1 + \mathbf{x}_i^{\mathsf{T}} \mathbf{x}_j)^2. \tag{3.34}$$

With the addition of a constant term, the dimension of $\boldsymbol{\phi}(\mathbf{x})$ in Eq. 3.34 increases to 6. This can be observed by writing $\boldsymbol{\phi}(\mathbf{u})^{\mathsf{T}} \boldsymbol{\phi}(\mathbf{v})$ for any vectors $\mathbf{u}$ and $\mathbf{v}$ in the input space as

$$\begin{aligned} \boldsymbol{\phi}(\mathbf{u})^{\mathsf{T}} \boldsymbol{\phi}(\mathbf{v}) &= (1 + \mathbf{u}^{\mathsf{T}} \mathbf{v})^2 \\ &= \left(1 + [u_1 \;\; u_2] \begin{bmatrix} v_1 \\ v_2 \end{bmatrix} \right)^2 \\ &= (1 + u_1 v_1 + u_2 v_2)(1 + u_1 v_1 + u_2 v_2) \\ &= 1 + 2u_1 v_1 + 2u_2 v_2 + 2u_1 v_1 u_2 v_2 + u_1^2 v_1^2 + u_2^2 v_2^2 \\ &= \left[1 \;\; \sqrt{2}u_1 \;\; \sqrt{2}u_2 \;\; \sqrt{2}u_1 u_2 \;\; u_1^2 \;\; u_2^2 \right] \begin{bmatrix} 1 \\ \sqrt{2}v_1 \\ \sqrt{2}v_2 \\ \sqrt{2}v_1 v_2 \\ v_1^2 \\ v_2^2 \end{bmatrix}. \end{aligned}$$

Therefore, vector $\mathbf{x}$ is mapped to $\boldsymbol{\phi}(x) = \left[1 \;\; \sqrt{2}x_1 \;\; \sqrt{2}x_2 \;\; \sqrt{2}x_1 x_2 \;\; x_1^2 \;\; x_2^2 \right]^{\mathsf{T}}$, which is a six-dimensional vector. The decision boundary in the $\boldsymbol{\phi}$-space is linear because the output of the SVM can now be written as:

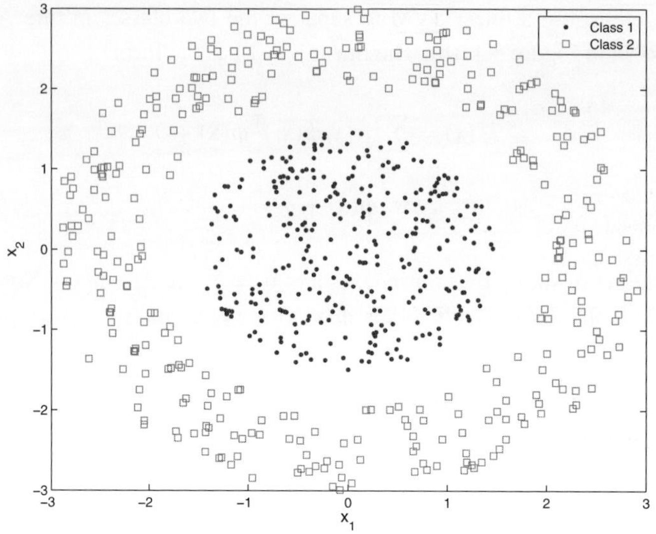

(a)

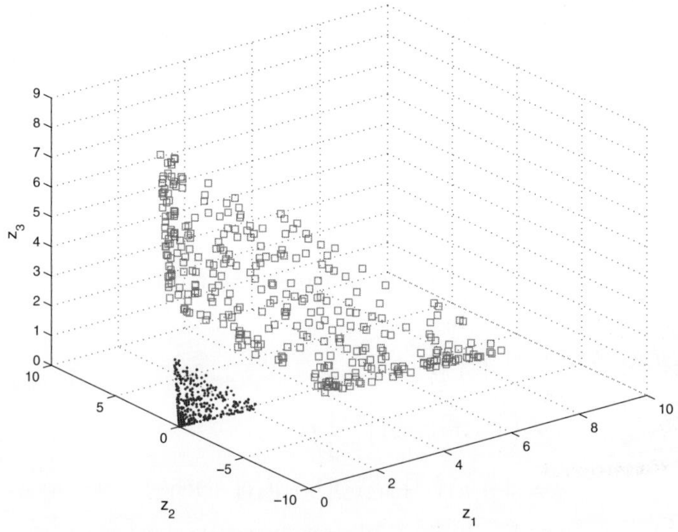

(b)

Figure 3.12 (a) A two-dimensional example that cannot be solved by linear SVMs. (b) By applying the mapping in Eq. 3.32, the same problem can be easily solved by a linear SVM on the three-dimensional space.

$$f(\mathbf{x}) = \sum_{i \in \mathcal{S}} \alpha_i y_i \boldsymbol{\phi}(\mathbf{x})^{\mathsf{T}} \boldsymbol{\phi}(\mathbf{x}_i) + b.$$

Increasing the dimension to 6, however, is still not sufficient to solve the two-D spiral problem, as shown in Figure 3.13. This is because the decision boundary produced by a second-order polynomial is too smooth. The problem can be alleviated by increasing the order to three. For example, if we define a mapping $\boldsymbol{\phi}(\mathbf{x})$ such that

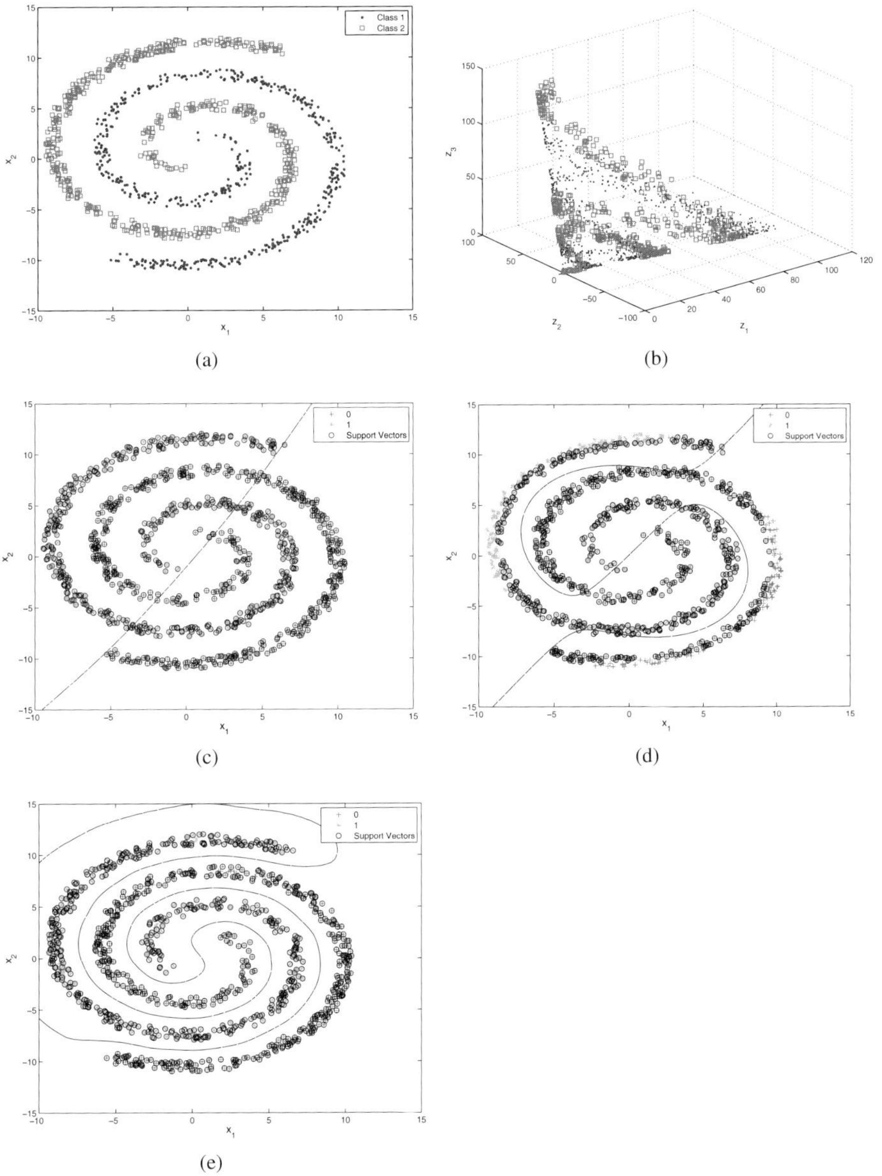

Figure 3.13 (a) A two-dimensional spiral example. (b) The mapping in Eq. 3.32 is not sufficient to convert the nonlinearly separable problem into a linearly separable one. (c) A decision boundary produced by a second-order polynomial SVM. (d) A decision boundary produced by a third-order polynomial SVM. (e) A decision boundary produced by an RBF-SVM with $\sigma = 1$. In (c)–(e), the penalty factor C was set to 1.

$$\boldsymbol{\phi}(\mathbf{x}_i)^{\mathsf{T}}\boldsymbol{\phi}(\mathbf{x}_j) = (1 + \mathbf{x}_i^{\mathsf{T}}\mathbf{x}_j)^3 \tag{3.35}$$

for any $\mathbf{x}_i$ and $\mathbf{x}_j$ in the input space, we have a better chance of solving (not perfectly) the two-spiral problem. This is because the polynomial in Eq. 3.35 has 10 terms,[4] meaning

[4] The number of terms in the polynomial expansion $\left(\sum_{i=1}^{k} x_i\right)^n$ is $\binom{n+k-1}{n}$.

that the transformed space has 10 dimensions. In such high dimensional space, the highly nonlinear two-spiral problem will become more linear. Figure 3.13(d) shows the decision boundary found by an SVM of the form

$$
\begin{aligned}
f(\mathbf{x}) &= \sum_{i \in \mathcal{S}} \alpha_i y_i \boldsymbol{\phi}(\mathbf{x}_i)^{\mathsf{T}} \boldsymbol{\phi}(\mathbf{x}) + b \\
&= \sum_{i \in \mathcal{S}} \alpha_i y_i (1 + \mathbf{x}_i^{\mathsf{T}} \mathbf{x})^3 + b.
\end{aligned}
$$

Evidently, the SVM can classify most of the data points correctly.

When the polynomial degree increases, the dimension of $\boldsymbol{\phi}(\mathbf{x})$ also increases. For example, for $\mathbf{x} \in \mathbb{R}^2$ and third-order polynomial (Eq. 3.35), the dimension is 10. However, increasing the degree of polynomial does not necessarily increase the complexity of the decision boundary. This may only increase the curvature of the boundary at some locations in the input space, as demonstrated in Figure 3.13(d).

The dimension of $\boldsymbol{\phi}$-space increases rapidly with the input dimension. For example, if the input dimension is 100, the dimension of $\boldsymbol{\phi}(\mathbf{x})$ becomes 176,851 for third-order polynomial. This will be too expensive to evaluate the dot products in such high-dimensional space. Fortunately, we may evaluate the right-hand side of Eq. 3.35 instead of the dot product on the left-hand side. The former is much cheaper and can be easily generalized to polynomials of any degree d:

$$
\boldsymbol{\phi}(\mathbf{x}_i)^{\mathsf{T}} \boldsymbol{\phi}(\mathbf{x}_j) = (1 + \mathbf{x}_i^{\mathsf{T}} \mathbf{x}_j)^d. \tag{3.36}
$$

To emphasize the fact that it is not necessary to evaluate the dot products, it is common to write it as a kernel:

$$
K(\mathbf{x}_i, \mathbf{x}_j) = (1 + \mathbf{x}_i^{\mathsf{T}} \mathbf{x}_j)^d.
$$

In additional to the polynomial kernel, the radial basis function (RBF) kernel:

$$
K(\mathbf{x}_i, \mathbf{x}_j) = \exp\left\{ -\frac{\|\mathbf{x}_i - \mathbf{x}_j\|^2}{2\sigma^2} \right\}
$$

is also commonly used. In this kernel, the parameter σ controls the kernel width, which in turn controls the curvature of the decision boundary. The smaller the value of σ, the sharper the curvature.

With this kernel notation, the decision function of an SVM can be written as

$$
f(\mathbf{x}) = \sum_{i \in \mathcal{S}} \alpha_i y_i K(\mathbf{x}_i, \mathbf{x}) + b.
$$

where $K(\cdot, \cdot)$ can be either a linear kernel, or a polynomial kernel, or an RBF kernel. Note that the linear SVM in Eq. 3.26 uses the linear kernel:

$$
K(\mathbf{x}_i, \mathbf{x}_j) = \mathbf{x}_i^{\mathsf{T}} \mathbf{x}_j.
$$

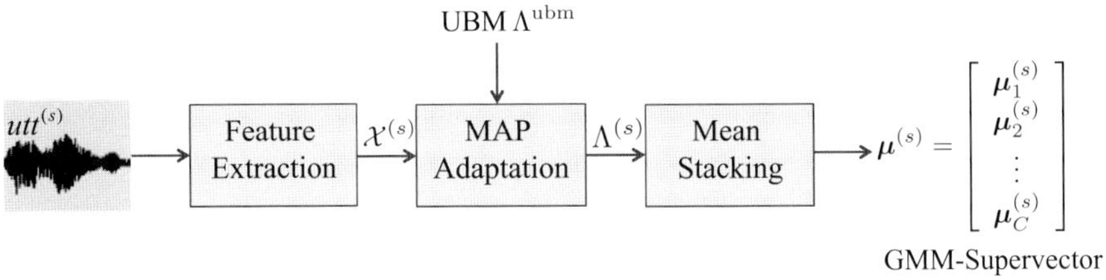

Figure 3.14 Extraction of a GMM-supervector from an utteracne.

3.2.2 GMM Supervectors

To apply SVMs for speaker verification, it is necessary to convert a variable-length utterance into a fixed-length vector, the so-called vectorization process. Campbell et al. [14] proposed to achieve such task by stacking the mean vectors of a MAP-adapted GMM as shown in Figure 3.14. Given the speech of a client speaker, MAP adaptation (Eq. 3.16) is applied to create his/her GMM model. Then, the mean vectors of the speaker model are stacked to form a supervector with dimension CF, where C is the number of Gaussians in the UBM and F is the dimension of the acoustic vectors. Typically, $C = 1024$ and $F = 60$, which result in supervectors with 61,440 dimensions.

Because of this high dimensionality, it is sensible to use linear SVMs to classify the supervectors. To incorporate the covariance matrices and the mixture coefficients of the UBM into the SVM, Campbell et al. suggest using a linear kernel of the form:[5]

$$K\left(\text{utt}^{(i)}, \text{utt}^{(j)}\right) = \sum_{c=1}^{C} \left(\sqrt{\pi_c}\boldsymbol{\Sigma}_c^{-\frac{1}{2}}\boldsymbol{\mu}_c^{(i)}\right)^{\mathsf{T}} \left(\sqrt{\pi_c}\boldsymbol{\Sigma}_c^{-\frac{1}{2}}\boldsymbol{\mu}_c^{(j)}\right), \tag{3.37}$$

where i and j index to the ith and jth utterances, respectively, and π_c, $\boldsymbol{\Sigma}_c$ and $\boldsymbol{\mu}_c$ are the weight, covariance matrix, and mean vector of the cth mixture, respectively. Eq. 3.37 can be written in a more compact form:

$$\begin{aligned} K\left(\text{utt}^{(i)}, \text{utt}^{(j)}\right) &= \left(\boldsymbol{\Omega}^{-\frac{1}{2}}\vec{\mu}^{(i)}\right)^{\mathsf{T}} \left(\boldsymbol{\Omega}^{-\frac{1}{2}}\vec{\mu}^{(j)}\right) \\ &\equiv K\left(\vec{\mu}^{(i)}, \vec{\mu}^{(j)}\right) \end{aligned} \tag{3.38}$$

where

$$\boldsymbol{\Omega} = \text{diag}\left\{\pi_1^{-1}\boldsymbol{\Sigma}_1, \ldots, \pi_C^{-1}\boldsymbol{\Sigma}_C\right\} \quad \text{and} \quad \vec{\mu} = \left[\boldsymbol{\mu}_1^{\mathsf{T}}, \ldots, \boldsymbol{\mu}_C^{\mathsf{T}}\right]^{\mathsf{T}}. \tag{3.39}$$

In practice, $\boldsymbol{\Sigma}_c$'s are assumed to be diagonal.

For each target speaker, a set of target-speaker's supervectors are obtained from his/her enrollment utterances, one for each utterance. Then, a linear SVM is trained to discriminate his/her supervector(s) from a set of supervectors derived from a number of background speakers. After training, for target-speaker s, we have

[5] To avoid cluttering with symbols, the superscript "ubm" in $\boldsymbol{\Sigma}_c$ and π_c is omitted.

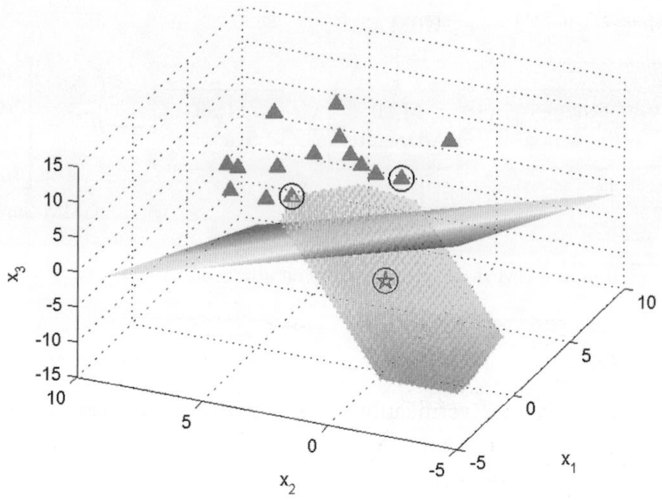

Figure 3.15 Illustration of a linear SVM in three-dimensional space when solving a data-imbalance problem in which the minority-class samples are represented by pentagrams and the majority-class samples are represented by filled triangles. The shaded area under the decision plane represents the possible locations of the minority-class samples. [Reprinted from *Acoustic vector resampling for GMMSVM-based speaker verification (Figure 1), M.W. Mak and W. Rao, Proceedings of Annual Conference of International Speech Communication Association, 2010, pp. 1449–1452, with permission of ISCA.*]

$$f^{(s)}(\vec{\mu}) = \sum_{i \in \mathcal{S}_s} \alpha_i^{(s)} K\left(\vec{\mu}^{(i)}, \vec{\mu}\right) - \sum_{i \in \mathcal{S}_{\text{bkg}}} \alpha_i^{(s)} K\left(\vec{\mu}^{(i)}, \vec{\mu}\right) + b^{(s)}, \tag{3.40}$$

where $\mathcal{S}_s$ and $\mathcal{S}_{\text{bkg}}$ are the support vector indexes corresponding to the target speaker and background speakers, respectively, and $\alpha_i^{(s)}$'s and $b^{(s)}$ are the Lagrange multipliers and the bias term of the target-speaker's SVM.

In practical situations, the number of target-speaker utterances is very small. This create a severe data-imbalance problem [35] in which the decision boundary is largely defined by the supervectors of the nontarget speakers. Another issue caused by data imbalance is that the SVM's decision boundary tends to skew toward the minority class [36, 37]. This will lead to a large number of false rejections unless the decision threshold has been adjusted to compensate for the bias. Figure 3.15 illustrates such situation. In the figure, there is a large region in the supervector space in which the target-speaker's supervector will not affect the orientation of the decision boundary.

The data-imbalance problem in GMM–SVM systems can be overcome by creating more supervectors from the utterance(s) of the target speakers [35, 38, 39]. Figure 3.16 illustrates the procedure, which is called utterance-partitioning with acoustic vector resampling (UP-VAR). The goal is to increase the number of sub-utterances without compromising their representation power. This is achieved by the following steps:

1. Randomly rearrange the sequence of acoustic vectors in an utterance;
2. Partition the acoustic vectors of an utterance into N segments;
3. Repeated Step 1 and Step 2 R times.

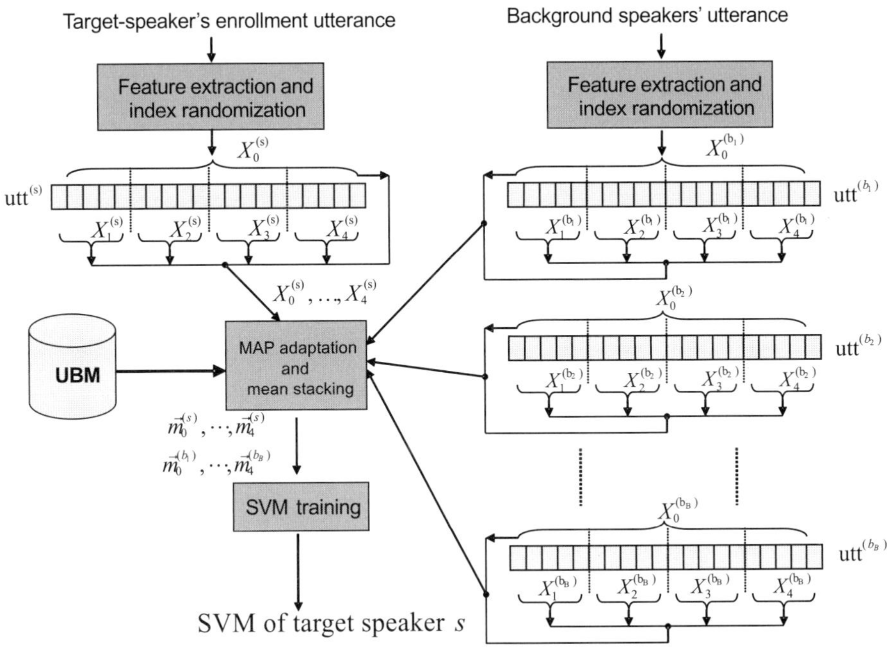

Figure 3.16 Illustration of the utterance partitioning with acoustic vector resampling (UP-AVR) process that increases the number of supervectors from a target speaker for training a speaker-dependent SVM. [Reprinted from *Utterance Partitioning with Acoustic Vector Resampling for GMM-SVM Speaker Verification (Figure 2), M.W. Mak and W. Rao, Speech Communication, vol. 53, no. 1, Jan. 2011, pp. 119–130, with permission of Elsevier.*]

By repeating Step 1 and Step 2 R times, we obtain $RN+1$ target-speaker's supervectors for training the speaker-dependent SVM.

3.2.3 GMM–SVM Scoring

During scoring, given a test utterance $\text{utt}^{(t)}$ and the SVM of a target-speaker s, we first apply MAP adaptation to the UBM to obtain the corresponding supervector $\boldsymbol{\mu}^{(t)}$ (see Figure 3.14). Then, we compute the GMM–SVM score as follows:

$$S_{\text{GMM-SVM}}(s,t) = \sum_{i\in\mathcal{S}_s} \alpha_i^{(s)} K\left(\vec{\mu}^{(i)}, \vec{\mu}^{(t)}\right) - \sum_{i\in\mathcal{S}_{\text{bkg}}} \alpha_i^{(s)} K\left(\vec{\mu}^{(i)}, \vec{\mu}^{(t)}\right) + b^{(s)}. \quad (3.41)$$

Therefore, GMM–SVM scoring amounts to finding the distance of the test supervector from the speaker-dependent hyperplane defined by the SVM of the target speaker.

Comparing the first line of Eq. 3.20 and Eq. 3.41, we can see the similarity between the GMM–UBM scoring and GMM–SVM scoring. In particular, both have two terms, one from the target speaker and one from the background speakers. However, there are also important differences. First, the two terms in GMM–UBM scoring are unweighted, meaning that they have equal contribution to the score. On the other hand, the two terms in GMM–SVM are weighted by Lagrange multipliers, which are optimized to produce the best discrimination between the target speaker and the background speakers. Second, the speaker model $\Lambda^{(s)}$ and Λ^{ubm} in GMM–UBM are separately trained, whereas the Lagrange multipliers in GMM–SVM are jointly trained by the SVM optimizer. This means that for each speaker, the optimizer will automatically find a set of *important* utterances from his/her enrollment sessions and the background speakers. This speaker-dependent selection of utterances make the GMM–SVM systems more flexible and perform much better than GMM–UBM systems.

3.2.4 Nuisance Attribute Projection

Because the vectorization process in Figure 3.14 will collect both the speaker and nonspeaker (e.g., channel) statistics from the input utterance through Eq. 3.3 and Eq. 3.16, the GMM-supervector $\vec{\mu}$ will contain not only speaker but also other nonspeaker information. Therefore, it is important to remove such unwanted information before performing GMM–SVM scoring. One of the most promising approaches is the nuisance attribute projection (NAP) [40–42].

Basic Idea of NAP

The idea of NAP is to find a subspace within the GMM-supervector space in which all nonspeaker variabilities occur. The method requires a training set comprising multiple speakers and multiple recording sessions per speaker. Assume that we are given a training set comprising N GMM-supervectors $\mathcal{X} = \{\vec{\mu}^{(1)}, \ldots, \vec{\mu}^{(N)}\}$. Our goal is to find a subspace defined by the column vectors in $\boldsymbol{U}$ and define a projection matrix

$$\mathbf{P} \equiv \mathbf{I} - \mathbf{U}\mathbf{U}^{\mathsf{T}}, \tag{3.42}$$

where $\mathbf{I}$ is an identity matrix, such that the projected vectors

$$\tilde{\mu}^{(i)} \equiv \mathbf{P}\vec{\mu}^{(i)} = (\mathbf{I} - \mathbf{U}\mathbf{U}^{\mathsf{T}})\vec{\mu}^{(i)}, \quad i = 1, \ldots, N \tag{3.43}$$

retain most of the speaker information but with nonspeaker information suppressed. For $\tilde{\mu}^{(i)}$'s to retain most of the speaker information, the rank of $\mathbf{U}$ should be much lower than the dimension of $\vec{\mu}^{(i)}$'s.

Figure 3.17 illustrates the concept of suppressing session variability in NAP. In the figure, the superscript h is the session index, and all sessions (supervectors) from the same speaker s – indexed by (s,h) – lie on the line defined by the single column in $\mathbf{U}$. By applying Eq. 3.43, we obtain a projected vector $\tilde{\mu}^{(s)}$ independent of the session index h.

Figure 3.18 demonstrates the capability of NAP in a three-dimensional toy problem in which nonspeaker variability occurs along the x_1-axis, i.e., $\boldsymbol{U} = [1\ 0\ 0]^{\mathsf{T}}$. In the figure, the feature vectors $\mathbf{x}$'s come from two different speakers, one for each cluster. Because $\boldsymbol{U} = [1\ \ 0\ \ 0]^{\mathsf{T}}$ is of rank 1, $\mathbf{U}\mathbf{U}^{\mathsf{T}}\mathbf{x}$'s vary along a straight line and $(\mathbf{I} - \mathbf{U}\mathbf{U}^{\mathsf{T}})\mathbf{x}$'s lie on a plane perpendicular to the line. As shown in the figure, after NAP projection, there is no direction in which the two clusters (black $\circ$) are indistinguishable.

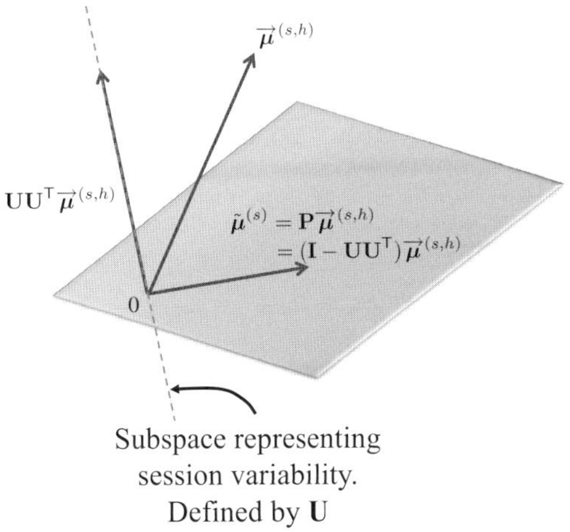

Figure 3.17 Suppressing session variability by nuisance attribute projection (NAP) when $\mathbf{U}$ is of rank 1. The superscripts s and h stand for speaker and session, respectively. The dashed line represents the nuisance direction and the shaped plane is orthogonal to it. All of the NAP-projected vectors lie on the shaped plane.

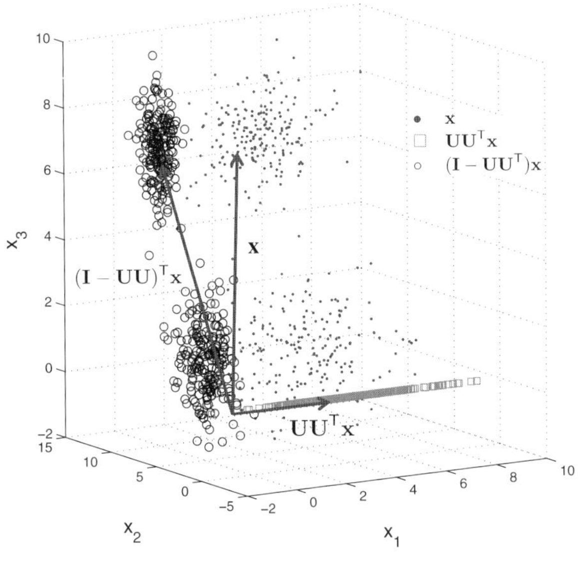

Figure 3.18 A three-dimensional example illustrating how NAP suppresses nonspeaker variability when $\mathbf{U} = [1\ \ 0\ \ 0]^{\mathsf{T}}$. Before the projection, the dots from two different speakers are indistinguishable along the x_1-axis, as indicated by the "□." After the projection, the black "○" of the two speakers lie on the plane perpendicular to the x_1-axis. As there is no variation along the x_1-axis after the projection, nonspeaker variability has been suppressed.

Objective Function of NAP

The subspace $\mathbf{U}$ can be found by minimizing the objective function:

$$\mathbf{U}^* = \underset{\mathbf{U}:\mathbf{U}^\mathsf{T}\mathbf{U}=\mathbf{I}}{\text{argmin}} \frac{1}{2}\sum_{ij} w_{ij} \left\| \mathbf{P}\left(\vec{\mu}^{(i)} - \vec{\mu}^{(j)}\right)\right\|^2, \tag{3.44}$$

where $\mathbf{P} = \mathbf{I} - \mathbf{U}\mathbf{U}^\mathsf{T}$ and

$$w_{ij} = \begin{cases} 1 & \vec{\mu}^{(i)} \text{ and } \vec{\mu}^{(j)} \text{ belong to the same speaker} \\ 0 & \text{otherwise.} \end{cases} \tag{3.45}$$

Eq. 3.44 and Eq. 3.45 suggest that we pull the projected supervectors belonging to the same speaker together and do not care about the vector pairs coming from different speakers.

To find $\mathbf{U}^*$ in Eq. 3.44, we define the data matrix $\mathbf{X} \equiv [\vec{\mu}^{(1)} \ \ldots \ \vec{\mu}^{(N)}]$ and the vector difference

$$\vec{\mathbf{d}}_{ij} \equiv \vec{\mu}^{(i)} - \vec{\mu}^{(j)}.$$

Then, the weighted projected distance in Eq. 3.44 can be expressed as

$$\begin{aligned} d_{\text{NAP}} &= \frac{1}{2}\sum_{ij} w_{ij} \left(\mathbf{P}\vec{\mathbf{d}}_{ij}\right)^\mathsf{T}\left(\mathbf{P}\vec{\mathbf{d}}_{ij}\right) \\ &= \frac{1}{2}\sum_{ij} w_{ij}\, \vec{\mathbf{d}}_{ij}^\mathsf{T}\mathbf{P}^\mathsf{T}\mathbf{P}\vec{\mathbf{d}}_{ij} \\ &= \frac{1}{2}\sum_{ij} w_{ij}\, \vec{\mathbf{d}}_{ij}^\mathsf{T}(\mathbf{I} - \mathbf{U}\mathbf{U}^\mathsf{T})^\mathsf{T}(\mathbf{I} - \mathbf{U}\mathbf{U}^\mathsf{T})\vec{\mathbf{d}}_{ij} \\ &= \frac{1}{2}\sum_{ij} w_{ij}\, \vec{\mathbf{d}}_{ij}^\mathsf{T}\vec{\mathbf{d}}_{ij} - \frac{1}{2}\sum_{ij} w_{ij}\, \vec{\mathbf{d}}_{ij}^\mathsf{T}\mathbf{U}\mathbf{U}^\mathsf{T}\vec{\mathbf{d}}_{ij}, \end{aligned} \tag{3.46}$$

where we have used the constraint $\mathbf{U}^\mathsf{T}\mathbf{U} = \mathbf{I}$. Dropping terms independent of $\mathbf{U}$, we have

$$\begin{aligned} d'_{\text{NAP}} &= -\frac{1}{2}\sum_{ij} w_{ij} \left(\vec{\mu}^{(i)} - \vec{\mu}^{(j)}\right)^\mathsf{T} \mathbf{U}\mathbf{U}^\mathsf{T} \left(\vec{\mu}^{(i)} - \vec{\mu}^{(j)}\right) \\ &= -\frac{1}{2}\sum_{ij} w_{ij}(\vec{\mu}^{(i)})^\mathsf{T}\mathbf{U}\mathbf{U}^\mathsf{T}\vec{\mu}^{(i)} - \frac{1}{2}\sum_{ij} w_{ij}(\vec{\mu}^{(j)})^\mathsf{T}\mathbf{U}\mathbf{U}^\mathsf{T}\mu^{(j)} \\ &\quad + \sum_{ij} w_{ij}(\vec{\mu}^{(i)})^\mathsf{T}\mathbf{U}\mathbf{U}^\mathsf{T}\vec{\mu}^{(j)} \\ &= -\sum_{ij} w_{ij}(\vec{\mu}^{(i)})^\mathsf{T}\mathbf{U}\mathbf{U}^\mathsf{T}\vec{\mu}^{(i)} + \sum_{ij} w_{ij}(\vec{\mu}^{(i)})^\mathsf{T}\mathbf{U}\mathbf{U}^\mathsf{T}\vec{\mu}^{(j)}, \end{aligned} \tag{3.47}$$

where we have used the property $w_{ij} = w_{ji}$. Using the identity $\mathbf{a}^\mathsf{T}\mathbf{B}\mathbf{B}^\mathsf{T}\mathbf{a} = \text{Tr}\{\mathbf{B}^\mathsf{T}\mathbf{a}\mathbf{a}^\mathsf{T}\mathbf{B}\}$, where Tr stands for matrix trace, Eq. 3.47 can be written as

$$
\begin{aligned}
d'_{\text{NAP}} &= -\text{Tr}\left\{\sum_{ij} w_{ij}\mathbf{U}^\mathsf{T}\vec{\mu}^{(i)}(\vec{\mu}^{(i)})^\mathsf{T}\mathbf{U}\right\} + \text{Tr}\left\{\sum_{ij} w_{ij}\mathbf{U}^\mathsf{T}\vec{\mu}^{(i)}(\vec{\mu}^{(j)})^\mathsf{T}\mathbf{U}\right\} \\
&= -\text{Tr}\left\{\mathbf{U}^\mathsf{T}\mathbf{X}\text{diag}(\mathbf{W}\mathbf{1})\mathbf{X}^\mathsf{T}\mathbf{U}\right\} + \text{Tr}\left\{\mathbf{U}^\mathsf{T}\mathbf{X}\mathbf{W}\mathbf{X}^\mathsf{T}\mathbf{U}\right\} \\
&= \text{Tr}\left\{\mathbf{U}^\mathsf{T}\mathbf{X}\left[\mathbf{W} - \text{diag}(\mathbf{W}\mathbf{1})\right]\mathbf{X}^\mathsf{T}\mathbf{U}\right\},
\end{aligned}
\tag{3.48}
$$

where $\mathbf{W}$ is the weight matrix in Eq. 3.45, diag($\mathbf{a}$) means converting $\mathbf{a}$ into a diagonal matrix, and $\mathbf{1}$ is a vector of all 1's. It can be shown [43] that minimizing d'_{NAP} in Eq. 3.48 with the constraint $\mathbf{U}^\mathsf{T}\mathbf{U} = \mathbf{I}$ is equivalent to finding the first K eigenvectors with the smallest eigenvalues of

$$
\mathbf{X}\left[\mathbf{W} - \text{diag}(\mathbf{W}\mathbf{1})\right]\mathbf{X}^\mathsf{T}\mathbf{U} = \mathbf{\Lambda}\mathbf{U},
$$

where $\mathbf{\Lambda}$ is a diagonal matrix containing the eigenvalues of $\mathbf{X}\left[\mathbf{W} - \text{diag}(\mathbf{W}\mathbf{1})\right]\mathbf{X}^\mathsf{T}$ in its diagonal. Once $\mathbf{U}$ has been found, we may apply Eq. 3.43 to all GMM-supervectors for both SVM training and GMM–SVM scoring.

Note that speakers with lots of sessions will dominate the objective function in Eq. 3.44. To balance the influence of individual speakers in the training set, we may change the weights in Eq. 3.45 to

$$
w_{ij} = \begin{cases} \frac{1}{N_k^2} & \text{both } \vec{\mu}^{(i)} \text{ and } \vec{\mu}^{(j)} \text{ belong to Speaker } k \\ 0 & \text{otherwise,} \end{cases}
\tag{3.49}
$$

where N_k is the number of utterances from Speaker k.

Relationship with Within-Class Covariance Matrix

With the weights in Eq. 3.49, the distance in Eq. 3.46 can be obtained from the within-class covariance matrix. To see this, let's assume that we have N training utterances spoken by K speakers, and for the kth speaker, we have N_k training utterances whose indexes are stored in the set $\mathcal{C}_k$. Then, we can write Eq. 3.46 as follows:

$$
\begin{aligned}
d_{\text{NAP}} &= \frac{1}{2}\sum_{k=1}^{K}\frac{1}{N_k^2}\sum_{i\in\mathcal{C}_k}\sum_{j\in\mathcal{C}_k}(\tilde{\mu}^{(i)} - \tilde{\mu}^{(j)})^\mathsf{T}(\tilde{\mu}^{(i)} - \tilde{\mu}^{(j)}) \\
&= \sum_{k=1}^{K}\frac{1}{N_k}\sum_{i\in\mathcal{C}_k}(\tilde{\mu}^{(i)})^\mathsf{T}\tilde{\mu}^{(i)} - \sum_{k=1}^{K}\frac{1}{N_k^2}\sum_{i\in\mathcal{C}_k}\sum_{j\in\mathcal{C}_k}(\tilde{\mu}^{(i)})^\mathsf{T}\tilde{\mu}^{(j)} \\
&= \sum_{k=1}^{K}\left[\frac{1}{N_k}\sum_{i\in\mathcal{C}_k}(\tilde{\mu}^{(i)})^\mathsf{T}\tilde{\mu}^{(i)} - \tilde{\mathbf{m}}_k^\mathsf{T}\tilde{\mathbf{m}}_k\right],
\end{aligned}
\tag{3.50}
$$

where $\tilde{\mathbf{m}}_k = \frac{1}{N_k}\sum_{i\in C_k}\tilde{\mu}^{(i)}$. Note that the within-class covariance (WCCN) matrix is given by

$$\begin{aligned}
\boldsymbol{\Sigma}_{\text{wccn}} &= \sum_{k=1}^{K}\frac{1}{N_k}\sum_{i\in C_k}(\tilde{\mu}^{(i)} - \tilde{\mathbf{m}}_k)(\tilde{\mu}^{(i)} - \tilde{\mathbf{m}}_k)^{\mathsf{T}} \\
&= \sum_{k=1}^{K}\frac{1}{N_k}\left[\sum_{i\in C_k}\tilde{\mu}^{(i)}(\tilde{\mu}^{(i)})^{\mathsf{T}} - \sum_{i\in C_k}\tilde{\mu}^{(i)}\tilde{\mathbf{m}}_k^{\mathsf{T}} - \sum_{i\in C_k}\tilde{\mathbf{m}}_k(\tilde{\mu}^{(i)})^{\mathsf{T}} + \sum_{i\in C_k}\tilde{\mathbf{m}}_k\tilde{\mathbf{m}}_k^{\mathsf{T}}\right] \\
&= \sum_{k=1}^{K}\frac{1}{N_k}\left[\sum_{i\in C_k}\tilde{\mu}^{(i)}(\tilde{\mu}^{(i)})^{\mathsf{T}} - 2N_k\tilde{\mathbf{m}}_k\tilde{\mathbf{m}}_k^{\mathsf{T}} + N_k\tilde{\mathbf{m}}_k\tilde{\mathbf{m}}_k^{\mathsf{T}}\right] \\
&= \sum_{k=1}^{K}\frac{1}{N_k}\left[\sum_{i\in C_k}\tilde{\mu}^{(i)}(\tilde{\mu}^{(i)})^{\mathsf{T}} - \tilde{\mathbf{m}}_k\tilde{\mathbf{m}}_k^{\mathsf{T}}\right].
\end{aligned} \tag{3.51}$$

Comparing Eq. 3.51 with Eq. 3.50 and using the property of matrix trace, $\text{Tr}\{\mathbf{a}\mathbf{a}^{\mathsf{T}}\} = \mathbf{a}^{\mathsf{T}}\mathbf{a}$, we obtain

$$\text{Tr}\{\boldsymbol{\Sigma}_{\text{wccn}}\} = d_{\text{NAP}}. \tag{3.52}$$

This means that the objective function in Eq. 3.44 using the weights in Eq. 3.49 is equivalent to minimizing the trace of the within-class covariance matrix.

3.3 Factor Analysis

Although NAP can reduce the session variability, it does not define a speaker subspace to model inter-speaker variability. Moreover, it does not take the prior distribution of the supervectors into account when projecting out the nuisance information. These two limitations, however, can be readily addressed by factor analysis models.

Factor analysis (FA) is a statistical method for modeling the variability and covariance structure of observed variables through a smaller number of unobserved variables called *factors*. It was originally introduced by psychologists to find the underlying latent factors that account for the correlations among a set of observations or measures.

For instance, the exam scores of 10 different subjects of 1000 students may be explained by two latent factors (also called common factors): language ability and math ability. Each student has their own values for these two factors across all of the 10 subjects and his/her exam score for each subject is a linear weighted sum of these two factors plus the score mean and an error. For each subject, all students share the same weights, which are referred to as the factor loadings, for this subject. Specifically, denote $\mathbf{x}_i = [x_{i,1} \ \cdots \ x_{i,10}]^{\mathsf{T}}$ and $\mathbf{z}_i = [z_{i,1} \ z_{i,2}]^{\mathsf{T}}$ as the exam scores and common factors of the ith student, respectively. Also, denote $\mathbf{m} = [m_1 \ \cdots \ m_{10}]^{\mathsf{T}}$ as the mean exam scores of these 10 subjects. Then, we have

$$x_{i,j} = m_j + v_{j,1}z_{i,1} + v_{j,2}z_{i,2} + \epsilon_{i,j}, \quad i = 1,\ldots,1000 \text{ and } j = 1,\ldots,10, \tag{3.53}$$

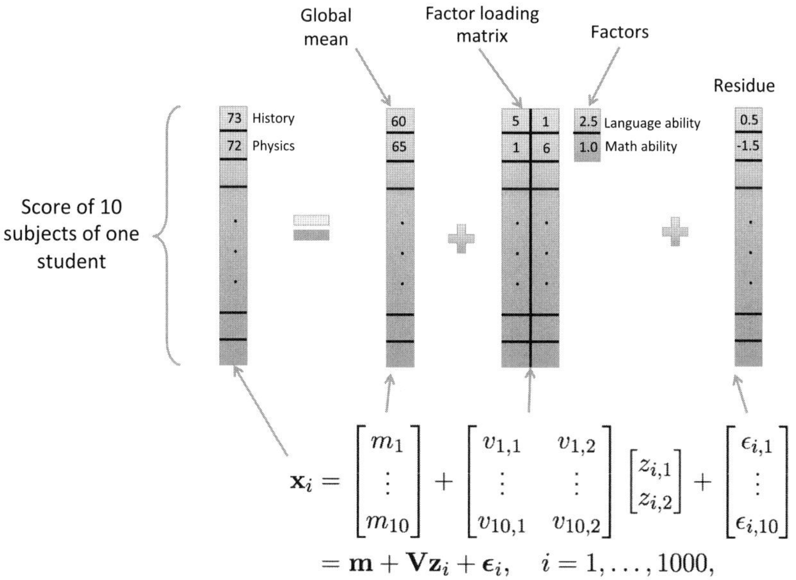

Figure 3.19 Illustrative example showing the idea of factor analysis. $\mathbf{x}_i$ represents the exam scores of ten subjects of a student who is strong in language but weak in math. The factor loading suggests that history demands language skill, whereas physics demands math skill.

where $v_{j,1}$ and $v_{j,2}$ are the factor loadings for the jth subject and $\epsilon_{i,j}$ is the error term. Eq. 3.53 can also be written as:

$$\mathbf{x}_i = \begin{bmatrix} x_{i1} \\ \vdots \\ x_{i,10} \end{bmatrix} = \begin{bmatrix} m_1 \\ \vdots \\ m_{10} \end{bmatrix} + \begin{bmatrix} v_{1,1} & v_{1,2} \\ \vdots & \vdots \\ v_{10,1} & v_{10,2} \end{bmatrix} \begin{bmatrix} z_{i,1} \\ z_{i,2} \end{bmatrix} + \begin{bmatrix} \epsilon_{i,1} \\ \vdots \\ \epsilon_{i,10} \end{bmatrix}$$
$$= \mathbf{m} + \mathbf{V}\mathbf{z}_i + \epsilon_i, \quad i = 1, \ldots, 1000, \tag{3.54}$$

where $\mathbf{V}$ is a 10×2 matrix comprising the factor loadings of the 10 subjects.

Figure 3.19 shows the exam scores of history and physics of a student who is strong in language but weak in math (as reflected by the latent factors). The first two rows of the loading matrix suggest that history requires language skill, whereas physics requires math skill, because of the large *factor loadings*. Should the latent factors of this student be swapped (i.e., 1.0 for language ability and 2.5 for math ability), he/she will obtain very a high score in physics but very low score in history.

In the context of speaker recognition, one may consider the elements of a GMM-supervector as the observed variables and the coordinates in the subspace in which the supervector lives are the factors. To simplify the discussions, in this subsection, we start from the simple case where the UBM only have one Gaussian component (so that the supervector reduces to ordinary vectors) and then extend the simple case to the more general case where the UBM has C mixtures.

3.3.1 Generative Model

Denote $\mathcal{X} = \{\mathbf{x}_1, \ldots, \mathbf{x}_N\}$ as a set of R-dimensional vectors. In factor analysis, $\mathbf{x}_i$'s are assume to follow a linear-Gaussian model:

$$\mathbf{x}_i = \mathbf{m} + \mathbf{V}\mathbf{z}_i + \boldsymbol{\epsilon}_i, \quad i = 1, \ldots, N, \tag{3.55}$$

where $\mathbf{m}$ is the global mean of vectors in $\mathcal{X}$, $\mathbf{V}$ is a low-rank $R \times D$ matrix, $\mathbf{z}_i$ is a D-dimensional latent vector with prior density $\mathcal{N}(\mathbf{0}, \mathbf{I})$, and $\boldsymbol{\epsilon}_i$ is the residual noise following a Gaussian density with zero mean and covariance matrix $\boldsymbol{\Sigma}$. Figure 3.20 shows the graphical model of factor analysis.

Given the factor analysis model in Eq. 3.55 and the graphical model in Figure 3.20, it can be shown that the marginal density of $\mathbf{x}$ is given by

$$\begin{aligned} p(\mathbf{x}) &= \int p(\mathbf{x}|\mathbf{z})p(\mathbf{z})d\mathbf{z} \\ &= \int \mathcal{N}(\mathbf{x}|\mathbf{m} + \mathbf{V}\mathbf{z}, \boldsymbol{\Sigma})\mathcal{N}(\mathbf{z}|\mathbf{0}, \mathbf{I})d\mathbf{z} \\ &= \mathcal{N}(\mathbf{x}|\mathbf{m}, \mathbf{V}\mathbf{V}^{\mathsf{T}} + \boldsymbol{\Sigma}). \end{aligned} \tag{3.56}$$

Eq. 3.56 can be obtained by convolution of Gaussian [44] or by noting that $p(\mathbf{x})$ is a Gaussian. For the latter, we take the expectation of $\mathbf{x}$ in Eq. 3.55 to obtain:

$$\mathbb{E}\{\mathbf{x}\} = \mathbf{m}.$$

Also, we take the expectation of $(\mathbf{x} - \mathbf{m})(\mathbf{x} - \mathbf{m})^{\mathsf{T}}$ in Eq. 3.55 to obtain:

$$\begin{aligned} \mathbb{E}\{(\mathbf{x} - \mathbf{m})(\mathbf{x} - \mathbf{m})^{\mathsf{T}}\} &= \mathbb{E}\{(\mathbf{V}\mathbf{z} + \boldsymbol{\epsilon})(\mathbf{V}\mathbf{z} + \boldsymbol{\epsilon})^{\mathsf{T}}\} \\ &= \mathbf{V}\mathbb{E}\{\mathbf{z}\mathbf{z}^{\mathsf{T}}\}\mathbf{V}^{\mathsf{T}} + \mathbb{E}\{\boldsymbol{\epsilon}\boldsymbol{\epsilon}^{\mathsf{T}}\} \\ &= \mathbf{V}\mathbf{I}\mathbf{V}^{\mathsf{T}} + \boldsymbol{\Sigma} \\ &= \mathbf{V}\mathbf{V}^{\mathsf{T}} + \boldsymbol{\Sigma}. \end{aligned} \tag{3.57}$$

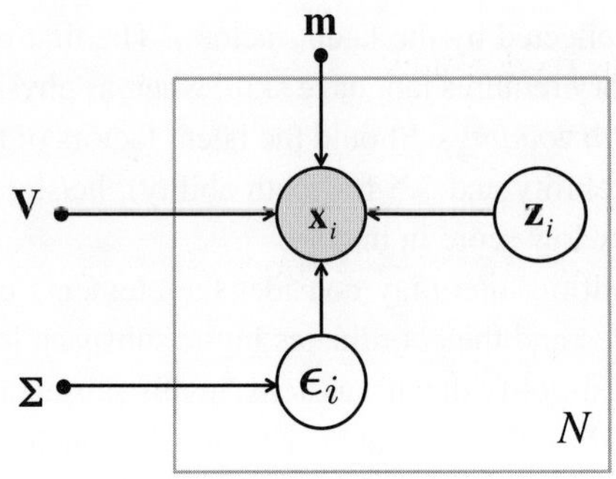

Figure 3.20 Graphical model of factor analysis.

Eq. 3.56 and Eq. 3.57 suggest that $\mathbf{x}$'s vary in a subspace of $\mathbb{R}^D$ with variability explained by the covariance matrix $\mathbf{V}\mathbf{V}^\mathsf{T}$. Any deviations away from this subspace are explained by $\boldsymbol{\Sigma}$.

3.3.2 EM Formulation

Because there is no closed-form solution for $\mathbf{V}$ and $\boldsymbol{\Sigma}$, they are estimated by the EM algorithm. Although each iteration of EM should be started with the E-step followed by the M-step, notationally, it is more convenient to describe the M-step first, assuming that the posterior expectations have already been computed in the E-step.

In the M-step, we maximize the expectation of the complete likelihood. Denote $\omega = \{\mathbf{m}, \mathbf{V}, \boldsymbol{\Sigma}\}$ and $\omega' = \{\mathbf{m}', \mathbf{V}', \boldsymbol{\Sigma}'\}$ as the old and new parameter sets, respectively. Also denote $\mathcal{Z} = \{\mathbf{z}_1, \ldots, \mathbf{z}_N\}$ as the latent factors of the FA model in Eq. 3.55. The E- and M-steps iteratively maximize the expectation of the complete likelihood with respect to the latent variables (see Section 2.2.2):

$$\begin{aligned} Q(\omega'|\omega) &= \mathbb{E}_{\mathbf{z}\sim p(\mathbf{z}|\mathbf{x})}\{\log p(\mathcal{X}, \mathcal{Z}|\omega')|\mathcal{X}, \omega\} \\ &= \mathbb{E}_{\mathbf{z}\sim p(\mathbf{z}|\mathbf{x})}\left\{\sum_i \log\left[p\left(\mathbf{x}_i|\mathbf{z}_i, \omega'\right) p(\mathbf{z}_i)\right]\Big|\mathcal{X}, \omega\right\} \\ &= \mathbb{E}_{\mathbf{z}\sim p(\mathbf{z}|\mathbf{x})}\left\{\sum_i \log\left[\mathcal{N}\left(\mathbf{x}_i|\mathbf{m}' + \mathbf{V}'\mathbf{z}_i, \boldsymbol{\Sigma}'\right)\mathcal{N}(\mathbf{z}_i|\mathbf{0}, \mathbf{I})\right]\Big|\mathcal{X}, \omega\right\}. \end{aligned} \tag{3.58}$$

To simplify notations, we drop the symbol $(')$ in Eq. 3.58, ignore the constant terms independent of the model parameters, and replace expectation notation $\mathbb{E}_{\mathbf{z}\sim p(\mathbf{z}|\mathbf{x})}\{\mathbf{z}|\mathbf{x}\}$ by $\langle\mathbf{z}|\mathbf{x}\rangle$, which results in

$$\begin{aligned} Q(\omega) &= -\sum_i \mathbb{E}_{\mathbf{z}\sim p(\mathbf{z}|\mathbf{x})}\left\{\frac{1}{2}\log|\boldsymbol{\Sigma}| + \frac{1}{2}(\mathbf{x}_i - \mathbf{m} - \mathbf{V}\mathbf{z}_i)^\mathsf{T}\boldsymbol{\Sigma}^{-1}(\mathbf{x}_i - \mathbf{m} - \mathbf{V}\mathbf{z}_i)\Big|\mathcal{X}, \omega\right\} \\ &= \sum_i\left[-\frac{1}{2}\log|\boldsymbol{\Sigma}| - \frac{1}{2}(\mathbf{x}_i - \mathbf{m})^\mathsf{T}\boldsymbol{\Sigma}^{-1}(\mathbf{x}_i - \mathbf{m})\right] \\ &\quad + \sum_i(\mathbf{x}_i - \mathbf{m})^\mathsf{T}\boldsymbol{\Sigma}^{-1}\mathbf{V}\langle\mathbf{z}_i|\mathbf{x}_i\rangle - \frac{1}{2}\left[\sum_i\left\langle\mathbf{z}_i^\mathsf{T}\mathbf{V}^\mathsf{T}\boldsymbol{\Sigma}^{-1}\mathbf{V}\mathbf{z}_i|\mathbf{x}_i\right\rangle\right]. \end{aligned} \tag{3.59}$$

Using the following properties of matrix derivatives

$$\begin{aligned} \frac{\partial\mathbf{a}^\mathsf{T}\mathbf{X}\mathbf{b}}{\partial\mathbf{X}} &= \mathbf{a}\mathbf{b}^\mathsf{T} \\ \frac{\partial\mathbf{b}^\mathsf{T}\mathbf{X}^\mathsf{T}\mathbf{B}\mathbf{X}\mathbf{c}}{\partial\mathbf{X}} &= \mathbf{B}^\mathsf{T}\mathbf{X}\mathbf{b}\mathbf{c}^\mathsf{T} + \mathbf{B}\mathbf{X}\mathbf{c}\mathbf{b}^\mathsf{T} \\ \frac{\partial}{\partial\mathbf{A}}\log|\mathbf{A}^{-1}| &= -(\mathbf{A}^{-1})^\mathsf{T} \end{aligned}$$

we have

$$\frac{\partial Q}{\partial\mathbf{V}} = \sum_i\boldsymbol{\Sigma}^{-1}(\mathbf{x}_i - \mathbf{m})\langle\mathbf{z}_i^\mathsf{T}|\mathbf{x}_i\rangle - \sum_i\boldsymbol{\Sigma}^{-1}\mathbf{V}\left\langle\mathbf{z}_i\mathbf{z}_i^\mathsf{T}|\mathbf{x}_i\right\rangle. \tag{3.60}$$

Setting $\frac{\partial Q}{\partial \mathbf{V}} = 0$, we have

$$\sum_i \mathbf{V} \left\langle \mathbf{z}_i \mathbf{z}_i^{\mathsf{T}} | \mathbf{x}_i \right\rangle = \sum_i (\mathbf{x}_i - \mathbf{m}) \langle \mathbf{z}_i^{\mathsf{T}} | \mathbf{x}_i \rangle \tag{3.61}$$

$$\mathbf{V} = \left[\sum_i (\mathbf{x}_i - \mathbf{m}) \langle \mathbf{z}_i | \mathbf{x}_i \rangle^{\mathsf{T}} \right] \left[\sum_i \left\langle \mathbf{z}_i \mathbf{z}_i^{\mathsf{T}} | \mathbf{x}_i \right\rangle \right]^{-1}. \tag{3.62}$$

To find $\boldsymbol{\Sigma}$, we evaluate

$$\frac{\partial Q}{\partial \boldsymbol{\Sigma}^{-1}} = \frac{1}{2} \sum_i \left[\boldsymbol{\Sigma} - (\mathbf{x}_i - \mathbf{m})(\mathbf{x}_i - \mathbf{m})^{\mathsf{T}} \right] + \sum_i (\mathbf{x}_i - \mathbf{m}) \langle \mathbf{z}_i^{\mathsf{T}} | \mathbf{x}_i \rangle \mathbf{V}^{\mathsf{T}}$$
$$- \frac{1}{2} \sum_i \mathbf{V} \langle \mathbf{z}_i \mathbf{z}_i^{\mathsf{T}} | \mathbf{x}_i \rangle \mathbf{V}^{\mathsf{T}}.$$

Note that according to Eq. 3.61, we have

$$\frac{\partial Q}{\partial \boldsymbol{\Sigma}^{-1}} = \frac{1}{2} \sum_i \left[\boldsymbol{\Sigma} - (\mathbf{x}_i - \mathbf{m})(\mathbf{x}_i - \mathbf{m})^{\mathsf{T}} \right] + \sum_i (\mathbf{x}_i - \mathbf{m}) \langle \mathbf{z}_i^{\mathsf{T}} | \mathbf{x}_i \rangle \mathbf{V}^{\mathsf{T}}$$
$$- \frac{1}{2} \sum_i (\mathbf{x}_i - \mathbf{m}) \langle \mathbf{z}_i^{\mathsf{T}} | \mathbf{x}_i \rangle \mathbf{V}^{\mathsf{T}}.$$

Therefore, setting $\frac{\partial Q}{\partial \boldsymbol{\Sigma}^{-1}} = 0$ we have

$$\sum_i \boldsymbol{\Sigma} = \sum_i \left[(\mathbf{x}_i - \mathbf{m})(\mathbf{x}_i - \mathbf{m})^{\mathsf{T}} - (\mathbf{x}_i - \mathbf{m}) \langle \mathbf{z}_i^{\mathsf{T}} | \mathbf{x}_i \rangle \mathbf{V}^{\mathsf{T}} \right].$$

Rearranging, we have

$$\boldsymbol{\Sigma} = \frac{1}{N} \sum_i \left[(\mathbf{x}_i - \mathbf{m})(\mathbf{x}_i - \mathbf{m})^{\mathsf{T}} - \mathbf{V} \langle \mathbf{z}_i | \mathbf{x}_i \rangle (\mathbf{x}_i - \mathbf{m})^{\mathsf{T}} \right].$$

To compute $\mathbf{m}$, we evaluate

$$\frac{\partial Q}{\partial \mathbf{m}} = - \sum_i \left(\boldsymbol{\Sigma}^{-1} \mathbf{m} - \boldsymbol{\Sigma}^{-1} \mathbf{x}_i \right) + \sum_i \boldsymbol{\Sigma}^{-1} \mathbf{V} \langle \mathbf{z}_i | \mathbf{x}_i \rangle.$$

Setting $\frac{\partial Q}{\partial \mathbf{m}} = 0$, we have

$$\mathbf{m} = \frac{1}{N} \sum_{i=1}^{N} \mathbf{x}_i,$$

where we have used the property $\sum_i \langle \mathbf{z}_i | \mathbf{x}_i \rangle \approx \mathbf{0}$ when N is sufficiently large. This property arises from the assumption that the prior of $\mathbf{z}$ follows a Gaussian distribution, i.e., $\mathbf{z} \sim \mathcal{N}(\mathbf{0}, \mathbf{I})$. Alternatively, we may take the expectation of $\langle \mathbf{z}_i | \mathbf{x}_i \rangle$ in Eq. 3.65(a) to get this result.

In the E-step, we compute the posterior means $\langle \mathbf{z}_i | \mathbf{x}_i \rangle$ and posterior moments $\langle \mathbf{z}_i \mathbf{z}_i^{\mathsf{T}} | \mathbf{x}_i \rangle$. Let's express the following posterior density in terms of its likelihood and prior [45]:

$$
\begin{aligned}
& p(\mathbf{z}_i|\mathbf{x}_i,\omega) \\
& \propto p(\mathbf{x}_i|\mathbf{z}_i,\omega)p(\mathbf{z}_i) \\
& \propto \exp\left\{-\frac{1}{2}(\mathbf{x}_i-\mathbf{m}-\mathbf{V}\mathbf{z}_i)^{\mathsf{T}}\boldsymbol{\Sigma}^{-1}(\mathbf{x}_i-\mathbf{m}-\mathbf{V}\mathbf{z}_i)-\frac{1}{2}\mathbf{z}_i^{\mathsf{T}}\mathbf{z}_i\right\} \\
& \propto \exp\left\{\mathbf{z}_i^{\mathsf{T}}\mathbf{V}^{\mathsf{T}}\boldsymbol{\Sigma}^{-1}(\mathbf{x}_i-\mathbf{m})-\frac{1}{2}\mathbf{z}_i^{\mathsf{T}}\left(\mathbf{I}+\mathbf{V}^{\mathsf{T}}\boldsymbol{\Sigma}^{-1}\mathbf{V}\right)\mathbf{z}_i\right\}.
\end{aligned}
\tag{3.63}
$$

Consider the following property of Gaussian distribution with mean $\boldsymbol{\mu}_z$ and covariance $\mathbf{C}_z$

$$
\begin{aligned}
\mathcal{N}(\mathbf{z}|\boldsymbol{\mu}_z,\mathbf{C}_z) & \propto \exp\left\{-\frac{1}{2}(\mathbf{z}-\boldsymbol{\mu}_z)^{\mathsf{T}}\mathbf{C}_z^{-1}(\mathbf{z}-\boldsymbol{\mu}_z)\right\} \\
& \propto \exp\left\{\mathbf{z}^{\mathsf{T}}\mathbf{C}_z^{-1}\boldsymbol{\mu}_z-\frac{1}{2}\mathbf{z}^{\mathsf{T}}\mathbf{C}_z^{-1}\mathbf{z}\right\}.
\end{aligned}
\tag{3.64}
$$

Comparing Eq. 3.63 and Eq. 3.64, we obtain the posterior mean and moment as follows:

$$
\begin{aligned}
\langle\mathbf{z}_i|\mathbf{x}_i\rangle &= \mathbf{L}^{-1}\mathbf{V}^{\mathsf{T}}\boldsymbol{\Sigma}^{-1}(\mathbf{x}_i-\mathbf{m}) \\
\langle\mathbf{z}_i\mathbf{z}_i^{\mathsf{T}}|\mathbf{x}_i\rangle &= \mathbf{L}^{-1}+\langle\mathbf{z}_i|\mathbf{x}_i\rangle\langle\mathbf{z}_i^{\mathsf{T}}|\mathbf{x}_i\rangle,
\end{aligned}
$$

where $\mathbf{L}^{-1}=(\mathbf{I}+\mathbf{V}^{\mathsf{T}}\boldsymbol{\Sigma}^{-1}\mathbf{V})^{-1}$ is the posterior covariance matrix of $\mathbf{z}_i$, $\forall i$.

In summary, we have the following EM algorithm for factor analysis:

E-step:

$$\langle\mathbf{z}_i|\mathbf{x}_i\rangle = \mathbf{L}^{-1}\mathbf{V}^{\mathsf{T}}\boldsymbol{\Sigma}^{-1}(\mathbf{x}_i-\mathbf{m}) \tag{3.65a}$$

$$\langle\mathbf{z}_i\mathbf{z}_i^{\mathsf{T}}|\mathbf{x}_i\rangle = \mathbf{L}^{-1}+\langle\mathbf{z}_i|\mathbf{x}_i\rangle\langle\mathbf{z}_i|\mathbf{x}_i\rangle^{\mathsf{T}} \tag{3.65b}$$

$$\mathbf{L}=\mathbf{I}+\mathbf{V}^{\mathsf{T}}\boldsymbol{\Sigma}^{-1}\mathbf{V} \tag{3.65c}$$

M-step:

$$\mathbf{m}'=\frac{1}{N}\sum_{i=1}^{N}\mathbf{x}_i \tag{3.65d}$$

$$\mathbf{V}'=\left[\sum_{i=1}^{N}(\mathbf{x}_i-\mathbf{m}')\langle\mathbf{z}_i|\mathbf{x}_i\rangle^{\mathsf{T}}\right]\left[\sum_{i=1}^{N}\langle\mathbf{z}_i\mathbf{z}_i^{\mathsf{T}}|\mathbf{x}_i\rangle\right]^{-1} \tag{3.65e}$$

$$\boldsymbol{\Sigma}'=\frac{1}{N}\left\{\sum_{i=1}^{N}\left[(\mathbf{x}_i-\mathbf{m}')(\mathbf{x}_i-\mathbf{m}')^{\mathsf{T}}-\mathbf{V}'\langle\mathbf{z}_i|\mathbf{x}_i\rangle(\mathbf{x}_i-\mathbf{m}')^{\mathsf{T}}\right]\right\} \tag{3.65f}$$

Note that Eq. 3.65(a) and the Eq. 3.65(c) are the posterior mean and posterior precision of the Bayesian general linear model [46, Eq.10.32 and Eq. 10.34].

3.3.3 Relationship with Principal Component Analysis

Both principal component analysis (PCA) and factor analysis (FA) are dimension reduction techniques. But they are fundamentally different. In particular, components in PCA are orthogonal to each other, whereas factor analysis does not require the column vectors in the loading matrix to be orthogonal, i.e., the correlation between the factors could be nonzero. Also, PCA finds a linear combination of the observed variables to preserve the variance of the data, whereas FA predicts observed variables from theoretical latent factors. Mathematically, if $\mathbf{W} = [w_1\ \mathbf{w}_2\ \cdots\ \mathbf{w}_M]$ is a PCA projection matrix in which $\mathbf{w}_j$'s are the eigenvectors of the data covariance matrix and $\mathbf{V}$ is an FA loading matrix, we have

$$\begin{aligned} \text{PCA}: \ \mathbf{y}_i &= \mathbf{W}^{\mathsf{T}}(\mathbf{x}_i - \mathbf{m}) \text{ and } \hat{\mathbf{x}}_i = \mathbf{m} + \mathbf{W}\mathbf{y}_i \\ \text{FA}: \ \mathbf{x}_i &= \mathbf{m} + \mathbf{V}\mathbf{z}_i + \epsilon_i, \end{aligned} \tag{3.66}$$

where $\mathbf{x}_i$'s are observed vectors, $\hat{\mathbf{x}}$'s are PCA reconstructed vectors, $\mathbf{y}_i$'s are PCA-projected vectors, $\mathbf{z}_i$'s are factors, and ϵ_i' are residues.

Under specific conditions, FA becomes PCA. Consider the special situation in which the covariance matrix of ϵ is diagonal and all elements in the diagonal are equals, i.e., $\boldsymbol{\Sigma} = \sigma^2\mathbf{I}$. Then, Eq. 3.65(c) and Eq. 3.65(a) become

$$\mathbf{L} = \mathbf{I} + \frac{1}{\sigma^2}\mathbf{V}^{\mathsf{T}}\mathbf{V} \quad \text{and} \quad \langle \mathbf{z}_i|\mathbf{x}_i\rangle = \frac{1}{\sigma^2}\mathbf{L}^{-1}\mathbf{V}^{\mathsf{T}}(\mathbf{x}_i - \mathbf{m}),$$

respectively. When $\sigma^2 \rightarrow 0$ and $\mathbf{V}$ is an orthogonal matrix, then $\mathbf{L} \rightarrow \frac{1}{\sigma^2}\mathbf{V}^{\mathsf{T}}\mathbf{V}$ and

$$\begin{aligned} \langle \mathbf{z}_i|\mathbf{x}_i\rangle &\rightarrow \frac{1}{\sigma^2}\sigma^2(\mathbf{V}^{\mathsf{T}}\mathbf{V})^{-1}\mathbf{V}^{\mathsf{T}}(\mathbf{x}_i - \mathbf{m}) \\ &= \mathbf{V}^{-1}(\mathbf{V}^{\mathsf{T}})^{-1}\mathbf{V}^{\mathsf{T}}(\mathbf{x}_i - \mathbf{m}) \\ &= \mathbf{V}^{\mathsf{T}}(\mathbf{x}_i - \mathbf{m}). \end{aligned} \tag{3.67}$$

Note that Eq. 3.67 has the same form as the PCA projection in Eq. 3.66 and that the posterior covariance ($\mathbf{L}^{-1}$) of $\mathbf{z}$ becomes 0.

The analysis above suggests that PCA is a specific case of factor analysis in which the covariances of residue noise ϵ collapse to zero. This means that in PCA, for a given $\mathbf{x}_i$ and $\mathbf{W}$ in Eq. 3.66, there is a *deterministic* $\mathbf{y}_i$. On the other hand, in factor analysis, an $\mathbf{x}_i$ can be generated by infinite combinations of $\mathbf{z}_i$ and ϵ_i, as evident in Eq. 3.56.

The covariance matrix ($\mathbf{V}\mathbf{V}^{\mathsf{T}} + \boldsymbol{\Sigma}$) of observed data in FA suggests that FA treats covariances and variances of observed data separately. More specifically, the D column vectors (factor loadings) of $\mathbf{V}$ in FA capture most of the covariances while the variances of the individual components of $\mathbf{x}$'s are captured by the diagonal elements of $\boldsymbol{\Sigma}$. Note that the diagonal elements in $\boldsymbol{\Sigma}$ can be different, providing extra flexibility in modeling variances. On the other hand, because $\boldsymbol{\Sigma} \rightarrow \mathbf{0}$ in PCA, the first M principal components attempt to capture most of the variabilities, including covariances and variances. As a result, FA is more flexible in terms of data modeling.

3.3.4 Relationship with Nuisance Attribute Projection

If we write the projected vectors in Eq. 3.43 as follows:

$$\begin{aligned} \vec{\mu}^{(s)} &= \tilde{\mu}^{(s)} + \mathbf{U}(\mathbf{U}^{\mathsf{T}}\vec{\mu}^{(s)}) \\ &= \tilde{\mu}^{(s)} + \mathbf{U}\mathbf{x}^{(s)}, \end{aligned} \tag{3.68}$$

where the superscript s stands for speaker s and $\mathbf{x}^{(s)} \equiv \mathbf{U}^{\mathsf{T}}\vec{\mu}^{(s)}$, then we may consider NAP as a special case of factor analysis in which the loading matrix $\mathbf{U}$ is determined by minimizing within-speaker covariance and $\mathbf{x}^{(s)}$ is the factor that gives the appropriate offset to produce $\vec{\mu}^{(s)}$. This factor analysis interpretation suggests that $\tilde{\mu}^{(s)}$ is considered fixed for speaker s and that there is no intra-speaker variability for speaker s, i.e., all variability in $\vec{\mu}^{(s)}$ is due to the channel. As a result, NAP does not consider the prior density of $\mathbf{x}^{(s)}$ and does not have a speaker subspace to model inter-speaker variability.

3.4 Probabilistic Linear Discriminant Analysis

3.4.1 Generative Model

The Gaussian probabilistic LDA (PLDA) is the supervised version of FA. Consider a data set comprising D-dimensional vectors $\mathcal{X} = \{\mathbf{x}_{ij}; i = 1, \ldots, N; j = 1, \ldots, H_i\}$ obtained from N classes such that the ith class contains H_i vectors (e.g., i-vectors). The i-vector $\mathbf{x}_{ij}$ is assumed to be generated by a factor analysis model:

$$\mathbf{x}_{ij} = \mathbf{m} + \mathbf{V}\mathbf{z}_i + \boldsymbol{\epsilon}_{ij}, \tag{3.69}$$

where $\mathbf{m}$ is the global mean of all i-vectors, $\mathbf{V}$ is the speaker loading matrix, $\mathbf{z}_i$ is the speaker factor (of the ith speaker), and $\boldsymbol{\epsilon}_{ij}$ is the residue that could not be captured by the factor analysis model.

The class-specific vectors should share the same latent factor:

$$\tilde{\mathbf{x}}_i = \tilde{\mathbf{m}} + \tilde{\mathbf{V}}\mathbf{z}_i + \tilde{\boldsymbol{\epsilon}}_i, \qquad \tilde{\mathbf{x}}_i, \tilde{\mathbf{m}} \in \mathbb{R}^{DH_i}, \; \tilde{\mathbf{V}} \in \mathbb{R}^{DH_i \times M}, \; \tilde{\boldsymbol{\epsilon}}_i \in \mathbb{R}^{DH_i}, \tag{3.70}$$

where $\tilde{\mathbf{x}}_i = [\mathbf{x}_{i1}^{\mathsf{T}} \;\cdots\; \mathbf{x}_{iH_i}^{\mathsf{T}}]^{\mathsf{T}}$, $\tilde{\mathbf{m}} = [\mathbf{m}^{\mathsf{T}} \;\cdots\; \mathbf{m}^{\mathsf{T}}]^{\mathsf{T}}$, $\tilde{\mathbf{V}} = [\mathbf{V}^{\mathsf{T}} \;\cdots\; \mathbf{V}^{\mathsf{T}}]^{\mathsf{T}}$, and $\tilde{\boldsymbol{\epsilon}}_i = [\boldsymbol{\epsilon}_{i1}^{\mathsf{T}} \;\cdots\; \boldsymbol{\epsilon}_{iH_i}^{\mathsf{T}}]^{\mathsf{T}}$. Figure 3.21 shows the graphical representation of a Gaussian PLDA model. Note that the key difference between the PLDA model in Figure 3.21 and the FA model in Figure 3.20 is that in the former, for each i we have H_i observed vectors; whereas for the latter, each i is associated with one observed vector only.

The E- and M-steps iteratively evaluate and maximize the expectation of the complete likelihood:

$$\begin{aligned} Q(\omega'|\omega) &= \mathbb{E}_{\mathcal{Z}}\{\ln p(\mathcal{X}, \mathcal{Z}|\omega')|\mathcal{X}, \omega\} \\ &= \mathbb{E}_{\mathcal{Z}}\left\{\sum\nolimits_{ij} \ln\left[p\left(\mathbf{x}_{ij}|\mathbf{z}_i, \omega'\right) p(\mathbf{z}_i)\right] \middle| \mathcal{X}, \omega\right\} \\ &= \mathbb{E}_{\mathcal{Z}}\left\{\sum\nolimits_{ij} \ln\left[\mathcal{N}\left(\mathbf{x}_{ij}|\mathbf{m}' + \mathbf{V}'\mathbf{z}_i, \boldsymbol{\Sigma}'\right) \mathcal{N}(\mathbf{z}_i|\mathbf{0}, \mathbf{I})\right] \middle| \mathcal{X}, \omega\right\}, \end{aligned} \tag{3.71}$$

where the notation $\sum_{ij}$ is a shorthand form of $\sum_{i=1}^{N}\sum_{j=1}^{H_i}$.

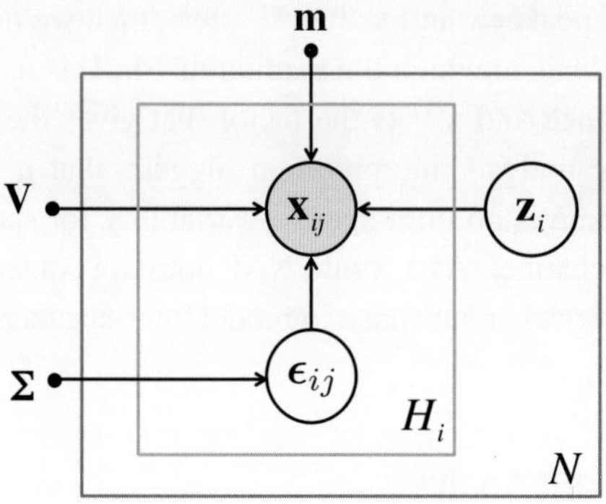

Figure 3.21 Graphical representation of a simplified Gaussian PLDA model in which the channel variability is modeled by the full covariance matrix $\mathbf{\Sigma}$. The prior of $\mathbf{z}_i$ follows a standard Gaussian distribution.

3.4.2 EM Formulations

Similar to Section 3.3.2, the auxiliary function can be simplified to

$$\begin{aligned} Q(\omega) &= -\sum_{ij} \mathbb{E}_{\mathcal{Z}} \left\{ \frac{1}{2} \log |\mathbf{\Sigma}| + \frac{1}{2} (\mathbf{x}_{ij} - \mathbf{m} - \mathbf{V}\mathbf{z}_i)^{\mathsf{T}} \mathbf{\Sigma}^{-1} (\mathbf{x}_{ij} - \mathbf{m} - \mathbf{V}\mathbf{z}_i) \right\} \\ &= \sum_{ij} \left[-\frac{1}{2} \log |\mathbf{\Sigma}| - \frac{1}{2} (\mathbf{x}_{ij} - \mathbf{m})^{\mathsf{T}} \mathbf{\Sigma}^{-1} (\mathbf{x}_{ij} - \mathbf{m}) \right] \\ &\quad + \sum_{ij} (\mathbf{x}_{ij} - \mathbf{m})^{\mathsf{T}} \mathbf{\Sigma}^{-1} \mathbf{V} \langle \mathbf{z}_i | \mathcal{X}_i \rangle - \frac{1}{2} \left[\sum_{ij} \left\langle \mathbf{z}_i^{\mathsf{T}} \mathbf{V}^{\mathsf{T}} \mathbf{\Sigma}^{-1} \mathbf{V} \mathbf{z}_i | \mathcal{X}_i \right\rangle \right]. \end{aligned} \tag{3.72}$$

Taking the derivative of Q with respect to $\mathbf{V}$, we have

$$\frac{\partial Q}{\partial \mathbf{V}} = \sum_{ij} \mathbf{\Sigma}^{-1} (\mathbf{x}_{ij} - \mathbf{m}) \langle \mathbf{z}_i^{\mathsf{T}} | \mathcal{X}_i \rangle - \sum_{ij} \mathbf{\Sigma}^{-1} \mathbf{V} \left\langle \mathbf{z}_i \mathbf{z}_i^{\mathsf{T}} | \mathcal{X}_i \right\rangle. \tag{3.73}$$

Setting $\frac{\partial Q}{\partial \mathbf{V}} = 0$, we have

$$\sum_{ij} \mathbf{V} \left\langle \mathbf{z}_i \mathbf{z}_i^{\mathsf{T}} | \mathcal{X}_i \right\rangle = \sum_{ij} (\mathbf{x}_{ij} - \mathbf{m}) \langle \mathbf{z}_i^{\mathsf{T}} | \mathcal{X}_i \rangle \tag{3.74}$$

$$\mathbf{V} = \left[\sum_{ij} (\mathbf{x}_{ij} - \mathbf{m}) \langle \mathbf{z}_i | \mathcal{X}_i \rangle^{\mathsf{T}} \right] \left[\sum_{ij} \left\langle \mathbf{z}_i \mathbf{z}_i^{\mathsf{T}} | \mathcal{X}_i \right\rangle \right]^{-1}. \tag{3.75}$$

To find $\mathbf{\Sigma}$, we evaluate

$$\begin{aligned} \frac{\partial Q}{\partial \mathbf{\Sigma}^{-1}} &= \frac{1}{2} \sum_{ij} \left[\mathbf{\Sigma} - (\mathbf{x}_{ij} - \mathbf{m})(\mathbf{x}_{ij} - \mathbf{m})^{\mathsf{T}} \right] + \sum_{ij} (\mathbf{x}_{ij} - \mathbf{m}) \langle \mathbf{z}_i^{\mathsf{T}} | \mathcal{X}_i \rangle \mathbf{V}^{\mathsf{T}} \\ &\quad - \frac{1}{2} \sum_{ij} \mathbf{V} \langle \mathbf{z}_i \mathbf{z}_i^{\mathsf{T}} | \mathcal{X}_i \rangle \mathbf{V}^{\mathsf{T}}. \end{aligned}$$

Note that according to Eq. 3.61, we have

$$\frac{\partial Q}{\partial \boldsymbol{\Sigma}^{-1}} = \frac{1}{2}\sum_{ij}\left[\boldsymbol{\Sigma} - (\mathbf{x}_{ij}-\mathbf{m})(\mathbf{x}_{ij}-\mathbf{m})^{\mathsf{T}}\right] + \sum_{ij}(\mathbf{x}_{ij}-\mathbf{m})\langle\mathbf{z}_i^{\mathsf{T}}|\mathcal{X}_i\rangle\mathbf{V}^{\mathsf{T}}$$
$$-\frac{1}{2}\sum_{ij}(\mathbf{x}_{ij}-\mathbf{m})\langle\mathbf{z}_i^{\mathsf{T}}|\mathcal{X}_i\rangle\mathbf{V}^{\mathsf{T}}.$$

Therefore, setting $\frac{\partial Q}{\partial \boldsymbol{\Sigma}^{-1}} = 0$ we have

$$\sum_{ij}\boldsymbol{\Sigma} = \sum_{ij}\left[(\mathbf{x}_{ij}-\mathbf{m})(\mathbf{x}_{ij}-\mathbf{m})^{\mathsf{T}} - (\mathbf{x}_{ij}-\mathbf{m})\langle\mathbf{z}_i^{\mathsf{T}}|\mathcal{X}_i\rangle\mathbf{V}^{\mathsf{T}}\right].$$

Rearranging, we have

$$\boldsymbol{\Sigma} = \frac{1}{\sum_{ij}1}\sum_{ij}\left[(\mathbf{x}_{ij}-\mathbf{m})(\mathbf{x}_{ij}-\mathbf{m})^{\mathsf{T}} - \mathbf{V}\langle\mathbf{z}_i|\mathcal{X}_i\rangle(\mathbf{x}_{ij}-\mathbf{m})^{\mathsf{T}}\right].$$

To compute $\mathbf{m}$, we evaluate

$$\frac{\partial Q}{\partial \mathbf{m}} = -\sum_{ij}\left(\boldsymbol{\Sigma}^{-1}\mathbf{m} - \boldsymbol{\Sigma}^{-1}\mathbf{x}_{ij}\right) + \sum_{ij}\boldsymbol{\Sigma}^{-1}\mathbf{V}\langle\mathbf{z}_i|\mathcal{X}_i\rangle.$$

Setting $\frac{\partial Q}{\partial \mathbf{m}} = 0$, we have

$$\mathbf{m} = \frac{\sum_{i=1}^{N}\sum_{j=1}^{H_i}\mathbf{x}_{ij}}{\sum_{i=1}^{N}\sum_{j=1}^{H_i}1},$$

where we have used the property $\sum_i\langle\mathbf{z}_i|\mathcal{X}_i\rangle \approx \mathbf{0}$ when the number of training speakers is sufficiently large. This property arises from the assumption that the factors $\mathbf{z} \sim \mathcal{N}(\mathbf{0},\mathbf{I})$.

In the E-step, we compute the posterior means $\langle\mathbf{z}_i|\mathcal{X}_i\rangle$ and posterior moments $\langle\mathbf{z}_i\mathbf{z}_i^{\mathsf{T}}|\mathcal{X}_i\rangle$, where $\mathcal{X}_i$ denotes the i-vectors of speaker i. Let's express the following posterior density in terms of its likelihood and prior:

$$\begin{aligned}
&p(\mathbf{z}_i|\mathbf{x}_{i1},\ldots,\mathbf{x}_{iH_i},\omega)\\
&\propto \prod_{j=1}^{H_i} p(\mathbf{x}_{ij}|\mathbf{z}_i,\omega)p(\mathbf{z}_i)\\
&\propto \exp\left\{-\frac{1}{2}\sum_j(\mathbf{x}_{ij}-\mathbf{m}-\mathbf{V}\mathbf{z}_i)^{\mathsf{T}}\boldsymbol{\Sigma}^{-1}(\mathbf{x}_{ij}-\mathbf{m}-\mathbf{V}\mathbf{z}_i) - \frac{1}{2}\mathbf{z}_i^{\mathsf{T}}\mathbf{z}_i\right\}\\
&= \exp\left\{\mathbf{z}_i^{\mathsf{T}}\mathbf{V}^{\mathsf{T}}\sum_j\boldsymbol{\Sigma}^{-1}(\mathbf{x}_{ij}-\mathbf{m}) - \frac{1}{2}\mathbf{z}_i^{\mathsf{T}}\left(\mathbf{I}+\sum_j\mathbf{V}^{\mathsf{T}}\boldsymbol{\Sigma}^{-1}\mathbf{V}\right)\mathbf{z}_i\right\}.
\end{aligned} \tag{3.76}$$

Consider the following property of Gaussian distribution with mean μ_z and covariance $\mathbf{C}_z$

$$\begin{aligned}\mathcal{N}(\mathbf{z}|\mu_z, \mathbf{C}_z) &\propto \exp\left\{-\frac{1}{2}(\mathbf{z}-\mu_z)^{\mathsf{T}}\mathbf{C}_z^{-1}(\mathbf{z}-\mu_z)\right\} \\ &\propto \exp\left\{\mathbf{z}^{\mathsf{T}}\mathbf{C}_z^{-1}\mu_z - \frac{1}{2}\mathbf{z}^{\mathsf{T}}\mathbf{C}_z^{-1}\mathbf{z}\right\}.\end{aligned} \tag{3.77}$$

Comparing Eq. 3.76 and Eq. 3.77, we obtain the posterior mean and moment as follows:

$$\langle \mathbf{z}_i|\mathcal{X}_i\rangle = \mathbf{L}_i^{-1}\mathbf{V}^{\mathsf{T}}\sum_{j=1}^{H_i}\mathbf{\Sigma}^{-1}(\mathbf{x}_{ij}-\mathbf{m}) \tag{3.78}$$

$$\langle \mathbf{z}_i\mathbf{z}_i^{\mathsf{T}}|\mathcal{X}_i\rangle = \mathbf{L}_i^{-1} + \langle \mathbf{z}_i|\mathcal{X}_i\rangle\langle \mathbf{z}_i^{\mathsf{T}}|\mathcal{X}_i\rangle \tag{3.79}$$

where $\mathbf{L}_i = \mathbf{I} + \sum_j \mathbf{V}^{\mathsf{T}}\mathbf{\Sigma}^{-1}\mathbf{V}$.

In summary, the EM algorithm for PLDA is as follows [47]:

E-step:

$\langle \mathbf{z}_i|\mathcal{X}_i\rangle = \mathbf{L}_i^{-1}\mathbf{V}^{\mathsf{T}}\mathbf{\Sigma}^{-1}\sum_{j=1}^{H_i}(\mathbf{x}_{ij}-\mathbf{m})$

$\langle \mathbf{z}_i\mathbf{z}_i^{\mathsf{T}}|\mathcal{X}_i\rangle = \mathbf{L}_i^{-1} + \langle \mathbf{z}_i|\mathcal{X}_i\rangle\langle \mathbf{z}_i|\mathcal{X}_i\rangle^{\mathsf{T}}$

$\mathbf{L}_i = \mathbf{I} + H_i\mathbf{V}^{\mathsf{T}}\mathbf{\Sigma}^{-1}\mathbf{V}$

M-step:

$$\mathbf{V}' = \left[\sum_{i=1}^{N}\sum_{j=1}^{H_i}(\mathbf{x}_{ij}-\mathbf{m}')\langle \mathbf{z}_i|\mathcal{X}_i\rangle^{\mathsf{T}}\right]\left[\sum_{i=1}^{N}\sum_{j=1}^{H_i}\langle \mathbf{z}_i\mathbf{z}_i^{\mathsf{T}}|\mathcal{X}_i\rangle\right]^{-1} \tag{3.80}$$

$$\mathbf{m}' = \frac{\sum_{ij}\mathbf{x}_{ij}}{\sum_i H_i}$$

$$\mathbf{\Sigma}' = \frac{1}{\sum_{i=1}^{N} H_i}\left\{\sum_{i=1}^{N}\sum_{j=1}^{H_i}\left[(\mathbf{x}_{ij}-\mathbf{m}')(\mathbf{x}_{ij}-\mathbf{m}')^{\mathsf{T}} - \mathbf{V}'\langle \mathbf{z}_i|\mathcal{X}_i\rangle(\mathbf{x}_{ij}-\mathbf{m}')^{\mathsf{T}}\right]\right\}$$

3.4.3 PLDA Scoring

PLDA is by far the most popular back end for speaker verification. Typically, it accepts an i-vector (see Section 3.6) pair or more recently x-vector (see Section 5.2.1) pairs as input and produces a score that quantifies how likely the pair comes from the same speaker against the likelihood that the pair is from two different speakers.

Given a test i-vector $\mathbf{x}_t$ and target-speaker's i-vector $\mathbf{x}_s$, we need to test the hypothesis H_0 that both i-vectors come from the same speaker against the alternative hypothesis that they belong to different speakers. Using the notations in Eq. 3.55, we have

$$H_0 : \text{Same speaker} \begin{cases} \mathbf{x}_s = \mathbf{m} + \mathbf{V}\mathbf{z} + \boldsymbol{\epsilon}_s \\ \mathbf{x}_t = \mathbf{m} + \mathbf{V}\mathbf{z} + \boldsymbol{\epsilon}_t \end{cases} \tag{3.81}$$

$$H_1 : \text{Different speakers} \begin{cases} \mathbf{x}_s = \mathbf{m} + \mathbf{V}\mathbf{z}_s + \boldsymbol{\epsilon}_s \\ \mathbf{x}_t = \mathbf{m} + \mathbf{V}\mathbf{z}_t + \boldsymbol{\epsilon}_t \end{cases} \tag{3.82}$$

Note that the speaker factor $\mathbf{z}$ is the same for both i-vectors in H_0, whereas they are different in H_1. This hypothesis testing problem can be solved by evaluating the likelihood ratio score:

$$\begin{aligned} S_{\text{LR}}(\mathbf{x}_s, \mathbf{x}_t) &= \frac{p(\mathbf{x}_s, \mathbf{x}_t | \text{same-speaker})}{p(\mathbf{x}_s, \mathbf{x}_t | \text{different-speakers})} \\ &= \frac{\int p(\mathbf{x}_s, \mathbf{x}_t, \mathbf{z} | \omega) d\mathbf{z}}{\int p(\mathbf{x}_s, \mathbf{z}_s | \omega) d\mathbf{z}_s \int p(\mathbf{x}_t, \mathbf{z}_t | \omega) d\mathbf{z}_t} \\ &= \frac{\int p(\mathbf{x}_s, \mathbf{x}_t | \mathbf{z}, \omega) p(\mathbf{z}) d\mathbf{z}}{\int p(\mathbf{x}_s | \mathbf{z}_s, \omega) p(\mathbf{z}_s) d\mathbf{z}_s \int p(\mathbf{x}_t | \mathbf{z}_t, \omega) p(\mathbf{z}_t) d\mathbf{z}_t} \\ &= \frac{\mathcal{N}\left(\left[\mathbf{x}_s^{\mathsf{T}} \ \mathbf{x}_t^{\mathsf{T}} \right]^{\mathsf{T}} \middle| \hat{\mathbf{m}}, \hat{\boldsymbol{\Lambda}} \right)}{\mathcal{N}\left(\left[\mathbf{x}_s^{\mathsf{T}} \ \mathbf{x}_t^{\mathsf{T}} \right]^{\mathsf{T}} \middle| \hat{\mathbf{m}}, \text{diag}\{\boldsymbol{\Lambda}, \boldsymbol{\Lambda}\} \right)} \end{aligned} \tag{3.83}$$

where $\hat{\mathbf{m}} = \left[\mathbf{m}^{\mathsf{T}}, \mathbf{m}^{\mathsf{T}} \right]^{\mathsf{T}}$, $\hat{\boldsymbol{\Lambda}} = \hat{\mathbf{V}}\hat{\mathbf{V}}^{\mathsf{T}} + \hat{\boldsymbol{\Sigma}}$, $\hat{\mathbf{V}} = \left[\mathbf{V}^{\mathsf{T}} \ \mathbf{V}^{\mathsf{T}} \right]^{\mathsf{T}}$, $\hat{\boldsymbol{\Sigma}} = \text{diag}\{\boldsymbol{\Sigma}, \boldsymbol{\Sigma}\}$, and $\boldsymbol{\Lambda} = \mathbf{V}\mathbf{V}^{\mathsf{T}} + \boldsymbol{\Sigma}$. Because both the numerator and denominator of Eq. 3.83 are Gaussians, it can be simplified as follows:

$$\begin{aligned} \log S_{\text{LR}}(\mathbf{x}_s, \mathbf{x}_t) = \log \mathcal{N} &\left(\begin{bmatrix} \mathbf{x}_s \\ \mathbf{x}_t \end{bmatrix} \middle| \begin{bmatrix} \mathbf{m} \\ \mathbf{m} \end{bmatrix}, \begin{bmatrix} \boldsymbol{\Sigma}_{tot} & \boldsymbol{\Sigma}_{ac} \\ \boldsymbol{\Sigma}_{ac} & \boldsymbol{\Sigma}_{tot} \end{bmatrix} \right) \\ &- \log \mathcal{N} \left(\begin{bmatrix} \mathbf{x}_s \\ \mathbf{x}_t \end{bmatrix} \middle| \begin{bmatrix} \mathbf{m} \\ \mathbf{m} \end{bmatrix}, \begin{bmatrix} \boldsymbol{\Sigma}_{tot} & \mathbf{0} \\ \mathbf{0} & \boldsymbol{\Sigma}_{tot} \end{bmatrix} \right), \end{aligned} \tag{3.84}$$

where $\boldsymbol{\Sigma}_{tot} = \mathbf{V}\mathbf{V}^{\mathsf{T}} + \boldsymbol{\Sigma}$ and $\boldsymbol{\Sigma}_{ac} = \mathbf{V}\mathbf{V}^{\mathsf{T}}$. Since $\mathbf{m}$ is a global offset that can be precomputed and removed from all i-vectors, we set it to zero and expand Eq. 3.84 to obtain:

$$\begin{aligned} \log S_{\text{LR}}(\mathbf{x}_s, \mathbf{x}_t) &= -\frac{1}{2} \begin{bmatrix} \mathbf{x}_s \\ \mathbf{x}_t \end{bmatrix}^{\mathsf{T}} \begin{bmatrix} \boldsymbol{\Sigma}_{tot} & \boldsymbol{\Sigma}_{ac} \\ \boldsymbol{\Sigma}_{ac} & \boldsymbol{\Sigma}_{tot} \end{bmatrix}^{-1} \begin{bmatrix} \mathbf{x}_s \\ \mathbf{x}_t \end{bmatrix} + \frac{1}{2} \begin{bmatrix} \mathbf{x}_s \\ \mathbf{x}_t \end{bmatrix}^{\mathsf{T}} \begin{bmatrix} \boldsymbol{\Sigma}_{tot} & 0 \\ 0 & \boldsymbol{\Sigma}_{tot} \end{bmatrix}^{-1} \begin{bmatrix} \mathbf{x}_s \\ \mathbf{x}_t \end{bmatrix} + \text{const} \\ &= -\frac{1}{2} \begin{bmatrix} \mathbf{x}_s \\ \mathbf{x}_t \end{bmatrix}^{\mathsf{T}} \begin{bmatrix} (\boldsymbol{\Sigma}_{tot} - \boldsymbol{\Sigma}_{ac}\boldsymbol{\Sigma}_{tot}^{-1}\boldsymbol{\Sigma}_{ac})^{-1} & -\boldsymbol{\Sigma}_{tot}^{-1}\boldsymbol{\Sigma}_{ac}(\boldsymbol{\Sigma}_{tot} - \boldsymbol{\Sigma}_{ac}\boldsymbol{\Sigma}_{tot}^{-1}\boldsymbol{\Sigma}_{ac})^{-1} \\ -(\boldsymbol{\Sigma}_{tot} - \boldsymbol{\Sigma}_{ac}\boldsymbol{\Sigma}_{tot}^{-1}\boldsymbol{\Sigma}_{ac})^{-1}\boldsymbol{\Sigma}_{ac}\boldsymbol{\Sigma}_{tot}^{-1} & (\boldsymbol{\Sigma}_{tot} - \boldsymbol{\Sigma}_{ac}\boldsymbol{\Sigma}_{tot}^{-1}\boldsymbol{\Sigma}_{ac})^{-1} \end{bmatrix} \begin{bmatrix} \mathbf{x}_s \\ \mathbf{x}_t \end{bmatrix} \\ &\quad + \frac{1}{2} \begin{bmatrix} \mathbf{x}_s \\ \mathbf{x}_t \end{bmatrix}^{\mathsf{T}} \begin{bmatrix} \boldsymbol{\Sigma}_{tot}^{-1} & 0 \\ 0 & \boldsymbol{\Sigma}_{tot}^{-1} \end{bmatrix} \begin{bmatrix} \mathbf{x}_s \\ \mathbf{x}_t \end{bmatrix} + \text{const} \\ &= \frac{1}{2} \begin{bmatrix} \mathbf{x}_s \\ \mathbf{x}_t \end{bmatrix}^{\mathsf{T}} \begin{bmatrix} \boldsymbol{\Sigma}_{tot}^{-1} - (\boldsymbol{\Sigma}_{tot} - \boldsymbol{\Sigma}_{ac}\boldsymbol{\Sigma}_{tot}^{-1}\boldsymbol{\Sigma}_{ac})^{-1} & \boldsymbol{\Sigma}_{tot}^{-1}\boldsymbol{\Sigma}_{ac}(\boldsymbol{\Sigma}_{tot} - \boldsymbol{\Sigma}_{ac}\boldsymbol{\Sigma}_{tot}^{-1}\boldsymbol{\Sigma}_{ac})^{-1} \\ \boldsymbol{\Sigma}_{tot}^{-1}\boldsymbol{\Sigma}_{ac}(\boldsymbol{\Sigma}_{tot} - \boldsymbol{\Sigma}_{ac}\boldsymbol{\Sigma}_{tot}^{-1}\boldsymbol{\Sigma}_{ac})^{-1} & \boldsymbol{\Sigma}_{tot}^{-1} - (\boldsymbol{\Sigma}_{tot} - \boldsymbol{\Sigma}_{ac}\boldsymbol{\Sigma}_{tot}^{-1}\boldsymbol{\Sigma}_{ac})^{-1} \end{bmatrix} \begin{bmatrix} \mathbf{x}_s \\ \mathbf{x}_t \end{bmatrix} + \text{const} \\ &= \frac{1}{2} \begin{bmatrix} \mathbf{x}_s^{\mathsf{T}} & \mathbf{x}_t^{\mathsf{T}} \end{bmatrix} \begin{bmatrix} \mathbf{Q} & \mathbf{P} \\ \mathbf{P} & \mathbf{Q} \end{bmatrix} \begin{bmatrix} \mathbf{x}_s \\ \mathbf{x}_t \end{bmatrix} + \text{const} \\ &= \frac{1}{2} [\mathbf{x}_s^{\mathsf{T}}\mathbf{Q}\mathbf{x}_s + \mathbf{x}_s^{\mathsf{T}}\mathbf{P}\mathbf{x}_t + \mathbf{x}_t^{\mathsf{T}}\mathbf{P}\mathbf{x}_s + \mathbf{x}_t^{\mathsf{T}}\mathbf{Q}\mathbf{x}_t] + \text{const} \\ &= \frac{1}{2} [\mathbf{x}_s^{\mathsf{T}}\mathbf{Q}\mathbf{x}_s + 2\mathbf{x}_s^{\mathsf{T}}\mathbf{P}\mathbf{x}_t + \mathbf{x}_t^{\mathsf{T}}\mathbf{Q}\mathbf{x}_t] + \text{const}, \end{aligned} \tag{3.85}$$

where

$$\mathbf{Q} = \boldsymbol{\Sigma}_{tot}^{-1} - (\boldsymbol{\Sigma}_{tot} - \boldsymbol{\Sigma}_{ac}\boldsymbol{\Sigma}_{tot}^{-1}\boldsymbol{\Sigma}_{ac})^{-1},$$
$$\mathbf{P} = \boldsymbol{\Sigma}_{tot}^{-1}\boldsymbol{\Sigma}_{ac}(\boldsymbol{\Sigma}_{tot} - \boldsymbol{\Sigma}_{ac}\boldsymbol{\Sigma}_{tot}^{-1}\boldsymbol{\Sigma}_{ac})^{-1}. \tag{3.86}$$

Eq. 3.83 suggests that instead of using a point estimate of the latent variable $\mathbf{z}$, the verification process marginalizes the joint density of the latent and observed variables over the latent variables. Instead of the actual value of the latent variable, the verification process considers the likelihood that the two i-vectors share the same latent variable [17]. Therefore, the verification score can be interpreted as the likelihood ratio between the test and target speakers sharing the same latent variable and the likelihood that they have different latent variables. As a result, a large score suggests that the two i-vectors come from the same speaker (i.e., sharing the same latent variable) rather than coming from two different speakers (i.e., having two distinct latent variables). Eq. 3.85 gives the PLDA score when there is only one enrollment i-vector per target speaker. When there are multiple enrollment utterances per target speaker, we may either average the corresponding enrollment i-vectors (so-called i-vector averaging) or average the scores obtained by applying Eq. 3.85 to each of the enrollment i-vectors (so-called score averaging).

While research has shown that in practice, i-vector averaging is a simple and effective way to process multiple enrollment utterances [48], a more elegant approach is to derive the exact likelihood ratio:

$$\begin{aligned} S_{\text{LR}}(\mathbf{x}_{s,1},\ldots,\mathbf{x}_{s,R},\mathbf{x}_t) &= \frac{p(\mathbf{x}_{s,1},\ldots,\mathbf{x}_{s,R},\mathbf{x}_t|H_0)}{p(\mathbf{x}_{s,1},\ldots,\mathbf{x}_{s,R},\mathbf{x}_t|H_1)} \\ &= \frac{p(\mathcal{X}_s,\mathbf{x}_t|H_0)}{p(\mathcal{X}_s|H_1)p(\mathbf{x}_t|H_1)} \\ &= \frac{p(\mathbf{x}_t|\mathcal{X}_s)}{p(\mathbf{x}_t)}, \end{aligned} \tag{3.87}$$

where $\mathcal{X}_s = \{\mathbf{x}_{s,1},\ldots,\mathbf{x}_{s,R}\}$ are R enrollment i-vectors from target-speaker s and $\mathbf{x}_t$ is an i-vector from a test speaker. Note that the numerator of Eq. 3.87 is a conditional likelihood [49]

$$p(\mathbf{x}_t|\mathcal{X}_s) = \mathcal{N}\left(\mathbf{x}_t\middle|\mathbf{m} + \mathbf{V}\langle\mathbf{z}_s|\mathcal{X}_s\rangle, \mathbf{V}\mathbf{L}_s^{-1}\mathbf{V}^{\mathsf{T}} + \boldsymbol{\Sigma}\right) \tag{3.88}$$

where $\langle\mathbf{z}_s|\mathcal{X}_s\rangle$ and $\mathbf{L}_s$ are the posterior mean and posterior precision given by

$$\langle z_s|\mathcal{X}_s\rangle = \mathbf{L}_s^{-1}\mathbf{V}\boldsymbol{\Sigma}^{-1}\sum_{r=1}^{R}(\mathbf{x}_{s,r} - \mathbf{m}) \tag{3.89a}$$

$$\mathbf{L}_s = \mathbf{I} + R\mathbf{V}^{\mathsf{T}}\boldsymbol{\Sigma}^{-1}\mathbf{V}. \tag{3.89b}$$

Eq. 3.88 and Eq. 3.89 can be derived by noting that

$$
\begin{aligned}
p(\mathbf{x}_t|\mathcal{X}_s) &= \int p(\mathbf{x}_t|\mathbf{z}_s)p(\mathbf{z}_s|\mathcal{X}_s)\mathrm{d}\mathbf{z}_s \\
&= \int \mathcal{N}(\mathbf{x}_t|\mathbf{m}+\mathbf{V}\mathbf{z}_s,\boldsymbol{\Sigma})\mathcal{N}(\mathbf{z}_s|\langle\mathbf{z}_s|\mathcal{X}_s\rangle,\mathbf{L}_s^{-1})\mathrm{d}\mathbf{z}_s \\
&= \mathcal{N}\left(\mathbf{x}_t\middle|\mathbf{m}+\mathbf{V}\langle\mathbf{z}_s|\mathcal{X}_s\rangle,\mathbf{V}\mathbf{L}_s^{-1}\mathbf{V}^{\mathsf{T}}+\boldsymbol{\Sigma}\right)
\end{aligned}
\tag{3.90}
$$

where the second Gaussian can be obtained by using Eq. 3.76. The marginal likelihood in the denominator of Eq. 3.87 can be obtained by using Eq. 3.56, i.e., $p(\mathbf{x}_t) = \mathcal{N}(\mathbf{x}_t|\mathbf{m},\mathbf{V}\mathbf{V}^{\mathsf{T}}+\boldsymbol{\Sigma})$.

Substituting Eq. 3.88 and the marginal likelihood $p(\mathbf{x}_t)$ into Eq. 3.87, we have

$$
S_{\mathrm{LR}}(\mathcal{X}_s,\mathbf{x}_t) = \frac{\mathcal{N}\left(\mathbf{x}_t\middle|\mathbf{m}+\mathbf{V}\langle\mathbf{z}_s|\mathcal{X}_s\rangle,\mathbf{V}\mathbf{L}_s^{-1}\mathbf{V}^{\mathsf{T}}+\boldsymbol{\Sigma}\right)}{\mathcal{N}(\mathbf{x}_t|\mathbf{m},\mathbf{V}\mathbf{V}^{\mathsf{T}}+\boldsymbol{\Sigma})},
$$

which can be viewed as the likelihood ratio between the speaker-adapted PLDA model and the universal PLDA model.

The above derivation assumes that the i-vectors from the target speaker are conditional independent, because the posterior mean and covariance matrix in Eq. 3.89 are derived from Eq. 3.76. If independency is not assumed, we may use the method described in [48]. But, the method will be computationally intensive when the number of enrollment utterances per speaker is large.

3.4.4 Enhancement of PLDA

A number of enhancements have been proposed to improve the performance of PLDA. For example, Cumani et al. [50] propose to map the acoustic vectors of an utterance to a posterior distribution of i-vectors rather than a single i-vector. The method is particularly suitable for short utterances. Burget et al. [51] and Vasilakakis et al. [52] consider pairs of i-vectors and formulate an SVM-style training procedure to train a PLDA model to discriminate between same-speaker pairs and different-speaker pairs. Other approaches include constraining the PLDA parameters [53], applying nonparametric discriminant analysis before PLDA [54–56], and the use of between-speaker scatter as well as within-speaker scatter in PLDA training [57].

3.4.5 Alternative to PLDA

As PLDA models are linear, attempts have been made to see if they can be replaced by nonlinear models. Research has shown that deep belief networks can be used for discriminating the i-vectors of target speakers from those of impostors [58]. In [59], a binary restricted Boltzmann machine (RBM) was used to model the joint density of target-speaker's i-vectors and test i-vectors. Because RBMs can model the distribution of arbitrary complexity, they are thought to be more powerful than the conventional PLDA models.

Instead of using PLDA for scoring i-vectors, [58, 60] use a deep belief network (DBN) with two output nodes to classify i-vectors into speaker-class and impostor-class. The key idea is to use unsupervised training to train a universal DBN for all speakers. Then, a layer comprising two output nodes is stacked on top of the universal DBN, followed by backpropagation fine-tuning to minimize the cross-entropy error.

3.5 Heavy-Tailed PLDA

In the simplified Gaussian PLDA discussed in Section 3.5, the prior of the latent factor $\mathbf{z}$ follows a standard Gaussian distribution. This means that the parameters of the prior are assumed deterministic. The idea of PLDA in [17] can be extended to a fully Bayesian model in which the parameters of the prior distribution are assumed random, with the distribution of the hyper-parameters follow Gamma distributions [61]. In that case, $\mathbf{z}$ follows a Student's t distribution rather than a Gaussian distribution and the resulting model is called heavy-tailed PLDA model.

3.5.1 Generative Model

Figure 3.22 shows the graphical model of the heavy-tailed PLDA. The generative model is given by

$$\mathbf{x}_{ij} = \mathbf{m} + \mathbf{V}\mathbf{z}_i + \boldsymbol{\epsilon}_{ij} \tag{3.91}$$

$$\mathbf{z}_i \sim \mathcal{N}(\mathbf{0}, u_i^{-1}\mathbf{I}) \tag{3.92}$$

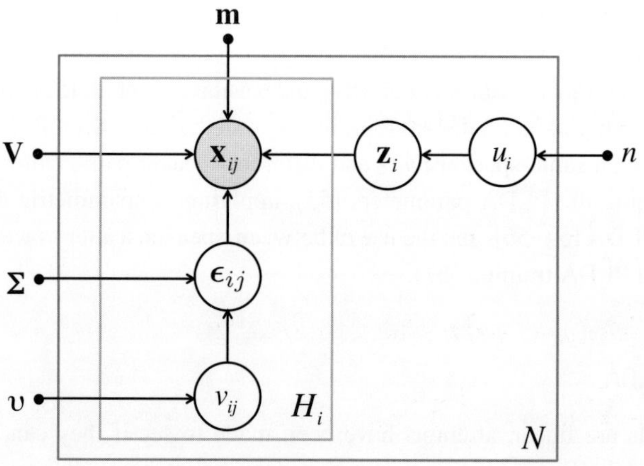

Figure 3.22 Graphical representation of a heavy-tailed PLDA model in which the channel variability is modeled by the full covariance matrix $\mathbf{\Sigma}$. The prior of $\mathbf{z}_i$ follows a Gaussian distribution $\mathcal{N}(\mathbf{0}, u_i^{-1}\mathbf{I})$, where u_i follows a Gamma distribution with n degree of freedom. The prior of ϵ_{ij} follows a Gaussian distribution $\mathcal{N}(\mathbf{0}, v_{ij}^{-1}\mathbf{\Sigma})$, where v_{ij} follows a Gamma distribution with ν degree of freedom.

$$\epsilon_{ij} \sim \mathcal{N}(\mathbf{0}, v_{ij}^{-1}\boldsymbol{\Sigma}) \quad (3.93)$$
$$u_i \sim \mathcal{G}(n/2, n/2) \quad (3.94)$$
$$v_{ij} \sim \mathcal{G}(v/2, v/2), \quad (3.95)$$

where $\mathcal{G}(\alpha/2, \alpha/2)$ is a Gamma distribution with α degree of freedom and identical scale and rate parameters. Note that because the latent variable u follows $\mathcal{G}(n/2, n/2)$, $\mathbf{z}_i$ follows a Student's t distribution with n degree of freedom [1, Eq. 2.161]. Similar argument applies to ϵ, i.e., $\epsilon \sim \mathcal{T}(\mathbf{0}, \boldsymbol{\Sigma}, v)$ [62].

3.5.2 Posteriors of Latent Variables

Figure 3.22 suggests that given a set of training vectors $\mathcal{X}_i = \{\mathbf{x}_{ij}; j = 1, \ldots, H_i\}$ from the ith speakers, the posteriors of the latent variables $\mathbf{z}_i$, u_i, ϵ_{ij}, and v_{ij} are dependent. This means that $p(z_i, u_i, \epsilon_{ij}, v_{ij}|\mathcal{X}_i)$ cannot be factorized; as a result, we resort to using variational Bayes [1]:

$$\log p(\mathbf{z}_i, u_i, \epsilon_{ij}, v_{ij}|\mathcal{X}_i) \approx \log q(\mathbf{z}_i) + \log q(u_i) + \log q(\epsilon_{ij}) + \log q(v_{ij}). \quad (3.96)$$

We denote the combined latent variables as $\mathbf{h}_i = \{\mathbf{z}_i, u_i, \mathbf{v}_i\}$,[6] where $\mathbf{v}_i = [v_{i1} \ \ldots \ v_{iH_i}]^{\mathsf{T}}$. To find $q()$'s, we apply variational Bayes:

$$\log q(\mathbf{z}_i) = \mathbb{E}_{q(\mathbf{h}_i \setminus \mathbf{z}_i)} \log p(\mathcal{X}_i, \mathbf{h}_i) + \text{const} \quad (3.97a)$$
$$\log q(u_i) = \mathbb{E}_{q(\mathbf{h}_i \setminus u_i)} \log p(\mathcal{X}_i, \mathbf{h}_i) + \text{const} \quad (3.97b)$$
$$\log q(v_{ij}) = \mathbb{E}_{q(\mathbf{h}_i \setminus v_{ij})} \log p(\mathcal{X}_i, \mathbf{h}_i) + \text{const}, \quad (3.97c)$$

where the notation $\mathbb{E}_{q(\mathbf{h}_i \setminus \mathbf{z}_i)}$ means taking expectation with respect to all latent variables except $\mathbf{z}_i$, using $q(u_i)q(\mathbf{v}_i)$ as the density. Similar meaning applies to other latent variables.

Using Figure 3.22, we may express $\log p(\mathcal{X}_i, \mathbf{h}_i)$ as follows:

$$\begin{aligned}
\log p(\mathcal{X}_i, \mathbf{h}_i) &= \log p(\mathcal{X}_i|\mathbf{z}_i, \epsilon_{ij}) p(\mathbf{z}_i|u_i) p(u_i) \prod_{j=1}^{H_i} p(\epsilon_{ij}|v_{ij}) p(v_{ij}) \\
&= -\frac{1}{2}\left[\sum_{j=1}^{H_i} v_{ij}(\mathbf{x}_{ij} - \mathbf{m} - \mathbf{V}\mathbf{z}_i)^{\mathsf{T}}\boldsymbol{\Sigma}^{-1}(\mathbf{x}_{ij} - \mathbf{m} - \mathbf{V}\mathbf{z}_i)\right] \\
&\quad + \frac{M}{2}\log u_i - \frac{u_i}{2}\mathbf{z}_i^{\mathsf{T}}\mathbf{z}_i + \log \mathcal{G}(u_i|n/2, n/2) \\
&\quad + \frac{D}{2}\sum_{j=1}^{H_i}\log v_{ij} - \frac{1}{2}\sum_{j=1}^{H_i} v_{ij}\epsilon_{ij}^{\mathsf{T}}\boldsymbol{\Sigma}^{-1}\epsilon_{ij} + \sum_{j=1}^{H_i}\log \mathcal{G}(v_{ij}|v/2, v/2) \\
&\quad + \text{const},
\end{aligned} \quad (3.98)$$

[6] Note that ϵ_{ij} is not part of the latent variables because it can be determined from $\mathbf{z}_i$, i.e., $\epsilon_{ij} = \mathbf{x}_{ij} - \mathbf{m} - \mathbf{V}\mathbf{z}_i$.

where M is the dimension of $\mathbf{z}_i$ and

$$\mathcal{G}(u_i|n/2,n/2) = \frac{1}{\Gamma(\frac{n}{2})}\left(\frac{n}{2}\right)^{\frac{n}{2}} u_i^{\frac{n}{2}-1} e^{-\frac{n}{2}u_i} \tag{3.99}$$

is a Gamma distribution. The constant term in Eq. 3.98 absorbs the terms that do not contain the latent variables.

Substituting Eq. 3.98 to Eq. 3.97(a), we have

$$\begin{aligned}\log q(\mathbf{z}_i) = &-\frac{1}{2}\left[\sum_{j=1}^{H_i}(\mathbf{x}_{ij} - \mathbf{m} - \mathbf{V}\mathbf{z}_i)^{\mathsf{T}}\langle v_{ij}\rangle\boldsymbol{\Sigma}^{-1}(\mathbf{x}_{ij} - \mathbf{m} - \mathbf{V}\mathbf{z}_i)\right] \\ &- \frac{1}{2}\langle u_i\rangle\mathbf{z}^{\mathsf{T}}\mathbf{z} + \text{const},\end{aligned} \tag{3.100}$$

where "const" absorbs all terms independent of $\mathbf{z}$ and $\langle x\rangle$ is the shorthand form for the expectation of x using the variational posterior $q(x)$ as the density function, i.e., $\langle x\rangle \equiv \mathbb{E}_{q(x)}\{x\}$. Note that we have used the factorization property of variational Bayes in Eq. 3.96 to derive Eq. 3.100. Using the same technique as in Eq. 3.76 and Eq. 3.77, we obtain the posterior mean and posterior covariance of $\mathbf{z}_i$ as follows:

$$\begin{aligned}\langle\mathbf{z}_i\rangle &= \left(\langle u_i\rangle\mathbf{I} + \sum_{j=1}^{H_i}\langle v_{ij}\rangle\mathbf{V}^{\mathsf{T}}\boldsymbol{\Sigma}^{-1}\mathbf{V}\right)^{-1}\left(\sum_{j=1}^{H_i}\langle v_{ij}\rangle\mathbf{V}^{\mathsf{T}}\boldsymbol{\Sigma}^{-1}(\mathbf{x}_{ij} - \mathbf{m})\right) \\ \text{Cov}(\mathbf{z}_i,\mathbf{z}_i) &= \left(\langle u_i\rangle\mathbf{I} + \sum_{j=1}^{H_i}\langle v_{ij}\rangle\mathbf{V}^{\mathsf{T}}\boldsymbol{\Sigma}^{-1}\mathbf{V}\right)^{-1},\end{aligned} \tag{3.101}$$

where $\langle u_i\rangle$ and $\langle v_{ij}\rangle$ are the posterior means of u_i and v_{ij}, respectively. Note that $q(\mathbf{z}_i)$ is a Gaussian distribution with mean $\langle\mathbf{z}_i\rangle$ and convariance matrix $\text{Cov}(\mathbf{z}_i,\mathbf{z}_i)$ even though u_i and v_{ij} follow Gamma distributions.

Substituting Eq. 3.98 to Eq. 3.97(b), we have

$$\begin{aligned}\log q(u_i) &= \frac{M}{2}\log u_i - \frac{u_i}{2}\langle\mathbf{z}_i^{\mathsf{T}}\mathbf{z}_i\rangle + \left(\frac{n}{2} - 1\right)\log u_i - \frac{n}{2}u_i + \text{const} \\ &= \left(\frac{n+M}{2} - 1\right)\log u_i - \left(\frac{n + \langle\mathbf{z}_i^{\mathsf{T}}\mathbf{z}_i\rangle}{2}\right)u_i + \text{const},\end{aligned} \tag{3.102}$$

where the const term absorbs all terms independent of u_i. Eq. 3.102 is the logarithm of a Gamma distribution with shape α and rate β given by

$$\alpha = \frac{n+M}{2} \quad\text{and}\quad \beta = \frac{n + \langle\mathbf{z}_i^{\mathsf{T}}\mathbf{z}_i\rangle}{2}. \tag{3.103}$$

As $q(u_i)$ is an approximation of the posterior density of u_i, the posterior mean $\langle u_i\rangle$ in Eq. 3.101 can be computed as follows:

$$\langle u_i\rangle = \mathbb{E}_{q(u_i)}\{u_i\} = \alpha\beta^{-1} = \frac{n+M}{n + \langle\mathbf{z}_i^{\mathsf{T}}\mathbf{z}_i\rangle}. \tag{3.104}$$

Substituting Eq. 3.98 to Eq. 3.97(c), we have

$$\begin{aligned}\log q(v_{ij}) &= \frac{D}{2}\log v_{ij} - \frac{1}{2}v_{ij}\langle \boldsymbol{\epsilon}_{ij}^{\mathsf{T}}\boldsymbol{\Sigma}^{-1}\boldsymbol{\epsilon}_{ij}\rangle_{\mathbf{z}_i} + \left(\frac{\nu}{2}-1\right)\log v_{ij} - \frac{\nu}{2}v_{ij} + \text{const} \\ &= \left(\frac{\nu+D}{2}-1\right)\log v_{ij} - \left(\frac{\nu + \langle \boldsymbol{\epsilon}_{ij}^{\mathsf{T}}\boldsymbol{\Sigma}^{-1}\boldsymbol{\epsilon}_{ij}\rangle_{\mathbf{z}_i}}{2}\right)v_{ij} + \text{const},\end{aligned} \tag{3.105}$$

where D is the dimensionality of $\mathbf{x}_{ij}$ and $\langle\cdot\rangle_{\mathbf{z}_i}$ means taking expectation with respect to the approximated posterior $q(\mathbf{z}_i)$. Eq. 3.105 is the logarithm of a Gamma distribution with shape α and rate β given by

$$\alpha = \frac{\nu+D}{2} \quad \text{and} \quad \beta = \frac{\nu + \langle \boldsymbol{\epsilon}_{ij}^{\mathsf{T}}\boldsymbol{\Sigma}^{-1}\boldsymbol{\epsilon}_{ij}\rangle_{\mathbf{z}_i}}{2}.$$

The posterior mean of v_{ij} in Eq. 3.101 is given by

$$\langle v_{ij}\rangle = \mathbb{E}_{q(v_{ij})}\{v_{ij}\} = \frac{\nu+D}{\nu + \langle \boldsymbol{\epsilon}_{ij}^{\mathsf{T}}\boldsymbol{\Sigma}^{-1}\boldsymbol{\epsilon}_{ij}\rangle_{\mathbf{z}_i}}. \tag{3.106}$$

The remaining posterior expectation is $\langle \boldsymbol{\epsilon}_{ij}^{\mathsf{T}}\boldsymbol{\Sigma}^{-1}\boldsymbol{\epsilon}_{ij}\rangle_{\mathbf{z}_i}$. To compute this posterior expectation, we leverage the symmetry of $\boldsymbol{\Sigma}^{-1}$ [63, Eq. 318]:

$$\langle \boldsymbol{\epsilon}_{ij}^{\mathsf{T}}\boldsymbol{\Sigma}^{-1}\boldsymbol{\epsilon}_{ij}\rangle_{\mathbf{z}_i} = \text{Tr}\{\boldsymbol{\Sigma}^{-1}\text{Cov}(\boldsymbol{\epsilon}_{ij},\boldsymbol{\epsilon}_{ij})\} + \langle \boldsymbol{\epsilon}_{ij}\rangle_{\mathbf{z}_i}^{\mathsf{T}}\boldsymbol{\Sigma}^{-1}\langle \boldsymbol{\epsilon}_{ij}\rangle_{\mathbf{z}_i}.$$

Because $\boldsymbol{\epsilon}_{ij} = \mathbf{x}_{ij} - \mathbf{m} - \mathbf{V}\mathbf{z}_i$, we have

$$\begin{aligned}\langle \boldsymbol{\epsilon}_{ij}\rangle_{\mathbf{z}_i} &= \mathbf{x}_{ij} - \mathbf{m} - \mathbf{V}\langle \mathbf{z}_i\rangle \\ \text{Cov}(\boldsymbol{\epsilon}_{ij},\boldsymbol{\epsilon}_{ij}) &= \mathbf{V}\text{Cov}(\mathbf{z}_i,\mathbf{z}_i)\mathbf{V}^{\mathsf{T}},\end{aligned}$$

where $\text{Cov}(\mathbf{z}_i,\mathbf{z}_i)$ and $\langle \mathbf{z}_i\rangle$ can be obtained from Eq. 3.101.

3.5.3 Model Parameter Estimation

Denote the model parameter of a heavy-tailed PLDA model as $\omega = \{\mathbf{m},\mathbf{V},\boldsymbol{\Sigma},n,\nu\}$. Also denote $\mathcal{H} = \{\mathbf{z}_i, u_i, v_{ij}; i = 1,\ldots,N; j = 1,\ldots,H_i\}$ as the set of latent variables associated with the training vectors $\mathcal{X} = \{\mathcal{X}_1,\ldots,\mathcal{X}_N\}$, where $\mathcal{X}_i = \{\mathbf{x}_{i1},\ldots,\mathbf{x}_{iH_i}\}$.

Similar to PLDA, the model parameter can be estimated by maximizing the following auxiliary function:

$$Q(\omega'|\omega) = \mathbb{E}_{\mathcal{H}}\{\ln p(\mathcal{X}, \mathcal{H}|\omega')\} \tag{3.107a}$$

$$= \mathbb{E}_{\mathcal{H}}\left\{\sum_{ij} \log p(\mathbf{x}_{ij}|\mathbf{z}_i, u_i, v_{ij}, \omega')p(\mathbf{z}_i|u_i, \omega')p(u_i|\omega')p(v_{ij}|\omega')\right\} \tag{3.107b}$$

$$= \mathbb{E}_{\mathcal{H}}\left\{\sum_{ij} \log \mathcal{N}(\mathbf{x}_{ij}|\mathbf{m}' + \mathbf{V}'\mathbf{z}_i, v_{ij}^{-1}\boldsymbol{\Sigma}')\right\} + \mathbb{E}_{\mathcal{H}}\left\{\sum_{ij} \log \mathcal{N}(\mathbf{z}_i|\mathbf{0}, u_i^{-1}\mathbf{I})\right\}$$

$$+ \mathbb{E}_{\mathcal{H}}\left\{\sum_{ij} \log \mathcal{G}(u_i|n/2, n/2)\mathcal{G}(v_{ij}|\nu/2, \nu/2)\right\}. \tag{3.107c}$$

When estimating $\mathbf{m}$, $\mathbf{V}$, and $\boldsymbol{\Sigma}$, the second and third terms in Eq. 3.107 can be dropped because they do not contain these model parameters. Therefore, we may follow the derivation in Section 3.4.2 to obtain the parameter update formulae (M-step) as follows:

$$\mathbf{V}' = \left[\sum_{i=1}^{N}\sum_{j=1}^{H_i} \langle v_{ij}\rangle(\mathbf{x}_{ij} - \mathbf{m}')\langle \mathbf{z}_i\rangle^{\mathsf{T}}\right]\left[\sum_{i=1}^{N}\sum_{j=1}^{H_i} \langle v_{ij}\rangle\langle \mathbf{z}_i\mathbf{z}_i^{\mathsf{T}}\rangle\right]^{-1} \tag{3.108a}$$

$$\mathbf{m}' = \frac{\sum_{ij}\langle v_{ij}\rangle \mathbf{x}_{ij}}{\sum_{ij}\langle v_{ij}\rangle} \tag{3.108b}$$

$$\boldsymbol{\Sigma}' = \frac{1}{\sum_{i=1}^{N} H_i}\left\{\sum_{i=1}^{N}\sum_{j=1}^{H_i} \langle v_{ij}\rangle \left[(\mathbf{x}_{ij} - \mathbf{m}')(\mathbf{x}_{ij} - \mathbf{m}')^{\mathsf{T}} - \mathbf{V}'\langle \mathbf{z}_i\rangle(\mathbf{x}_{ij} - \mathbf{m}')^{\mathsf{T}}\right]\right\} \tag{3.108c}$$

The degrees of freedom n can be estimating by minimizing the sum of the KL-divergence:

$$d(n) = \sum_{i=1}^{N} \mathcal{D}_{\text{KL}}(\mathcal{G}(\alpha_i, \beta_i)\|\mathcal{G}(n/2, n/2))$$

$$= \sum_{i=1}^{N}\left[\log \frac{\Gamma(n/2)}{\Gamma(\alpha_i)} + \alpha_i \log \beta_i + \left(\alpha_i - \frac{n}{2}\right)(\psi(\alpha_i) - \log \beta_i) + \frac{\alpha_i}{\beta_i}\left(\frac{n}{2} - \beta_i\right)\right]$$

$$- \frac{nN}{2}\log \frac{n}{2},$$

where α_i and β_i are Gamma distribution parameters obtained from Eq. 3.103, $\Gamma(\alpha_i)$ is the gamma function, and

$$\psi(x) = \frac{\partial}{\partial x}\log \Gamma(x) = \frac{\Gamma'(x)}{\Gamma(x)}$$

is the digamma function. We may use the following properties to simplify $d(n)$:

$$\langle u_i\rangle = \alpha_i\beta_i^{-1} \quad \text{and} \quad \langle \log u_i\rangle = \psi(\alpha_i) - \log \beta_i. \tag{3.109}$$

Also, consolidating terms in $d(n)$ that are independent of n as a constant term, we have

$$d(n) = N \log \Gamma(n/2) + \sum_{i=1}^{N} \left[-\frac{n}{2}\langle \log u_i \rangle + \langle u_i \rangle \frac{n}{2} \right] - \frac{nN}{2} \log \frac{n}{2} + \text{const.}$$

Setting the derivative of $d(n)$ to 0, we have

$$N \frac{\Gamma'(n/2)}{\Gamma(n/2)} + \sum_{i=1}^{N} \left[-\langle \log u_i \rangle + \langle u_i \rangle \right] - N \log \frac{n}{2} - N = 0,$$

which is equivalent to

$$N\psi\left(\frac{n}{2}\right) + \sum_{i=1}^{N} \left[-\langle \log u_i \rangle + \langle u_i \rangle \right] - N \log \frac{n}{2} - N = 0. \tag{3.110}$$

The optimal value of n can be found by applying line search on Eq. 3.110. The value of ν can be found similarly. Specifically, line search is applied to

$$N'\psi\left(\frac{\nu}{2}\right) + \sum_{i=1}^{N} \sum_{j=1}^{H_i} \left[-\langle \log v_{ij} \rangle + \langle v_{ij} \rangle \right] - N' \log \frac{\nu}{2} - N' = 0, \tag{3.111}$$

where $N' = \sum_{i=1}^{N} H_i$.

3.5.4 Scoring in Heavy-Tailed PLDA

Similar to the Gaussian PLDA in Section 3.4.3, scoring in heavy-tailed PLDA also involves the computation of a likelihood ratio score

$$S_{\text{LR}}(\mathbf{x}_s, \mathbf{x}_t) = \frac{p(\mathbf{x}_s, \mathbf{x}_t | H_1)}{p(\mathbf{x}_s | H_0) p(\mathbf{x}_t | H_0)}, \tag{3.112}$$

where H_1 and H_0 are the hypotheses that the input vectors $\mathbf{x}_s$ and $\mathbf{x}_t$ are from the same speaker and from different speakers, respectively. But unlike the Gaussian PLDA in Eq. 3.83, in heavy-tailed PLDA, there is no close form solution to the marginal likelihoods in Eq. 3.112. Therefore, we need to define a tractable solution and use it as a proxy to compute the marginal likelihoods. Let's define

$$\begin{aligned}
\mathcal{L} &= \mathbb{E}_{\mathbf{h} \sim q(\mathbf{h})} \left\{ \log \frac{p(\mathbf{x}, \mathbf{h})}{q(\mathbf{h})} \right\} \\
&= \mathbb{E}_{\mathbf{h} \sim q(\mathbf{h})} \{ \log p(\mathbf{x} | \mathbf{h}) \} - \mathbb{E}_{\mathbf{h} \sim q(\mathbf{h})} \left\{ \log \frac{q(\mathbf{h})}{p(\mathbf{h})} \right\} \\
&= \mathcal{L}_1 + \mathcal{L}_2,
\end{aligned} \tag{3.113}$$

where $\mathbf{x}$ can be $\mathbf{x}_s$ (target), $\mathbf{x}_t$ (test), or $[\mathbf{x}_s^{\mathsf{T}}\ \mathbf{x}_t^{\mathsf{T}}]^{\mathsf{T}}$; $\mathbf{h}$ comprises all latent variables, i.e., $\mathbf{h} = \{\mathbf{z}, u, v\}$; and $q(\mathbf{h})$ is the variational posterior of $\mathbf{h}$. Note that $\mathcal{L} \leq \log p(\mathbf{x})$ with equality holds when $q(\mathbf{h})$ is equal to the true posterior $p(\mathbf{h}|\mathbf{x})$. Therefore, we may use $\mathcal{L}$ as an approximation to the marginal likelihood $p(\mathbf{x})$.

$\mathcal{L}_1$ can be obtained from Eq. 3.98 by dropping all the prior over the latent variables:

$$\begin{aligned}\mathcal{L}_1 &= \mathbb{E}_{\mathbf{h}\sim q(\mathbf{h})}\{\log p(\mathbf{x}|\mathbf{z},u,v,\boldsymbol{\omega})\}\\ &= \mathbb{E}_{\mathbf{h}\sim q(\mathbf{h})}\{\log \mathcal{N}(\mathbf{x}|\mathbf{m}+\mathbf{V}\mathbf{z}, v^{-1}\boldsymbol{\Sigma})\}\\ &= \frac{D}{2}\langle\log v\rangle - \frac{D}{2}\log 2\pi - \frac{1}{2}\log|\boldsymbol{\Sigma}| - \frac{1}{2}\langle v\rangle\langle\boldsymbol{\epsilon}^{\mathsf{T}}\boldsymbol{\Sigma}^{-1}\boldsymbol{\epsilon}\rangle,\end{aligned} \tag{3.114}$$

where $\boldsymbol{\epsilon} = \mathbf{x} - \mathbf{m} - \mathbf{V}\mathbf{z}$ and $\langle\cdot\rangle$ stands for expectation with respect to the latent variables using $q(\mathbf{h})$ as the density. The posterior expectation $\langle v\rangle$ can be obtained from Eq. 3.106 and $\langle\log v\rangle$ can be computed by replacing u by v in Eq. 3.109 as follows:

$$\langle\log v\rangle = \psi\left(\frac{v+D}{2}\right) - \log\left(\frac{v+\boldsymbol{\epsilon}^{\mathsf{T}}\boldsymbol{\Sigma}^{-1}\boldsymbol{\epsilon}}{2}\right).$$

As $\mathcal{L}_2$ is the negative of the Kullback–Leibler divergence from $q(\mathbf{h})$ to $p(\mathbf{h})$ and the variational posteriors are factorizable, we have

$$\begin{aligned}\mathcal{L}_2 &= -\mathcal{D}_{\mathrm{KL}}(q(\mathbf{z},u,v)\|p(\mathbf{z},u,v))\\ &= -\mathcal{D}_{\mathrm{KL}}(q(\mathbf{z},u)q(v)||p(\mathbf{z},u)p(v))\\ &= -\left[\mathcal{D}_{\mathrm{KL}}(q(\mathbf{z},u)\|p(\mathbf{z},u)) + \mathcal{D}_{\mathrm{KL}}(q(v)\|p(v))\right]\\ &= -\left[\mathcal{D}_{\mathrm{KL}}(q(u)\|p(u)) + \mathbb{E}_{u\sim p(u)}\{\mathcal{D}_{\mathrm{KL}}(q(\mathbf{z}|u)\|p(\mathbf{z}|u))\} + \mathcal{D}_{\mathrm{KL}}(q(v)\|p(v))\right]\\ &= -\left[\mathcal{D}_{\mathrm{KL}}(q(u)\|p(u)) + \mathbb{E}_{u\sim p(u)}\{\mathcal{D}_{\mathrm{KL}}(q(\mathbf{z})\|p(\mathbf{z}|u))\} + \mathcal{D}_{\mathrm{KL}}(q(v)\|p(v))\right]\end{aligned} \tag{3.115}$$

where we have used the chain rule of KL-divergence[7] and the notation $q(\mathbf{z}) \equiv q(\mathbf{z}|u)$. Note that we have also used the fact $p(v|\mathbf{z},u) = p(v)$ and $\mathbb{E}_{p(\mathbf{z},u)}p(v) = p(v)$ in Line 3 of Eq. 3.115.

The distributions in Eq. 3.115 can be obtained from the posteriors and priors of the latent variables as follows:

$$q(u) = \mathcal{G}\left(\frac{n+M}{2}, \frac{n+\langle\mathbf{z}_i^{\mathsf{T}}\mathbf{z}_i\rangle}{2}\right) \tag{3.116a}$$

$$p(u) = \mathcal{G}\left(\frac{n}{2}, \frac{n}{2}\right) \tag{3.116b}$$

$$q(v) = \mathcal{G}\left(\frac{v+D}{2}, \frac{v+\langle\boldsymbol{\epsilon}_{ij}^{\mathsf{T}}\boldsymbol{\Sigma}^{-1}\boldsymbol{\epsilon}_{ij}\rangle}{2}\right) \tag{3.116c}$$

$$p(v) = \mathcal{G}\left(\frac{v}{2}, \frac{v}{2}\right) \tag{3.116d}$$

$$q(\mathbf{z}) = \mathcal{N}\left(\langle\mathbf{z}\rangle, \mathrm{Cov}(\mathbf{z},\mathbf{z})\right) \tag{3.116e}$$

$$p(\mathbf{z}|u) = \mathcal{N}\left(\mathbf{0}, u^{-1}\mathbf{I}\right). \tag{3.116f}$$

Note that the KL-divergence between two Gamma distributions and between two Gaussian distributions are [64, 65]:

[7] $\mathcal{D}_{\mathrm{KL}}(q(x,y)\|p(x,y)) = \mathcal{D}_{\mathrm{KL}}(q(x)\|p(x)) + \mathbb{E}_{x\sim p(x)}\{\mathcal{D}_{\mathrm{KL}}(q(y|x)\|p(y|x))\}$.

$$\mathcal{D}_{\text{KL}}\left(\mathcal{N}(\boldsymbol{\mu}_q, \boldsymbol{\Sigma}_q)\|\mathcal{N}(\boldsymbol{\mu}_p, \boldsymbol{\Sigma}_p)\right)$$
$$= \frac{1}{2}\left[\log\frac{|\boldsymbol{\Sigma}_p|}{|\boldsymbol{\Sigma}_q|} + \text{Tr}\{\boldsymbol{\Sigma}_p^{-1}\boldsymbol{\Sigma}_q\} + (\boldsymbol{\mu}_q - \boldsymbol{\mu}_p)^{\mathsf{T}}\boldsymbol{\Sigma}_p^{-1}(\boldsymbol{\mu}_q - \boldsymbol{\mu}_p) - D\right] \quad (3.117a)$$
$$\mathcal{D}_{\text{KL}}\left(\mathcal{G}(\alpha_q, \beta_q)\|\mathcal{G}(\alpha_p, \beta_p)\right)$$
$$= \alpha_p \log\frac{\beta_q}{\beta_p} - \log\frac{\Gamma(\alpha_q)}{\Gamma(\alpha_p)} + (\alpha_q - \alpha_p)\psi(\alpha_q) - (\beta_q - \beta_p)\frac{\alpha_q}{\beta_q}. \quad (3.117b)$$

Using these KL-divergence, the second term in Eq. 3.115 is given by

$$\mathbb{E}_{u\sim p(u)}\{\mathcal{D}_{\text{KL}}(q(\mathbf{z})\|p(\mathbf{z}|u))\}$$
$$= \frac{1}{2}\left[-D\langle\log u\rangle - \log|\text{Cov}(\mathbf{z},\mathbf{z})| + \text{Tr}\{\langle u\rangle\text{Cov}(\mathbf{z},\mathbf{z})\} + \langle u\rangle\langle\mathbf{z}\rangle^{\mathsf{T}}\langle\mathbf{z}\rangle - D\right].$$

The first and the third terms in Eq. 3.115 can be computed similarly using Eq. 3.117(b) and Eq. 3.116.

3.5.5 Heavy-Tailed PLDA versus Gaussian PLDA

While the training and scoring of heavy-tailed PLDA are more complicated and computationally expensive than Gaussian PLDA, in speaker recognition, the former is proposed earlier than the latter. In fact, Kenny has shown in [61] that heavy-tailed PLDA performs better than Gaussian PLDA if no preprocessing is applied to the i-vectors. However, in 2011, Garcia-Romero and Espy-Wilson [66] discovered that if i-vectors are Gaussianized by length-normalization, Gaussian PLDA can perform as good as heavy-tailed PLDA. Since then, Gaussian PLDA became more popular than heavy-tailed PLDA.

In 2018, Silnova et al. [67] proposed a method that leverage the more exact modeling of heavy-tailed PLDA but with the computation advantage of Gaussian PLDA. With heavy-tailed PLDA, length-normalization, which may remove some speaker information, is no longer needed. The main idea is based on the observation that the conditional distribution of i-vectors given the speaker factor z is proportional to another t-distribution with degrees of freedom $n' = n + D - M$. If n' is large, the distribution is practically a Gaussian. In practice, D is about 500 and M is 150. As a result, n' is large enough to make the conditional distribution of $\mathbf{x}$ Gaussian. This approximation allows the computation of the joint variational posterior $q(\mathbf{v}_i)$ in Eq. 3.105 independent of the speaker factor z_i, which results in a very fast E-step.

3.6 I-Vectors

I-vectors are based on factor analysis in which the acoustic features (typically MFCC and log-energy plus their first and second derivatives) are generated by a Gaussian mixture model (GMM). Since its first appeared [16], it has become the de facto method for many text-independent speaker verification systems. Although i-vectors were originally proposed for speaker verification, they have been found useful in many other applications, including speaker diarization [68], voice activity detection [69], language recognition [70], and ECG classification [71].

3.6.1 Generative Model

Given the ith utterance, we extract the F-dimensional acoustic vectors $\mathcal{O}_i = \{\mathbf{o}_{i1}, \ldots, \mathbf{o}_{iT_i}\}$ from the utterance, where T_i is the number of frames in the utterance. We assume that the acoustic vectors are generated by an utterance-dependent GMM with parameters $\Lambda_i = \{\pi_c, \boldsymbol{\mu}_{ic}, \boldsymbol{\Sigma}_c\}_{c=1}^{C}$, i.e.,

$$p(\mathbf{o}_{it}) = \sum_{c=1}^{C} \pi_c^{(b)} \mathcal{N}(\mathbf{o}_{it} | \boldsymbol{\mu}_{ic}, \boldsymbol{\Sigma}_c^{(b)}), \quad t = 1, \ldots, T_i, \tag{3.118}$$

where C is the number of mixtures in the GMM. For simplicity, we assume that $\lambda_c^{(b)}$ and $\boldsymbol{\Sigma}_c^{(b)}$ are tied across all utterances and are equal to the mixture weights and covariance matrices of the universal background model (UBM), respectively.

In the i-vector framework [16], the supervector representing the ith utterance is assumed to be generated by the following factor analysis model [1]:[8]

$$\boldsymbol{\mu}_i = \boldsymbol{\mu}^{(b)} + \mathbf{T}\mathbf{w}_i \tag{3.119}$$

where $\boldsymbol{\mu}^{(b)}$ is obtained by stacking the mean vectors of the UBM, $\mathbf{T}$ is a $CF \times D$ low-rank total variability matrix modeling the speaker and channel variability, and $\mathbf{w}_i \in \mathbb{R}^D$ comprises the latent (total) factors. Eq. 3.119 suggests that the generated supervectors $\boldsymbol{\mu}_i$'s have mean $\boldsymbol{\mu}^{(b)}$ and convariance matrix $\mathbf{T}\mathbf{T}^{\mathsf{T}}$. Eq. 3.119 can also be written in a component-wise form:

$$\boldsymbol{\mu}_{ic} = \boldsymbol{\mu}_c^{(b)} + \mathbf{T}_c\mathbf{w}_i, \quad c = 1, \ldots, C \tag{3.120}$$

where $\boldsymbol{\mu}_{ic} \in \mathbb{R}^F$ is the cth sub-vector of $\boldsymbol{\mu}_i$ (similarly for $\boldsymbol{\mu}_c^{(b)}$) and $\mathbf{T}_c$ is an $F \times D$ sub-matrix of $\mathbf{T}$. Figure 3.23 shows the graphical representation of this generative model.

In [45], it was assumed that for every speaker s, the acoustic vectors (frames) aligning to the cth mixture component have mean $\boldsymbol{\mu}_{sc}$ and covariance $\boldsymbol{\Sigma}_c^{(b)}$. In the case of i-vectors, every utterance is assumed to be spoken by a different speaker. As a result, the frames of utterance i aligning to mixture c have mean $\boldsymbol{\mu}_{ic}$ and covariance matrix $\boldsymbol{\Sigma}_c^{(b)}$. This matrix measures the deviation of the acoustic vectors – which are aligned to the cth mixture – from the utterance-dependent mean $\boldsymbol{\mu}_{ic}$. A super-covariance matrix, denoted as $\boldsymbol{\Sigma}^{(b)}$, can be formed by packing the C covariance matrices, i.e., $\boldsymbol{\Sigma}^{(b)} = \text{diag}\{\boldsymbol{\Sigma}_1^{(b)}, \ldots, \boldsymbol{\Sigma}_C^{(b)}\}$. In the i-vector framework, model estimation amounts to finding the parameter set $\boldsymbol{\Lambda} = \{\boldsymbol{\mu}^{(b)}, \mathbf{T}, \boldsymbol{\Sigma}^{(b)}\}$. In practice, $\boldsymbol{\mu}^{(b)}$ and $\boldsymbol{\Sigma}^{(b)}$ are the mean vectors and covariance matrices of the UBM.

Note that unlike Eq. 3.55, Eq. 3.119 and Eq. 3.120 do not have the residue terms $\boldsymbol{\epsilon}_i$ and $\boldsymbol{\epsilon}_{ic}$, respectively. This is because the FA model in Eq. 3.119 aims to generate the supervector $\boldsymbol{\mu}_i$. If the supervector is observed, e.g., an GMM-supervector obtained from

[8] To simplify notations, from now on, we drop the over-arrow symbol ($\rightarrow$) in the supervectors.

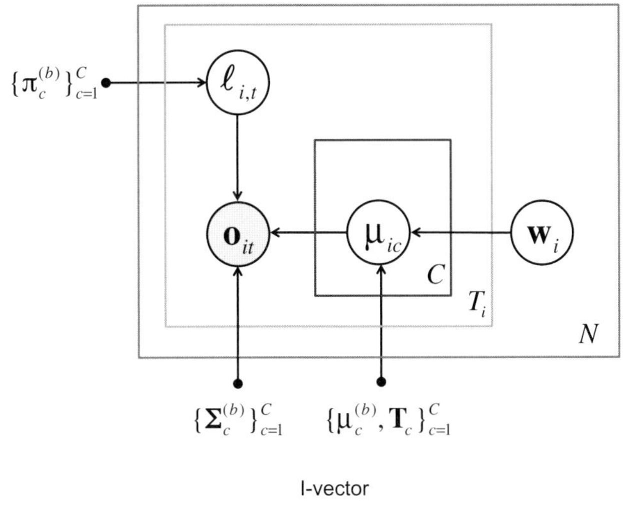

Figure 3.23 Graphical representation of the factor analysis model for i-vector extraction. $\boldsymbol{\ell}_{i,t} = [\ell_{i,t,1} \ \cdots \ \ell_{i,t,c} \ \cdots \ \ell_{i,t,C}]^{\mathsf{T}}$ are the mixture labels in one-hot encoding. The prior of $\mathbf{w}_i$ is assumed to be a standard Gaussian.

the MAP adaptation, an residue term will need to be added to Eq. 3.119 to represent the difference between the observed and the generated supervectors.

There is a close relationship between the residue covariance matrix $\boldsymbol{\Sigma}$ in Eq. 3.56 and the component-wise covariance matrices $\boldsymbol{\Sigma}_c^{(b)}$ described above. Given the acoustic vectors $\mathcal{O}_i$ of the ith utterance, we can compute the posterior mean (see Section 3.6.2), $\langle \mathbf{w}_i | \mathcal{O}_i \rangle$. Then, the density of $\boldsymbol{\mu}_{ic}$ in Eq. 3.120 is $\mathcal{N}(\boldsymbol{\mu}_{ic} | \boldsymbol{\mu}_c^{(b)} + \mathbf{T}_c \langle \mathbf{w}_i | \mathcal{O}_i \rangle, \boldsymbol{\Sigma}_c^{(b)})$. Here, the deviations of the acoustic vectors from the mean $(\boldsymbol{\mu}_c^{(b)} + \mathbf{T}_c \langle \mathbf{w}_i | \mathcal{O}_i \rangle)$ are specified by $\boldsymbol{\Sigma}_c^{(b)}$. In Eq. 3.55, these deviations are represented by $\boldsymbol{\epsilon}$ whose covariance matrix is $\boldsymbol{\Sigma}$.

3.6.2 Posterior Distributions of Total Factors

Given an utterance with acoustic vectors $\mathcal{O}_i$, the i-vector $\mathbf{x}_i$ representing the utterance is the posterior mean of $\mathbf{w}_i$, i.e., $\mathbf{x}_i = \langle \mathbf{w}_i | \mathcal{O}_i \rangle$. To determine $\mathbf{x}_i$, we may express the joint posterior density of $\mathbf{w}_i$ and indicator variables $\ell_{i,t,c}$'s, where $t = 1, \ldots, T_i$ and $c = 1, \ldots, C$. The indicator variable $\ell_{i,t,c}$ specifies which of the C Gaussians generates $\mathbf{o}_{it}$. More specifically, $\ell_{i,t,c} = 1$ if the cth mixture generates $\mathbf{o}_{it}$; otherwise, $\ell_{i,t,c} = 0$. The joint posterior density can be expressed in terms of $\mathbf{w}_i$ using the Bayes rule:

$$
\begin{aligned}
p(\mathbf{w}_i, \ell_{i,\cdot,\cdot} | \mathcal{O}_i) &\propto p(\mathcal{O}_i | \mathbf{w}_i, \ell_{i,\cdot,\cdot} = 1) p(\ell_{i,\cdot,\cdot}) p(\mathbf{w}_i) \\
&= \prod_{t=1}^{T} \prod_{c=1}^{C} \left[\pi_c p(\mathbf{o}_{it} | \ell_{i,t,c} = 1, \mathbf{w}_i) \right]^{\ell_{i,t,c}} p(\mathbf{w}_i) \quad [1, \text{Eq. } 9.38] \\
&= \underbrace{p(\mathbf{w}_i) \prod_{t=1}^{T} \prod_{c=1}^{C} \left[\mathcal{N}(\mathbf{o}_{it} | \boldsymbol{\mu}_c^{(b)} + \mathbf{T}_c \mathbf{w}_i, \boldsymbol{\Sigma}_c^{(b)}) \right]^{\ell_{i,t,c}}}_{\propto\, p(\mathbf{w}_i | \mathcal{O}_i)} \pi_c^{\ell_{i,t,c}}, \qquad (3.121)
\end{aligned}
$$

where $\ell_{i,\cdot,\cdot}$ means all possible combinations of t and c for utterance i. In Eq. 3.121, we have used the fact that for each (i,t)-pair, only one of the $\ell_{i,t,c}$ for $c = 1,\dots,C$ equals to 1, the rest are zeros. Extracting terms depending on $\mathbf{w}_i$ from Eq. 3.121, we obtain

$$\begin{aligned}
& p(\mathbf{w}_i|\mathcal{O}_i) \\
& \propto \exp\left\{-\frac{1}{2}\sum_{c=1}^{C}\sum_{t\in\mathcal{H}_{ic}}(\mathbf{o}_{it}-\boldsymbol{\mu}_c^{(b)}-\mathbf{T}_c\mathbf{w}_i)^{\mathsf{T}}(\boldsymbol{\Sigma}_c^{(b)})^{-1}(\mathbf{o}_{it}-\boldsymbol{\mu}_c^{(b)}-\mathbf{T}_c\mathbf{w}_i)-\frac{1}{2}\mathbf{w}_i^{\mathsf{T}}\mathbf{w}_i\right\} \\
& = \exp\left\{\mathbf{w}_i^{\mathsf{T}}\sum_{c=1}^{C}\sum_{t\in\mathcal{H}_{ic}}\mathbf{T}_c^{\mathsf{T}}(\boldsymbol{\Sigma}_c^{(b)})^{-1}(\mathbf{o}_{it}-\boldsymbol{\mu}_c^{(b)})\right. \\
& \qquad \left.-\frac{1}{2}\mathbf{w}_i^{\mathsf{T}}\left(\mathbf{I}+\sum_{c=1}^{C}\sum_{t\in\mathcal{H}_{ic}}\mathbf{T}_c^{\mathsf{T}}(\boldsymbol{\Sigma}_c^{(b)})^{-1}\mathbf{T}_c\right)\mathbf{w}_i\right\},
\end{aligned} \tag{3.122}$$

where $\mathcal{H}_{ic}$ comprises the frame indexes for which $\mathbf{o}_{it}$ aligned to mixture c. Comparing Eq. 3.122 with Eq. 3.77, we obtain the following posterior expectations:

$$\begin{aligned}
\langle\mathbf{w}_i|\mathcal{O}_i\rangle &= \mathbf{L}_i^{-1}\sum_{c=1}^{C}\sum_{t\in\mathcal{H}_{ic}}\mathbf{T}_c^{\mathsf{T}}\left(\boldsymbol{\Sigma}_c^{(b)}\right)^{-1}(\mathbf{o}_{it}-\boldsymbol{\mu}_c^{(b)}) \\
&= \mathbf{L}_i^{-1}\sum_{c=1}^{C}\mathbf{T}_c^{\mathsf{T}}\left(\boldsymbol{\Sigma}_c^{(b)}\right)^{-1}\sum_{t\in\mathcal{H}_{ic}}(\mathbf{o}_{it}-\boldsymbol{\mu}_c^{(b)})
\end{aligned} \tag{3.123}$$

$$\langle\mathbf{w}_i\mathbf{w}_i^{\mathsf{T}}|\mathcal{O}_i\rangle = \mathbf{L}_i^{-1} + \langle\mathbf{w}_i|\mathcal{O}_i\rangle\langle\mathbf{w}_i^{\mathsf{T}}|\mathcal{O}_i\rangle, \tag{3.124}$$

where

$$\mathbf{L}_i = \mathbf{I}+\sum_{c=1}^{C}\sum_{t\in\mathcal{H}_{ic}}\mathbf{T}_c^{\mathsf{T}}(\boldsymbol{\Sigma}_c^{(b)})^{-1}\mathbf{T}_c. \tag{3.125}$$

Note that when $C = 1$ (i.e., the UBM has one Gaussian only), Eq. 3.123 reduces to the posterior mean of the latent factor in the PLDA model in Eq. 3.78. This means that the factor analysis model in PLDA is a special case of the more general factor analysis model in i-vectors.

In Eq. 3.123 and Eq. 3.125, the sum over $\mathcal{H}_{ic}$ can be evaluated in two ways:

(1) *Hard Decisions*. For each t, the posterior probabilities of $\ell_{i,t,c}$, for $c = 1,\dots,C$, are computed. Then, $\mathbf{o}_{it}$ is aligned to mixture c^* when

$$c^* = \underset{c}{\operatorname{argmax}}\,\gamma(\ell_{i,t,c}),$$

where

$$\begin{aligned}
\gamma(\ell_{i,t,c}) &\equiv \Pr(\text{Mixture} = c|\mathbf{o}_{it}) \\
&= \frac{\pi_c^{(b)}\mathcal{N}(\mathbf{o}_{it}|\boldsymbol{\mu}_c^{(b)},\boldsymbol{\Sigma}_c^{(b)})}{\sum_{j=1}^{C}\pi_j^{(b)}\mathcal{N}(\mathbf{o}_{it}|\boldsymbol{\mu}_j^{(b)},\boldsymbol{\Sigma}_j^{(b)})}, \quad c = 1,\dots,C
\end{aligned} \tag{3.126}$$

are the posterior probabilities of mixture c given $\mathbf{o}_{it}$.

(2) *Soft Decisions*. Each frame is aligned to all of the mixtures with degree of alignment according to the posterior probabilities $\gamma(\ell_{i,t,c})$. Then, we have

$$\begin{aligned}\sum_{t\in\mathcal{H}_{ic}} 1 &\rightarrow \sum_{t=1}^{T_i} \gamma(\ell_{i,t,c}) \\ \sum_{t\in\mathcal{H}_{ic}} (\mathbf{o}_{it} - \boldsymbol{\mu}_c^{(b)}) &\rightarrow \sum_{t=1}^{T_i} \gamma(\ell_{i,t,c})(\mathbf{o}_{it} - \boldsymbol{\mu}_c^{(b)}).\end{aligned} \tag{3.127}$$

Eq. 3.127 results in the following Baum–Welch statistics:

$$N_{ic} \equiv \sum_{t=1}^{T_i} \gamma(\ell_{i,t,c}) \text{ and } \tilde{\mathbf{f}}_{ic} \equiv \sum_{t=1}^{T_i} \gamma(\ell_{i,t,c})(\mathbf{o}_{it} - \boldsymbol{\mu}_c^{(b)}). \tag{3.128}$$

3.6.3 I-Vector Extractor

Training of the total variability matrix involves estimating the total variability matrix $\mathbf{T}$ using the EM algorithm. The formulation can be derived based on the EM steps in Section 3.3. Specifically, using Eqs. 3.65, 3.123, and 3.124, the M-step for estimating $\mathbf{T}$ is

$$\mathbf{T}_c = \left[\sum\nolimits_i \tilde{\mathbf{f}}_{ic}\langle \mathbf{w}_i|\mathcal{O}_i\rangle^{\mathsf{T}}\right]\left[\sum\nolimits_i N_{ic}\langle \mathbf{w}_i\mathbf{w}_i^{\mathsf{T}}|\mathcal{O}_i\rangle\right]^{-1}, \quad c = 1,\ldots,C. \tag{3.129}$$

Figure 3.24 shows the pseudo-code of training an i-vector extractor.

When estimating the TV matrix $\mathbf{T}$, the likelihood of training data can also be maximized through minimizing the divergence [72]. The idea is to force the i-vectors to follow the standard Gaussian prior. This can be achieved by performing the following transformation after computing $\mathbf{T}_c$ in Eq. 3.129:

$$\mathbf{T}_c \leftarrow \mathbf{T}_c\mathbf{L} \text{ and } \langle \mathbf{w}_i|\mathcal{O}_i\rangle \leftarrow \mathbf{L}^{-1}\langle \mathbf{w}_i|\mathcal{O}_i\rangle, \tag{3.130}$$

where $\mathbf{L}\mathbf{L}^{\mathsf{T}}$ is the Cholesky decomposition of $\frac{1}{M}\sum_{i=1}^{M}\langle \mathbf{w}_i\mathbf{w}_i^{\mathsf{T}}|\mathcal{O}_i\rangle$, where M is the number of training utterances. The advantage of performing these transformations is the speedup of the convergence [73].

To extract the i-vector $\mathbf{x}_i$ from an utterance with acoustic vectors $\mathcal{O}_i$, we substitute Eq. 3.128 into Eq. 3.123 and Eq. 3.125:

$$\begin{aligned}\mathbf{x}_i \equiv \langle \mathbf{w}_i|\mathcal{O}_i\rangle &= \mathbf{L}_i^{-1}\sum_{c=1}^{C}\mathbf{T}_c^{\mathsf{T}}\left(\boldsymbol{\Sigma}_c^{(b)}\right)^{-1}\tilde{\mathbf{f}}_{ic} \\ &= \mathbf{L}_i^{-1}\mathbf{T}^{\mathsf{T}}(\boldsymbol{\Sigma}^{(b)})^{-1}\tilde{\mathbf{f}}_i,\end{aligned} \tag{3.131}$$

where $\tilde{\mathbf{f}}_i = [\tilde{\mathbf{f}}_{i1}^{\mathsf{T}} \cdots \tilde{\mathbf{f}}_{iC}^{\mathsf{T}}]^{\mathsf{T}}$ and

$$\mathbf{L}_i = \mathbf{I} + \sum_{c=1}^{C} N_{ic}\mathbf{T}_c^{\mathsf{T}}(\boldsymbol{\Sigma}_c^{(b)})^{-1}\mathbf{T}_c = \mathbf{I} + \mathbf{T}^{\mathsf{T}}(\boldsymbol{\Sigma}^{(b)})^{-1}\mathbf{N}_i\mathbf{T}, \tag{3.132}$$

$M =$ no. of training utterances

Randomize $\mathbf{T} \in \Re^{CF \times R}$ with submatrices denoted by $\mathbf{T}_c \in \Re^{F \times R}$, $c = 1,\ldots,C$

foreach **iteration**

for $i = 1$: no. of training utterances

$$\mathbf{N}_i = \begin{bmatrix} N_{i1}\mathbf{I}_{F\times F} & \mathbf{0} & \cdots & \mathbf{0} \\ \mathbf{0} & N_{i2}\mathbf{I}_{F\times F} & \mathbf{0} & \mathbf{0} \\ \mathbf{0} & \mathbf{0} & \ddots & \mathbf{0} \\ \mathbf{0} & \mathbf{0} & \cdots & N_{iC}\mathbf{I}_{F\times F} \end{bmatrix}_{CF\times CF}$$

$$\mathbf{L}_i = \mathbf{I} + \mathbf{T}^T \boldsymbol{\Sigma}^{-1} \mathbf{N}_i \mathbf{T} \qquad \mathbf{L}_i \in \Re^{R\times R} \text{ and } \mathbf{T} \in \Re^{CF\times R}$$

$$\mathbf{x}_i = \mathbf{L}_i^{-1} \mathbf{T}^T \boldsymbol{\Sigma}^{-1} \tilde{\mathbf{f}}_i \qquad \mathbf{x}_i \in \Re^{R}$$

end

$$\boldsymbol{\Omega}_c = \sum_{i=1}^{M} N_{ic} \left[\mathbf{L}_i^{-1} + \mathbf{x}_i \mathbf{x}_i^T \right] \qquad \boldsymbol{\Omega}_c \in \Re^{R\times R}$$

$$\boldsymbol{\Psi} = \sum_{i=1}^{M} \tilde{\mathbf{f}}_i \mathbf{x}_i^T \qquad \boldsymbol{\Psi} \in \Re^{CF\times R}$$

$$\mathbf{T}_c = \boldsymbol{\Psi}_c \boldsymbol{\Omega}_c^{-1}$$

end

Figure 3.24 The pseudo-code for computing the total variability matrix.

where $\mathbf{N}_i$ is a $CF \times CF$ block diagonal matrix containing $N_{ic}\mathbf{I}$, $c = 1,\ldots,C$, as its block diagonal elements.

Eq. 3.131 and Eq. 3.132 suggest that given a total variability matrix $\mathbf{T}$, an UBM, and an utterance with acoustic vectors $\mathcal{O}_i$, the i-vector of the utterance can be extracted by firstly aligning every frame in $\mathcal{O}_i$ with the UBM using Eq. 3.126 to compute the mixture posteriors $\gamma(\ell_{i,t,c})$'s. These posteriors, together with the mean vectors in the UBM, allow us to compute the zeroth and first-order sufficient statistics. Substituting these statistics into Eq. 3.132 and Eq. 3.131 leads to the i-vector. Figure 3.25 shows the procedure of i-vector extraction.

As will be discussed in Section 3.6.10, the posteriors $\gamma(\ell_{i,t,c})$'s do not have to be computed by aligning individual frames with the UBM. In the senone i-vectors, the mixture components c are replaced by senones and $\gamma(\ell_{i,t,c})$'s are replaced by senone posteriors. The latter can be estimated by a phonetic-aware deep neural network. Note that in senone i-vectors, a UBM is still required because Eq. 3.131 and Eq. 3.132 involve C covariance matrices. However, instead of computing these matrices through the EM algorithm, we may compute these matrices using the senone posteriors and the acoustic vectors in one iteration. Figure 3.26 shows the procedure of senone i-vector extraction.

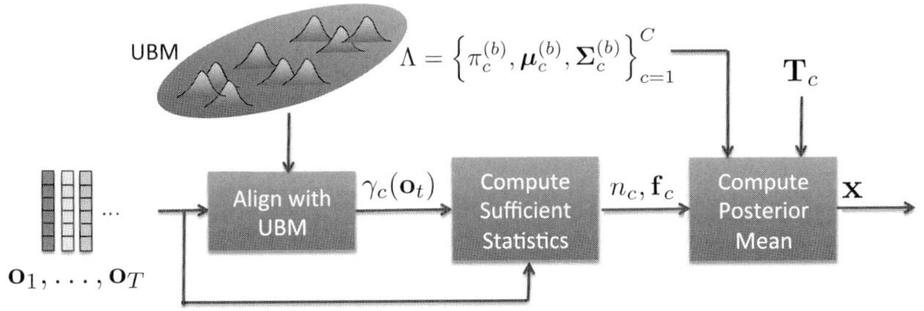

Figure 3.25 The procedure of GMM i-vector extraction. For clarity, the utterance index i is omitted.

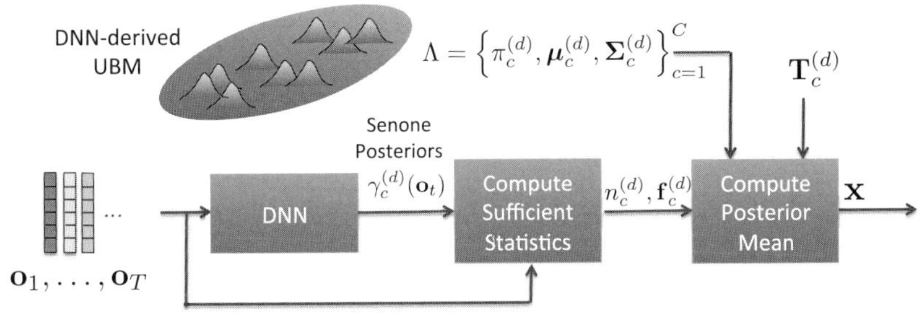

Figure 3.26 The procedure of DNN senone i-vector extraction. For clarity, the utterance index i is omitted.

3.6.4 Relation with MAP Adaptation in GMM–UBM

It can be shown that MAP adaptation in GMM–UBM [7] is a special case of the factor analysis model in Eq. 3.119. Specifically, when $\mathbf{T}$ is a $CF \times CF$ diagonal matrix denoted as $\mathbf{D}$, then Eq. 3.119 can be rewritten as

$$\boldsymbol{\mu}_i = \boldsymbol{\mu}^{(b)} + \mathbf{D}\mathbf{z}_i. \tag{3.133}$$

Using Eq. 3.131, we obtain the posterior mean of latent factor $\mathbf{z}_i$

$$\langle \mathbf{z}_i | \mathcal{O}_i \rangle = \mathbf{L}_i^{-1}\mathbf{D}^{\mathsf{T}}(\boldsymbol{\Sigma}^{(b)})^{-1}\tilde{\mathbf{f}}_i,$$

where $\mathcal{O}_i$ comprises the MFCC vectors of utterance i. Define r as the relevance factor in MAP adaptation such that $r\mathbf{D}^{\mathsf{T}}(\boldsymbol{\Sigma}^{(b)})^{-1}\mathbf{D} = \mathbf{I}$. Then, Eq. 3.132 becomes

$$\begin{aligned}\mathbf{L}_i &= r\mathbf{D}^{\mathsf{T}}(\boldsymbol{\Sigma}^{(b)})^{-1}\mathbf{D} + \mathbf{D}^{\mathsf{T}}(\boldsymbol{\Sigma}^{(b)})^{-1}\mathbf{N}_i\mathbf{D} \\ &= \mathbf{D}^{\mathsf{T}}(\boldsymbol{\Sigma}^{(b)})^{-1}(r\mathbf{I} + \mathbf{N}_i)\mathbf{D}.\end{aligned}$$

As a result, the offset to the UBM's supervector $\mu^{(b)}$ for utterance i is

$$\begin{aligned}\mathbf{D}\langle\mathbf{z}_i|\mathcal{O}_i\rangle &= \mathbf{D}\left[\mathbf{D}^{\mathsf{T}}(\mathbf{\Sigma}^{(b)})^{-1}(r\mathbf{I}+\mathbf{N}_i)\mathbf{D}\right]^{-1}\mathbf{D}^{\mathsf{T}}(\mathbf{\Sigma}^{(b)})^{-1}\tilde{\mathbf{f}}_i \\ &= \mathbf{D}\left[\mathbf{D}^{-1}(r\mathbf{I}+\mathbf{N}_i)^{-1}(\mathbf{\Sigma}^{(b)})\mathbf{D}^{-\mathsf{T}}\right]\mathbf{D}^{\mathsf{T}}(\mathbf{\Sigma}^{(b)})^{-1}\tilde{\mathbf{f}}_i \\ &= (r\mathbf{I}+\mathbf{N}_i)^{-1}\tilde{\mathbf{f}}_i.\end{aligned} \tag{3.134}$$

Recall from Eq. 3.16 that given utterance i, the MAP adaptation of mixture c in GMM–UBM is

$$\begin{aligned}\boldsymbol{\mu}_{i,c} &= \alpha_{i,c}E_c(\mathcal{O}_i) + (1-\alpha_{i,c})\boldsymbol{\mu}_c^{(b)} \\ &= \frac{N_{i,c}}{N_{i,c}+r}\frac{\mathbf{f}_{i,c}}{N_{i,c}} + \boldsymbol{\mu}_c^{(b)} - \frac{N_{i,c}}{N_{i,c}+r}\boldsymbol{\mu}_c^{(b)} \\ &= \boldsymbol{\mu}_c^{(b)} + \frac{1}{N_{i,c}+r}(\mathbf{f}_{i,c} - N_{i,c}\boldsymbol{\mu}_c^{(b)}) \\ &= \boldsymbol{\mu}_c^{(b)} + (r\mathbf{I}+\mathbf{N}_{i,c})^{-1}\tilde{\mathbf{f}}_{i,c},\end{aligned} \tag{3.135}$$

where r is the relevance factor, $N_{i,c}$ is the zeroth order sufficient statistics of mixture c, and $\mathbf{N}_{i,c} = \text{diag}\{N_{i,c},\ldots,N_{i,c}\}$ is the cth block of matrix $\mathbf{N}_i$. Note that the offset to UBM's means in Eq. 3.135 is equivalent to Eq. 3.134.

3.6.5 I-Vector Preprocessing for Gaussian PLDA

It is necessary to preprocess the i-vectors before using the Gaussian PLDA because it requires the i-vectors to follow a Gaussian distribution. Gaussianization of i-vectors involves two steps. First, a whitening transform is applied to i-vectors:

$$\mathbf{x}^{\text{wht}} = \mathbf{W}^{\mathsf{T}}(\mathbf{x}-\bar{\mathbf{x}}), \tag{3.136}$$

where $\mathbf{W}$ is the Cholesky decomposition of the within-class covariance matrix of i-vectors [74], $\bar{\mathbf{x}}$ is the global mean of i-vectors, and $\mathbf{x}^{\text{wht}}$ is the whitened i-vector. In the second step, we apply a simple length-normalization to the whitened i-vectors:

$$\mathbf{x}^{\text{len-norm}} = \frac{\mathbf{x}^{\text{wht}}}{\|\mathbf{x}^{\text{wht}}\|}. \tag{3.137}$$

It is a common practice to include LDA (or NDA) and within-class covariance normalization (WCCN) [74] in the preprocessing steps. The whole preprocessing can be written in a more concise form:

$$\mathbf{x} \leftarrow \frac{\mathbf{P}(\mathbf{x}-\bar{\mathbf{x}})}{\|\mathbf{x}^{\text{wht}}\|}, \tag{3.138}$$

where $\mathbf{P}$ denotes the transformation matrix that combines whitening, LDA and WCCN, and $\mathbf{x}$ on the left of Eq. 3.138 is the preprocessed i-vector that is ready for PLDA modeling.

3.6.6 Session Variability Suppression

Because the total variability matrix $\mathbf{T}$ models not only the speaker variability but also other variabilities in the acoustic vectors, it is important to remove the unwanted variabilities in the i-vectors. This can be done by minimizing within-speaker scatter and maximizing between-speaker scatter.

LDA+WCCN

A classical approach to session variability suppression is to apply linear discriminant analysis (LDA) followed by within-class covariance normalization (WCCN). As shown in Figure 3.27, given the utterances of a target speaker s and a test speaker t, the verification score is the cosine of the angle between the two transformed vectors:

$$S_{\text{CD}}(\mathbf{x}_s, \mathbf{x}_t) = \frac{(\mathbf{W}^{\mathsf{T}}\mathbf{x}_s) \cdot (\mathbf{W}^{\mathsf{T}}\mathbf{x}_t)}{\|\mathbf{W}^{\mathsf{T}}\mathbf{x}_s\| \, \|\mathbf{W}^{\mathsf{T}}\mathbf{x}_t\|}, \tag{3.139}$$

where $\mathbf{W}$ is the combined LDA+WCCN transformation matrix.

The LDA+WCCN transformation matrix can be estimated by supervised learning as follows. Denote $\mathbf{A}$ and $\mathbf{B}$ as the LDA and WCCN transformation matrices, respectively. Therefore, the combined matrix is $\mathbf{W} = \mathbf{BA}$. To find $\mathbf{A}$ and $\mathbf{B}$, we need a number of training speakers, each providing a number of utterances (i-vectors). Assume that we have K training speakers whose i-vectors are denoted as $\mathbf{x}_n$'s. To find $\mathbf{A}$, we maximize the between-speaker scatter and minimize the within-speaker scatter after the transformation. This can be achieved by maximizing the LDA criterion:

$$J(\mathbf{A}) = \frac{\text{Between-speaker scatter}}{\text{Within-speaker scatter}} = \text{Tr}\left\{\left(\mathbf{A}^{\mathsf{T}}\mathbf{S}_b\mathbf{A}\right)\left(\mathbf{A}^{\mathsf{T}}\mathbf{S}_w\mathbf{A}\right)^{-1}\right\}, \tag{3.140}$$

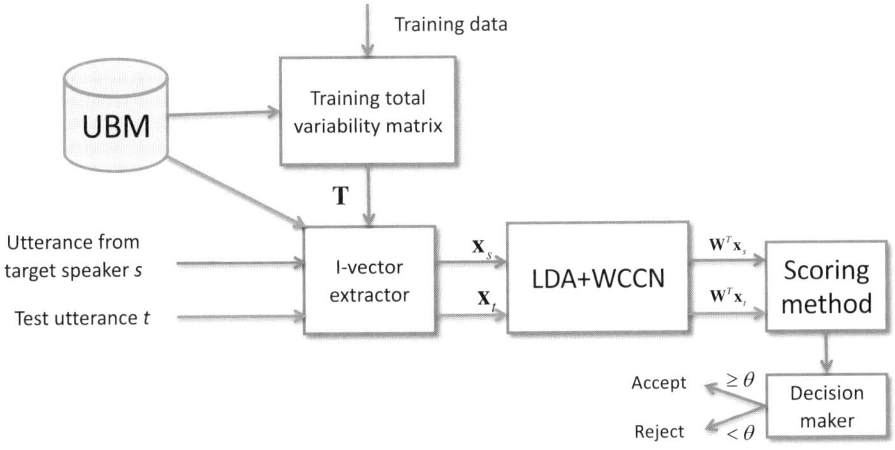

Figure 3.27 Preprocessing of i-vectors for cosine-distance or PLDA scoring.

where $\mathbf{S}_b$ and $\mathbf{S}_w$ are respectively the between-speaker and within-speaker scatter matrices

$$\mathbf{S}_b = \sum_{k=1}^{K} N_k(\bar{\mathbf{x}}_k - \bar{\mathbf{x}})(\bar{\mathbf{x}}_k - \bar{\mathbf{x}})^{\mathsf{T}} \tag{3.141a}$$

$$\mathbf{S}_w = \sum_{k=1}^{K} \sum_{n \in \mathcal{C}_k} (\mathbf{x}_n - \bar{\mathbf{x}}_k)(\mathbf{x}_n - \bar{\mathbf{x}}_k)^{\mathsf{T}}. \tag{3.141b}$$

In Eq. 3.141, $\bar{\mathbf{x}}_k$ is the mean i-vector of speaker k, $\bar{\mathbf{x}}$ is the global mean of i-vectors, $\mathcal{C}_k$ comprises the utterance indexes of speaker k, and N_k is the number of samples in the class k, i.e., $N_k = |\mathcal{C}_k|$.

Eq. 3.140 can be framed as a constrained optimization problem:

$$\begin{aligned} \max_{\mathbf{A}} \quad & \mathrm{Tr}\{\mathbf{A}^{\mathsf{T}}\mathbf{S}_b\mathbf{A}\} \\ \text{subject to} \quad & \mathbf{A}^{\mathsf{T}}\mathbf{S}_w\mathbf{A} = \mathbf{I}, \end{aligned} \tag{3.142}$$

where $\mathbf{I}$ is an $M \times M$ identity matrix. Denote $\mathbf{A} = [\mathbf{a}_1, \ldots, \mathbf{a}_{K'}]$. To find $\mathbf{a}_j$, $j = 1, \ldots, K'$, we write the Lagrangian function as:

$$L(\mathbf{a}_j, \lambda_j) = \mathbf{a}_j^{\mathsf{T}}\mathbf{S}_b\mathbf{a}_j - \lambda_j(\mathbf{a}_j^{\mathsf{T}}\mathbf{S}_w\mathbf{a}_j - 1),$$

where λ_j is a Lagrange multiplier. Setting $\frac{\partial L}{\partial \mathbf{w}_j} = 0$, we obtain the optimal solution of $\mathbf{a}_j$, which satisfies

$$(\mathbf{S}_w^{-1}\mathbf{S}_b)\mathbf{a}_j = \lambda_j\mathbf{a}_j.$$

Therefore, $\mathbf{A}$ comprises the first K' eigenvectors of $S_w^{-1}\mathbf{S}_b$. A more formal proof can be find in [75]. As the maximum rank of $\mathbf{S}_b$ is $K - 1$, $S_w^{-1}\mathbf{S}_b$ has at most $K - 1$ nonzero eigenvalues. As a result, K' can be at most $K - 1$. This suggests that for practical systems, a few hundred training speakers are required. Figure 3.28 shows the effect of LDA+WCCN on i-vectors. The diagram shows the projection of the original i-vectors and the LDA+WCCN-projected i-vectors on a three-dimensional t-SNE space [4]. Evidently, the LDA is very effective in separating the i-vectors of different speakers.

After finding $\mathbf{A}$, all training i-vectors are transformed, i.e., $\mathbf{A}^{\mathsf{T}}\mathbf{x}_n$, $\forall n$. Then, a within-class covariance matrix is computed based on the LDA-transformed i-vectors:

$$\mathbf{W}_{wccn} = \sum_{k=1}^{K} \frac{1}{N_k} \sum_{n \in \mathcal{C}_k} \mathbf{A}^{\mathsf{T}}(\mathbf{x}_n - \bar{\mathbf{x}}_k)(\mathbf{x}_n - \bar{\mathbf{x}}_k)^{\mathsf{T}}\mathbf{A}. \tag{3.143}$$

Then, the matrix $\mathbf{B}$ can be obtained from the Cholesky decomposition of $\mathbf{W}_{wccn}$, i.e., $\mathbf{B}\mathbf{B}^{\mathsf{T}} = \mathbf{W}_{wccn}$.

Once we have matrices $\mathbf{A}$ and $\mathbf{B}$, we may apply LDA+WCCN transformation on all i-vectors, including test i-vectors as follows:

$$\mathbf{x} \leftarrow \mathbf{W}^{\mathsf{T}}\mathbf{x} = \mathbf{B}^{\mathsf{T}}\mathbf{A}^{\mathsf{T}}\mathbf{x}.$$

As Eq. 3.142 suggests, LDA aims to maximize speaker discrimination. WCCN, on the other hand, aims to inversely scale the LDA-projected space according to the

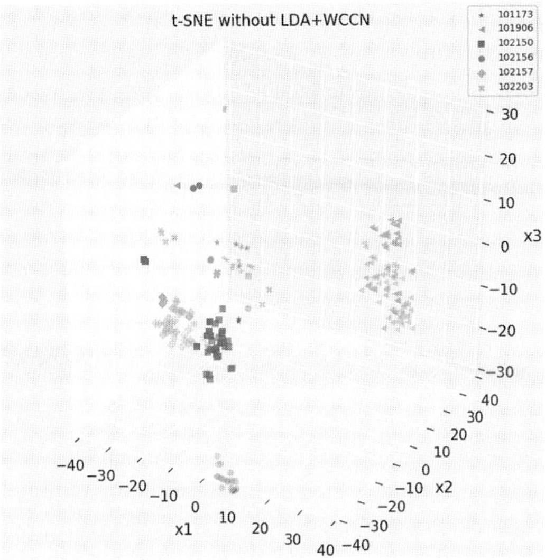

(a)

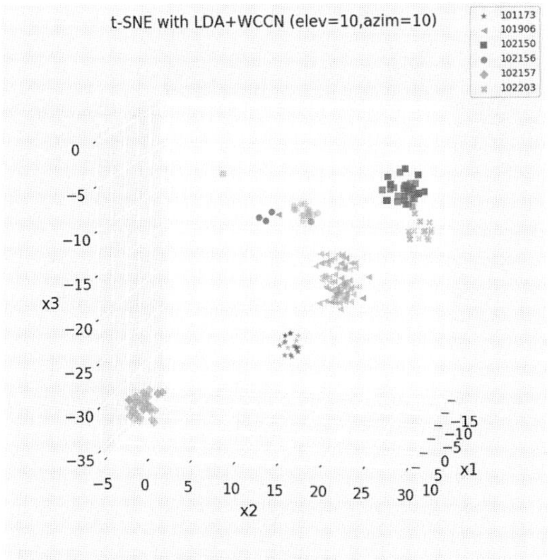

(b)

Figure 3.28 T-SNE plot of I-vectors (a) before and (b) after LDA+WCCN transformation. Each marker type represents a speaker. The legend shows the speaker IDs.

within-speaker covariance matrix (Eq. 3.143). After the scaling, the directions of high intra-speaker variability will be de-emphasized. Thus, WCCN has the effect of variance normalization.

NDA

In LDA, it is assumed that samples in individual classes follow a Gaussian distribution. It is also assumed that these samples share the same covariance matrix. Because the within-class and between-class scatter matrices are estimated empirically from training data, their accuracy depends on a number of factors. One important factor is the accuracy of the class-dependent means. Because in practical situation, the number of sessions per training speaker is not very large and some speakers may only have a few sessions, some of the class-dependent means could be very inaccurate. If the effect of these inaccurate means is not suppressed, the performance of LDA will suffer.

To address the limitation of LDA, a nonparametric discriminant analysis (NDA) [76] – which is originally proposed for face recognition – has successfully been applied to speaker verification [55, 56].

As mentioned earlier, in LDA, the class-dependent means are used for computing the within-class scatter matrix. On the other hand, in NDA, the class-dependent means are replaced by the nearest neighbors to each training vector. Moreover, instead of using the global mean and class-dependent means to compute the between-class scatter matrix, NDA uses all of the training vectors and their nearest neighbors from other classes to compute the between-class scatter matrix. It was found that this strategy is more effective than LDA in capturing the structural information of speaker boundaries.

Denote $\mathbf{x}_{ij}$ as the jth length-normalized i-vector from speaker i. In NDA, the nonparametric between-class and within-class scatter matrices are given by

$$\mathbf{S}_b^{\text{NDA}} = \sum_{i=1}^{S} \sum_{\substack{r=1 \\ r \neq i}}^{S} \sum_{l=1}^{K_b} \sum_{j=1}^{N_i} \omega(i,r,l,j)(\mathbf{x}_{ij} - \psi_l(\mathbf{x}_{ij},r))(\mathbf{x}_{ij} - \psi_l(\mathbf{x}_{ij},r))^{\mathsf{T}} \tag{3.144}$$

and

$$\mathbf{S}_w^{\text{NDA}} = \sum_{i=1}^{S} \sum_{l=1}^{K_w} \sum_{j=1}^{N_i} (\mathbf{x}_{ij} - \psi_l(\mathbf{x}_{ij},i))(\mathbf{x}_{ij} - \psi_l(\mathbf{x}_{ij},i))^{\mathsf{T}}, \tag{3.145}$$

respectively. In Eq. 3.144 and Eq. 3.145, N_i is the number of i-vectors from the ith speaker, S is the total number of speakers in the training set, $\psi_l(\mathbf{x}_{ij},r)$ is the lth nearest neighbor from speaker r to $\mathbf{x}_{ij}$, and K_b and K_w are the number of nearest neighbors selected for computing $\mathbf{S}_b^{\text{NDA}}$ and $\mathbf{S}_w^{\text{NDA}}$, respectively.

The weighting term $\omega(i,r,l,j)$ in Eq. 3.144 is defined as:

$$\begin{aligned} \omega(i,r,l,j) &= \frac{\min\{d^{\alpha}(\mathbf{x}_{ij},\psi_l(\mathbf{x}_{ij},i)), d^{\alpha}(\mathbf{x}_{ij},\psi_l(\mathbf{x}_{ij},r))\}}{d^{\alpha}(\mathbf{x}_{ij},\psi_l(\mathbf{x}_{ij},i)) + d^{\alpha}(\mathbf{x}_{ij},\psi_l(\mathbf{x}_{ij},r))} \\ &= \frac{1}{1+g(\mathbf{x}_{ij})^{\alpha}}, \qquad i \neq r \end{aligned} \tag{3.146}$$

where α controls the rate of change of the weight with respect to the distance ratio $g(\mathbf{x}_{ij})$ and $d(\mathbf{x}_p,\mathbf{x}_q)$ is the Euclidean distance between vector $\mathbf{x}_p$ and vector $\mathbf{x}_q$.

In Eq. 3.146, if $d^{\alpha}(\mathbf{x}_{ij},\psi_l(\mathbf{x}_{ij},i)) < d^{\alpha}(\mathbf{x}_{ij},\psi_l(\mathbf{x}_{ij},r))$, then

$$g(\mathbf{x}_{ij}) = \frac{d(\mathbf{x}_{ij},\psi_l(\mathbf{x}_{ij},r))}{d(\mathbf{x}_{ij},\psi_l(\mathbf{x}_{ij},i))},$$

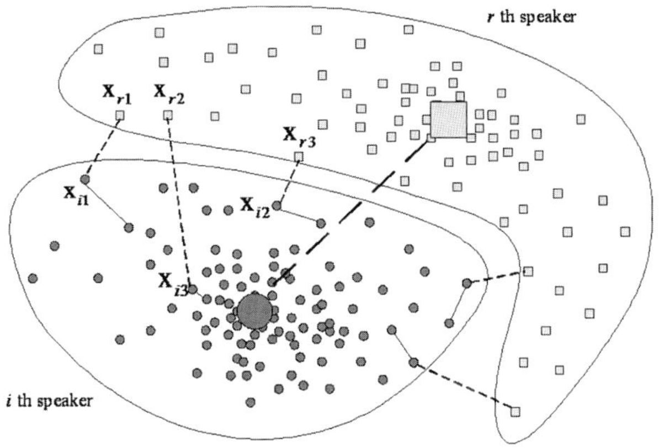

Figure 3.29 A two-dimensional example showing the within-class distance $d(\mathbf{x}_{ij}, \psi_l(\mathbf{x}_{ij}, i))$ [solid lines] and the between-class distance $d(\mathbf{x}_{ij}, \psi_l(\mathbf{x}_{ij}, r))$ [dashed lines] and for the training i-vectors $\mathbf{x}_{ij}$ from the ith speaker. For $\mathbf{x}_{i1}$ and $\mathbf{x}_{i2}$, the weighting function $\omega(\cdot)$ in Eq. 3.146 approaches 0.5. For $\mathbf{x}_{i3}$, $\omega(\cdot)$ approaches 0.0. [Reprinted from *SNR-Invariant PLDA Modeling in Nonparametric Subspace for Robust Speaker Verification (Figure 1), N. Li and M.W. Mak, IEEE/ACM Trans. on Audio Speech and Language Processing, vol. 23, no. 10, pp. 1648–1659, Oct. 2015, with permission of IEEE.*]

otherwise

$$g(\mathbf{x}_{ij}) = \frac{d(\mathbf{x}_{ij}, \psi_l(\mathbf{x}_{ij}, i))}{d(\mathbf{x}_{ij}, \psi_l(\mathbf{x}_{ij}, r))}. \tag{3.147}$$

If a selected i-vector $\mathbf{x}_{ij}$ is close to a speaker boundary (e.g., $\mathbf{x}_{i1}$ and $\mathbf{x}_{i2}$ in Figure 3.29), the between-class distance $d(\mathbf{x}_{ij}, \psi_l(\mathbf{x}_{ij}, r))$ [dashed line] will be comparable to the within-class distance $d(\mathbf{x}_{ij}, \psi_l(\mathbf{x}_{ij}, i))$ [solid line]. This will cause $g(\mathbf{x}_{ij})$ in Eq. 3.147 approaching 1.0. As a result, the weighting function $\omega(\cdot)$ in Eq. 3.146 will approach 0.5. On the other hand, if the selected i-vector $\mathbf{x}_{ij}$ is far away from a speaker boundary, e.g., $\mathbf{x}_{i3}$, the between-class distance $d(\mathbf{x}_{ij}, \psi_l(\mathbf{x}_{ij}, i))$ will be much smaller than the between-class distance $d(\mathbf{x}_{ij}, \psi_l(\mathbf{x}_{ij}, r))$. This will make the weighting function $\omega(\cdot)$ approaching 0.0. Consequently, the weighting function is able to emphasize the discriminative speaker boundary information in the training set. In [55], α was set to 2.

In LDA, the class-dependent means ($\bar{\mathbf{x}}_k$ in Eq. 3.141) are assumed to be the parameters of normal distributions. More precisely, each class has its own class-dependent mean but all classes share the same covariance matrix. In that sense, LDA is parametric. On the other hand, NDA is nonparametric because the terms in Eqs. 3.145 and 3.144 are vectors near the decision boundaries. They are not the parameters of normal distributions.

Similar to the standard LDA, the NDA projection matrix comprises the eigenvectors of $(\mathbf{S}_w^{\text{NDA}})^{-1}\mathbf{S}_b^{\text{NDA}}$.

SVDA

SVDA (support vector discriminant analysis) [77] is a variant of LDA in that its objective function is the same as Eq. 3.140. The difference is that the between-speaker and within-speaker scatter matrices are computed by using the support vectors of linear SVMs that optimally separate any combinations of two speakers.

Given C speakers in a training set, there are $\frac{1}{2}C(C-1)$ combinations of speaker pairs. For each pair, we train a linear SVM to classify the two speakers in the pair. The linear SVM that separates the speakers pair (i,j), where $i < j$, has a decision function

$$\begin{aligned} f_{ij}(\mathbf{x}) &= \sum_{k \in \mathcal{S}_{ij}} \alpha_k y_k \mathbf{x}_k^{\mathsf{T}} \mathbf{x} + b_{ij} \\ &= \mathbf{w}_{ij}^{\mathsf{T}} \mathbf{x} + b_{ij}, \end{aligned}$$

where $\mathcal{S}_{ij}$ contains the support vector indexes for this speaker pair, α_k's and $y_k \in \{-1,+1\}$ are Lagrange multipliers and speaker labels, respectively, and b_{ij} is a bias term. As shown in Figure 3.8, $\mathbf{w}_{ij}$ is parallel to the optimal direction to classify speakers i and j. Therefore, we may use it to compute the within-speaker scatter matrix as follows:

$$\mathbf{S}_b = \sum_{1 \le i \le j \le C} \mathbf{w}_{ij} \mathbf{w}_{ij}^{\mathsf{T}}.$$

Similarly, the within-speaker scatter matrix can be computed from the support vectors only:

$$\mathbf{S}_w = \sum_{c=1}^{C} \sum_{k \in \mathcal{S}_c} (\mathbf{x}_k - \boldsymbol{\mu}_c)(\mathbf{x}_k - \boldsymbol{\mu}_c)^{\mathsf{T}},$$

where $\mathcal{S}_c$ contains the support vector indexes of speaker c and

$$\boldsymbol{\mu}_c = \frac{1}{|\mathcal{S}_c|} \sum_{k \in \mathcal{S}_c} \mathbf{x}_k$$

is the mean of support vectors from speaker c. Then, the SVDA projection matrix is obtained from the eigenvectors of $\mathbf{S}_w^{-1}\mathbf{S}_b$.

It was found in [77] that replacing LDA by SVDA can improve the performance by up to 32 percent.

3.6.7 PLDA versus Cosine-Distance Scoring

The cosine distance scoring in Eq. 3.139 simply computes the similarity between two i-vectors, regardless of their variability. In fact, all nonspeaker variabilities are assumed to be removed by the LDA+WCCN projection. One may think of this projection as creating a speaker subspace on which all projected i-vectors depend on speaker characteristics only. The PLDA discussed in Section 3.5, however, finds the speaker subspace by maximizing the likelihood of observed i-vectors using a factor analysis model and the speaker labels of the observed i-vectors. The advantage of PLDA is that it produces a probabilistic model through which true likelihood-ratio can be computed. As a result, in theory, score normalization [13] is not necessary.

3.6.8 Effect of Utterance Length

An interesting question to ask is "will the modeling capability of i-vectors keeps on improving when the utterance duration increases." The answer is "no." This is because the information captured by i-vectors saturates when the utterances reach certain duration.

To demonstrate this saturation effect, we computed the intra- and inter-speaker cosine-distance scores of 272 speakers extracted from the telephone calls (phonecall_tel) and interview sessions (interview_mic) of SRE10. Each conversation was divided into a number of equal-length segments. By concatenating variable numbers of equal-length segments, sub-utterances of variable length were obtained. A voice activity detector (VAD) [78] was applied to remove the nonspeech regions in the sub-utterances. An i-vector was then estimated by using the acoustic vectors of each sub-utterance. Then, LDA and WCCN were applied to reduce the dimension of the vectors to 150 for computing cosine distance scores.

The mean intra- and inter-speaker scores (with error bars indicating two standard deviations) of the three types of speech are shown in Figure 3.30. The scores in "8-min interview_mic" were acquired from the 8-min interview sessions of 29 male speakers; each of these speakers give four interview-style conversations. Totally, for each utterance-length, there are 6496 inter-speaker scores and 174 intra-speaker scores. The scores in "3-min interview_mic" were acquired from the 3-minute interview sessions of 196 male speakers; each of them provide four interview sessions. This gives rise to 305,760 inter-speaker scores and 1176 intra-speaker scores for each utterance-length. For the "5-min phonecall_tel," the scores were acquired from the 5-minute telephone conversations of 47 male speakers, each giving four conversations. For each segment-length, there are 282 intra-speaker scores and 17,296 inter-speaker scores. Figure 3.30 clearly shows that both types of scores level off when the segment length is longer than a certain threshold.

Figure 3.31 shows the minimum decision cost (minDCF) versus the utterance length for estimating the i-vectors using the intra- and inter-speaker scores shown in Figure 3.30. The results further demonstrate the discriminative power of i-vectors with respect to the utterance duration. Evidently, the capability of discriminating speakers get saturated when the utterance duration is longer than two minutes. This result indicates that it may not be necessary to record very long utterances. From another perspective, when an utterance is sufficiently long, we may be able to leverage its long duration by dividing it into a number of sub-utterances so that more i-vectors can be derived, as illustrated in Figure 3.32 [79].

3.6.9 Gaussian PLDA with Uncertainty Propagation

While the i-vector/PLDA framework has been a great success, its performance drops rapidly when both the enrollment and test utterances have a wide range of durations. There are several reasons for this performance degradation. First, the duration of utterances is totally ignored in i-vector extraction, which means that the utterances are represented by vectors of fixed dimension without taking the utterance duration into account.

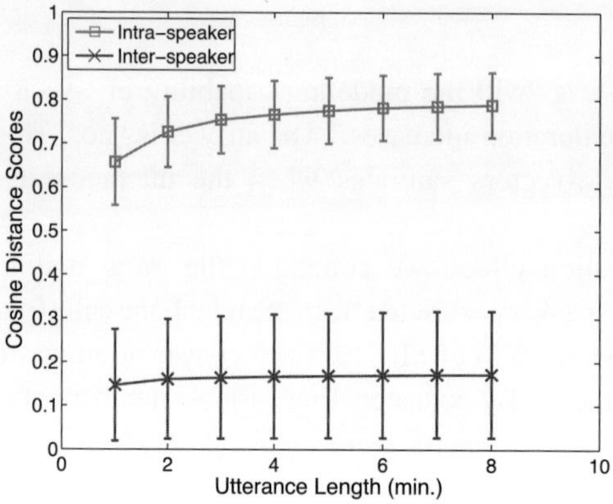

(a) 8-min interview_mic

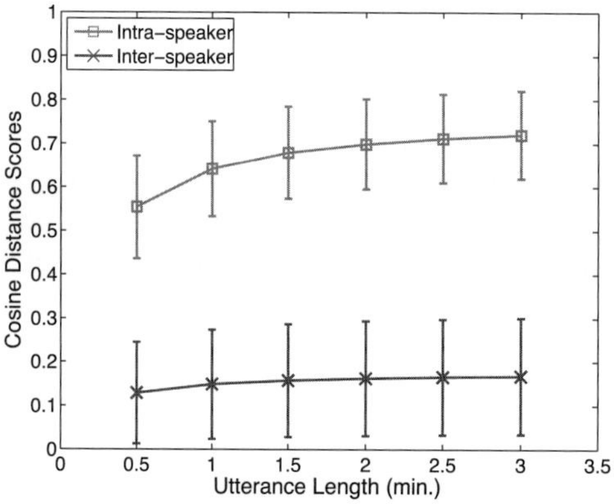

(b) 3-min interview_mic

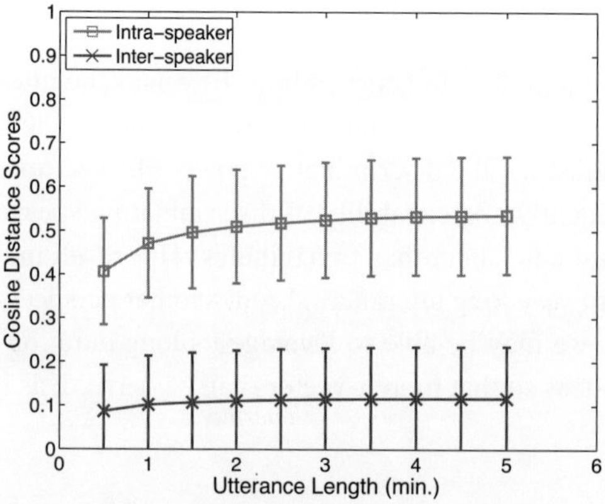

(c) 5-min phonecall_tel

Figure 3.30 Inter- and intra-speaker cosine-distance scores against utterance duration. (a) 8-min interview conversation; (b) 3-min interview conversation; (c) 5-min telephone conversation. [Reprinted from *Boosting the Performance of I-Vector Based Speaker Verification via Utterance Partitioning (Figure 1), W. Rao and M.W. Mak, IEEE Trans. on Audio Speech and Language Processing, vol. 21, no. 5, pp. 1012–1022, May 2013, with permission of IEEE.*]

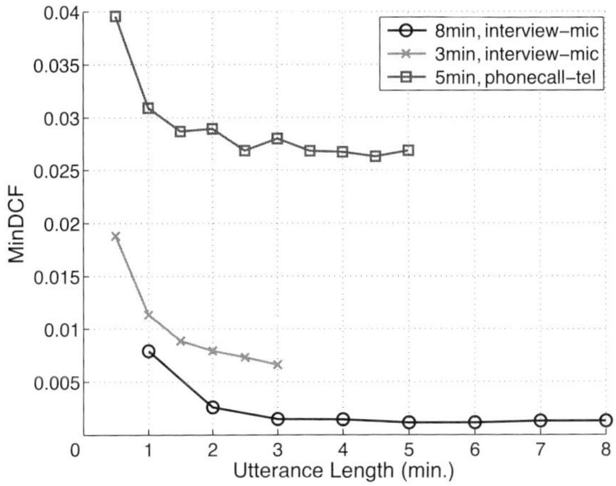

Figure 3.31 Minimum decision costs (MinDCF) versus utterance duration. The costs were based on the inter- and intra-speaker cosine-distances shown in Figure 3.30. [Reprinted from *Boosting the Performance of I-Vector Based Speaker Verification via Utterance Partitioning (Figure 2), W. Rao and M.W. Mak, IEEE Trans. on Audio Speech and Language Processing, vol. 21, no. 5, pp. 1012–1022, May 2013, with permission of IEEE.*]

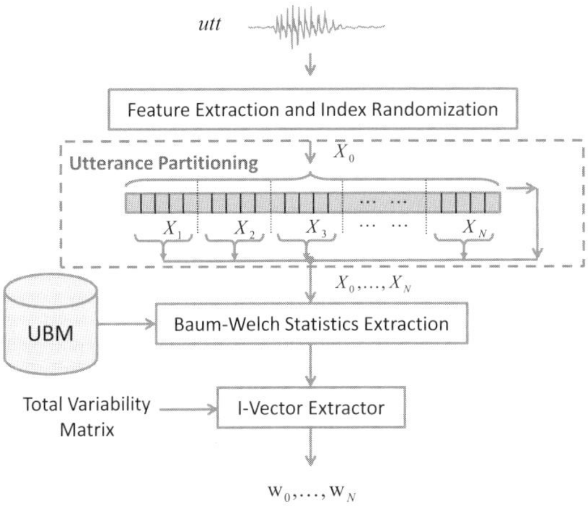

Figure 3.32 The utterance partitioning with acoustic vector resampling (UP-AVR) process. [Reprinted from *Boosting the Performance of I-Vector Based Speaker Verification via Utterance Partitioning (Figure 3), W. Rao and M.W. Mak, IEEE Trans. on Audio Speech and Language Processing, vol. 21, no. 5, pp. 1012–1022, May 2013, with permission of IEEE.*]

The accuracy of i-vectors depends on the number of acoustic vectors used for computing the posterior mean of the latent variable $\mathbf{w}$. If we do not take the utterance duration into consideration, we essentially treat all i-vectors to be equally reliable. Second, PLDA assumes that the intra-speaker variability (represented by $\boldsymbol{\Sigma}$ in Eq. 3.80) is the same across all i-vectors. This does not really reflect the real situation because intra-speaker variability in short utterances is larger than that of long utterances.

Kenny et al. [80] proposed to modify the standard PLDA to better accommodate utterance-length variability. The method aims to tie the i-vector extraction and PLDA modeling through propagating the uncertainty of i-vectors into the PLDA model. As shown in Eq. 3.132, the uncertainty (posterior covariance) of i-vectors depends on the number of acoustic vectors involves in computing the sufficient statistics. The longer the utterances, the smaller the uncertainty. By propagating the uncertainty to the PLDA model through the loading matrix, we will be able to model the variability in i-vectors due to difference in utterance duration. The resulting model will be better in handling the duration variability in utterances than the conventional PLDA model.

Preprocessing for Gaussian PLDA with UP

To apply UP, we need to apply the preprocessing steps in Section. 3.6.5 to the posterior covariance matrices as well. Assume that only a linear transformation $\mathbf{P}$ is applied to an i-vector, the corresponding preprocessed covariance matrix is

$$\text{cov}(\mathbf{Pw}, \mathbf{Pw}) = \mathbf{PL}^{-1}\mathbf{P}^{\mathsf{T}}, \tag{3.148}$$

where $\mathbf{L}$ is the precision matrix in Eq. 3.132. If length-normalization is also applied, the preprocessed covariance matrix can be estimated as follows [80]:

$$\boldsymbol{\Lambda} \leftarrow \frac{\mathbf{PL}^{-1}\mathbf{P}^{\mathsf{T}}}{\|\mathbf{x}^{\text{wht}}\|}. \tag{3.149}$$

Other methods to deal with this nonlinear transformation on the posterior matrix can be found in [80, 81].

Generative Model for Gaussian PLDA with UP

In uncertainty propagation, the i-vectors and the PLDA model are considered as a whole. Given an i-vector $\mathbf{x}_r$, its uncertainty is propagated to the PLDA model by adding an utterance-dependent loading matrix $\mathbf{U}_r$ to the generative model as follows:

$$\mathbf{x}_r = \mathbf{m} + \mathbf{Vz} + \mathbf{U}_r\mathbf{q}_r + \boldsymbol{\epsilon}_r. \tag{3.150}$$

In this model, $\mathbf{U}_r$ is the Cholesky decomposition of the posterior covariance matrix $\boldsymbol{\Lambda}_r$ of the i-vector, $\mathbf{q}_r$ is a latent variable with prior $\mathcal{N}(\mathbf{0}, \mathbf{I})$, and $\boldsymbol{\epsilon}_r$ is the residue that follows $\mathcal{N}(\mathbf{0}, \boldsymbol{\Sigma})$. The intra-speaker variability of $\mathbf{x}_r$ in Eq. 3.150 is:

$$\text{cov}(\mathbf{x}_r, \mathbf{x}_r | \mathbf{z}) = \boldsymbol{\Lambda}_r + \boldsymbol{\Sigma}, \tag{3.151}$$

where $\boldsymbol{\Lambda}_r$ varies from utterances to utterances, thus reflecting the reliability of i-vector $\mathbf{x}_r$.

Given a target speaker's i-vector $\mathbf{x}_s$ together with its posterior covariance matrix $\boldsymbol{\Lambda}_s$ and a test i-vector $\mathbf{x}_t$ together with its posterior covariance matrix $\boldsymbol{\Lambda}_t$, the log-likelihood ratio can be written as:

$$
\begin{aligned}
S_{LR}(\mathbf{x}_s, \mathbf{x}_t; \boldsymbol{\Lambda}_s, \boldsymbol{\Lambda}_t) &= \log \frac{p(\mathbf{x}_s, \mathbf{x}_t; \boldsymbol{\Lambda}_s, \boldsymbol{\Lambda}_t | \text{same-speaker})}{p(\mathbf{x}_s, \mathbf{x}_t; \boldsymbol{\Lambda}_s, \boldsymbol{\Lambda}_t | \text{different-speaker})} \\
&= \log p\left(\begin{bmatrix} \mathbf{x}_s \\ \mathbf{x}_t \end{bmatrix} \middle| \begin{bmatrix} \mathbf{0} \\ \mathbf{0} \end{bmatrix}, \begin{bmatrix} \boldsymbol{\Sigma}_s & \boldsymbol{\Sigma}_{ac} \\ \boldsymbol{\Sigma}_{ac} & \boldsymbol{\Sigma}_t \end{bmatrix} \right) \\
&\quad - \log p\left(\begin{bmatrix} \mathbf{x}_s \\ \mathbf{x}_t \end{bmatrix} \middle| \begin{bmatrix} \mathbf{0} \\ \mathbf{0} \end{bmatrix}, \begin{bmatrix} \boldsymbol{\Sigma}_s & \mathbf{0} \\ \mathbf{0} & \boldsymbol{\Sigma}_t \end{bmatrix} \right) \\
&= \frac{1}{2}\mathbf{x}_s^{\mathsf{T}} \mathbf{A}_{x,t} \mathbf{x}_s + \mathbf{x}_s^{\mathsf{T}} \mathbf{B}_{s,t} \mathbf{x}_t + \frac{1}{2}\mathbf{x}_t^{\mathsf{T}} \mathbf{C}_{s,t} \mathbf{x}_t + D_{s,t}
\end{aligned}
\tag{3.152}
$$

where

$$\mathbf{A}_{s,t} = \boldsymbol{\Sigma}_s^{-1} - (\boldsymbol{\Sigma}_s - \boldsymbol{\Sigma}_t^{-1}\boldsymbol{\Sigma}_{ac})^{-1} \tag{3.153}$$

$$\mathbf{B}_{s,t} = \boldsymbol{\Sigma}_s^{-1}\boldsymbol{\Sigma}_{ac}(\boldsymbol{\Sigma}_t - \boldsymbol{\Sigma}_{ac}\boldsymbol{\Sigma}_s^{-1}\boldsymbol{\Sigma}_{ac})^{-1} \tag{3.154}$$

$$\mathbf{C}_{s,t} = \boldsymbol{\Sigma}_t^{-1} - (\boldsymbol{\Sigma}_t - \boldsymbol{\Sigma}_s^{-1}\boldsymbol{\Sigma}_{ac})^{-1} \tag{3.155}$$

$$D_{s,t} = -\frac{1}{2}\log \begin{vmatrix} \boldsymbol{\Sigma}_s & \boldsymbol{\Sigma}_{ac} \\ \boldsymbol{\Sigma}_{ac} & \boldsymbol{\Sigma}_t \end{vmatrix} + \frac{1}{2}\log \begin{vmatrix} \boldsymbol{\Sigma}_s & \mathbf{0} \\ \mathbf{0} & \boldsymbol{\Sigma}_t \end{vmatrix} \tag{3.156}$$

$$\boldsymbol{\Sigma}_t = \mathbf{V}\mathbf{V}^{\mathsf{T}} + \boldsymbol{\Lambda}_t + \boldsymbol{\Sigma} \tag{3.157}$$

$$\boldsymbol{\Sigma}_s = \mathbf{V}\mathbf{V}^{\mathsf{T}} + \boldsymbol{\Lambda}_s + \boldsymbol{\Sigma} \tag{3.158}$$

$$\boldsymbol{\Sigma}_{ac} = \mathbf{V}\mathbf{V}^{\mathsf{T}}. \tag{3.159}$$

Note that Eqs. 3.153–3.156 involve terms dependent on both the target speaker's utterance and the test utterance, which means that these terms need to be evaluated during scoring. This means that uncertainty propagation is computationally expansive.

Fortunately, fast scoring methods [82] have been proposed to reduce the complexity of UP. The basic idea in [82] is to group i-vectors with similar reliability so that for each group the utterance-dependent loading matrices are replaced by a representative one. The procedure is as follows:

1. Compute the posterior covariance matrices from development data
2. For the kth group, select the representative $\mathbf{U}_k\mathbf{U}_k^{\mathsf{T}}$ as shown in Figure 3.33.

In [83], three approaches to finding the representative posterior covariances were investigated: (1) utterance duration, (2) mean of diagonal elements of posterior covariance matrices, and (3) based on the largest eigenvalue of posterior covariance matrices. With the above procedure, a set of representative matrices that cover all possible i-vectors can be precomputed. It has been demonstrated that by replacing the group

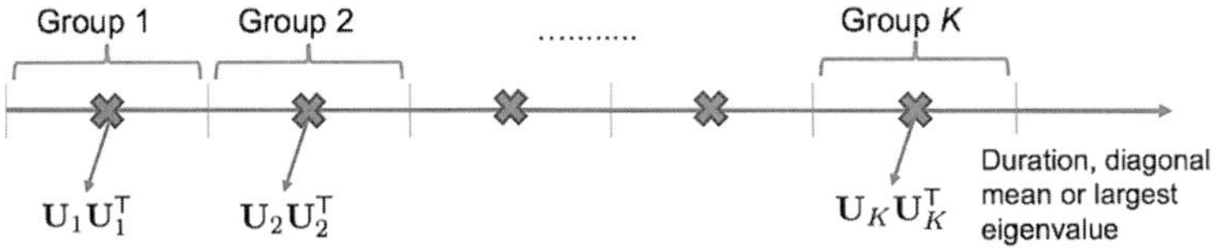

Figure 3.33 Representing the posterior covariance groups by one posterior covariance matrix.

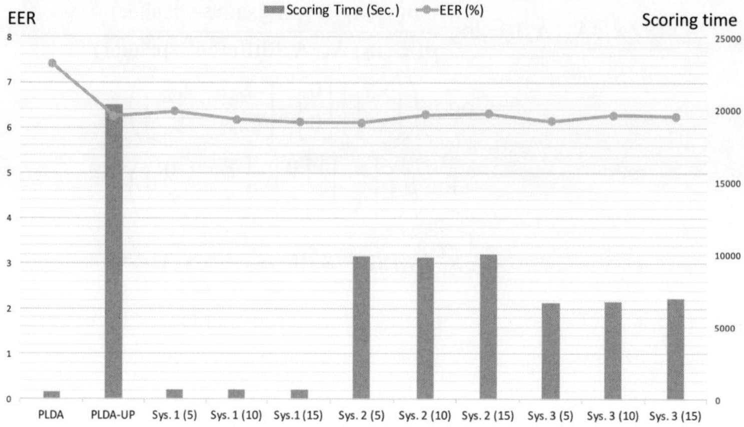

Figure 3.34 Equal error rates (EERs) and scoring time of PLDA, PLDA-UP, and fast PLDA-UP. For fast PLDA-UP, the EERs and scoring time of three systems, each with a different number of i-vector groups, were shown. The numbers within the parenthesis in the x-axis labels denote the numbers of i-vector groups. [Reprinted from *Fast Scoring for PLDA with Uncertainty Propagation via I-vector Grouping (Figure 2), W.W. Lin, M.W. Mak and J.T. Chien, Computer Speech and Language, vol. 45, pp. 503–515, 2017, with permission of Elsevier.*]

members by their representative, scoring time can be reduced by 33 times with negligible performance degradation. Figure 3.34 shows the dramatic reduction in the scoring time when this strategy is used.

3.6.10 Senone I-Vectors

Recently, Lei et al. [84] and Ferrer et al. [85] proposed to replace the UBM in the i-vector extractor by a deep neural network (DNN). The idea is to replace the zeroth order sufficient statistics (Eq. 3.126) by the outputs of a DNN trained for speech recognition. An advantage of this method is that it enables each test frame to be compared with the training frames for the same phonetic content.

One important characteristic of this method is that the acoustic features for speech recognition in the DNN are not necessarily the same as the features for the i-vector extractor. Specifically, denote the acoustic features of utterance i for the DNN and i-vector extractor as $\mathbf{a}_{it}$ and $\mathbf{o}_{it}$, respectively, where $t = 1, \ldots, T_i$. Then, the Baum–Welch statistics in Eq. 3.128 can be rewritten as:

$$N_{ic} \equiv \sum_{t=1}^{T_i} \gamma_c^{\text{DNN}}(\mathbf{a}_{it}) \quad \text{and} \quad \tilde{\mathbf{f}}_{ic} \equiv \sum_{t=1}^{T_i} \gamma_c^{\text{DNN}}(\mathbf{a}_{it})(\mathbf{o}_{it} - \boldsymbol{\mu}_c), \tag{3.160}$$

where $\gamma_c^{\text{DNN}}(\mathbf{a}_{it})$'s are the senone posterior probabilities obtained from the DNN's outputs and $\boldsymbol{\mu}_c$'s are the probabilistic weighted sum of speaker features:

$$\boldsymbol{\mu}_c = \frac{\sum_{i=1}^{N} \sum_{t=1}^{T_i} \gamma_c^{\text{DNN}}(\mathbf{a}_{it})\mathbf{o}_{it}}{\sum_{i=1}^{N} \sum_{t=1}^{T_i} \gamma_c^{\text{DNN}}(\mathbf{a}_{it})}. \tag{3.161}$$

Similar strategy can also be applied to Eq. 3.132:

$$\mathbf{L}_i = \mathbf{I} + \sum_{c=1}^{C} N_{ic}\mathbf{T}_c^{\mathsf{T}}(\boldsymbol{\Sigma}_c)^{-1}\mathbf{T}_c = \mathbf{I} + \mathbf{T}^{\mathsf{T}}\boldsymbol{\Sigma}^{-1}\mathbf{N}_i\mathbf{T}, \tag{3.162}$$

where N_{ic} is obtained in Eq. 3.160 and

$$\boldsymbol{\Sigma}_c = \frac{\sum_{i=1}^{N}\sum_{t=1}^{T_i}\gamma_c^{\text{DNN}}(\mathbf{a}_{it})\mathbf{o}_{it}\mathbf{o}_{it}^{\mathsf{T}}}{\sum_{i=1}^{N}\sum_{t=1}^{T_i}\gamma_c^{\text{DNN}}(\mathbf{a}_{it})} - \boldsymbol{\mu}_c\boldsymbol{\mu}_c^{\mathsf{T}}. \tag{3.163}$$

Note that $\mathbf{T}_c$ in Eq. 3.162 should be estimated by replacing $\tilde{\mathbf{f}}_{ic}$ in Eq. 3.129 with the first-order statistics in Eq. 3.160. Similarly, the posterior expectation in Eq. 3.123 should also be computed as follows:

$$\langle \mathbf{w}_i|\mathcal{O}_i\rangle = \mathbf{L}_i^{-1}\sum_{c=1}^{C}\mathbf{T}_c^{\mathsf{T}}\boldsymbol{\Sigma}_c^{-1}\tilde{\mathbf{f}}_{ic}, \tag{3.164}$$

where $\mathbf{L}_i$ and $\boldsymbol{\Sigma}_c$ are obtained from Eqs. 3.162 and 3.163, respectively. Because the mixture posteriors are now replaced by senone posteriors, the resulting i-vectors are termed senone i-vectors [3]. In some literature, they are also called DNN i-vectors.

Figure 3.35 shows an example of senone i-vector extraction [3]. The input $\mathbf{a}_{it}$ to the DNN comprises a context frames of acoustic features (could be MFCCs or filterbank outputs) centered at frame t and the vector $\mathbf{o}_{it}$ can be obtained from the bottleneck layer of the DNN.

3.7 Joint Factor Analysis

Before the introduction of i-vectors in 2010, joint factor analysis (JFA) [15] was the de facto standard for text-independent speaker verification. JFA is based on the assumption that a speaker- and channel-dependent supervector can be decomposed into the sum of two statistically independent and normally distributed supervectors: a speaker-dependent supervector and a channel-dependent supervector. Because a JFA model comprises three sets of latent factors, it is the most complex among all FA models in speaker recognition.

3.7.1 Generative Model of JFA

Denote $\mathbf{V}$ and $\mathbf{U}$ as the loading matrices representing the speaker- and channel subspaces, respectively. In JFA, the speaker- and channel-dependent supervectors $\mathbf{m}_h(s)$ can be generated by the following generative model:[9]

$$\mathbf{m}_h(s) = \mathbf{m} + \mathbf{V}\mathbf{y}(s) + \mathbf{U}\mathbf{x}_h(s) + \mathbf{D}\mathbf{z}(s), \tag{3.165}$$

[9] We follow the notations in [15, 86] as much as possible.

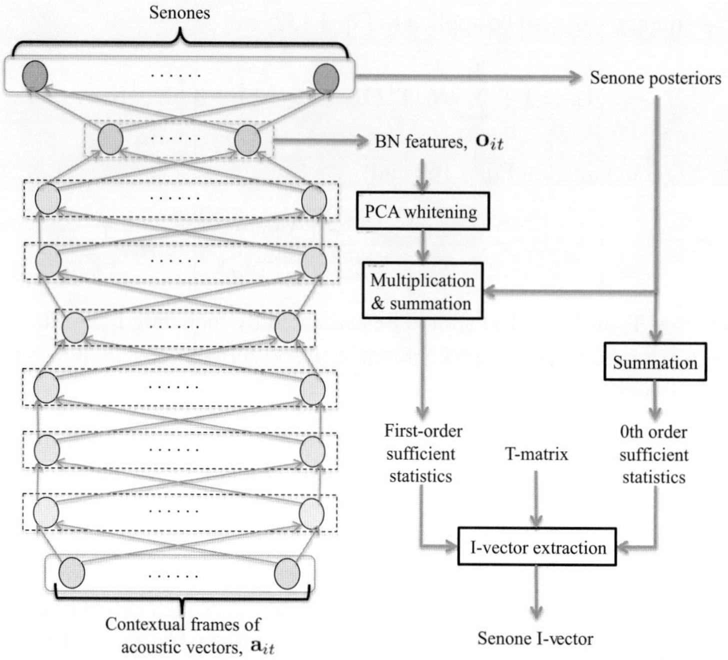

Figure 3.35 The procedure of extracting senone i-vectors. [Reprinted from *Denoised Senone I-Vectors for Robust Speaker Verification (Figure 2), Z.L. Tan, M.W. Mak, et al., IEEE/ACM Trans. on Audio Speech and Language Processing, vol. 26, no. 4, pp. 820–830, April 2018, with permission of IEEE.*]

where h and s stand for the recording (session) and speaker, respectively. In this model, $\mathbf{y}(s)$ is a speaker factor that depends on the speaker s and $\mathbf{x}_h(s)$ is a channel factor that depends on both the speaker and the session. Also, $\mathbf{D}$ is a diagonal matrix and the priors of $\mathbf{y}(s)$, $\mathbf{x}_h(s)$ and $\mathbf{z}(s)$ follow the standard Gaussian distribution. Figure 3.36 shows the graphical model of JFA.

Without the second and the third terms, Eq. 3.165 reduces to the classical MAP adaptation [7] where the adapted model is the mode of the posterior distribution of $\mathbf{m}_h(s)$. Also, without the third and the fourth terms, Eq. 3.165 reduces to the eigenvoice MAP [45]. Unlike Eq. 3.55, Eq. 3.165 does not have the residue term ϵ. This is because Eq. 3.165 is to generate the supervector $\mathbf{m}_h(s)$. If the supervector is observed, an residue term will need to be added to Eq. 3.165 to represent the difference between the observed and the generated supervectors.

3.7.2 Posterior Distributions of Latent Factors

To find the posterior distribution of all latent factors jointly, we may stack all the latent factors of speaker s with recording $h = 1, \ldots, H_s$ and pack the loading matrices as follows:

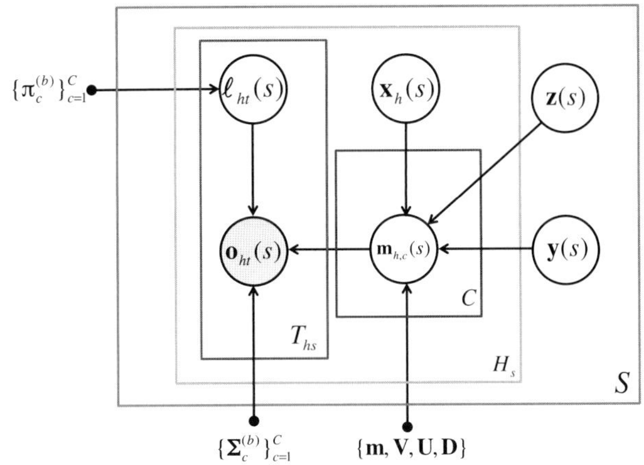

Figure 3.36 Graphical representation of the joint factor analysis model. $\boldsymbol{\ell}_{ht} = [\ell_{h,t,1} \ldots \ell_{h,t,c} \ldots \ell_{h,t,C}]^{\mathsf{T}}$ are the mixture labels in one-hot encoding. The priors of $\mathbf{x}_h(s)$, $\mathbf{z}(s)$, and $\mathbf{y}(s)$ are assumed to be standard Gaussians.

$$\begin{bmatrix} \mathbf{m}_1(s) \\ \vdots \\ \mathbf{m}_{H_s}(s) \end{bmatrix} = \begin{bmatrix} \mathbf{m} \\ \vdots \\ \mathbf{m} \end{bmatrix} + \begin{bmatrix} \mathbf{V} & \mathbf{U} & \mathbf{0} & \mathbf{0} & \cdots & \mathbf{D} \\ \mathbf{V} & \mathbf{0} & \mathbf{U} & \mathbf{0} & \cdots & \mathbf{D} \\ \vdots & \vdots & \ddots & \ddots & \cdots & \vdots \\ \mathbf{V} & \mathbf{0} & \cdots & \mathbf{0} & \mathbf{U} & \mathbf{D} \end{bmatrix} \begin{bmatrix} \mathbf{y}(s) \\ \mathbf{x}_1(s) \\ \vdots \\ \mathbf{x}_{H_s}(s) \\ \mathbf{z}(s) \end{bmatrix}. \tag{3.166}$$

Eq. 3.166 can be written in a compact form:[10]

$$\underline{\mathbf{m}}(s) = \underline{\mathbf{m}} + \underline{\mathbf{V}}(s)\underline{\mathbf{z}}(s), \tag{3.167}$$

which has the same form as Eq. 3.119. This means that we can use Eq. 3.132 to compute the posterior precision matrix of $\underline{\mathbf{z}}(s)$ for speaker s as follows:

$$\underline{\mathbf{L}}(s) = \underline{\mathbf{I}} + \underline{\mathbf{V}}(s)^{\mathsf{T}}\underline{\boldsymbol{\Sigma}}^{-1}(s)\underline{\mathbf{N}}(s)\underline{\mathbf{V}}(s), \tag{3.168}$$

In Eq. 3.168, $\underline{\mathbf{N}}(s)$ is a $CFH_s \times CFH_s$ block-diagonal matrix

$$\underline{\mathbf{N}}(s) = \begin{bmatrix} \mathbf{N}_1(s) & & & & \\ & \ddots & & & \\ & & \mathbf{N}_h(s) & & \\ & & & \ddots & \\ & & & & \mathbf{N}_{H(s)} \end{bmatrix},$$

where $\mathbf{N}_h(s)$ is a $CF \times CF$ diagonal matrix with elements

$$N_{hc}(s) = \sum_{\mathbf{o} \in \mathcal{O}_h(s)} \gamma_c(\mathbf{o}), \tag{3.169}$$

[10] While $\mathbf{V}$, $\mathbf{U}$, $\mathbf{D}$, and $\boldsymbol{\Sigma}$ are independent of s, $\underline{\mathbf{V}}(s)$ and $\underline{\boldsymbol{\Sigma}}(s)$ do, because their size depends on H_s.

where $\mathcal{O}_h(s)$ comprises the acoustic vectors of the hth recording of speaker s. Similarly, $\underline{\boldsymbol{\Sigma}}$ is a $CFH_s \times CFH_s$ block-diagonal matrix whose hth diagonal block is a $CF \times CF$ matrix $\boldsymbol{\Sigma}$. In practice, $\boldsymbol{\Sigma}$ can be substituted by the super-covariance matrix of the UBM $\boldsymbol{\Sigma}^{(b)}$.

Using Eq. 3.131, the posterior mean and the posterior moment of $\underline{\mathbf{z}}(s)$ are given by

$$\left\langle \underline{\mathbf{z}}(s) \middle| \{\mathcal{O}_h(s)\}_{h=1}^{H_s} \right\rangle = \underline{\mathbf{L}}^{-1}(s)\underline{\mathbf{V}}^{\mathsf{T}}(s)\underline{\boldsymbol{\Sigma}}^{-1}(s)\tilde{\underline{\mathbf{f}}}(s), \tag{3.170}$$

and

$$\left\langle \underline{\mathbf{z}}(s)\underline{\mathbf{z}}(s)^{\mathsf{T}} \middle| \{\mathcal{O}_h(s)\}_{h=1}^{H_s} \right\rangle = \underline{\mathbf{L}}^{-1}(s) + \left\langle \underline{\mathbf{z}}(s) \middle| \{\mathcal{O}_h(s)\}_{h=1}^{H_s} \right\rangle \left\langle \underline{\mathbf{z}}(s) \middle| \{\mathcal{O}_h(s)\}_{h=1}^{H_s} \right\rangle^{\mathsf{T}}, \tag{3.171}$$

respectively, where $\tilde{\underline{\mathbf{f}}}(s)$ is a CFH_s-dimensional vector whose cth segment of the hth block is an F-dimension vector

$$\tilde{\mathbf{f}}_{hc}(s) = \sum_{\mathbf{o} \in \mathcal{O}_h(s)} \gamma_c(\mathbf{o})(\mathbf{o} - \mathbf{m}_c), \quad h = 1, \ldots, H_s \text{ and } c = 1, \ldots, C. \tag{3.172}$$

3.7.3 Model Parameter Estimation

The training in JFA amounts to finding the model parameters $\boldsymbol{\Lambda} = \{\mathbf{m}, \mathbf{V}, \mathbf{U}, \mathbf{D}, \boldsymbol{\Sigma}\}$, where $\boldsymbol{\Sigma}$ is a diagonal covariance matrix. For each speaker s, we assume that there are H_s recording sessions. Define the joint latent factor

$$\hat{\mathbf{z}}_h(s) \equiv [\mathbf{y}(s) \;\; \mathbf{x}_h(s) \;\; \mathbf{z}(s)]^{\mathsf{T}}, \quad h = 1, \ldots, H_s. \tag{3.173}$$

Also define the joint loading matrix $\hat{\mathbf{V}} \equiv [\mathbf{V} \;\; \mathbf{U} \;\; \mathbf{D}]$. Then, for recording h of speaker s, we have

$$\mathbf{m}_h(s) = \mathbf{m} + \hat{\mathbf{V}}\hat{\mathbf{z}}_h(s), \quad h = 1, \ldots, H_s.$$

Using Eq. 3.170 and Eq. 3.171, we obtain the posterior mean and posterior covariance of $\hat{\mathbf{z}}_h(s)$ as follows:

$$\langle \hat{\mathbf{z}}_h(s) | \mathcal{O}_h(s) \rangle = \hat{\mathbf{L}}_h^{-1}(s)\hat{\mathbf{V}}^{\mathsf{T}}\boldsymbol{\Sigma}^{-1}\tilde{\mathbf{f}}_h(s) \tag{3.174}$$

and

$$\left\langle \hat{\mathbf{z}}_h(s)\hat{\mathbf{z}}_h(s)^{\mathsf{T}} | \mathcal{O}_h(s) \right\rangle = \hat{\mathbf{L}}_h^{-1}(s) + \langle \hat{\mathbf{z}}_h(s) | \mathcal{O}_h(s) \rangle \langle \hat{\mathbf{z}}_h(s) | \mathcal{O}_h(s) \rangle^{\mathsf{T}}. \tag{3.175}$$

In Eq. 3.174, $\tilde{\mathbf{f}}_h(s)$ is a $CF \times 1$ vector formed by stacking the C components of Eq. 3.172 and the precision matrix $\hat{\mathbf{L}}_h(s)$ is an $(R_v + R_u + CF) \times (R_v + R_u + CF)$ matrix

$$\hat{\mathbf{L}}_h(s) = \hat{\mathbf{I}} + \hat{\mathbf{V}}^{\mathsf{T}}\boldsymbol{\Sigma}^{-1}\mathbf{N}_h(s)\hat{\mathbf{V}}, \tag{3.176}$$

where R_v and R_u are the rank of $\mathbf{U}$ and $\mathbf{V}$, respectively, and $\hat{\mathbf{I}}$ is an $(R_v + R_u + CF) \times (R_v + R_u + CF)$ identity matrix.

Using Eq. 3.80 and Eq. 3.129, the maximum-likelihood solution of $\hat{\mathbf{V}}$ is

$$\hat{\mathbf{V}}_c = \left[\sum_s \sum_{h=1}^{H_s} \tilde{\mathbf{f}}_{hc}(s) \langle \hat{\mathbf{z}}_h(s) | \mathcal{O}_h(s) \rangle^{\mathsf{T}}\right] \left[\sum_s \sum_{h=1}^{H_s} N_{hc}(s) \left\langle \hat{\mathbf{z}}_h(s) \hat{\mathbf{z}}_h(s)^{\mathsf{T}} | \mathcal{O}_h(s) \right\rangle\right]^{-1}. \tag{3.177}$$

Concatenating the latent factors as in Eq. 3.166 and Eq. 3.173 means that these factors are correlated in the posterior, i.e., the precision matrices $\underline{\mathbf{L}}(s)$ in Eq. 3.168 and $\hat{\mathbf{L}}_h(s)$ in Eq. 3.176 are non-diagonal. Because the dimension of $\mathbf{z}(s)$ is very high ($= 61,440$ for 1024 Gaussians and 60-dimensional acoustic vectors), using Eqs. 3.177, 3.174, and 3.175 to compute the model parameters is impractical.

A more practical approach is to estimate the loading matrices separately, as suggested in [87]. Specifically, the loading matrices are estimated using the following steps:

1. Estimate the eigenvoice matrix $\mathbf{V}$ by assuming that $\mathbf{U} = \mathbf{D} = \mathbf{0}$.
2. Given $\mathbf{V}$, estimate the eigenchannel matrix $\mathbf{U}$ by assuming that $\mathbf{D} = \mathbf{0}$.
3. Given $\mathbf{V}$ and $\mathbf{U}$, estimate $\mathbf{D}$.

With this simplification, the formulations in Step 1 are

$$\mathbf{m}_h(s) = \mathbf{m} + \mathbf{V}\mathbf{y}(s), \qquad h = 1, \ldots, H_s \tag{3.178a}$$

$$\mathbf{L}(s) = \mathbf{I} + \mathbf{V}^{\mathsf{T}}\boldsymbol{\Sigma}^{-1}\mathbf{N}(s)\mathbf{V} \tag{3.178b}$$

$$\mathbf{N}(s) = \text{diag}\{N_1(s)\mathbf{I}, \ldots, N_C(s)\mathbf{I}\} \tag{3.178c}$$

$$N_c(s) = \sum_{h=1}^{H_s} \sum_{\mathbf{o} \in \mathcal{O}_h(s)} \gamma_c(\mathbf{o}) \tag{3.178d}$$

$$\tilde{\mathbf{f}}_c(s) = \sum_{h=1}^{H_s} \sum_{\mathbf{o} \in \mathcal{O}_h(s)} \gamma_c(\mathbf{o})(\mathbf{o} - \mathbf{m}_c), \qquad c = 1, \ldots, C \tag{3.178e}$$

$$\left\langle \mathbf{y}(s) \middle| \{\mathcal{O}_h(s)\}_{h=1}^{H_s} \right\rangle = \mathbf{L}^{-1}(s)\mathbf{V}^{\mathsf{T}}\boldsymbol{\Sigma}^{-1}\tilde{\mathbf{f}}(s) \tag{3.178f}$$

$$\left\langle \mathbf{y}(s)\mathbf{y}(s)^{\mathsf{T}} \middle| \{\mathcal{O}_h(s)\}_{h=1}^{H_s} \right\rangle = \mathbf{L}^{-1}(s) + \left\langle \mathbf{y}(s) \middle| \{\mathcal{O}_h(s)\}_{h=1}^{H_s} \right\rangle \left\langle \mathbf{y}(s) \middle| \{\mathcal{O}_h(s)\}_{h=1}^{H_s} \right\rangle^{\mathsf{T}} \tag{3.178g}$$

$$\mathbf{V}_c = \left[\sum_s \tilde{\mathbf{f}}_c(s) \left\langle \mathbf{y}(s) \middle| \{\mathcal{O}_h(s)\}_{h=1}^{H_s} \right\rangle^{\mathsf{T}}\right] \left[\sum_s N_c(s) \left\langle \mathbf{y}(s)\mathbf{y}(s)^{\mathsf{T}} \middle| \{\mathcal{O}_h(s)\}_{h=1}^{H_s} \right\rangle\right]^{-1}. \tag{3.178h}$$

In Step 2, for all of the recordings from speaker s, we account for his/her speaker shift using the estimated $\mathbf{V}$ and speaker factor $\mathbf{y}(s)$, which results in the first-order statistics:

$$\tilde{\mathbf{f}}_{hc}(s) = \sum_{\mathbf{o} \in \mathcal{O}_h(s)} \gamma_c(\mathbf{o})(\mathbf{o} - \mathbf{V}_c\mathbf{y}(s) - \mathbf{m}_c), \quad h = 1, \ldots, H_s \text{ and } c = 1, \ldots, C,$$

where $\mathbf{y}(s)$ can be obtained from the posterior mean in Eq. 3.178f. Because we are modeling channel variability, there is no need to sum over the recordings of speaker s to compute the zeroth order statistic, i.e., $N_c(s)$ in Eq. 3.178d becomes

$$N_{hc}(s) = \sum_{\mathbf{o}\in\mathcal{O}_h(s)} \gamma_c(\mathbf{o}).$$

Also, the posterior precision matrix of $\mathbf{x}_h(s)$ is recording- and speaker-dependent:

$$\mathbf{L}_h(s) = \mathbf{I} + \mathbf{U}^{\mathsf{T}}\boldsymbol{\Sigma}^{-1}\mathbf{N}_h(s)\mathbf{U}.$$

Finally, we compute the posterior mean and posterior moment of $\mathbf{x}_h(s)$ and estimate the eigenchannel matrix $\mathbf{U}$ as follows:

$$\langle \mathbf{x}_h(s)|\mathcal{O}_h(s)\rangle = \mathbf{L}_h^{-1}(s)\mathbf{U}^{\mathsf{T}}\boldsymbol{\Sigma}^{-1}\tilde{\mathbf{f}}_h(s) \tag{3.179a}$$

$$\left\langle \mathbf{x}_h(s)\hat{\mathbf{x}}_h(s)^{\mathsf{T}}|\mathcal{O}_h(s)\right\rangle = \mathbf{L}_h^{-1}(s) + \langle \mathbf{x}_h(s)|\mathcal{O}_h(s)\rangle \langle \mathbf{x}_h(s)|\mathcal{O}_h(s)\rangle^{\mathsf{T}} \tag{3.179b}$$

$$\mathbf{U}_c = \left[\sum_s\sum_{h=1}^{H_s}\tilde{\mathbf{f}}_{hc}(s)\langle \mathbf{x}_h(s)|\mathcal{O}_h(s)\rangle^{\mathsf{T}}\right]\left[\sum_s\sum_{h=1}^{H_s}N_{hc}(s)\left\langle \mathbf{x}_h(s)\mathbf{x}_h(s)^{\mathsf{T}}|\mathcal{O}_h(s)\right\rangle\right]^{-1} \tag{3.179c}$$

In Step 3, for each recording from speaker s, we need to account for the speaker shift and the channel shift using the estimated $\mathbf{V}$ and $\mathbf{U}$, respectively. Using the same principle as in Step 2, we have the following zeroth- and first-order statistics:

$$N_c(s) = \sum_{h=1}^{H_s}\sum_{\mathbf{o}\in\mathcal{O}_h(s)} \gamma_c(\mathbf{o})$$

$$\tilde{\mathbf{f}}_c(s) = \sum_{h=1}^{H_s}\sum_{\mathbf{o}\in\mathcal{O}_h(s)} \gamma_c(\mathbf{o})(\mathbf{o} - \mathbf{V}_c\mathbf{y}(s) - \mathbf{U}_c\mathbf{x}_h(s) - \mathbf{m}_c), \quad c = 1,\ldots,C.$$

The posterior precision matrix becomes

$$\mathbf{L}(s) = \mathbf{I} + \mathbf{D}^2\boldsymbol{\Sigma}^{-1}\mathbf{N}(s),$$

where $\mathbf{N}(s) = \text{diag}\{N_1(s)\mathbf{I},\ldots,N_C(s)\mathbf{I}\}$. Then, the diagonal elements of $\mathbf{D}$ can be estimated as follows:

$$\text{diag}\{\mathbf{D}_c\} = \frac{\text{diag}\left\{\sum_s \tilde{\mathbf{f}}_c(s)\left\langle \mathbf{z}(s)|\{\mathcal{O}_h(s)\}_{h=1}^{H_s}\right\rangle^{\mathsf{T}}\right\}}{\text{diag}\left\{\sum_s N_c(s)\left\langle \mathbf{z}(s)\mathbf{z}(s)^{\mathsf{T}}|\{\mathcal{O}_h(s)\}_{h=1}^{H_s}\right\rangle\right\}}, \tag{3.180}$$

where diag{**A**} means converting the diagonal elements of **A** into a vector and the division is applied element-wise, and

$$\left\langle \mathbf{z}(s) \middle| \{\mathcal{O}_h(s)\}_{h=1}^{H_s} \right\rangle = \mathbf{L}^{-1}(s)\mathbf{D}^{\mathsf{T}}\boldsymbol{\Sigma}^{-1}\tilde{\mathbf{f}}(s) \tag{3.181a}$$

$$\left\langle \mathbf{z}(s)\mathbf{z}(s)^{\mathsf{T}} \middle| \{\mathcal{O}_h(s)\}_{h=1}^{H_s} \right\rangle = \mathbf{L}^{-1}(s) + \left\langle \mathbf{z}(s) \middle| \{\mathcal{O}_h(s)\}_{h=1}^{H_s} \right\rangle \left\langle \mathbf{z}(s) \middle| \{\mathcal{O}_h(s)\}_{h=1}^{H_s} \right\rangle^{\mathsf{T}}. \tag{3.181b}$$

3.7.4 JFA Scoring

Scoring in JFA involves computing the likelihoods of acoustic vectors with respect to the target speaker and the UBM. There are several approaches to computing these likelihoods [88]. Here, we focus on the one based on the point estimates of the channel factors.

Point Estimate of Channel Factors

To express the likelihood of acoustic vectors, we need some notations. Let's define the channel-independent supervector of target-speaker s as

$$\mathbf{m}(s) = \mathbf{m} + \mathbf{V}\mathbf{y}(s) + \mathbf{D}\mathbf{z}(s), \tag{3.182}$$

where $\mathbf{m}$, $\mathbf{V}$, $\mathbf{D}$, $\mathbf{y}(s)$, and $\mathbf{z}(s)$ have the same meaning as those in Eq. 3.165. Note that while we use the same symbol s in both Eq. 3.182 and Eq. 3.165, the target speaker does not necessarily to be one of the training speakers in Eq. 3.165. Enrollment in JFA amounts to estimating $\mathbf{m}(s)$ for each target speaker using his/her enrollment utterances. The point estimates of $\mathbf{y}(s)$ and $\mathbf{z}(s)$ can be obtained from their posterior means.

Denote $\mathcal{O}_{\text{tst}}$ as the set of acoustic vectors from a test utterance. We compute the posterior mean of the channel factor given $\mathcal{O}$ and $\mathbf{m}(s)$:

$$\mathbf{x}_s \equiv \langle \mathbf{x} | \mathcal{O}_{\text{tst}}, \mathbf{m}(s) \rangle = (\mathbf{I} + \mathbf{U}^{\mathsf{T}}\boldsymbol{\Sigma}^{-1}\mathbf{N}\mathbf{U})^{-1}\mathbf{U}^{\mathsf{T}}\boldsymbol{\Sigma}^{-1}\tilde{\mathbf{f}}(s),$$

where $\mathbf{N}$ comprises the zeroth order statistics along its diagonal and $\tilde{\mathbf{f}}(s)$ contains the first-order statistics, i.e., the cth component of $\tilde{\mathbf{f}}(s)$ is

$$\tilde{\mathbf{f}}_c(s) = \sum_{\mathbf{o} \in \mathcal{O}_{\text{tst}}} \gamma_c(\mathbf{o})(\mathbf{o} - \mathbf{m}_c(s)) = \mathbf{f}_c - \mathbf{N}_c\mathbf{m}_c(s).$$

Similarly, we also compute the posterior mean of the channel factor based on the supervector of the UBM:

$$\mathbf{x}_{\text{ubm}} \equiv \langle \mathbf{x} | \mathcal{O}_{\text{tst}}, \mathbf{m} \rangle = (\mathbf{I} + \mathbf{U}^{\mathsf{T}}\boldsymbol{\Sigma}^{-1}\mathbf{N}\mathbf{U})^{-1}\mathbf{U}^{\mathsf{T}}\boldsymbol{\Sigma}^{-1}\tilde{\mathbf{f}}(\text{ubm}),$$

where

$$\tilde{\mathbf{f}}_c(\text{ubm}) = \sum_{\mathbf{o} \in \mathcal{O}_{\text{tst}}} \gamma_c(\mathbf{o})(\mathbf{o} - \mathbf{m}_c) = \mathbf{f}_c - \mathbf{N}_c\mathbf{m}_c.$$

Using Theorem 1 of [86], the log-likelihood of $\mathcal{O}_{\text{tst}}$ with respect to the target speaker s is

$$\log p(\mathcal{O}_{\text{tst}}|\mathbf{m}(s), \mathbf{x}_s) = \sum_{c=1}^{C} N_c \log \frac{1}{(2\pi)^{\frac{F}{2}}|\boldsymbol{\Sigma}_c|^{\frac{1}{2}}} - \frac{1}{2}\text{Tr}\{\boldsymbol{\Sigma}^{-1}\mathbf{S}\}$$
$$+ \mathbf{r}(s)^{\mathsf{T}}\mathbf{G}^{\mathsf{T}}\boldsymbol{\Sigma}^{-1}\tilde{\mathbf{f}} - \frac{1}{2}\mathbf{r}(s)^{\mathsf{T}}\mathbf{G}^{\mathsf{T}}\mathbf{N}\boldsymbol{\Sigma}^{-1}\mathbf{G}\mathbf{r}(s) \tag{3.183}$$

where

$$\mathbf{r}(s) \equiv \begin{bmatrix} \mathbf{y}(s) \\ \mathbf{z}(s) \\ \mathbf{x}_s \end{bmatrix}, \qquad \mathbf{G} \equiv [\mathbf{V} \ \ \mathbf{D} \ \ \mathbf{U}]$$

and the cth component of $\mathbf{S}$ is

$$\mathbf{S}_c = \text{diag}\left\{ \sum_{\mathbf{o}\in\mathcal{O}_{\text{tst}}} (\mathbf{o} - \mathbf{m}_c)(\mathbf{o} - \mathbf{m}_c)^{\mathsf{T}} \right\}.$$

Note that $\mathbf{y}(s)$ and $\mathbf{z}(s)$ – which are approximated by their posterior means – have been computed during enrollment, whereas $\mathbf{x}_s$ is computed during verification. Similarly, the the log-likelihood of $\mathcal{O}_{\text{tst}}$ with respect to the UBM is

$$\log p(\mathcal{O}|\mathbf{m}, \mathbf{x}_{\text{ubm}}) = \sum_{c=1}^{C} N_c \log \frac{1}{(2\pi)^{\frac{F}{2}}|\boldsymbol{\Sigma}_c|^{\frac{1}{2}}} - \frac{1}{2}\text{Tr}\{\boldsymbol{\Sigma}^{-1}\mathbf{S}\}$$
$$+ \mathbf{r}(\text{ubm})^{\mathsf{T}}\mathbf{G}^{\mathsf{T}}\boldsymbol{\Sigma}^{-1}\tilde{\mathbf{f}}(\text{ubm}) - \frac{1}{2}\mathbf{r}(\text{ubm})^{\mathsf{T}}\mathbf{G}^{\mathsf{T}}\mathbf{N}\boldsymbol{\Sigma}^{-1}\mathbf{G}\mathbf{r}(\text{ubm}) \tag{3.184}$$

where

$$\mathbf{r}(\text{ubm}) = \begin{bmatrix} \mathbf{0} \\ \mathbf{0} \\ \mathbf{x}_{\text{ubm}} \end{bmatrix}.$$

Using Eq. 3.183 and Eq. 3.184, the log-likelihood ratio (LLR) score for speaker verification can be computed as follows:

$$\begin{aligned} S_{\text{LLR}}(\mathcal{O}_{\text{tst}}|\mathbf{m}(s)) &= \log p(\mathcal{O}_{\text{tst}}|\mathbf{m}(s), \mathbf{x}_s) - \log p(\mathcal{O}|\mathbf{m}, \mathbf{x}_{\text{ubm}}) \\ &= \mathbf{r}(s)^{\mathsf{T}}\mathbf{G}^{\mathsf{T}}\boldsymbol{\Sigma}^{-1}\tilde{\mathbf{f}}(s) - \frac{1}{2}\mathbf{r}(s)^{\mathsf{T}}\mathbf{G}^{\mathsf{T}}\mathbf{N}\boldsymbol{\Sigma}^{-1}\mathbf{G}\mathbf{r}(s) \\ &\quad - \mathbf{r}(\text{ubm})^{\mathsf{T}}\mathbf{G}^{\mathsf{T}}\boldsymbol{\Sigma}^{-1}\tilde{\mathbf{f}}(\text{ubm}) + \frac{1}{2}\mathbf{r}(\text{ubm})^{\mathsf{T}}\mathbf{G}^{\mathsf{T}}\mathbf{N}\boldsymbol{\Sigma}^{-1}\mathbf{G}\mathbf{r}(\text{ubm}). \end{aligned} \tag{3.185}$$

Linear Approximation

Define $\mathbf{q} \equiv \mathbf{G}\mathbf{r}(s)$ and consider the first-order Taylor expansion of

$$f(\mathbf{q}) = \mathbf{q}^{\mathsf{T}}\boldsymbol{\Sigma}^{-1}\tilde{\mathbf{f}}(s) - \frac{1}{2}\mathbf{q}^{\mathsf{T}}\mathbf{N}\boldsymbol{\Sigma}^{-1}\mathbf{q}.$$

We have $f(\mathbf{q}) \approx \mathbf{q}^{\mathsf{T}}\boldsymbol{\Sigma}^{-1}\tilde{\mathbf{f}}(s)$ and therefore only the first and the third term in Eq. 3.185 remain:

$$\begin{aligned} S_{\text{LLR}}(\mathcal{O}_{\text{tst}}|\mathbf{m}(s)) &\approx \mathbf{r}(s)^{\mathsf{T}}\mathbf{G}^{\mathsf{T}}\boldsymbol{\Sigma}^{-1}\tilde{\mathbf{f}}(s) - \mathbf{r}(\text{ubm})^{\mathsf{T}}\mathbf{G}^{\mathsf{T}}\boldsymbol{\Sigma}^{-1}\tilde{\mathbf{f}}(\text{ubm}) \\ &= (\mathbf{V}\mathbf{y}(s) + \mathbf{D}\mathbf{z}(s) + \mathbf{U}\mathbf{x}_s)^{\mathsf{T}}\boldsymbol{\Sigma}^{-1}\tilde{\mathbf{f}}(s) - \mathbf{x}_{\text{ubm}}^{\mathsf{T}}\mathbf{U}^{\mathsf{T}}\boldsymbol{\Sigma}^{-1}\tilde{\mathbf{f}}(\text{ubm}). \end{aligned} \tag{3.186}$$

If we further assume that the first-order statistics are obtained from the target speaker $\mathbf{m}(s)$ for both the target speaker and the UBM and that $\mathbf{x}_s = \mathbf{x}_{\text{ubm}}$, we have

$$\begin{aligned} S_{\text{LLR}}(\mathcal{O}_{\text{tst}}|\mathbf{m}(s)) &\approx (\mathbf{V}\mathbf{y}(s) + \mathbf{D}\mathbf{z}(s) + \mathbf{U}\mathbf{x}_{\text{ubm}})^{\mathsf{T}}\boldsymbol{\Sigma}^{-1}\tilde{\mathbf{f}}(s) - (\mathbf{U}\mathbf{x}_{\text{ubm}})^{\mathsf{T}}\boldsymbol{\Sigma}^{-1}\tilde{\mathbf{f}}(s) \\ &= (\mathbf{V}\mathbf{y}(s) + \mathbf{D}\mathbf{z}(s))^{\mathsf{T}}\boldsymbol{\Sigma}^{-1}\tilde{\mathbf{f}}(s) \\ &= (\mathbf{V}\mathbf{y}(s) + \mathbf{D}\mathbf{z}(s))^{\mathsf{T}}\boldsymbol{\Sigma}^{-1}(\mathbf{f} - \mathbf{N}\mathbf{m}(s)) \\ &= (\mathbf{V}\mathbf{y}(s) + \mathbf{D}\mathbf{z}(s))^{\mathsf{T}}\boldsymbol{\Sigma}^{-1}(\mathbf{f} - \mathbf{N}\mathbf{m} - \mathbf{N}\mathbf{U}\mathbf{x}_{\text{ubm}}). \end{aligned} \tag{3.187}$$

3.7.5 From JFA to I-Vectors

One problem of JFA is its insufficiency in distinguishing between speaker and channel information. It was found in [16] that the channel factors also contain speaker information, causing imprecise modeling if the speaker subspace and channel subspace are modeled separately by two loading matrices. This finding motivates the development of i-vectors, which combines the two subspaces into one subspace called total variability space. Modeling in i-vectors involves two steps: (1) use low-dimensional vectors (called i-vectors) that comprise both speaker and channel information to represent utterances (see Section 3.6.3); and (2) model the channel variabilities of the i-vectors during scoring (see Sections 3.4.3 and 3.6.6).

Part II

Advanced Studies

4 Deep Learning Models

This chapter addresses the background knowledge of deep learning theories and algorithms that is used as a foundation for developing speaker recognition solutions to meet different issues in implementation of real-world systems. We will start from an unsupervised learning machine called the restricted Boltzmann machine (RBM), which serves as a building block for deep belief networks and deep Boltzmann machines, which will be introduced in Section 4.3. Discriminative fine-tuning will be applied to build a supervised model with good performance for classification. RBM will be described in Section 4.1. Then, the algorithm for constructing and training a supervised deep neural network (DNN) will be addressed in Section 4.2. However, training a reliable DNN is challenging. Section 4.4 mentions the algorithm of stacking autoencoder, which is performed to build a deep model based on a two-layer structural module in a layer-by-layer fashion. Furthermore, in Section 4.3, the deep belief network is introduced to carry out a procedure of estimating a reliable deep model based on the building block of RBM. Nevertheless, DNN is not only feasible to build a supervised model for classification or regression problems in speaker recognition but also applicable to construct an unsupervised model that serves as generative model for data generation to deal with data sparseness problem. From this perspective, we introduce two deep learning paradigms. One is variational auto-encoder as mentioned in Section 4.5 while the other is the generative adversarial network as provided in Section 4.6. Such general neural network models can be merged in solving different issues in a speaker recognition system. At last, this chapter will end in Section 4.7, which addresses the fundamentals of transfer learning and the solutions based on deep learning. The deep transfer learning is developed to deal with domain mismatch problem between training and test data.

4.1 Restricted Boltzmann Machine

Restricted Boltzmann machine (RBM) [89, 90] plays a crucial role in building deep belief networks that are seen as the foundations in deep learning. Basically, RBM is seen as a bipartite graph that can be represented by an undirected graphical model or equivalently a bidirectional graphical model in a two-layer structure. One is the visible layer and the other is the hidden layer. Each connection must connect visible units $\mathbf{v} = \{v_i\}_{i=1}^{V}$ with hidden units $\mathbf{h} = \{h_j\}_{j=1}^{H}$ using weight parameters $\mathbf{W}$ where bias parameters $\{\mathbf{a}, \mathbf{b}\}$ are usually assumed to be zero for compact notation. Hidden variable

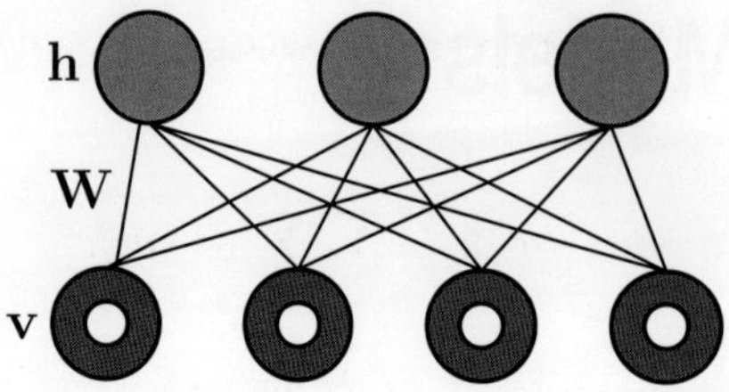

Figure 4.1 A two-layer structure of the restricted Boltzmann machine.

h extracts high-level information from observations **v**. Unlike the Boltzmann machines, the name "restricted" is added because there are no connections between two neurons in the same layer, namely visible-visible or hidden-hidden connections do not exist. Sparsity is controlled in RBM. RBM structure is depicted in Figure 4.1. There is no arrows in between-layer connections because these connections are bidirectional. RBM meets the property that visible units are conditionally independent given hidden units and vice versa. Or equivalently, the following probabilities are met

$$p(\mathbf{v}|\mathbf{h}, \boldsymbol{\theta}) = \prod_i p(v_i|\mathbf{h}, \boldsymbol{\theta}) \tag{4.1}$$

$$p(\mathbf{h}|\mathbf{v}, \boldsymbol{\theta}) = \prod_j p(h_j|\mathbf{v}, \boldsymbol{\theta}) \tag{4.2}$$

where model parameters $\boldsymbol{\theta} = \{\mathbf{a}, \mathbf{b}, \mathbf{W}\}$ are used.

4.1.1 Distribution Functions

In general, there are two different types of RBM that we usually use in DNN pretraining. The difference between two types depends on the input data values. First, the Bernoulli-Bernoulli RBM considers that both visible and hidden units are binary, i.e., either 0 or 1. Bernoulli distribution is used to represent the binary units. The conditional distribution in Bernoulli-Bernoulli RBM is expressed by

$$p(v_i = 1|\mathbf{h}, \boldsymbol{\theta}) = \sigma\left(a_i + \sum_j w_{ij} h_j\right) \tag{4.3}$$

$$p(h_j = 1|\mathbf{v}, \boldsymbol{\theta}) = \sigma\left(b_j + \sum_i w_{ij} v_i\right) \tag{4.4}$$

where w_{ij} denotes the weight between visible units v_i and between hidden units h_j, a_i denotes the visible unit bias, b_j denotes the hidden unit bias, $\boldsymbol{\theta} = \{a_i, b_j, w_{ij}\}$ denotes the model parameters and

$$\sigma(x) = \frac{1}{1 + e^{-x}} \tag{4.5}$$

denotes a logistic sigmoid function. The energy [91] of such a joint configuration of visible units and hidden units is given by

$$E(\mathbf{v},\mathbf{h}|\boldsymbol{\theta}) = -\sum_{i=1}^{V} a_i v_i - \sum_{j=1}^{H} b_j h_j - \sum_{i=1}^{V}\sum_{j=1}^{H} w_{ij} v_i h_j, \tag{4.6}$$

which is accumulated from V units v_i in visible layer, H units h_j in hidden layer and $V \cdot H$ mutual connections between two layers by using the corresponding parameters a_i, b_j, and w_{ij}.

On the other hand, the Gaussian-Bernoulli RBM is developed for the case that hidden units are binary while input units follow a linear model with Gaussian noise. This RBM deals with the real-valued data. Speech features are treated as real-valued inputs to neural networks. Namely, for Gaussian-Bernoulli RBM, the conditional distribution is given by

$$p(h_j = 1|\mathbf{v},\boldsymbol{\theta}) = \sigma\left(b_j + \sum_i \frac{v_i}{\sigma_i} w_{ij}\right) \tag{4.7}$$

$$p(v_i = r|\mathbf{h},\boldsymbol{\theta}) = \mathcal{N}\left(a_i + \sigma_i \sum_j w_{ij} h_j, \sigma_i^2\right) \tag{4.8}$$

where r is a real value and σ_i^2 denotes the the variance of Gaussian noise for visible unit v_i. We usually set $\sigma_i^2 = 1$ to simplify the implementation. Using Gaussian-Bernoulli RBM, the energy of a joint configuration is calculated by

$$E(\mathbf{v},\mathbf{h}|\boldsymbol{\theta}) = \sum_{i=1}^{V} \frac{(v_i - a_i)^2}{2\sigma_i^2} - \sum_{j=1}^{H} b_j h_j - \sum_{i=1}^{V}\sum_{j=1}^{H} \frac{v_i}{\sigma_i} w_{ij} h_j. \tag{4.9}$$

Given the energy functions $E(\mathbf{v},\mathbf{h}|\boldsymbol{\theta})$ in Eqs. 4.6 and 4.9, the energy-based distribution of visible units $\mathbf{v}$ and latent units $\mathbf{h}$ is defined in the form

$$p(\mathbf{v},\mathbf{h}|\boldsymbol{\theta}) = \frac{e^{-E(\mathbf{v},\mathbf{h}|\boldsymbol{\theta})}}{\sum_{\mathbf{v},\mathbf{h}} e^{-E(\mathbf{v},\mathbf{h}|\boldsymbol{\theta})}}, \tag{4.10}$$

where the normalization constant or the *partition* function is defined as

$$Z(\boldsymbol{\theta}) = \sum_{\mathbf{v},\mathbf{h}} e^{-E(\mathbf{v},\mathbf{h}|\boldsymbol{\theta})}. \tag{4.11}$$

This joint distribution of $\mathbf{v}$ and $\mathbf{h}$ is seen as a realization of exponential family.

The marginal likelihood and conditional likelihood are obtained by

$$p(\mathbf{v}|\boldsymbol{\theta}) = \frac{1}{Z(\boldsymbol{\theta})} \sum_{\mathbf{h}} e^{-E(\mathbf{v},\mathbf{h}|\boldsymbol{\theta})} \tag{4.12}$$

$$p(\mathbf{h}|\mathbf{v},\boldsymbol{\theta}) = \frac{p(\mathbf{v},\mathbf{h}|\boldsymbol{\theta})}{p(\mathbf{v}|\boldsymbol{\theta})} = \frac{e^{-E(\mathbf{v},\mathbf{h}|\boldsymbol{\theta})}}{\sum_{\mathbf{h}} e^{-E(\mathbf{v},\mathbf{h}|\boldsymbol{\theta})}}. \tag{4.13}$$

In general, $p(\mathbf{h}|\mathbf{v},\boldsymbol{\theta})$ is an easy and exact calculation for an RBM and $p(\mathbf{v}|\boldsymbol{\theta})$ is functioned as an empirical likelihood. An RBM is seen as a generative model for observations $\mathbf{v}$, which is driven by the marginal distribution $p(\mathbf{v}|\boldsymbol{\theta})$. Correlations between nodes in $\mathbf{v}$ are present in the marginal distribution $p(\mathbf{v}|\boldsymbol{\theta})$ as shown in Eq. 4.6 or Eq. 4.9.

4.1.2 Learning Algorithm

According to the maximum likelihood (ML) principle, we maximize the logarithm of the likelihood function of visible data $\mathbf{v}$ to estimate the ML parameters by

$$\boldsymbol{\theta}_{\text{ML}} = \underset{\boldsymbol{\theta}}{\operatorname{argmax}} \log p(\mathbf{v}|\boldsymbol{\theta}). \tag{4.14}$$

To solve this estimation problem, we differentiate the log-likelihood function with respect to individual parameters in $\boldsymbol{\theta}$ and derive the updating formulas. The derivative is derived as follows:

$$\begin{aligned}
\frac{\partial \log p(\mathbf{v}|\boldsymbol{\theta})}{\partial \boldsymbol{\theta}} &= \frac{\partial \log\left(\frac{\sum_{\mathbf{h}} e^{-E(\mathbf{v},\mathbf{h})}}{Z}\right)}{\partial \boldsymbol{\theta}} \\
&= \frac{Z}{\sum_{\mathbf{h}} e^{-E(\mathbf{v},\mathbf{h})}} \frac{1}{Z^2}\left(\frac{Z\partial\left(\sum_{\mathbf{h}} Ze^{-E(\mathbf{v},\mathbf{h})}\right)}{\partial \boldsymbol{\theta}} - \sum_{\mathbf{h}} e^{-E(\mathbf{v},\mathbf{h})}\frac{\partial Z}{\partial \boldsymbol{\theta}}\right) \\
&= \frac{1}{\sum_{\mathbf{h}} e^{-E(\mathbf{v},\mathbf{h})}} \frac{\partial\left(\sum_{\mathbf{h}} e^{-E(\mathbf{v},\mathbf{h})}\right)}{\partial \boldsymbol{\theta}} - \frac{1}{Z}\frac{\partial Z}{\partial \boldsymbol{\theta}} \\
&= \frac{-\sum_{\mathbf{h}} e^{-E(\mathbf{v},\mathbf{h})}\frac{\partial E(\mathbf{v},\mathbf{h})}{\partial \boldsymbol{\theta}}}{\sum_{\mathbf{h}} e^{-E(\mathbf{v},\mathbf{h})}} - \frac{-\sum_{\mathbf{v},\mathbf{h}} e^{-E(\mathbf{v},\mathbf{h})}\frac{\partial E(\mathbf{v},\mathbf{h})}{\partial \boldsymbol{\theta}}}{\sum_{\mathbf{v},\mathbf{h}} e^{-E(\mathbf{v},\mathbf{h})}} \\
&= \sum_{\mathbf{h}} p(\mathbf{h}|\mathbf{v},\boldsymbol{\theta})\left(-\frac{\partial E(\mathbf{v},\mathbf{h})}{\partial \boldsymbol{\theta}}\right) - \sum_{\mathbf{v},\mathbf{h}} p(\mathbf{v},\mathbf{h}|\boldsymbol{\theta})\left(-\frac{\partial E(\mathbf{v},\mathbf{h})}{\partial \boldsymbol{\theta}}\right) \\
&= \mathbb{E}_{\mathbf{h}\sim p(\mathbf{h}|\mathbf{v},\boldsymbol{\theta})}\left\{-\frac{\partial E(\mathbf{v},\mathbf{h})}{\partial \boldsymbol{\theta}}\middle|\mathbf{v}\right\} - \mathbb{E}_{\mathbf{v},\mathbf{h}\sim p(\mathbf{v},\mathbf{h}|\boldsymbol{\theta})}\left\{-\frac{\partial E(\mathbf{v},\mathbf{h})}{\partial \boldsymbol{\theta}}\right\}.
\end{aligned}$$

As a result, by referring to Eq. 4.6, the updating of the three parameters w_{ij}, a_i, and b_j in Bernoulli-Bernoulli RBMs can be formulated as

$$\frac{\partial \log p(\mathbf{v}|\boldsymbol{\theta})}{\partial w_{ij}} = \langle v_i h_j \rangle_0 - \langle v_i h_j \rangle_\infty \tag{4.15}$$

$$\frac{\partial \log p(\mathbf{v}|\boldsymbol{\theta})}{\partial a_i} = \langle v_i \rangle_0 - \langle v_i \rangle_\infty \tag{4.16}$$

$$\frac{\partial \log p(\mathbf{v}|\boldsymbol{\theta})}{\partial b_j} = \langle h_j \rangle_0 - \langle h_j \rangle_\infty \tag{4.17}$$

where the angle brackets $\langle\rangle$ denote the expectations under the distribution either $p(\mathbf{h}|\mathbf{v},\boldsymbol{\theta})$ in the first term or $p(\mathbf{v},\mathbf{h}|\boldsymbol{\theta})$ in the second term. The subscript 0 means the step number from the original data and ∞ means the step number after doing the Gibbs sampling steps. The difference between two terms in Eqs. 4.15, 4.16, and 4.17 is seen as a contrastive divergence (CD) between the expectations of $v_i h_j$, v_i or h_j with initialization from visible data $\mathbf{v}$ and with calculation after infinite Gibbs sampling. A Gibbs chain is run continuously with initialization from the data.

The contrastive divergence algorithm [92, 93] is performed to train an RBM. Because the parameter updating in exact implementation is time-consuming, the so-called k-step contrastive divergence (CD-k) [92, 94] is applied to speed up the implementation procedure. In practice, the case $k = 1$ is used to carry out one step of Gibbs sampling or Markov-chain Monte Carlo sampling (CD-1), as addressed in Section 2.4, instead of $k = \infty$. More specifically, the blocked-Gibbs sampling is performed. This trick practically works well. Correspondingly, the formulas for parameter updating are modified as

$$\begin{aligned}\frac{\partial \log p(\mathbf{v}|\boldsymbol{\theta})}{\partial w_{ij}} &= \langle v_i h_j\rangle_0 - \langle v_i h_j\rangle_1 \\ &= \mathbb{E}_{p_{\text{data}}}[v_i h_j] - \mathbb{E}_{p_{\text{model}}}[v_i h_j] \triangleq \Delta w_{ij}\end{aligned} \tag{4.18}$$

$$\frac{\partial \log p(\mathbf{v}|\boldsymbol{\theta})}{\partial a_i} = \langle v_i\rangle_0 - \langle v_i\rangle_1 \tag{4.19}$$

$$\frac{\partial \log p(\mathbf{v}|\boldsymbol{\theta})}{\partial b_j} = \langle h_j\rangle_0 - \langle h_j\rangle_1, \tag{4.20}$$

where the subscript 1 means that one step of Gibbs sampling is performed. Figure 4.2 shows the training procedure of RBM based on CD-1 and CD-k. The arrows between visible layer and hidden layers are used to indicate the sampling steps and directions. Calculation of $\langle v_i h_j\rangle_\infty$ using each individual neuron i in visible layer and neuron j in hidden layer is performed step by step by starting from $\langle v_i h_j\rangle_0$. Step k is seen as a kind of state k in a Markov chain. A kind of Markov switching is realized in the CD algorithm.

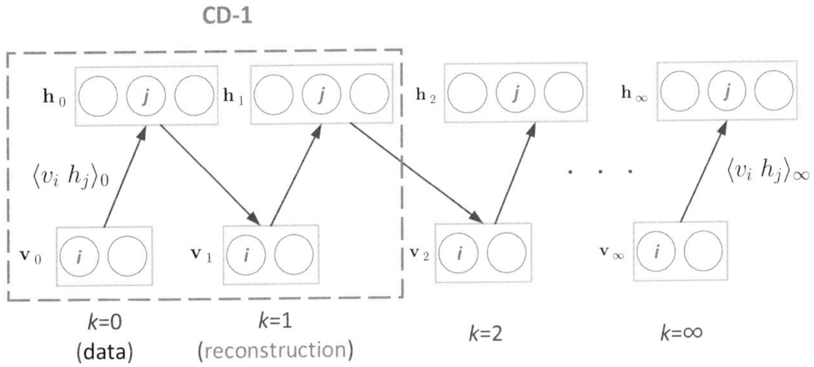

Figure 4.2 Training procedure for a restricted Boltzmann machine using the contrastive divergence algorithm.

After finding the partial derivatives of log likelihood with respect to three individual parameters $\{w_{ij}, a_i, b_j\}$ using the CD algorithm, the stochastic gradient descent (SGD) algorithm with momentum is implemented to construct the RBM model as shown in Algorithm 1. When running the SGD algorithm, the training data are split into minibatches. SGD algorithm proceeds with one minibatch at a time and runs a number of epochs. Meaningfully, the expectation from data $\langle v_i h_j \rangle_0$ using data distribution p_{data} is also denoted by $\langle v_i h_j \rangle_{\text{data}}$ or $\mathbb{E}_{p_{\text{data}}}[v_i h_j]$. The other expectations $\langle v_i h_j \rangle_1$ and $\langle v_i h_j \rangle_\infty$ are seen as those from model, i.e. $\langle v_i h_j \rangle_{\text{model}}$ or $\mathbb{E}_{p_{\text{model}}}[v_i h_j]$ using model distribution p_{model}. Calculation from model in step $k = 1$ is also viewed as a kind of reconstruction from data in step $k = 0$.

Algorithm 1 Training procedure for the restricted Boltzmann machine

$\Delta w_{ij} = \langle v_i h_j \rangle_{\text{data}} - \langle v_i h_j \rangle_{\text{model}}$
Use CD-1 to approximate $\langle v_i h_j \rangle_{\text{model}}$
  Initialize $\mathbf{v}_0$ with observation data
  Sample $\mathbf{h}_0 \sim p(\mathbf{h}|\mathbf{v}_0, \boldsymbol{\theta})$
  Sample $\mathbf{v}_1 \sim p(\mathbf{v}|\mathbf{h}_0, \boldsymbol{\theta})$
  Sample $\mathbf{h}_1 \sim p(\mathbf{h}|\mathbf{v}_1, \boldsymbol{\theta})$
  Call $(\mathbf{v}_1, \mathbf{h}_1)$ a sample from the model
$(\mathbf{v}_\infty, \mathbf{h}_\infty)$ is a true sample from the model
$(\mathbf{v}_1, \mathbf{h}_1)$ is a very rough estimate but worked

An RBM acts as an unsupervised two-layer learning machine that is a shallow neural network with only one hidden layer. Nevertheless, a number of RBMs can act as building blocks to construct a deep belief network that will be addressed in Section 4.3. Deep model is equipped with strong modeling capability by increasing the depth of hidden layers so as to represent the high-level abstract meaning of data. In what follows, we address the principle of deep neural networks and their training procedure.

4.2 Deep Neural Networks

A deep neural network (DNN) has a hierarchy or deep architecture that is composed of a number of hidden layers. This architecture aims to learn an abstract representation or high-level model from observation data, which is used to find the highly nonlinear mapping between input data and their target values. DNNs have been widely developed to carry out different classification and regression applications, e.g., speaker recognition [95], image classification [96], speech recognition [97–99], natural language processing [100], and music or audio information retrieval, etc.

4.2.1 Structural Data Representation

Real-world applications involve different kinds of technical data that are inherent with structural features. It is crucial to capture the hierarchical information and conduct representation learning when building the information systems in presence of various observation data. System performance is highly affected by the modeling capability of a learning machine. To assure the learning capability, it is essential to increase the level of abstraction extracted from observation data and carry out a hierarchy of representations based on a deep learning machine. For example, image data can be learned via a representation process with multiple stages:

$$\text{pixel} \rightarrow \text{edge} \rightarrow \text{texton} \rightarrow \text{motif} \rightarrow \text{part} \rightarrow \text{object}.$$

Each stage is run as a kind of trainable feature transform that captures the hierarchical features corresponding to different levels of observed evidence including pixel, edge, texton, motif, part, and object. Such a trainable feature hierarchy can be also extended to represent other technical data, e.g., text and speech, in a hierarchical style of

$$\text{character} \rightarrow \text{word} \rightarrow \text{word group} \rightarrow \text{clause} \rightarrow \text{sentence} \rightarrow \text{story}$$

and

$$\text{sample} \rightarrow \text{spectral band} \rightarrow \text{sound} \rightarrow \text{phoneme} \rightarrow \text{phone} \rightarrow \text{word}$$

based on the hierarchies from character to story and from time sample to word sequence, respectively. A deep structural model is required to perform delicate representation so as to achieve desirable system performance in heterogeneous environments.

Figure 4.3 represents a bottom-up fully connected neural network that can characterize the multilevel abstraction in the hidden layers given the observation data in the bottom layer. The calculation starts from the input signals and then propagates with feedforward layers toward the abstraction outputs for regression or classification. In layer-wise calculation, we sum up the inputs in low-level layer and produce the activation outputs in high-level layer. The regression or classification error is determined at the top, and it is minimized and passed backward from the top to the bottom layer. Deep neural networks are structural models with three kinds of realizations as shown in Figure 4.4.

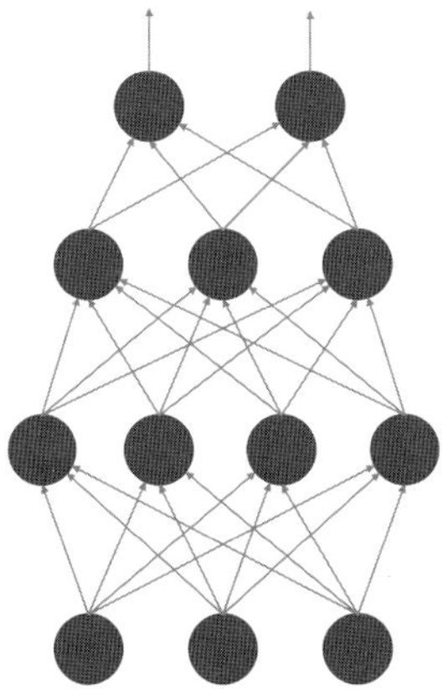

Figure 4.3 Bottom-up layer-wise representation for structural data.

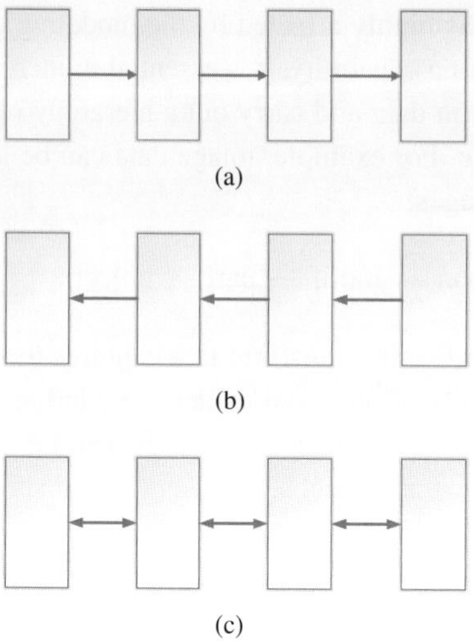

Figure 4.4 Three types of structural model (a) feedforward model, (b) feedback model, and (c) bidirectional model.

The multilayer neural networks and the convolutional neural networks correspond to the feedforward model where the inputs are summed, activated, and feedforwarded toward outputs. Without loss of generality, the bottom-up network has an alternative graphical representation using the left-to-right model topology. In addition to feedforward model, the layer-wise network can also be realized as a feedback model as well as a bidirectional model according to the direction of data flow in the model. Deconvolutional neural network [101] and stacked sparse coding [102] are seen as the feedback models. Deep Boltzmann machine [103], stacked autoencoder [104], and bidirectional recurrent neural network [105] are known as the bidirectional models. This book mainly addresses the feedforward model for deep learning based on multilayer perceptrons that will be introduced next.

4.2.2 Multilayer Perceptron

Figure 4.5 depicts a multilayer perceptron (MLP) [106] where $\mathbf{x}_t$ denotes the input with D dimensions, $\mathbf{z}_t$ denotes the hidden feature with M dimensions, $\mathbf{y}_t$ denotes the output with K dimensions, and t denotes the time index. A hierarchy of hidden layers is constructed by stacking a number of hidden layers. The mapping or relation between an input vector $\mathbf{x} = \{x_{td}\}$ and an output vector $\mathbf{y} = \{y_{tk}\}$ is formulated as

$$\begin{aligned} y_{tk} &= y_k(\mathbf{x}_t, \mathbf{w}) \\ &= f\left(\sum_{m=0}^{M} w_{mk}^{(2)} f\left(\sum_{d=0}^{D} w_{dm}^{(1)} x_{td}\right)\right) \\ &= f\left(\sum_{m=0}^{M} w_{mk}^{(2)} z_{tm}\right) \triangleq f(a_{tk}). \end{aligned} \tag{4.21}$$

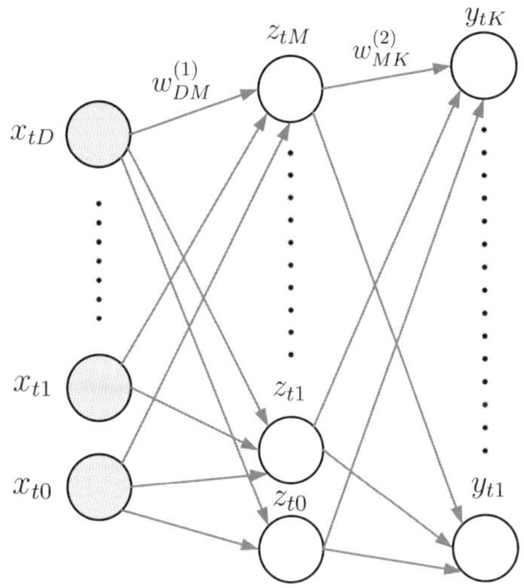

Figure 4.5 A multilayer perceptron with one input layer, one hidden layer, and one output layer. [Based on *Pattern Recognition and Machine Learning (Figure 5.1), by C. M. Bishop, 2006, Springer.*]

Here, we introduce the weight parameters of the first two layers, which are expressed by $\mathbf{w} = \{w^{(1)}_{dm}, w^{(2)}_{mk}\}$. We have $x_{t0} = z_{t0} = 1$ and the bias parameters that are denoted by $\{w^{(1)}_{0m}, w^{(2)}_{0k}\}$. This is a feedforward neural network (FNN) with two layer-wise calculations in forward pass. The first calculation is devoted to the layer-wise affine transformation with multiplication by using the layered parameters $\{w^{(1)}_{dm}\}$ and $\{w^{(2)}_{mk}\}$. The transformation is calculated to find the activation from input layer to output layer in an order of

$$a_{tm} = \sum_{d=0}^{D} w^{(1)}_{dm} x_{td} \quad \Longrightarrow \quad a_{tk} = \sum_{m=0}^{M} w^{(2)}_{mk} f(a_{tm}). \tag{4.22}$$

The second calculation is performed with the nonlinear activation function $f(\cdot)$. Figure 4.6 shows different activation functions, including the logistic sigmoid function in Eq. 4.5, the rectified linear unit (ReLU)

$$f(a) = \text{ReLU}(a) = \max\{0, a\}, \tag{4.23}$$

and the hyperbolic tangent function

$$f(a) = \tanh(a) = \frac{e^a - e^{-a}}{e^a + e^{-a}}. \tag{4.24}$$

The logic sigmoid function has value between 0 and 1. Differently, the hyperbolic tangent function has value between -1 and 1. However, the most popular activation in the implementation is ReLU, although others could be just as good for specific tasks [107]. For the case of classification network, the output vector $\mathbf{y}_t = \{y_{tk}\}$ is additionally calculated as a softmax function

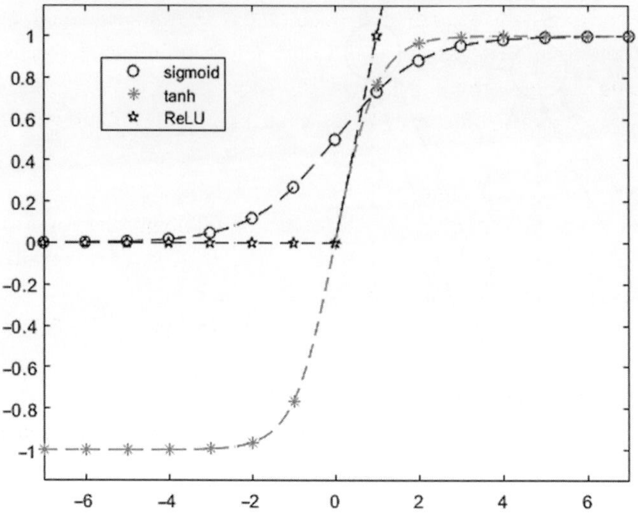

Figure 4.6 Three common activation functions in deep neural networks.

$$y_{tk} = \frac{\exp(a_{tk})}{\sum_m \exp(a_{tm})}. \tag{4.25}$$

In general, a DNN is constructed as a fully connected multilayer perceptron consisting of multiple hidden layers. Given an input vector, it is simple to compute the DNN's output in the forward pass, which includes the nonlinearity in the hidden layers and softmax activations in the output layer.

4.2.3 Error Backpropagation Algorithm

In implementation of DNN training, we collect a set of training samples $\{\mathbf{X}, \mathbf{R}\} = \{\mathbf{x}_t, \mathbf{r}_t\}_{t=1}^T$ in the form of input-output pairs. In the context of speech applications, $\mathbf{x}_t$ can be a speech vector or a number of consecutive speech vectors, whereas $\mathbf{r}_t$ is the desirable network output. In case $\mathbf{r}_t$'s are also speech vectors, a regression problem is formulated in an optimization procedure. We therefore estimate DNN parameters $\mathbf{w}$ by optimizing an objective based on the sum-of-squares error function, which is computed in DNN output layer in a form of

$$E(\mathbf{w}) = \frac{1}{2} \sum_{t=1}^{T} \|\mathbf{y}(\mathbf{x}_t, \mathbf{w}) - \mathbf{r}_t\|^2, \tag{4.26}$$

where $\mathbf{y}(\mathbf{x}_t, \mathbf{w}) = \{y_k(\mathbf{x}_t, \mathbf{w})\}$. However, the closed-form solution to solve this nonlinear regression problem is not analytical. The optimization is then implemented according to the stochastic gradient descent (SGD) algorithm

$$\mathbf{w}^{(\tau+1)} = \mathbf{w}^{(\tau)} - \eta \nabla E_n(\mathbf{w}^{(\tau)}), \tag{4.27}$$

where τ denotes the iteration index, η denotes the learning rate and $E_n(\cdot)$ denotes the error function computed from the nth minibatch in training set $\{\mathbf{X}_n, \mathbf{R}_n\}$, which is sampled by using the entire training data $\{\mathbf{X}, \mathbf{R}\}$. A training epoch is run by scanning the whole set of minibatches denoted by $\{\mathbf{X}, \mathbf{R}\} = \{\mathbf{X}_n, \mathbf{R}_n\}_{n=1}^N$. In the implementation, we initialize the parameters from $\mathbf{w}^{(0)}$ and continuously perform SGD algorithm to reduce the value of error function E_n until the iteration procedure converges. Practically, the SGD algorithm is implemented by randomizing all minibatches in a learning epoch. Also, a large number of epochs are run to assure that the convergence condition is met after DNN training. In general, SGD training can achieve better regression or classification performance when compared with the batch training where all of the training data are put into a single batch.

The training of DNNs using the backpropagation algorithm involves two passes. In forward pass, an affine transformation followed by nonlinear activation is computed layer by layer until the output layer. In the background passes, the error (typically cross-entropy for classification and mean squared error for regression) between the desired output and the actual output is computed and the error gradient with respect to the network weights are computed from the output layer back to the input layer. That is, we find the gradients for updating the weights in different layers in the following order

$$\frac{\partial E_n(\mathbf{w}^{(\tau)})}{\partial w_{mk}^{(2)}} \quad \rightarrow \quad \frac{\partial E_n(\mathbf{w}^{(\tau)})}{\partial w_{dm}^{(1)}}, \tag{4.28}$$

where the minibatch samples $\{\mathbf{X}_n, \mathbf{R}_n\}_{n=1}^N$ are used. Figures 4.7(a) and (b) illustrate the computation in the forward pass and the backward pass, respectively. To fulfill the error backpropagation algorithm in a backward pass, an essential trick is to compute the local gradient of neuron m in a hidden layer by using the input vector at each time $\mathbf{x}_t$

$$\begin{aligned} \delta_{tm} &\triangleq \frac{\partial E_t}{\partial a_{tm}} \\ &= \sum_k \frac{\partial E_t}{\partial a_{tk}} \frac{\partial a_{tk}}{\partial a_{tm}} \\ &= \sum_k \delta_{tk} \frac{\partial a_{tk}}{\partial a_{tm}}, \end{aligned} \tag{4.29}$$

which is updated and recalculated by integrating local gradients δ_{tk} from all neurons k in output layer. The order of updating for local gradient is shown by $\delta_{tk} \rightarrow \delta_{tm}$. Given the local gradients in output layer and hidden layer, the SGD updating involves the calculation of differentiations in a way of

$$\frac{\partial E_n(\mathbf{w}^{(\tau)})}{\partial w_{dm}^{(1)}} = \sum_{t \in \{\mathbf{X}_n, \mathbf{R}_n\}} \delta_{tm} x_{td} \tag{4.30}$$

$$\frac{\partial E_n(\mathbf{w}^{(\tau)})}{\partial w_{mk}^{(2)}} = \sum_{t \in \{\mathbf{X}_n, \mathbf{R}_n\}} \delta_{tk} z_{tm}, \tag{4.31}$$

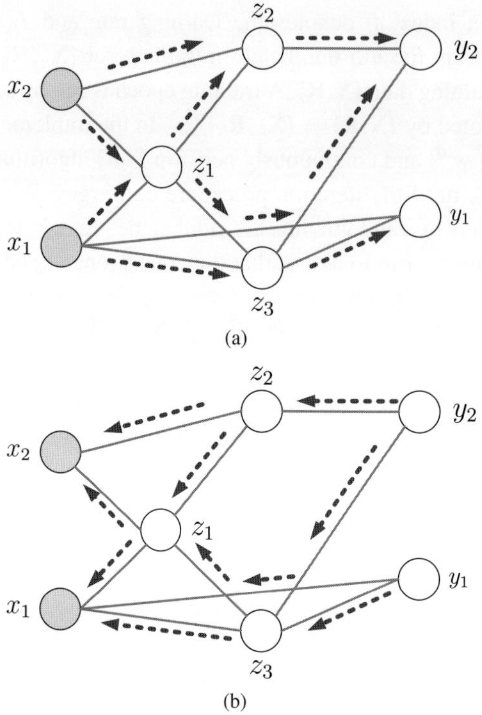

Figure 4.7 Illustration for computations in error backpropagation algorithm including those in (a) forward pass and (b) backward pass. The propagation directions are displayed by arrows. In a forward pass, each individual node involves the computation and propagation of activations a_t and outputs z_t. In a backward pass, each individual node involves the computation and propagation of local gradients δ_t. [Adapted from *Pattern Recognition and Machine Learning (Figure 5.2), by C. M. Bishop, 2006, Springer.*]

where the error function E_t is accumulated over $t \in \{\mathbf{X}_n, \mathbf{R}_n\}$, which is an calculation using the input-target pairs in a minibatch $E_n = \sum_{t \in \{\mathbf{X}_n, \mathbf{R}_n\}} E_t$. At each time t, we simply express the gradient for correcting a weight parameter $w_{mk}^{(2)}$ by using a multiplication of the output of neuron m, z_{tm}, in the hidden layer and the local gradient of neuron k, δ_{tk}, in the output layer. This type of computation is similarly employed in updating the weight parameter $w_{dm}^{(1)}$ for the neurons in input layer d and hidden layer m. In Figure 4.8, we illustrate a general procedure for error backpropagation algorithm that is used to train a l-layer multilayer perceptron. The forward calculation of error function and the backward calculations of local gradients or error gradients from layer l to layer $l-1$ until layer 1.

4.2.4 Interpretation and Implementation

The deep hierarchy in DNN is efficient in learning representation. The computation units based on linear transformation and nonlinear activation are modulated and adopted in

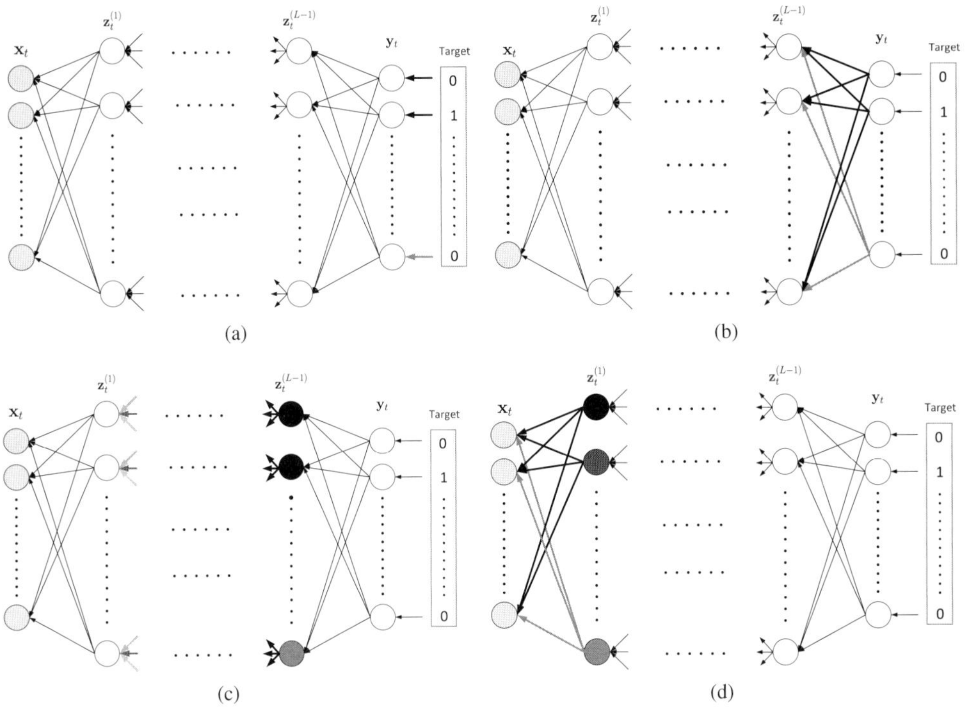

Figure 4.8 Procedure in the backward pass based on an error backpropagation algorithm. (a) Calculate the error function, (b) propagate the local gradient from output layer L to (c) hidden layer $L - 1$ (d) until input layer 1.

different neurons and different layers. DNN is naturally seen as a discriminative model for hierarchical learning, which works for regression and classification tasks. It is crucial to explain why deep hierarchy does help when training a DNN for speaker recognition. Basically, the nonlinear and deep architecture provides a vehicle to deal with the weakness of the bounded performance due to a complicated regression mapping. Deep learning aims to conduct structural representation and pursue the comprehensibility in hierarchical learning. A hierarchical model with multiple layers of hidden neurons opens an avenue to conduct combinational sharing over the synapse of statistics. Structural learning in DNN provides a monitoring platform to analyze what has been learned and what subspace has been projected to obtain a better prediction for regression or classification. A hierarchical representation consisting of different levels of abstraction is trained for future prediction. Each level is trainable and corresponds to a specific feature transformation. A deep model is trained to find a high-level representation that is generalized for different tasks.

On the other hand, DNN is built with a huge parameter space and is spanned by a number of fully connected layers. The convergence condition in training procedure is usually hard to meet. Backpropagation does not work well if parameters are randomly initialized. The performance of deep model could not be theoretically guaranteed.

Basically, the weights of a DNN without unsupervised pretraining are randomly initialized. The resulting system performance is even worse than that of a shallow model [108]. It is crucial that DNN training procedure requires a reliable initialization and a rapid convergence to learn a desirable "deep" model for different regression or classification problems in a speaker recognition system.

The problems and solvers with the error backpropagation algorithm are discussed here. Basically, the gradient in backpropagation is progressively getting diluted. Below a top few layers, the correction signal for the connection weights is too weak. Backpropagation accordingly gets stuck in a local minimum, especially in case of random initialization where the starting point is far from good regions. On the other hand, in usual settings, we can use only labeled data to train a supervised neural network. But, almost all data are unlabeled in practical applications. Nevertheless, a human brain can efficiently learn from unlabeled data. To tackle this issue, we may introduce unsupervised learning via a greedy layer-wise training procedure. This procedure allows abstraction to be developed naturally from one layer to another and helps the network initialize with good parameters. After that, the supervised top-down training is performed as a final step to refine the features in intermediate layers that are directly relevant for the task. An elegant solution based on deep belief network is therefore constructed and addressed in what follows.

4.3 Deep Belief Networks

Deep belief network (DBN) [109] is a probabilistic generative model that provides a meaningful initialization or pretraining for constructing a neural network with deep structure consisting of multiple hidden layers. DBN conducts an unsupervised learning where the outputs are learned to reconstruct the original inputs. Different layers in DBN are viewed as the feature extractors. Such an unsupervised learning could be further merged with the supervised retraining for different regression or classification problems in speaker recognition. DBN is seen as a theoretical tool to pretrain or initialize different layers in DNN training. In the training procedure, the RBM, addressed in Section 4.1.2, is treated as a building component to construct multiple layers of hidden neurons for DBN. The building method is based on a stack-wise and bottom-up style. Each stack is composed of a pair of layers that is trained by RBM. After training each stack, the hidden layer of RBM is subsequently used as an observable layer to train the RMB in next stack for a deeper hidden layer. Following this style, we eventually train a bottom-up deep machine in accordance with a stack-wise and tandem-based training algorithm. DBN obtained great results in [109] due to good *initialization* and *deep* model structure. A lower bound of log likelihood of observation data $\mathbf{v}$ is maximized to estimate DBN parameters [94].

4.3.1 Training Procedure

As illustrated in Figure 4.9, a stack-wise and tandem-based training procedure is performed to learn DBN parameters. In this training procedure, the first RBM is estimated from a set of training samples $\{\mathbf{x}\}$ and then used to transform each individual speech token $\mathbf{x}$ into a latent variable $\mathbf{z}^{(1)}$ using the trained RBM parameters $\mathbf{w}^{(1)}$. The latent units $\{\mathbf{z}^{(1)}\}$ are subsequently treated as the observation data to learn the next RBM, which transforms each sample $\mathbf{z}^{(1)}$ into a deeper sample $\mathbf{z}^{(2)}$ in the next layer. RBM is used as a learning representation for a pair of layers. According to this procedure, a deep hierarchy using DBN is built to explore the observed and hidden variables from input layer to deep layers in a chain of

$$\mathbf{x} \to \mathbf{z}^{(1)} \to \mathbf{z}^{(2)} \to \mathbf{z}^{(3)} \to \cdots . \tag{4.32}$$

We develop the level of abstraction in a layer-wise manner. The progression of model structure from low level to high level represents the natural complexity. An unsupervised learning is performed to estimate a DBN by using a greedy and layer-wise training procedure. After training the DBN, we use DBN parameters as the initial parameters to train a DNN that performs much better than that based on the random initialization. The chance of going stuck in a local minimum point is much smaller. Finally, a supervised top-down fine-tuning procedure is executed to adjust the features in middle layers based on the labels of training data $\mathbf{r}$. Compared to the original features, the adapted features are better fitted to produce the target values $\mathbf{y}$ in output layer. The supervised learning using error backpropagation algorithm is employed in a fine-tuning process. The resulting method is also called DBN-DNN, which has been successfully developed for deep learning. This DBN-DNN is seen as *generative* model as well as *discriminative* model due to this two-step procedure, including one step of unsupervised stack-wise training and the other step of supervised fine-tuning and processing. Specifically, the unlabeled samples $\{\mathbf{x}\}$ are collected to build a generative model in a stack-wise and bottom-up manner. Then, a small set of labeled samples $\{\mathbf{x}, \mathbf{r}\}$ is used to adjust DBN parameters to final DNN parameters based on an error backpropagation algorithm.

In general, the greedy and layer-wise training of DNN performs well from the perspectives of optimization and regularization due to twofold reasons. First, the pretraining step in each layer in a bottom-up way helps constraining the learning process around

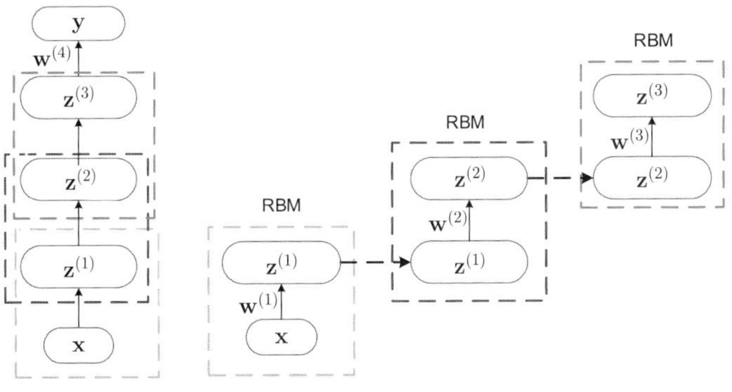

Figure 4.9 A stack-wise training procedure for deep belief network based on the restricted Boltzmann machine. [Based on *Source Separation and Machine Learning (Figure 2.16), by J.T. Chien, 2018, Academic Press.*]

the region of parameters, which is relevant to represent the unlabeled data. The learning representation is performed to describe the unlabeled data in a way of providing discriminative representation for labeled data. DNN is accordingly regularized to improve model generalization for prediction of unseen data. In addition, the second reason is caused by the fact that the unsupervised learning starts from lower-layer parameters that are close to the localities and near to the minimum in optimization. Random initialization is hard to attain these properties. Such a perspective sufficiently explains why the optimization for DNN based on DBN initialization works better than that based on random initialization.

There are three training strategies for deep neural networks or deep belief networks developed in different application domains under various training conditions. First, the deep model can be purely trained in supervised mode. The model parameters can be initialized randomly and estimated according to SGD using the backpropagation method to compute gradients. Most practical systems for speech, text, and image applications have been built by using this strategy. Second, the unsupervised learning is combined with supervised training for constructing a deep model. Using this strategy, each layer is trained in an unsupervised way using RBM. The layer-wise hierarchy is grown one layer after the other until the top layer. Then, a supervised classifier on top layers is trained while keeping the parameters of the other layers fixed. This strategy is suitable especially when very few labeled samples are available. A semi-supervised learning is carried out from a large set of unlabeled data and a small set of labeled data. Similar to the second strategy, the third strategy builds a deep structure in an unsupervised mode layer by layer based on RBMs. A classifier layer is added on the top for classification task. The key difference is to retrain the whole model in a supervised way. This strategy works well especially when the labels are poorly transcribed. Typically, the unsupervised pretraining in second and third strategies is often performed by applying the regularized stacked autoencoders that will be addressed later in Section 4.4. In what follows, we further detail the procedure and the meaning of greedy training for deep belief networks.

4.3.2 Greedy Training

In deep learning, it is essential to carry out the greedy training procedure for stack-wise construction of a bottom-up neural network model. Greedy training is performed in an unsupervised style. Figure 4.10 depicts a building block of an RBM and an approach to stack a number of RBMs toward a deep model. The undirectional two-layer RBM functions as described in Section 4.1. The hidden units in $\mathbf{h}$ are characterized from visible speech data $\mathbf{v}$ using the weight parameters $\mathbf{W}^1$ in the first RBM. The construction of DBN with one visible layer $\mathbf{v}$ and three hidden layers $\mathbf{h}^1$, $\mathbf{h}^2$, and $\mathbf{h}^3$ in this example can be decomposed into three steps as illustrated in Figures 4.10(b) and (c).

At the first step, a two-layer RBM is constructed with an input layer $\mathbf{v}$ and a hidden layer $\mathbf{h}^1$. RBM parameters $\mathbf{W}^1$ are trained by maximizing the log likelihood $\log p(\mathbf{v})$ or alternatively the lower bound (or variational bound) of log likelihood, which is derived by

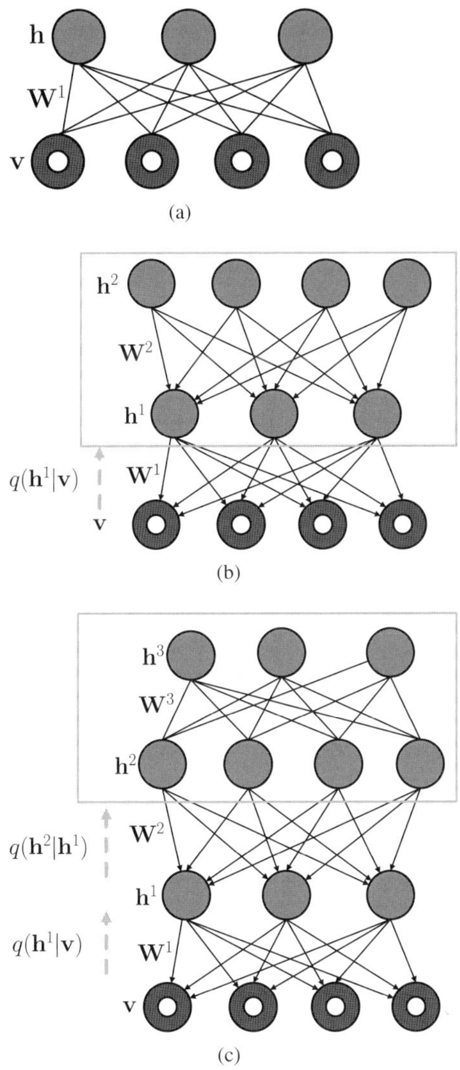

Figure 4.10 (a) A building block of two-layer RBM. Construction of deep belief network using (b) the second step of RBM and (c) the third step of RBM. The variational distributions $q(\mathbf{h}^1|\mathbf{v})$ and $q(\mathbf{h}^2|\mathbf{h}^1)$ are introduced to infer hidden variables $\mathbf{h}^1$ and $\mathbf{h}^2$ and model parameters $\mathbf{W}^1$ and $\mathbf{W}^2$ in different layers, respectively.

$$\log p(\mathbf{v}) \geq \sum_{\mathbf{h}^1} q(\mathbf{h}^1|\mathbf{v}) \Big(\log p(\mathbf{h}^1) + \log p(\mathbf{v}|\mathbf{h}^1) \Big) + \mathbb{H}_{q(\mathbf{h}^1|\mathbf{v})}[\mathbf{h}^1] \tag{4.33}$$

according to variational inference where the variational distribution $q(\mathbf{h}^1|\mathbf{v})$ for hidden variable $\mathbf{h}^1$ is used. $\mathbb{H}[\cdot]$ denotes an entropy function. The derivation is similar to Eq. 2.33 in Section 2.3 when finding the evidence lower bound. At the second step, another hidden layer is stacked on top of the RBM to form a new RBM. When training this second RBM, the parameters $\mathbf{W}^1$ of the first RBM are fixed and the units of the first hidden layer $\mathbf{h}^1$ are sampled by variational distribution $q(\mathbf{h}^1|\mathbf{v})$ and treated as visible data for training the second RBM with parameters $\mathbf{W}^2$. At the third step, this stack-wise process is continued in order to stack layers on top of the network and to train the third RBM parameter $\mathbf{W}^3$ as in the previous step, with the sample $\mathbf{h}^2$ drawn from $q(\mathbf{h}^2|\mathbf{h}^1)$ and treated as the visible data to explore the latent variable $\mathbf{h}^3$ in the third hidden layer. The joint distribution of visible data $\mathbf{v}$ and l hidden variables $\{\mathbf{h}^1, \ldots, \mathbf{h}^l\}$ is therefore expressed in accordance with the property of Markov switching where the probability of a layer k depends only on layer $k+1$ as

$$p(\mathbf{v}, \mathbf{h}^1, \mathbf{h}^2, \ldots, \mathbf{h}^l) = p(\mathbf{v}|\mathbf{h}^1)p(\mathbf{h}^1|\mathbf{h}^2) \ldots p(\mathbf{h}^{l-2}|\mathbf{h}^{l-1})p(\mathbf{h}^{l-1}|\mathbf{h}^l). \tag{4.34}$$

Figure 4.11(a) shows the construction of DBN with one visible layer and three hidden layers. RBM is on the top while the directed belief network is on the bottom. The arrows indicate the direction of data generation. We basically estimate the model parameters $\{\mathbf{W}^1, \ldots, \mathbf{W}^l\}$ by maximizing the variational lower bound of log likelihood $p(\mathbf{v})$ where all possible configurations of the higher variables are integrated to get the prior for lower variables toward the likelihood of visible variables. In addition, as illustrated in Figure 4.11(b), if the first RBM with parameter $\mathbf{W}^1$ is stacked by an inverse RBM (the second RBM) with parameters $(\mathbf{W}^1)^\top$, the hidden units $\mathbf{h}^2$ of the second RBM with input units $\mathbf{h}^2$ are seen as the reconstruction of visible data $\mathbf{v}$ in the first RBM.

It is important to explain why greedy training works for construction of a deep belief network. In general, DBN is constructed via RBM where the joint distribution of a two-layer model $p(\mathbf{v}, \mathbf{h})$ is calculated by using the conditional distributions $p(\mathbf{v}|\mathbf{h})$ and $p(\mathbf{h}|\mathbf{v})$ that implicitly reflects the marginal distributions $p(\mathbf{v})$ and $p(\mathbf{h})$. The key idea behind DBN is originated from the stacking of RBM, which preserves the conditional likelihood $p(\mathbf{v}|\mathbf{h}^1)$ from the first-level RBM and replaces the distribution of hidden units by using the distribution $p(\mathbf{h}^1|\mathbf{h}^2)$ generated by the second-level RBM. Furthermore, DBN performs an easy approximate inference where $p(\mathbf{h}^{k+1}|\mathbf{h}^k)$ is approximated by the associated RBM in layer k using $q(\mathbf{h}^{k+1}|\mathbf{h}^k)$. This is an approximation because $p(\mathbf{h}^{k+1})$ differs between RBM and DBN.

During training time [110], variational bound is maximized to justify greedy layer-wise training of RBMs. In the whole training procedure, we basically initialize from the stacked RBMs in a pretraining stage, which is seen as two-step fine-tuning. First, the *generative* fine-tuning is run to construct or fine-tune a deep model based on the variational or mean-field approximation where the persistent chain and the stochastic approximation are performed. Second, the *discriminative* fine-tuning is executed by minimizing the classification loss in the backpropagation procedure. In addition, greedy training meets the *regularization* hypothesis where the pretraining is performed to constrain the parameters in a region relevant to an unsupervised dataset. Greedy training also follows the *optimization* hypothesis where the unsupervised training initializes the lower-level parameters near the localities of better minima. Such a property is *not* held by using the scheme of random initialization. Nevertheless, the greedy procedure of stacking RBMs is suboptimal. A very approximate inference procedure is performed in

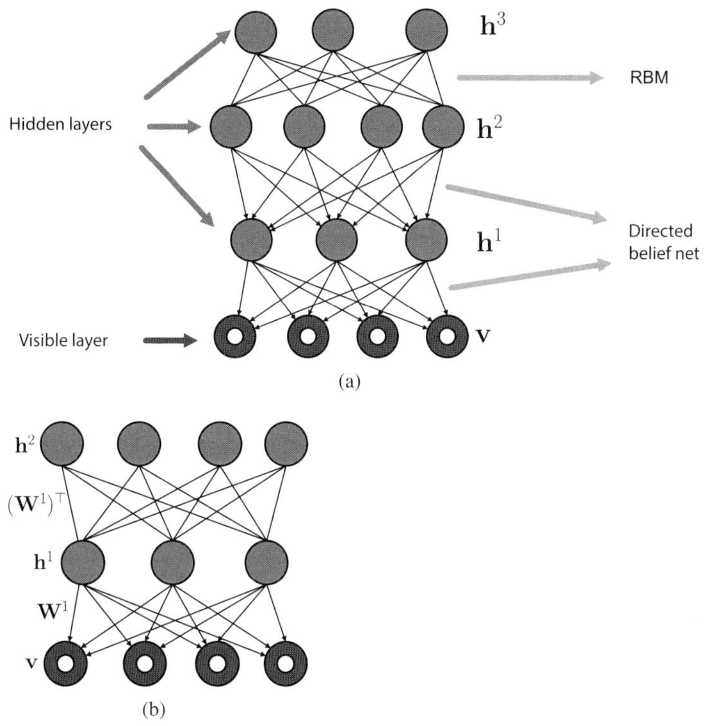

Figure 4.11 (a) Construction of deep belief network with three hidden layers. The top two layers form an undirectional bipartite graph using a RBM while the remaining layers form a sigmoid belief net with directed and top-down connections. (b) RBMs are stacked as an invertible learning machine for data reconstruction.

DBN. These weaknesses were tackled by using a different type of hierarchical probabilistic model, called the deep Boltzmann machine, which is addressed in what follows.

4.3.3 Deep Boltzmann Machine

Deep Boltzmann machine (DBM) [103, 111] is a type of Markov random field where all connections between layers are undirected as depicted in Figure 4.12, which differs from Figure 4.11 for deep belief network. The undirected connection between the layers make a complete Boltzmann machine. Similar to RBM and DBN, using this unsupervised DBM, there are no connections between the nodes in the same layer. High-level representations are built from the unlabeled inputs. Labeled data are only used to slightly fine-tune the model.

Considering the example of DBM with three hidden layers, the marginal likelihood function is given by

$$p(\mathbf{v}|\boldsymbol{\theta}) = \sum_{\mathbf{h}^1,\mathbf{h}^2,\mathbf{h}^3} \frac{1}{Z(\boldsymbol{\theta})} \exp\left(\mathbf{v}^\top \mathbf{W}^1 \mathbf{h}^1 + (\mathbf{h}^1)^\top \mathbf{W}^2 \mathbf{h}^2 + (\mathbf{h}^2)^\top \mathbf{W}^3 \mathbf{h}^3\right) \tag{4.35}$$

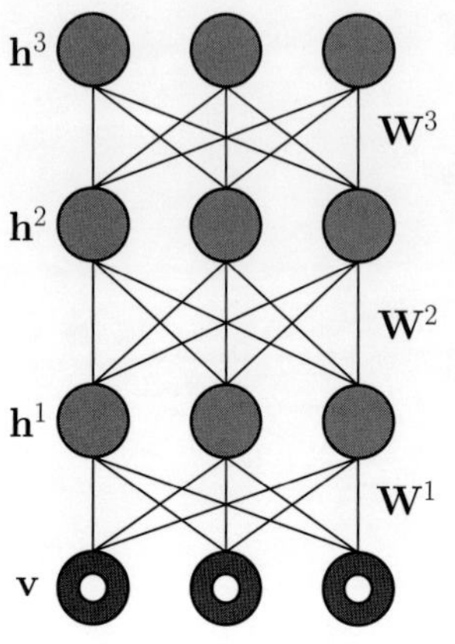

Figure 4.12 A deep Boltzmann machine with three hidden layers. All connections between layers are undirected but with no within-layer connections.

where $\{\mathbf{h}^1, \mathbf{h}^2, \mathbf{h}^3\}$ are the hidden units, $\boldsymbol{\theta} = \{\mathbf{W}^1, \mathbf{W}^2, \mathbf{W}^3\}$ are the model parameters, and $Z(\boldsymbol{\theta})$ is a normalization constant. Similar to RBM in Section 4.1, the energy-based likelihood function in DBM is defined by using the energy function

$$E(\mathbf{v}, \mathbf{h}|\boldsymbol{\theta}) = -\mathbf{v}^\top \mathbf{W}^1 \mathbf{h}^1 - (\mathbf{h}^1)^\top \mathbf{W}^2 \mathbf{h}^2 - (\mathbf{h}^2)^\top \mathbf{W}^3 \mathbf{h}^3, \tag{4.36}$$

which is accumulated from three pairs of weight connections $\{\mathbf{v}, \mathbf{h}^1\}$, $\{\mathbf{h}^1, \mathbf{h}^2\}$, and $\{\mathbf{h}^2, \mathbf{h}^3\}$. For example, the updating of parameter vector $\mathbf{W}^1$ is performed according to the following derivative

$$\frac{\partial p(\mathbf{v}|\boldsymbol{\theta})}{\partial \mathbf{W}^1} = \mathbb{E}_{p_{\text{data}}}\left[\mathbf{v}(\mathbf{h}^1)^\top\right] - \mathbb{E}_{p_{\text{model}}}\left[\mathbf{v}(\mathbf{h}^1)^\top\right] \tag{4.37}$$

Similar updating can be also found for $\mathbf{W}^2$ and $\mathbf{W}^3$.

The mean-field theory is developed for variational inference of DBM parameters. A factorized variational distribution is introduced for a set of hidden variables $\mathbf{h} = \{\mathbf{h}^1, \mathbf{h}^2, \mathbf{h}^3\}$ in a form of

$$q(\mathbf{h}|\mathbf{v}, \boldsymbol{\mu}) = \prod_j \prod_k \prod_m q(h_j^1) q(h_k^2) q(h_m^3) \tag{4.38}$$

where $\boldsymbol{\mu} = \{\boldsymbol{\mu}^1, \boldsymbol{\mu}^2, \boldsymbol{\mu}^3\}$ denote the mean-field parameters with $q(h_i^k = 1) = \mu_i^k$ for $k = 1, 2, 3$. The lower bound of log likelihood in DBM is then derived by referring Eq. 4.33 using the variational distribution in Eq. 4.38. The bound is expressed by

$$\begin{aligned}\log p(\mathbf{v}|\boldsymbol{\theta}) &\geq \mathbf{v}^\top \mathbf{W}^1 \boldsymbol{\mu}^1 + (\boldsymbol{\mu}^1)^\top \mathbf{W}^2 \boldsymbol{\mu}^2 + (\boldsymbol{\mu}^2)^\top \mathbf{W}^3 \boldsymbol{\mu}^2 \\ &\quad - \log Z(\boldsymbol{\theta}) + \mathbb{H}_{\mathbf{q}(\mathbf{h}|\mathbf{v})}[\mathbf{h}] \\ &\triangleq \mathcal{L}(q(\mathbf{h}|\mathbf{v}, \boldsymbol{\mu})).\end{aligned} \tag{4.39}$$

An approximate inference, by following Section 2.3, is performed to fulfill a learning procedure of RBM that corresponds to a type of VB-EM algorithm as addressed in

Section 2.3.3. For each training sample, we first estimate the mean-field parameters $\boldsymbol{\mu}$ by maximizing the variational lower bound $\mathcal{L}(q(\mathbf{h}|\mathbf{v},\boldsymbol{\mu}))$ given the current value of model parameters $\boldsymbol{\theta}$. This optimal variational parameters $\boldsymbol{\mu}$ are estimated to meet the mean-field conditions

$$\mu_j^1 = q(h_j^1 = 1) \leftarrow p(h_j^1 = 1|\mathbf{v}, \boldsymbol{\mu}^2, \boldsymbol{\theta}) = \sigma\left(\sum_i W_{ij}^1 v_i + \sum_k W_{jk}^2 \mu_k^2\right) \quad (4.40)$$

$$\mu_k^2 = q(h_k^2 = 1) \leftarrow p(h_k^2 = 1|\boldsymbol{\mu}^1, \boldsymbol{\mu}^3, \boldsymbol{\theta}) = \sigma\left(\sum_j W_{jk}^2 \mu_j^1 + \sum_m W_{km}^3 \mu_m^3\right) \quad (4.41)$$

$$\mu_m^3 = q(h_m^3 = 1) \leftarrow p(h_m^3 = 1|\boldsymbol{\mu}^2, \boldsymbol{\theta}) = \sigma\left(\sum_k W_{km}^3 \mu_k^2\right) \quad (4.42)$$

where the biases in sigmoid function are considered and merged in the summation terms. The VB-E step is completed. Given the variational parameters $\boldsymbol{\mu}$, the lower bound $\mathcal{L}(q(\mathbf{h}|\mathbf{v},\boldsymbol{\mu}))$ is updated and then maximized to find model parameters $\boldsymbol{\theta}$ in the VB-M step. Let $\boldsymbol{\theta}_t$ and $\mathcal{H}_t = \{\mathbf{v}_t, \mathbf{h}_t^1, \mathbf{h}_t^2, \mathbf{h}_t^3\}$ denote the current parameters and the state, respectively. These two variables are sequentially updated and sampled. Continuous calculating $\mathcal{H}_{t+1} \leftarrow \mathcal{H}_t$ is similar to performing the Gibbs sampling. The new parameter $\boldsymbol{\theta}_{t+1}$ is updated by calculating a gradient step where the intractable model expectation $\mathbb{E}_{p_{\text{model}}}[\cdot]$ in the gradient is approximated by a point estimate at sample $\mathcal{H}_{t+1}$ or more precisely calculated from a set of L samples or particles $\{\mathbf{x}_t^{(1)}, \ldots, \mathbf{x}_t^{(L)}\}$. This MCMC-based optimization assures the asymptotic convergence in the learning process. Next, we further address the scheme of stacking autoencoder, which is employed to build a deep neural network model.

4.4 Stacked Autoencoder

It is crucial to explore a meaningful strategy and learn a useful representation for layer-wise structure in a deep neural network. Stacking layers of denoising autoencoder provides an effective approach to deep unsupervised learning [104]. In what follows, we will first address the principle of denoising autoencoder and then introduce the procedure of greedy layer-wise learning for a deep network with the stacked autoencoder.

4.4.1 Denoising Autoencoder

Denoising autoencoder (DAE) [107, 112] is an autoencoder that is trained to produce the original and uncorrupted data point as output when given a corrupted data point as its input. Figure 4.13 illustrates a procedure of finding a reconstructed data $\widehat{\mathbf{x}}$ from a corrupted sample $\widetilde{\mathbf{x}}$ given its original data $\mathbf{x}$ via a deterministic hidden code $\mathbf{h}$. The corrupted version $\widetilde{\mathbf{x}}$ is obtained by means of a stochastic mapping function for data corruption under a predefined probability distribution $\widetilde{\mathbf{x}} \sim q_c(\widetilde{\mathbf{x}}|\mathbf{x})$. DAE aims to learn an autoencoder that maps the corrupted sample $\widetilde{\mathbf{x}}$ to a hidden representation $\mathbf{h}$ via an

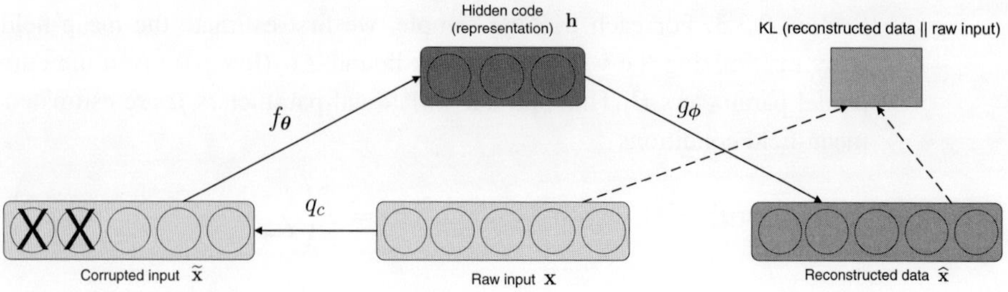

Figure 4.13 An illustration of denosing autoencoder for finding a hidden representation **h** from a corrupted input $\widetilde{\mathbf{x}}$. [Based on *Stacked Denoising Autoencoders: Learning Useful Representations in a Deep Network with a Local Denoising Criterion (Figure 1), by P. Vincent et al., J. of Machine Learning Research, vol. 11, 2010, pp. 3371–3408, MIT Press.*]

encoder $f_\theta(\widetilde{\mathbf{x}})$ with parameters $\boldsymbol{\theta}$ and attempts to estimate the reconstructed sample $\widehat{\mathbf{x}}$ via a decoder $\widehat{\mathbf{x}} = g_\phi(\mathbf{h})$ with parameters $\boldsymbol{\phi}$. DAE is constructed by learning the parameters of encoder $\boldsymbol{\theta}$ and decoder $\boldsymbol{\phi}$ by minimizing the reconstruction error or the negative log likelihood of decoder output or reconstructed data, i.e., $-\log p(\widehat{\mathbf{x}}|\mathbf{h})$, from the minibatches of training pairs $\{\mathbf{x}, \widetilde{\mathbf{x}}\}$. More specifically, we perform the stochastic gradient descent and minimize the following expectation function

$$\mathcal{L}_{\text{DAE}}(\boldsymbol{\theta}, \boldsymbol{\phi}) = -\mathbb{E}_{\mathbf{x}\sim p_{\text{data}}(\mathbf{x}), \widetilde{\mathbf{x}}\sim q_c(\widetilde{\mathbf{x}}|\mathbf{x})} \left[\log p(\widehat{\mathbf{x}} = g_\phi(\mathbf{h})|\mathbf{h} = f_\theta(\widetilde{\mathbf{x}}))\right], \tag{4.43}$$

where $p_{\text{data}}(\mathbf{x})$ denotes the distribution of training data. The fulfillment of DAE learning procedure is done by running three steps.

(1) Sample a training sample $\mathbf{x}$ from a collection of training data.
(2) Sample a corrupted version $\widetilde{\mathbf{x}}$ from the corruption model $q_c(\widetilde{\mathbf{x}}|\mathbf{x})$.
(3) Use this data pair $\{\mathbf{x}, \widetilde{\mathbf{x}}\}$ as a training example for estimating the distribution of decoder for reconstruction $p(\widehat{\mathbf{x}}|\mathbf{h})$ or equivalently finding the encoder and decoder parameters $\{\boldsymbol{\theta}, \boldsymbol{\phi}\}$ by minimizing $\mathcal{L}_{\text{DAE}}(\boldsymbol{\theta}, \boldsymbol{\phi})$.

In the implementation, the encoder and decoder are expressed by a feedforward network that is trained by the standard error backpropagation algorithm as mentioned in Section 4.2.3. The corruption model is represented by an isotropic Gaussian distribution

$$q_c(\widetilde{\mathbf{x}}|\mathbf{x}) = \mathcal{N}(\widetilde{\mathbf{x}}|\mathbf{x}, \sigma^2\mathbf{I}), \tag{4.44}$$

where the corrupted sample $\widetilde{\mathbf{x}}$ is around the the original data $\mathbf{x}$ with a shared variance σ^2 across different dimensions of $\mathbf{x}$.

Figure 4.14 depicts the concept of DAE from a manifold learning perspective. The corrupted sample $\widetilde{\mathbf{x}}$ is shown with a circle of equiprobable corruption. Basically, DAE is trained to learn a *mean vector field*, i.e., $g_\phi(f_\theta(\widetilde{\mathbf{x}})) - \mathbf{x}$, toward the regions with high probability or low reconstruction error. Different corrupted samples $\widetilde{\mathbf{x}}$ are constructed with vector fields shown by the dashed arrows. A generative model based on DAE is accordingly trained by minimizing the loss function in Eq. 4.43. This objective function corresponds to the negative variational lower bound by using the variational autoencoder, which will be addressed in Section 4.5. In fact, DAE loss function is minimized

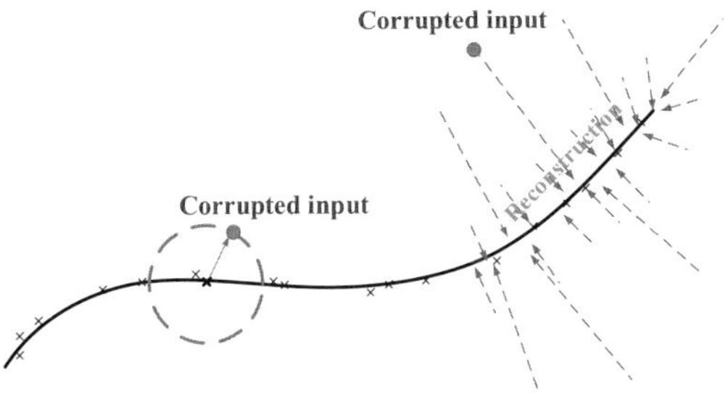

Figure 4.14 A denoising autoencoder that maps corrupted data $\widetilde{\mathbf{x}}$'s to their raw data $\mathbf{x}$'s along a data manifold. [Based on *Stacked Denoising Autoencoders: Learning Useful Representations in a Deep Network with a Local Denoising Criterion (Figure 2), by P. Vincent et al., J. of Machine Learning Research, vol. 11, 2010, pp. 3371–3408, MIT Press.*]

to realize a maximum likelihood (ML) estimate of DAE parameters $\{\boldsymbol{\theta}, \boldsymbol{\phi}, \sigma^2\}$. Such an estimation in DAE is similar to fulfill the regularized score matching [113] or the denoising score matching [107] on an RBM with Gaussian visible unites. The score matching aims to find the model distribution p_{model}, which is encouraged to produce the same likelihood score as the data distribution p_{data} at every training point $\mathbf{x}$. The robustness to noise corruption is assured in this specialized and regularized RBM. In the next section, the denoising autoencoder is treated as a building block to perform unsupervised stacking for a deep neural network.

4.4.2 Greedy Layer-Wise Learning

This section addresses how the autoencoder or RBM is stacked to carry out the greedy layer-wise learning for a deep structural model. Uncorrupted encoding is acted as the inputs for next level of stacking. Uncorrupted inputs are reconstructed in each stacking. Briefly speaking, we first start with the lowest level and stack upwards. Second, each layer of autoencoder is trained by using the intermediate codes or features from the layer below. Third, top layer can have a different output, e.g., softmax nonlinearity, to provide an output for classification. Figures 4.15(a), (b), and (c) demonstrate three steps toward constructing a deep classification network.

(1) Autoencoder or RBM is trained from visible data $\mathbf{x}$ to explore the hidden units $\mathbf{h}^1$ in the first hidden layer by using the first-level parameter $\mathbf{W}^1$. The transpose of parameters $(\mathbf{W}^1)^\top$ is used to implement an inverse of model or equivalently to find the reconstructed data $\widehat{\mathbf{x}}$.

(2) Reconstructed data $\widehat{\mathbf{x}}$ are masked by fixing the trained parameter $\mathbf{W}^1$ or $(\mathbf{W}^1)^\top$. The second hidden layer with hidden units $\mathbf{h}^2$ is explored by training the second-level parameter $\mathbf{W}^2$ or its transpose $(\mathbf{W}^2)^\top$ to reconstruct the hidden units $\widehat{\mathbf{h}}^1$ in the second hidden layer.

(3) Labels $\mathbf{y}$ of the visible data $\mathbf{x}$ are then used to train the third-level parameter $\mathbf{U}$ where the reconstructions $\widehat{\mathbf{x}}$ and $\widehat{\mathbf{h}}^1$ are masked by fixing the inverse parameters $\{(\mathbf{W}^1)^\top, (\mathbf{W}^2)^\top\}$. Three levels of parameters $\{\mathbf{W}^1, \mathbf{W}^2, \mathbf{U}\}$ are further fine-tuned by using error backpropagation algorithm via minimization of classification loss.

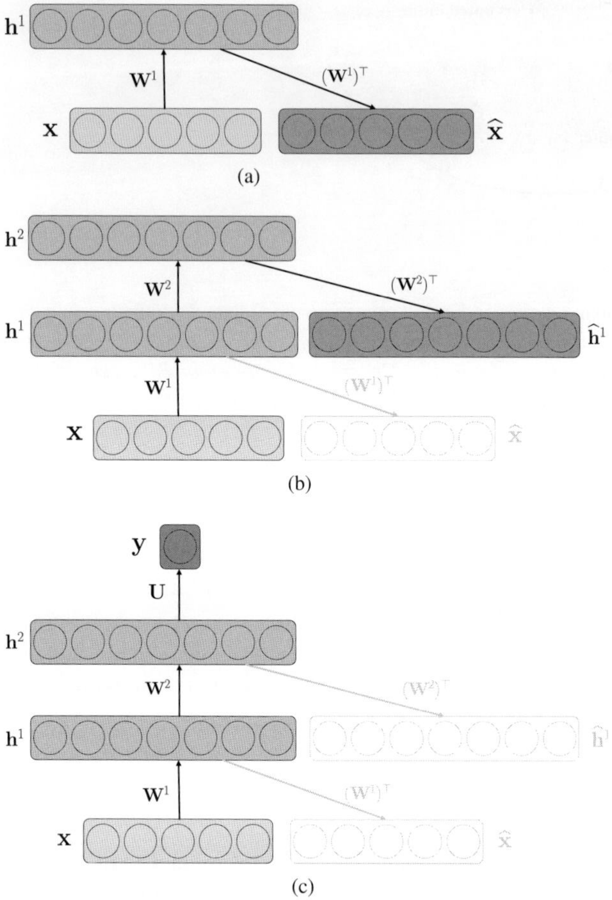

Figure 4.15 Procedure of stacking autoencoders for the construction of a deep classification network with input **x** and classification output **y**.

Alternatively, Figures 4.16(a), (b), and (c) show a different greedy procedure to construct a deep classification network where the output is predicted directly for class label $\widehat{\mathbf{y}}$ rather than indirectly for training sample $\widehat{\mathbf{x}}$. In the first step, the hidden units $\mathbf{h}^1$ are discovered from visible data $\mathbf{v}$ by using the first-level parameter $\mathbf{W}^1$. The reconstructed class label $\widehat{\mathbf{y}}$ is then calculated by using transformation parameter $\mathbf{U}^1$. However, the two-layer model is insufficient to estimate a good class label $\widehat{\mathbf{y}}$. The second step is presented to explore a deeper model to $\mathbf{h}^2$ and use this higher abstraction to predict class label $\widehat{\mathbf{y}}$ via an updated higher-level parameter $\mathbf{U}^2$ while the estimated class label $\widehat{\mathbf{y}}$ using lower-level parameter $\mathbf{U}^1$ is masked. The whole model with three layers is fine-tuned by using error backpropagation.

In general, using DAE, there is no partition function in the training criterion. The encoder and decoder can be represented by any parametric functions although feedforward neural networks have been widely adopted. The experiments on classification tasks in [104] showed that the unsupervised pretraining based on greedy layer-wise learning using RBM and DAE worked well for the construction of deep classification networks.

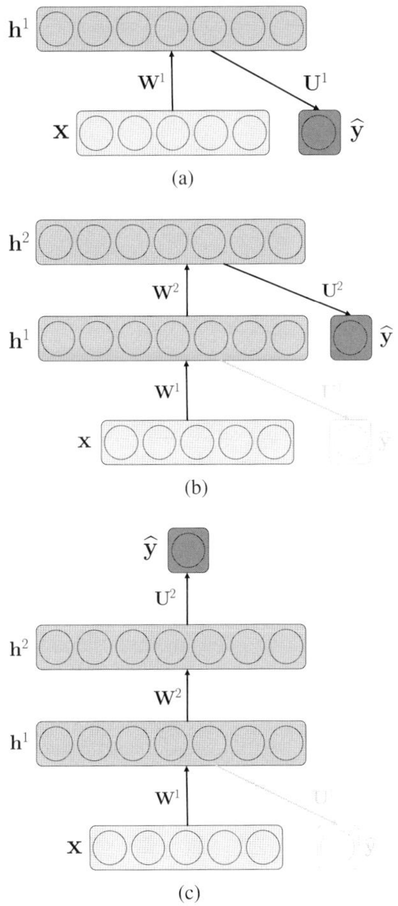

Figure 4.16 An alternative procedure for greedy layer-wise learning in a deep classification network where the output layer produces the reconstructed class label $\widehat{\mathbf{y}}$.

When the number of training examples is increased, the stacking DAE performs better than stacking RBM.

Although the restricted Boltzmann machine and denoising autoencoder are developed as the building block for greedy layer-wise learning of a deep neural network, the hidden units in different layers $\{\mathbf{h}^k\}_{k=1}^l$ are seen as the deterministic variables without uncertainty modeling. The interpretation of model capacity is therefore constrained. For example, in real-world speaker recognition systems, the utterances or feature vectors of a target speaker are usually collected in adverse environments with different contents and lengths in presence of various noise types and noise levels. It is challenging to build a robust deep model that can accommodate the variations of speech patterns from abundant as well as sparse training utterances. Uncertainty modeling or Bayesian learning (as described in Section 2.5) becomes crucial in the era of deep learning. In the next section, we will introduce the variational autoencoder where the hidden units are considered random.

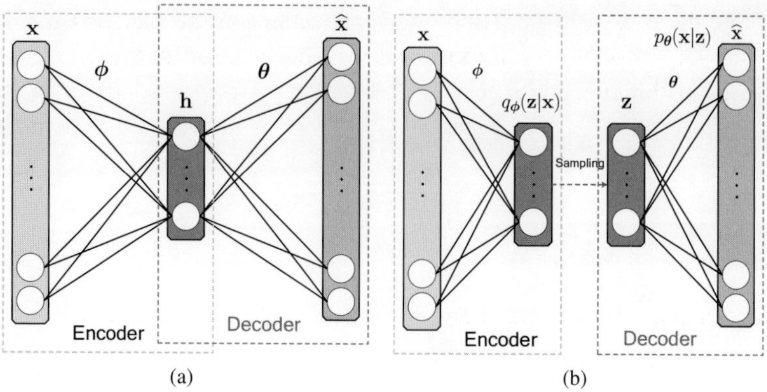

Figure 4.17 Network structures for (a) autoencoder and (b) variational autoencoder. [Adapted from *Variational Recurrent Neural Networks for Speech Separation (Figure 2), by J.T. Chien and K.T. Kuo, Proceedings Interspeech, 2017, pp. 1193–1197, ISCA.*]

4.5 Variational Autoencoder

Figures 4.17(a) and (b) show the model structures for an autoencoder and a variational autoencoder, respectively, which are designed to find the reconstructed data $\widehat{\mathbf{x}}$ corresponding to its original data $\mathbf{x}$ based on an encoder and decoder. A traditional autoencoder is seen as a deterministic model to extract the deterministic latent features $\mathbf{h}$. In [114], the variational autoencoder (VAE) was developed by incorporating the variational Bayesian learning into the construction of an autoencoder that is used as a building block for different stochastic neural networks [115, 116]. Uncertainty modeling is performed to build a latent variable model for data generation through stochastic hidden features $\mathbf{z}$. As addressed in Section 2.3, the variational inference involves the optimization of an approximation to an intractable posterior distribution. The standard mean-field approach requires analytical derivations to calculate the expectations with respect to the approximate posterior. Such a calculation is obviously intractable in the case of neural networks. A VAE provides an approach to illustrate how the reparameterization of variational lower bound is manipulated to yield a simple differentiable and unbiased estimator, also called the stochastic gradient variational Bayes method. This estimator is used to calculate the posterior distribution and carry out the approximate inference in presence of continuous latent variables. The optimization is implemented according to a standard stochastic gradient ascent technique as detailed in what follows.

4.5.1 Model Construction

Given a data collection $\mathbf{X} = \{\mathbf{x}_t\}_{t=1}^T$ consisting of T i.i.d. (identically, independently distributed) samples of continuous observation variable $\mathbf{x}$, we aim to train a generative model to synthesize a new value of $\mathbf{x}$ from its latent variable $\mathbf{z}$. The latent variable $\mathbf{z}$ and observation data $\mathbf{x}$ are generated by prior distribution $p_\theta(\mathbf{z})$ and likelihood function $p_\theta(\mathbf{x}|\mathbf{z})$, respectively, but the true parameters $\boldsymbol{\theta}$, as well as the values of $\mathbf{z}$, are unknown.

Notably, there is no assumption or simplification on the marginal distribution $p(\mathbf{x})$ or posterior distribution $p_\theta(\mathbf{z}|\mathbf{x})$. To deal with the issue that marginal likelihood or posterior distribution is intractable, the variational Bayes algorithm is introduced to approximate the logarithm of marginal likelihood of individual data points $\log p(\mathbf{x})$ or

$$\log p(\mathbf{x}_1, \ldots, \mathbf{x}_T) = \sum_{t=1}^{T} \log p(\mathbf{x}_t) \tag{4.45}$$

through maximizing the variational lower bound $\mathcal{L}(\theta, \phi)$ derived by

$$\begin{aligned}
\log p(\mathbf{x}) &= \log \sum_{\mathbf{z}} p_\theta(\mathbf{x}|\mathbf{z}) p_\theta(\mathbf{z}) \\
&= \log \sum_{\mathbf{z}} p_\theta(\mathbf{x}|\mathbf{z}) \frac{q_\phi(\mathbf{z}|\mathbf{x})}{q_\phi(\mathbf{z}|\mathbf{x})} p_\theta(\mathbf{z}) \\
&= \log \mathbb{E}_{q_\phi(\mathbf{z}|\mathbf{x})}\left[\frac{p_\theta(\mathbf{x}|\mathbf{z}) p_\theta(\mathbf{z})}{q_\phi(\mathbf{z}|\mathbf{x})} \right] \\
&\geq \mathbb{E}_{q_\phi(\mathbf{z}|\mathbf{x})}\Bigg[\underbrace{\log \left(p_\theta(\mathbf{x}|\mathbf{z}) \frac{p_\theta(\mathbf{z})}{q_\phi(\mathbf{z}|\mathbf{x})} \right)}_{f_\Theta(\mathbf{x},\mathbf{z})} \Bigg] \\
&= \mathbb{E}_{q_\phi(\mathbf{z}|\mathbf{x})}\left[\log p_\theta(\mathbf{x}|\mathbf{z}) \right] - \mathcal{D}_{\text{KL}}(q_\phi(\mathbf{z}|\mathbf{x}) \| p_\theta(\mathbf{z})) \triangleq \mathcal{L}(\theta, \phi)
\end{aligned} \tag{4.46}$$

where $\mathbf{\Theta} = \{\theta, \phi\}$ denote the whole parameter set of a VAE. By subtracting this lower bound from the log-marginal distribution, as referred to Eqs. 2.32 and 2.33, we obtain the relation

$$\log p(\mathbf{x}) = \mathcal{D}_{\text{KL}}(q_\phi(\mathbf{z}|\mathbf{x}) \| p_\theta(\mathbf{z}|\mathbf{x})) + \mathcal{L}(\theta, \phi). \tag{4.47}$$

Maximizing the likelihood function $p(\mathbf{x})$ turns out to maximizing lower bound $\mathcal{L}(\theta, \phi)$ while the KL divergence term is minimized, or equivalently, the variational distribution $q_\phi(\mathbf{z}|\mathbf{x})$ is estimated to be maximally close to the true posterior $p_\theta(\mathbf{z}|\mathbf{x})$.

Figure 4.18 shows the graphical representation of generative model based on a VAE. The variational autoencoder is composed of two parts. One is the recognition model or probabilistic encoder $q_\phi(\mathbf{z}|\mathbf{x})$, which is an approximation to the intractable true posterior $p_\theta(\mathbf{z}|\mathbf{x})$. The other part is the generative model or probabilistic decoder $p_\theta(\mathbf{x}|\mathbf{z})$, which generates the data sample $\mathbf{x}$ from latent code $\mathbf{z}$. Note that in contrast with the approximate posterior in mean-field variational inference, the factorized inference is not required in a VAE. Variational parameter ϕ and model parameter θ are not computed via a closed-form expectation. All parameters are jointly learned with this network structure by maximizing the variational lower bound $\mathcal{L}(\theta, \phi)$, as shown in Eq. 4.46, which consists of two terms: the KL divergence term corresponds to the optimization of the parameters for recognition model ϕ, and the other term corresponds to maximizing the log-likelihood of training samples given the generative model θ.

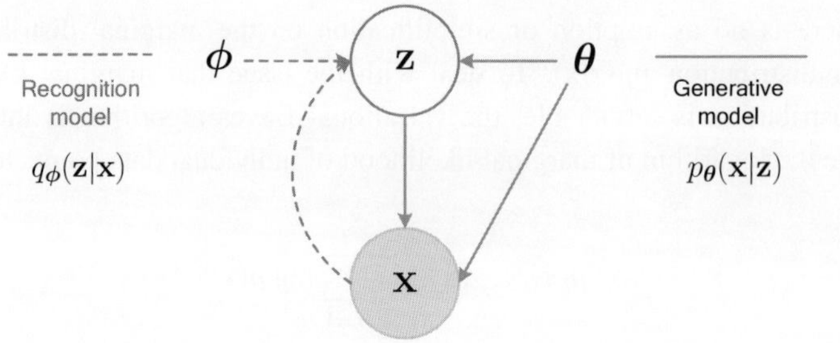

Figure 4.18 Graphical model for variational autoencoder. Solid lines denote the generative model $p_\theta(\mathbf{x}|\mathbf{z})$ and dash lines denote the variational approximation $q_\phi(\mathbf{z}|\mathbf{x})$ to the intractable posterior $p_\theta(\mathbf{z}|\mathbf{x})$. [Adapted from *Variational Recurrent Neural Networks for Speech Separation (Figure 2), by J.T. Chien and K.T. Kuo, Proceedings Interspeech, 2017, pp. 1193–1197, ISCA.*]

4.5.2 Model Optimization

Different from traditional error backpropagation algorithm developed for training neural networks with deterministic hidden variables $\mathbf{h}$, the stochastic neural network, i.e., the variational autoencoder in this section, is trained according to the stochastic backpropagation [117] where the random hidden variable $\mathbf{z}$ is considered in the optimization of the variational lower bound.

Stochastic Backpropagation

There is no closed-form solution for the expectation in the objective function of a VAE in Eq 4.46. The approximate inference based on sampling method, as introduced in Section 2.4, was employed in model inference for a VAE [114]. To implement the stochastic backpropagation, the expectation with respect to variational distribution $q_\phi(\mathbf{z}|\mathbf{x})$ in a VAE is simply approximated by sampling a hidden variable $\mathbf{z}^{(l)}$ and using this random sample to calculate the gradient for parameter updating. For example, the Gaussian distribution with mean vector $\boldsymbol{\mu}_z$ and diagonal covariance matrix $\sigma_z^2\mathbf{I}$

$$\mathbf{z}^{(l)} \sim q_\phi(\mathbf{z}|\mathbf{x}) = \mathcal{N}(\mathbf{z}|\boldsymbol{\mu}_z(\mathbf{x}), \sigma_z^2(\mathbf{x})\mathbf{I}) \tag{4.48}$$

is popularly assumed and used to find the M-dimensional random sample of continuous-density latent variable $\mathbf{z}^{(l)}$. Given the sample $\mathbf{z}^{(l)}$, the lower bound can be approximated by a *single sample* using

$$\begin{aligned}\mathcal{L}(\theta,\phi) &= \mathbb{E}_{q_\phi(\mathbf{z}|\mathbf{x})}\left[f_\Theta(\mathbf{x},\mathbf{z})\right]\\ &\simeq f_\Theta(\mathbf{x},\mathbf{z}^{(l)}).\end{aligned} \tag{4.49}$$

Figure 4.19(a) shows three steps toward fulfilling the stochastic gradient estimator for model parameter and variational parameter $\Theta = \{\theta,\phi\}$ where the single sample $\mathbf{z}^{(l)}$ is used in the expectation.

(1) Sample the latent variable $\mathbf{z}^{(l)}$ from $q_\phi(\mathbf{z}|\mathbf{x})$.
(2) Replace the variable $\mathbf{z}$ by $\mathbf{z}^{(l)}$ to obtain the objective $\mathcal{L} \simeq f_\Theta(\mathbf{x},\mathbf{z}^{(l)})$.

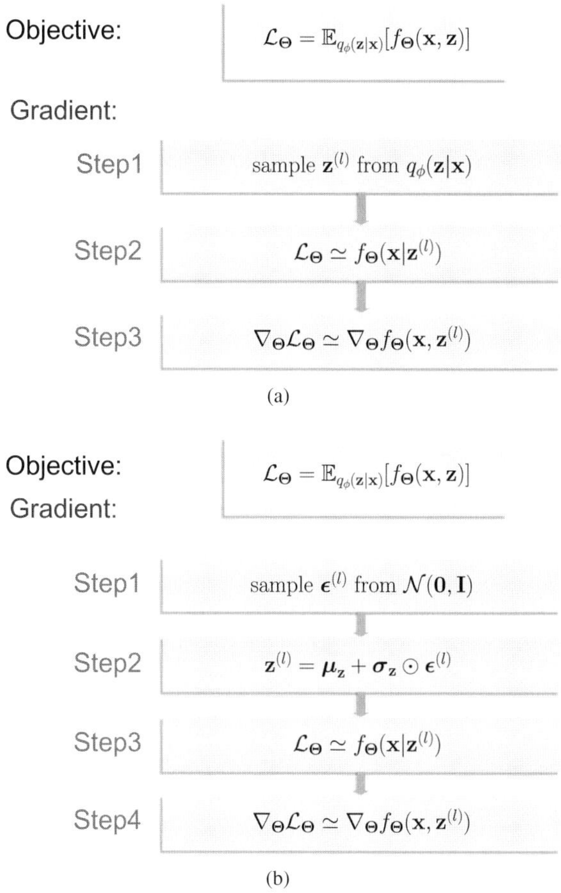

Figure 4.19 Objectives and implementation steps for (a) stochastic backpropagation and (b) stochastic gradient variational Bayes.

(3) Individually take gradient of $\mathcal{L}$ with respect to parameters $\boldsymbol{\theta}$ and $\boldsymbol{\phi}$ to obtain $\nabla_\Theta \mathcal{L} \simeq \nabla_\Theta f_\Theta(\mathbf{x}, \mathbf{z}^{(l)})$.

As shown in Figure 4.17(b), the encoder of a VAE is configured as a two-layer neural network with variational parameter $\boldsymbol{\phi}$ where the D-dimensional input data $\mathbf{x}$ is used to estimate the variational posterior $q_\phi(\mathbf{z}|\mathbf{x})$. Practically, the hidden layer is arranged with M units for mean vector $\boldsymbol{\mu}_z(\mathbf{x})$ and M units for diagonal covariance matrix $\sigma_z^2(\mathbf{x})\mathbf{I}$, which are both functions of input data $\mathbf{x}$. Hidden variable $\mathbf{z}$ corresponding to input $\mathbf{x}$ is then sampled from this Gaussian distribution.

Although the standard gradient estimator is valid by using variational posterior $q_\phi(\mathbf{z}|\mathbf{x})$, the problem of *high variance* will happen [117] if latent variable $\mathbf{z}$ is *directly* sampled from a Gaussian distribution $\mathcal{N}(\mathbf{z}|\boldsymbol{\mu}_z(\mathbf{x}), \sigma_z^2(\mathbf{x})\mathbf{I})$. To deal with this problem, a reparameterization trick [114] is introduced and applied as the variance reduction method, which will be detailed in the following section.

Stochastic Gradient Variational Bayes

Stochastic gradient variational Bayes (SGVB) is a method to address the high variance issue in a VAE. In this implementation, the sampling based on a chosen distribution $q_\phi(\mathbf{z}|\mathbf{x})$ is executed via a reparameterization trick. This trick is performed by reparameterizing the random variable $\mathbf{z} \sim q_\phi(\mathbf{z}|\mathbf{x})$ using a differentiable transformation function $g_\phi(\mathbf{x}, \epsilon)$ with an auxiliary random noise ϵ

$$\widetilde{\mathbf{z}} = g_\phi(\mathbf{x}, \epsilon) \tag{4.50}$$

with $\epsilon \sim p(\epsilon)$ where $p(\epsilon)$ is chosen for an appropriate distribution. Basically, the M-dimensional standard Gaussian distribution

$$p(\epsilon) = \mathcal{N}(\epsilon|\mathbf{0}, \mathbf{I}) \tag{4.51}$$

is adopted. In this way, the latent variable $\mathbf{z}$ is viewed as a deterministic mapping function with deterministic parameter ϕ but driven by a standard Gaussian variable ϵ. A stable *random* variable sampled by the variational posterior $q_\phi(\mathbf{z}|\mathbf{x})$ is then approximated by the mapping function $g_\phi(\mathbf{x}, \epsilon)$. The reduction of variance in sampling can be assured. More specially, we can calculate the Monte Carlo estimate of the expectation of a target function $f(\mathbf{z})$ with respect to $q_\phi(\mathbf{z}|\mathbf{x})$ by using L samples of $\mathbf{z}$ drawn from $q_\phi(\mathbf{z}|\mathbf{x})$, by referring to Eqs. 2.50 and 2.51, via the reparameterization trick

$$\begin{aligned}\mathbb{E}_{q_\phi(\mathbf{z}|\mathbf{x})}[f(\mathbf{z})] &= \mathbb{E}_{q_\phi(\mathbf{z}|\mathbf{x})}[f(g_\phi(\mathbf{x}, \epsilon))] \\ &\simeq \frac{1}{L}\sum_{l=1}^{L} f(g_\phi(\mathbf{x}, \epsilon^{(l)}))\end{aligned} \tag{4.52}$$

where $\epsilon^{(l)} \sim \mathcal{N}(\epsilon|\mathbf{0}, \mathbf{I})$. An SGVB estimator is then developed by maximizing the variational lower bound in Eq. 4.46, which is written in a form of

$$\mathcal{L}(\theta, \phi) = \frac{1}{L}\sum_{l=1}^{L} \log p_\theta(\mathbf{x}|\mathbf{z}^{(l)}) - \mathcal{D}_{\mathrm{KL}}(q_\phi(\mathbf{z}|\mathbf{x})\|p_\theta(\mathbf{z})) \tag{4.53}$$

where $\mathbf{z}^{(l)} = g_\phi(\mathbf{x}, \epsilon^{(l)})$ and $\epsilon^{(l)} \sim p(\epsilon)$. In this way, the variance of the gradient simply depends on the noise term ϵ, which is sampled from distribution $p(\epsilon)$. When compared with directly sampling of $\mathbf{z}$, the variance of sampling $p(\epsilon)$ is considerably reduced.

Algorithm 2 Autoencoding variational Bayes algorithm

Initialize hidden state parameters θ, ϕ
For parameters (θ, ϕ) until convergence
  $\mathbf{X}_n \leftarrow$ Random minibatch with T_n data points from full dataset $\mathbf{X}$
  $\epsilon \leftarrow$ Random samples from noise distribution $p(\epsilon)$
  $\mathbf{g} \leftarrow \nabla_{\theta,\phi}\widetilde{\mathcal{L}}(\theta, \phi; \mathbf{X}_n, \epsilon)$ (Gradient of Eq. (4.54))
  $\theta, \phi \leftarrow$ Update parameters using $\mathbf{g}$
End For
Return θ, ϕ

4.5.3 Autoencoding Variational Bayes

Based on the model construction in Section 4.5.1 and the model optimization in Section 4.5.2, the autoencoding variational Bayes (AEVB) algorithm is developed to fulfill the building block of variational autoencoder for deep learning. AEVB learning for variational autoencoder is shown in Algorithm 2. From the full data set $\mathbf{X} = \{\mathbf{x}_t\}_{t=1}^{T}$ with T data points, we first randomly sample a minibatch $\mathbf{X}_n = \{\mathbf{x}_t\}_{t=1}^{T_n}$ with T_n data points and calculate an estimator of variational lower bound of log marginal likelihood using a VAE as

$$\mathcal{L}(\boldsymbol{\theta}, \boldsymbol{\phi}; \mathbf{X}) \simeq \frac{T}{T_n} \widetilde{\mathcal{L}}(\boldsymbol{\theta}, \boldsymbol{\phi}; \mathbf{X}_n, \boldsymbol{\epsilon}) = \frac{T}{T_n} \sum_{t=1}^{T_n} \widetilde{\mathcal{L}}(\boldsymbol{\theta}, \boldsymbol{\phi}; \mathbf{x}_t, \boldsymbol{\epsilon}), \tag{4.54}$$

where

$$T = \sum_n T_n \tag{4.55}$$

and a random noise $\boldsymbol{\epsilon}$ is used by sampling from a standard Gaussian distribution ϵ. Then, the gradients of variational lower bound with respect to a VAE parameters $\{\boldsymbol{\theta}, \boldsymbol{\phi}\}$ are calculated to perform the stochastic optimization based on the stochastic gradient ascent or the adaptive gradient (AdaGrad) algorithm [107] via

$$\{\boldsymbol{\theta}, \boldsymbol{\phi}\} \leftarrow \{\boldsymbol{\theta}, \boldsymbol{\phi}\} + \eta \underbrace{\nabla_{\boldsymbol{\theta}, \boldsymbol{\phi}} \widetilde{\mathcal{L}}(\boldsymbol{\theta}, \boldsymbol{\phi}; \mathbf{X}_n, \boldsymbol{\epsilon})}_{\triangleq \mathbf{g}}. \tag{4.56}$$

In the implementation, an isotropic Gaussian distribution is adopted as the prior density of latent variable

$$p_{\boldsymbol{\theta}}(\mathbf{z}) = \mathcal{N}(\mathbf{z}|\mathbf{0}, \mathbf{I}). \tag{4.57}$$

Likelihood function $p_{\boldsymbol{\theta}}(\mathbf{x}_t|\mathbf{z})$ for the reconstructed data $\widehat{\mathbf{x}}_t$ is either modeled by a multivariate Gaussian distribution for real-valued data or a Bernoulli distribution for binary data where the distribution parameters are computed from $\mathbf{z}$ by using a decoder neural network with parameter $\boldsymbol{\theta}$. More specifically, AEVB algorithm is performed to maximize the variational lower bound jointly for the variational parameter $\boldsymbol{\theta}$ in the encoder and the model parameter $\boldsymbol{\phi}$ in the decoder. The variational posterior distribution $q_{\boldsymbol{\phi}}(\mathbf{z}|\mathbf{x}_t)$ is approximated by a multivariate Gaussian with parameters calculated by an encoder neural network $[\boldsymbol{\mu}_z(\mathbf{x}_t), \boldsymbol{\sigma}_z^2(\mathbf{x}_t)] = f_{\boldsymbol{\phi}}^{\text{enc}}(\mathbf{x}_t)$. As explained in Section 4.5.2, we sample the M-dimensional posterior latent variable by

$$\mathbf{z}^{(l)} = g_{\boldsymbol{\phi}}(\mathbf{x}_t, \boldsymbol{\epsilon}^{(l)}) = \boldsymbol{\mu}_z(\mathbf{x}_t) + \boldsymbol{\sigma}_z(\mathbf{x}_t) \odot \boldsymbol{\epsilon}^{(l)}, \tag{4.58}$$

where $\odot$ denotes the element-wise product of two vectors and $\boldsymbol{\epsilon}^{(l)}$ is a standard Gaussian variable with $\mathcal{N}(\mathbf{0}, \mathbf{I})$. The resulting estimator for the variational lower bound in Eq. 4.54 using a speech observation $\mathbf{x}_t$ is expressed by

$$
\begin{aligned}
\widetilde{\mathcal{L}}(\boldsymbol{\theta}, \boldsymbol{\phi}; \mathbf{x}_t, \boldsymbol{\epsilon}) \simeq & \frac{1}{L} \sum_{l=1}^{L} \log p_{\boldsymbol{\theta}}(\mathbf{x}_t | \mathbf{z}^{(l)}) \\
& + \underbrace{\frac{1}{2} \sum_{j=1}^{M} \left(1 + \log(\sigma_{z,j}^2(\mathbf{x}_t)) - \mu_{z,j}^2(\mathbf{x}_t) - \sigma_{z,j}^2(\mathbf{x}_t)\right)}_{-\mathcal{D}_{\mathrm{KL}}(q_{\boldsymbol{\phi}}(\mathbf{z}|\mathbf{x}) \| p_{\boldsymbol{\theta}}(\mathbf{z}))},
\end{aligned}
\tag{4.59}
$$

where the second term on the right-hand side is calculated as a negative KL divergence between the variational posterior distribution $q_{\boldsymbol{\phi}}(\mathbf{z}|\mathbf{x})$ and the standard Gaussian distribution $p_{\boldsymbol{\theta}}(\mathbf{z})$. The KL term can be derived as

$$
-\mathcal{D}_{\mathrm{KL}}(q_{\boldsymbol{\phi}}(\mathbf{z}|\mathbf{x}) \| p_{\boldsymbol{\theta}}(\mathbf{z})) = \int q_{\boldsymbol{\phi}}(\mathbf{z}|\mathbf{x}) \left(\log p_{\boldsymbol{\theta}}(\mathbf{z}) - \log q_{\boldsymbol{\phi}}(\mathbf{z}|\mathbf{x})\right) d\mathbf{z} \tag{4.60}
$$

where

$$
\begin{aligned}
\int q_{\boldsymbol{\phi}}(\mathbf{z}|\mathbf{x}) \log p_{\boldsymbol{\theta}}(\mathbf{z}) d\mathbf{z} &= \int \mathcal{N}(\mathbf{z}|\boldsymbol{\mu}_z(\mathbf{x}_t), \boldsymbol{\sigma}_z^2(\mathbf{x}_t)) \log \mathcal{N}(\mathbf{z}|\mathbf{0}, \mathbf{I}) d\mathbf{z} \\
&= -\frac{M}{2} \log(2\pi) - \frac{1}{2} \sum_{j=1}^{M} \left(\mu_{z,j}^2(\mathbf{x}_t) + \sigma_{z,j}^2(\mathbf{x}_t)\right)
\end{aligned}
\tag{4.61}
$$

and

$$
\begin{aligned}
\int q_{\boldsymbol{\phi}}(\mathbf{z}|\mathbf{x}) \log q_{\boldsymbol{\phi}}(\mathbf{z}|\mathbf{x}) d\mathbf{z} &= \int \mathcal{N}(\mathbf{z}|\boldsymbol{\mu}_z(\mathbf{x}_t), \boldsymbol{\sigma}_z^2(\mathbf{x}_t)) \log \mathcal{N}(\mathbf{z}|\boldsymbol{\mu}_z(\mathbf{x}_t), \boldsymbol{\sigma}_z^2(\mathbf{x}_t)) d\mathbf{z} \\
&= -\frac{M}{2} \log(2\pi) - \frac{1}{2} \sum_{j=1}^{M} (1 + \log \sigma_j^2).
\end{aligned}
\tag{4.62}
$$

The AEVB algorithm is accordingly implemented to construct the variational autoencoder with parameters $\boldsymbol{\theta}$ and $\boldsymbol{\phi}$. This type of autoencoder can be stacked in a way as addressed in Section 4.4 to build the deep regression or classification network for speaker recognition. In general, a VAE is sufficiently supported by exploring the randomness of latent code $\mathbf{z}$. The capability as a generative model is assured to synthesize an artificial sample $\widehat{\mathbf{x}}$ from a latent variable $\mathbf{z}$ where the global structure of observation data are preserved to minimize the reconstruction error. In what follows, we introduce another paradigm of generative model, called the generative adversarial network, which produces artificial samples based on adversarial learning.

4.6 Generative Adversarial Networks

This section addresses the pros and cons of the generative models based on the generative adversarial network (GAN) [118] that has been extensively studied and developed for different applications including adversarial learning for speaker recognition [119, 120]. There are two competing neural networks that are estimated mutually through adversarial learning. Several generative models based on deep learning and maximum likelihood estimation have been discussed in Sections 4.1 and 4.5. Before explaining the details of GANs, we would like to point out the key idea and property where GANs perform differently when compared with the other generative models.

4.6.1 Generative Models

Figure 4.20 depicts the taxonomy of generative models. In general, the parameter of a generative model is estimated according to the maximum likelihood (ML) principle where a probabilistic model or distribution $p_{\text{model}}(\mathbf{x}|\boldsymbol{\theta})$ with parameter $\boldsymbol{\theta}$ characterize how a speech observation $\mathbf{x}$ is obtained, namely $\mathbf{x} \sim p_{\text{model}}(\mathbf{x}|\boldsymbol{\theta})$. Training samples are collected or drawn from an unknown data distribution by $\mathbf{x} \sim p_{\text{data}}(\mathbf{x})$. The ML method assumes that there exists a parameter $\boldsymbol{\theta}$ that make the model distribution exactly the same as the true data distribution

$$p_{\text{model}}(\mathbf{x}|\boldsymbol{\theta}) = p_{\text{data}}(\mathbf{x}). \tag{4.63}$$

The ML parameter $\boldsymbol{\theta}$ of a model distribution p_{model} is then estimated by maximizing the log likelihood function of training data collected from data distribution p_{data}, i.e., solving the optimization problem

$$\boldsymbol{\theta}_{\text{ML}} = \underset{\theta}{\text{argmax}}\, \mathbb{E}_{\mathbf{x}\sim p_{\text{data}}}[\log p_{\text{model}}(\mathbf{x}|\boldsymbol{\theta})]. \tag{4.64}$$

Traditionally, the ML solution is derived through an *explicit* model distribution where a specific distribution is assumed. Such a model distribution is possibly intractable. An approximate distribution is accordingly required. The approximation can be either based on the variational inference or driven by the Markov chain. The variational autoencoder in Section 4.5 adopts the variational distribution, which is a Gaussian, to approximate a true posterior distribution in variational inference. The restricted Boltzmann machine in Section 4.1 employs the explicit distributions of Bernoulli and Gaussian in a style of distribution approximation based on a Markov chain using the contrastive divergence

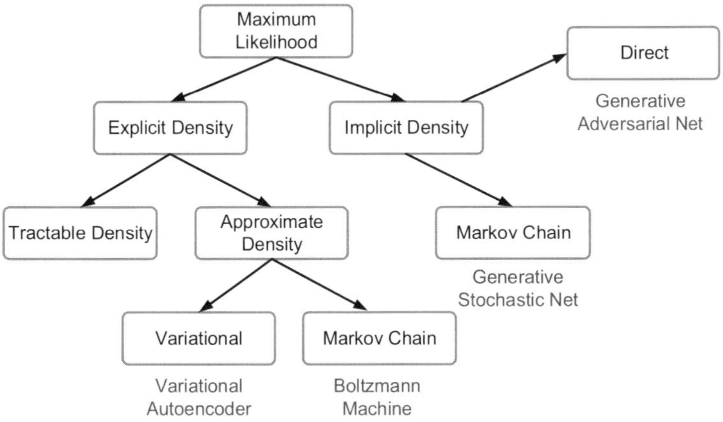

Figure 4.20 Taxonomy of generative models based on maximum likelihood estimation.

algorithm. On the other hand, the ML model can also be constructed without assuming any explicit distribution. In [121], the generative stochastic network was proposed to carry out a Markov chain to find state transitions for updating a generative model. An implicit distribution was actually adopted to drive the construction of the generative model.

In what follows, we are introducing the generative model based on the generative adversarial network. GAN presents a general idea to learn a sampling mechanism for data generation. The generation process is based on the sampling of a latent variable $\mathbf{z}$ by using a generator network G. The generated sample is distinguished from the true data sample by using a discriminator network D. There is no explicit density assumed in the optimization procedure. There is no Markov chain involved. The issue of *mixing* between the updated models of two states is mitigated. The property of asymptotically consistent performance is assured. GAN is likely to produce the best samples through the adversarial learning. GAN is seen as a *direct* model that directly estimates the data distribution without assuming distribution function. Some theoretical justification is provided to illustrate why GAN is powerful to estimate true data distribution as addressed in what follows.

4.6.2 Adversarial Learning

Generative adversarial network is a generative model consisting of two competing neural networks as illustrated in Figure 4.21. One of the competing neural networks is the generator G that takes an random hidden variable $\mathbf{z}$ from $p(\mathbf{z})$ where $\mathbf{z}$ is a sample from probability distribution $p(\mathbf{z})$ and is used to generate an artificial or fake sample. The other model is the discriminator D that receives samples from both the generator and the real samples. The discriminator is to take the input either from the real data or from the generator and tries to predict whether the input is real or generated. It takes an input $\mathbf{x}$

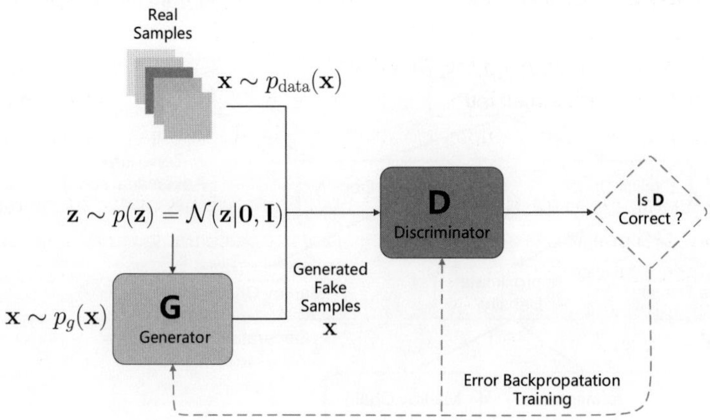

Figure 4.21 Illustration for generative adversarial network. [Adapted from *Adversarial Learning and Augmentation for Speaker Recognition (Figure 1), by J.T. Chien and K.T. Peng, Proceedings Odyssey 2018 The Speaker and Language Recognition Workshop, pp. 342–348, ISCA.*]

from real data distribution $p_{\text{data}}(\mathbf{x})$. The discriminator then solves a binary classification problem using a sigmoid function giving output in the range 0 to 1. More specifically, GAN is formulated as a two-player game between a generator with mapping function

$$G(\mathbf{z}) : \mathbf{z} \rightarrow \mathbf{x} \tag{4.65}$$

and a binary discriminator with the mapping

$$D(\mathbf{x}) : \mathbf{x} \rightarrow [0, 1] \tag{4.66}$$

via a *minimax* optimization over a value function $V(G, D)$ or an adversarial loss $\mathcal{L}_{\text{adv}}$. This minimax game is expressed by a joint optimization over G and D according to

$$\min_G \max_D V(G, D). \tag{4.67}$$

GAN aims to pursue a discriminator D, which optimally classifies if the input comes from true sample $\mathbf{x}$ (with discriminator output $D(\mathbf{x})$) or the fake sample $G(\mathbf{z})$ using a random code $\mathbf{z}$ for generator G (with discriminator output $D(G(\mathbf{z}))$). Meaningfully, the value function is formed as a *negative cross-entropy error* function for binary classification in a two-player game

$$\begin{aligned} V(G, D) = {} & \mathbb{E}_{\mathbf{x} \sim p_{\text{data}}(\mathbf{x})}[\log D(\mathbf{x})] \\ & + \mathbb{E}_{\mathbf{z} \sim p(\mathbf{z})}[\log(1 - D(G(\mathbf{z})))] \triangleq \mathcal{L}_{\text{adv}}, \end{aligned} \tag{4.68}$$

where $p_{\text{data}}(\mathbf{x})$ is data distribution and $p(\mathbf{z})$ is a distribution that draws a noise sample or latent code $\mathbf{z}$. Similar to a VAE in Section 4.5, GAN adopts the standard Gaussian distribution as the prior density of $\mathbf{z}$

$$\mathbf{z} \sim p(\mathbf{z}) = \mathcal{N}(\mathbf{z}|\mathbf{0}, \mathbf{I}). \tag{4.69}$$

Using GAN, $G(\mathbf{z})$ captures the data distribution and generates a sample $\mathbf{x}$ from $\mathbf{z}$. The discriminator produces a probability $0 \leq D(\mathbf{x}) \leq 1$, which measures how likely a data point $\mathbf{x}$ is sampled from the data distribution $p_{\text{data}}(\mathbf{x})$ as a real sample or from the generator $G(\mathbf{z})$ (with a distribution of generator $p_g(\mathbf{x})$ or equivalently a model distribution $p_{\text{model}}(\mathbf{x})$) as a fake sample. Correspondingly, discriminator D classifies a sample coming from the training data rather than G.

The generator G is trained via maximizing the probability of discriminator D to make a mistake. Generator G minimizes the classification accuracy of D. Generator $G(\mathbf{z})$ and discriminator $D(\mathbf{x})$ are both realized as the fully connected neural network models. This minimax optimization assures the *worst* performance in which a fake sample is classified as a real sample. A powerful generator model $G(\mathbf{z})$ is therefore implemented.

4.6.3 Optimization Procedure

There are two steps in the optimization procedure for training the discriminator and the generator.

First Step: Finding the Optimal Discriminator

To derive the formula for an optimal discriminator, we first rewrite the value function in Eq. 4.67 by

$$\begin{aligned} V(G,D) &= \mathbb{E}_{\mathbf{x}\sim p_{\text{data}}(\mathbf{x})}[\log D(\mathbf{x})] + \mathbb{E}_{\mathbf{z}\sim p(\mathbf{z})}[\log\left(1 - D(G(\mathbf{z}))\right)] \\ &= \int p_{\text{data}}(\mathbf{x}) \log D(\mathbf{x}) d\mathbf{x} + \int p(\mathbf{z}) \log\left(1 - D(G(\mathbf{z}))\right) d\mathbf{z} \\ &= \int \left\{ p_{\text{data}}(\mathbf{x}) \log(D(\mathbf{x})) + p_g(\mathbf{x}) \log(1 - D(\mathbf{x})) \right\} d\mathbf{x}, \end{aligned} \tag{4.70}$$

where the variable $\mathbf{z}$ in the second term in the right-hand side is manipulated and expressed by using the variable $\mathbf{x}$ so that $p(\mathbf{z})$ is expressed by $p_g(\mathbf{x})$ and $D(G(\mathbf{z})) = D(\mathbf{x})$. Taking the derivative of Eq. 4.70 with respect to the discriminator $D(\mathbf{x})$, the optimal discriminator is derived for any given generator G with a distribution $p_g(\mathbf{x})$ in a form of

$$D^*(\mathbf{x}) = \frac{p_{\text{data}}(\mathbf{x})}{p_{\text{data}}(\mathbf{x}) + p_g(\mathbf{x})} \tag{4.71}$$

A global optimum happens in the condition that the generator or model distribution is exactly the same as data distribution

$$p_g(\mathbf{x}) \text{ (or } p_{\text{model}}(\mathbf{x})) = p_{\text{data}}(\mathbf{x}), \tag{4.72}$$

which results in the worst performance in binary classification using the discriminator. Namely, no matter the input $\mathbf{x}$ to the discriminator is either drawn from the data distribution $p_{\text{data}}(\mathbf{x})$ or from the model distribution $p_g(\mathbf{x})$, the classification accuracy is always 50 percent because

$$D^*(\mathbf{x})\Big|_{p_g(\mathbf{x}) = p_{\text{data}}(\mathbf{x})} = 0.5. \tag{4.73}$$

Second Step: Finding the Optimal Generator

By substituting this discriminator into value function, the second step of GAN is to find the optimal generator G by solving a minimization problem

$$\begin{aligned} \min_G V(G, D^*) &= \min_G C(G) \\ &= \min_G \; \{2\mathcal{D}_{\text{JS}}(p_{\text{data}}(\mathbf{x}) \| p_g(\mathbf{x})) - \log 4\}, \end{aligned} \tag{4.74}$$

where $\mathcal{D}_{\text{JS}}(\cdot\|\cdot)$ denotes the Jensen–Shannon (JS) divergence. JS divergence quantifies how distinguishable two distributions p and q are from each other. It is related to the KL divergence by

$$\mathcal{D}_{\text{JS}}(p\|q) = \frac{1}{2}\mathcal{D}_{\text{KL}}\left(p \middle\| \frac{p+q}{2}\right) + \frac{1}{2}\mathcal{D}_{\text{KL}}\left(q \middle\| \frac{p+q}{2}\right). \tag{4.75}$$

In Eq. 4.74, the detailed derivation of the cost function $C(G)$ for finding optimal generator is shown as

$$\begin{aligned}
C(G) &= \max_D V(G,D) \\
&= \max_D \int \left\{ p_{\text{data}}(\mathbf{x}) \log(D(\mathbf{x})) + p_g(\mathbf{x}) \log(1 - D(\mathbf{x})) \right\} d\mathbf{x} \\
&= \int \left\{ p_{\text{data}}(\mathbf{x}) \log(D^*(\mathbf{x})) + p_g(\mathbf{x}) \log(1 - D^*(\mathbf{x})) \right\} d\mathbf{x} \\
&= \int \left\{ p_{\text{data}}(\mathbf{x}) \log\left(\frac{p_{\text{data}}(\mathbf{x})}{p_{\text{data}}(\mathbf{x}) + p_g(\mathbf{x})} \right) + p_g(\mathbf{x}) \log\left(\frac{p_g(\mathbf{x})}{p_{\text{data}}(\mathbf{x}) + p_g(\mathbf{x})} \right) \right\} d\mathbf{x} \\
&= \int \left\{ p_{\text{data}}(\mathbf{x}) \log\left(\frac{2p_{\text{data}}(\mathbf{x})}{p_{\text{data}}(\mathbf{x}) + p_g(\mathbf{x})} \right) + p_g(\mathbf{x}) \log\left(\frac{2p_g(\mathbf{x})}{p_{\text{data}}(\mathbf{x}) + p_g(\mathbf{x})} \right) \right\} d\mathbf{x} \\
&\quad - \log 4 \\
&= \mathcal{D}_{\text{KL}}\left(p_{\text{data}}(\mathbf{x}) \middle\| \frac{p_{\text{data}}(\mathbf{x}) + p_g(\mathbf{x})}{2} \right) + \mathcal{D}_{\text{KL}}\left(p_g(\mathbf{x}) \middle\| \frac{p_{\text{data}}(\mathbf{x}) + p_g(\mathbf{x})}{2} \right) - \log 4.
\end{aligned} \tag{4.76}$$

Eq. 4.74 is therefore obtained by substituting Eq. 4.75 into Eq. 4.76. According to Eq. 4.74, the optimal generator G^* is calculated to reflect the generator distribution $p_g(\mathbf{x})$, which is closest to real data distribution $p_{\text{data}}(\mathbf{x})$. GAN encourages G with $p_g(\mathbf{x})$ to fit $p_{\text{data}}(\mathbf{x})$ so as to fool D with its generated samples. The global minimum or the lower bound of $C(G)$

$$C^*(G) = -\log 4 \tag{4.77}$$

is achieved if and only if $p_g(\mathbf{x}) = p_{\text{data}}(\mathbf{x})$. G and D are trained to update the parameters of both models by the *error backpropagation* algorithm.

Training Algorithm

Algorithm 3 presents the stochastic training algorithm for estimating the parameters $\{\boldsymbol{\theta}_d, \boldsymbol{\theta}_g\}$ of discriminator D and generator G for the construction of generative adversarial network. Starting from an initial set of parameters, GAN is implemented by running K updating steps for discriminator parameter $\boldsymbol{\theta}_d$ before one updating step for generator parameter $\boldsymbol{\theta}_g$. The minibatches of noise samples $\mathbf{Z}_n = \{\mathbf{z}_t\}_{t=1}^{T_n}$ and training samples $\mathbf{X}_n = \{\mathbf{x}_t\}_{t=1}^{T_n}$ are drawn from $p(\mathbf{z})$ and $p_{\text{data}}(\mathbf{x})$, respectively, and then employed in different updating steps for $\boldsymbol{\theta}_d$ and $\boldsymbol{\theta}_g$. Gradient-based updates with momentum can be used [118]. Minimax optimization is performed because the value function is maximized to run the ascending of stochastic gradient for discriminator parameter $\boldsymbol{\theta}_d$ in Eq. 4.78 and is minimized to run the descending of stochastic gradient for generator $\boldsymbol{\theta}_g$ in Eq. 4.79. Notably, Eq. 4.79 does not include the first term of Eq. 4.78 because this term is independent of generator parameter $\boldsymbol{\theta}_g$.

Algorithm 3 SGD training for generative adversarial net

Initialize discriminator and generator parameters $\boldsymbol{\theta}_d, \boldsymbol{\theta}_g$

For number of training iterations

For K steps

Sample a minibatch of T_n noise samples $\{\mathbf{z}_1, \ldots, \mathbf{z}_{T_n}\}$ from prior $p(\mathbf{z})$

Sample a minibatch of T_n training examples $\{\mathbf{x}_1, \ldots, \mathbf{x}_{T_n}\}$ from $p_{\text{data}}(\mathbf{x})$

Update the discriminator by *ascending* its stochastic gradient

$$\boldsymbol{\theta}_d \leftarrow \nabla_{\boldsymbol{\theta}_d} \frac{1}{T_n} \sum_{t=1}^{T_n} \left[\log D(\mathbf{x}_t) + \log\left(1 - D(G(\mathbf{z}_t))\right) \right] \tag{4.78}$$

End For

Sample a minibatch of T_n noise samples $\{\mathbf{z}_1, \ldots, \mathbf{z}_{T_n}\}$ from prior $p(\mathbf{z})$

Update the generator by *descending* its stochastic gradient

$$\boldsymbol{\theta}_g \leftarrow \nabla_{\boldsymbol{\theta}_g} \frac{1}{T_n} \sum_{t=1}^{T_n} \log\left(1 - D(G(\mathbf{z}_t))\right) \tag{4.79}$$

End For

Return $\boldsymbol{\theta}_d, \boldsymbol{\theta}_g$

Interpretation for Learning Procedure

Figure 4.22(a) interprets how the discriminator $D(\mathbf{x})$ and generator $G(\mathbf{x})$ are run to estimate the distribution of generated data $p_g(\mathbf{x})$ relative to the unknown distribution of real data $p_{\text{data}}(\mathbf{x})$. Actually, the distribution of generated data $p_g(\mathbf{x})$ is obtained from the samples $\mathbf{x}$ generated by the generator $G(\mathbf{z})$ using the samples $\mathbf{z}$ drawn by prior density

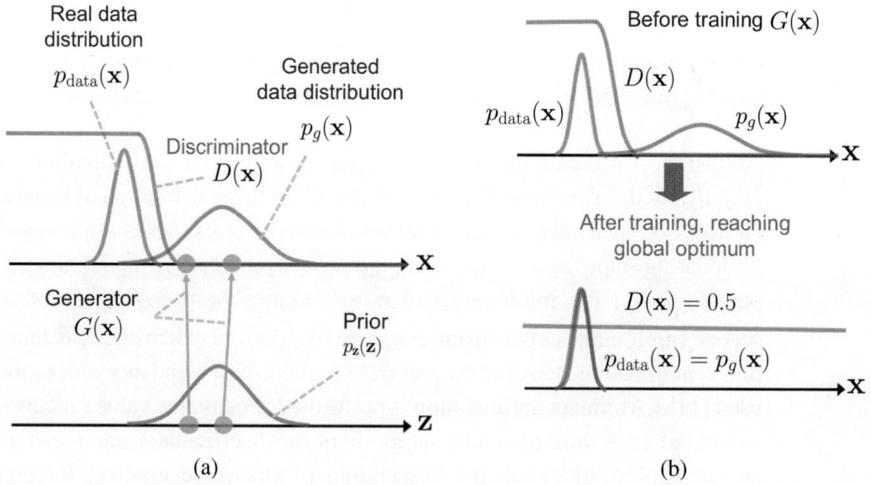

Figure 4.22 Interpretation of discriminator, generator, data distribution, and model distribution for generative adversarial networks.

$p(\mathbf{z})$ with a standard Gaussian distribution. The discriminator D will correctly determine whether the input data $\mathbf{x}$ comes from the real data distribution or from the generator distribution. The output of D is likely to be one for true data in low-value range or to be zero for fake data in high-value range.

Figure 4.22(b) shows the situation of data distribution $p_{\text{data}}(\mathbf{x})$ and model distribution $p_g(\mathbf{x})$ before and after GAN training. In the beginning of GAN training, the estimated model distribution is different from the original data distribution. The resulting discriminator classifies the data in region of $p_{\text{data}}(\mathbf{x})$ as $D(\mathbf{x}) = 1$ and the data in region of $p_g(\mathbf{x})$ as $D(\mathbf{x}) = 0$. After a number of updating steps and learning epochs, we may achieve the global optimum at $p_{\text{data}} = p_g$ where the output of discriminator becomes $D(\mathbf{x}) = 0.5$ in various data region. Nevertheless, the joint training of discriminator and generator may be difficult and easily stuck in local optimum.

The training procedure for GAN can be further demonstrated by Figures 4.23(a) and (b) where the optimal discriminator and generator are trained and searched in the spaces of D and G, respectively. In estimating the discriminator, the maximization problem

$$D^* = \underset{D}{\operatorname{argmax}}\, V(G, D) \tag{4.80}$$

is solved to find the optimum D^*. Using the estimated discriminator D^*, the resulting value function is minimized to estimate the optimal generator G^* by

$$G^* = \underset{G}{\operatorname{argmax}}\, V(G, D^*). \tag{4.81}$$

The global optimum happens in an *equilibrium* condition for minimax optimization where the *saddle point* of the value function $V(G, D)$ is reached in a hybrid space of the discriminator and the generator. However, the value manifold in real-world applications is much more complicated than this example. The training procedure becomes difficult for GAN that uses neural networks as the discriminator and the generator in presence of heterogeneous training data. Although GAN is theoretically meaningful from the optimization perspective, in the next section, we discuss a number of challenges in the training procedure that will affect the performance of data generation.

4.6.4 Gradient Vanishing and Mode Collapse

Generative adversarial networks are generally difficult to optimize. Their training procedure is prone to be unstable. A careful setting and design of network architecture is required to balance the model capacities between discriminator and generator so that a successful training with convergence can be achieved. This section addresses two main issues in the training procedure that are caused by the unbalance in model capacity. One is the gradient vanishing and the other is the mode collapse.

First of all, the *gradient vanishing* problem may easily happen in the optimization procedure in accordance with a minimax game where the discriminator minimizes a cross-entropy error function, but the generator maximizes the same cross-entropy error function. This is unfortunate for the generator because when the discriminator

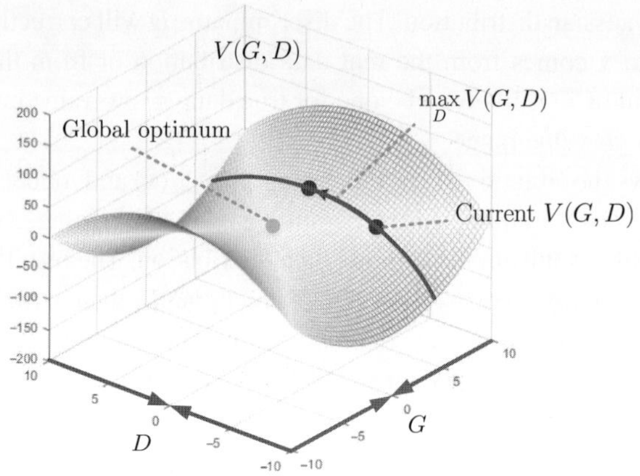

(a)

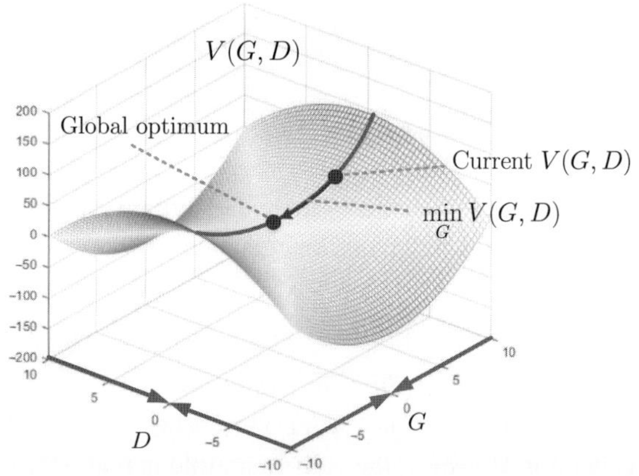

(b)

Figure 4.23 Optimization steps for finding (a) optimal discriminator D and then (b) optimal generator G. The D- and G-axis represent the parameter space of D and G, respectively.

successfully rejects the generated data with high confidence, the gradient of cross-entropy error or value function with respect to the generator will vanish. No updating or slow updating is performed for the generator. Only the discriminator is updated continuously. Therefore, the generator is too weak and the discriminator is too strong. The discriminator is completely optimal and achieves 100 percent accuracy in binary classification, which means that the classification probability is $D^*(\mathbf{x})$ for real data $\mathbf{x} \sim p_{\text{data}}(\mathbf{x})$ and $(1 - D^*(\mathbf{x}))$ for generated data $\mathbf{x} \sim p_g(\mathbf{x})$ are surely 1 and 0, respectively. This shows that $\nabla_{\theta_g} D^*(G(\mathbf{z})) = 0$ because $\log(1 - D^*(G(\mathbf{z}))) \approx \log 1$. Such a problem usually happens when the generator is poorly designed or built with weak capacity. To deal with this problem, a common practice is to reformulate the optimization for the generator from Eq. 4.79 to a minimization of the generator loss

$$\mathbb{E}_{\mathbf{z}\sim p(\mathbf{z})}[-\log D^*(G(\mathbf{z}))] \quad \text{or} \quad \mathbb{E}_{\mathbf{x}\sim p_g(\mathbf{x})}[-\log D^*(\mathbf{x})], \tag{4.82}$$

which yields the same solution but exhibits much stronger gradient for updating.

However, the *mode collapse* problem is further exacerbated in the training procedure if the generator loss in Eq. 4.82 is minimized. The key issue in the implementation of GAN is the instability when calculating the gradients of the value function with respect to $\boldsymbol{\theta}_d$ and $\boldsymbol{\theta}_g$. This issue results in mode collapse in the trained model where the generator or the model could not produce a variety of samples with sufficient randomness. To address this issue, we first express the loss function of a standard GAN given by an optimal discriminator $D^*(G(\mathbf{z}))$ or $D^*(\mathbf{x})$ as

$$\begin{aligned}&\mathbb{E}_{\mathbf{x}\sim p_{\text{data}}(\mathbf{x})}\big[\log D^*(\mathbf{x})\big] + \mathbb{E}_{\mathbf{x}\sim p_g(\mathbf{x})}\big[\log(1 - D^*(\mathbf{x}))\big]\\ &\quad = 2\mathcal{D}_{\text{JS}}(p_{\text{data}}(\mathbf{x})\|p_g(\mathbf{x})) - \log 4,\end{aligned} \tag{4.83}$$

which is minimized to estimate the optimal generator G or equivalently the optimal generator distribution $p_g(\mathbf{x})$. From a representation of Kullback–Leibler divergence between $p_g(\mathbf{x})$ and $p_{\text{data}}(\mathbf{x})$

$$\begin{aligned}\mathcal{D}_{\text{KL}}(p_g(\mathbf{x})\|p_{\text{data}}(\mathbf{x})) &= \mathbb{E}_{\mathbf{x}\sim p_g(\mathbf{x})}\left[\log\frac{p_g(\mathbf{x})}{p_{\text{data}}(\mathbf{x})}\right]\\ &= \mathbb{E}_{\mathbf{x}\sim p_g(\mathbf{x})}\left[\log\frac{\frac{p_g(\mathbf{x})}{p_{\text{data}}(\mathbf{x})+p_g(\mathbf{x})}}{\frac{p_{\text{data}}(\mathbf{x})}{p_{\text{data}}(\mathbf{x})+p_g(\mathbf{x})}}\right]\\ &= \mathbb{E}_{\mathbf{x}\sim p_g(\mathbf{x})}\left[\log\frac{1 - D^*(\mathbf{x})}{D^*(\mathbf{x})}\right]\\ &= \mathbb{E}_{\mathbf{x}\sim p_g(\mathbf{x})}[\log(1 - D^*(\mathbf{x}))] - \mathbb{E}_{\mathbf{x}\sim p_g}[\log D^*(\mathbf{x})].\end{aligned} \tag{4.84}$$

We derive an equation for the modified generator loss in Eq. 4.82 by

$$\begin{aligned}&\mathbb{E}_{\mathbf{x}\sim p_g(\mathbf{x})}[-\log D^*(\mathbf{x})]\\ &\quad = \mathcal{D}_{\text{KL}}(p_g(\mathbf{x})\|p_{\text{data}}(\mathbf{x})) - \mathbb{E}_{\mathbf{x}\sim p_g(\mathbf{x})}[\log(1 - D^*(\mathbf{x}))]\\ &\quad = \mathcal{D}_{\text{KL}}(p_g(\mathbf{x})\|p_{\text{data}}(\mathbf{x})) - 2\mathcal{D}_{\text{JS}}(p_{\text{data}}(\mathbf{x})\|p_g(\mathbf{x}))\\ &\qquad + \mathbb{E}_{\mathbf{x}\sim p_{\text{data}}(\mathbf{x})}[\log D^*(\mathbf{x})] + \log 4,\end{aligned} \tag{4.85}$$

which is obtained by substituting Eqs. 4.84 and 4.83 into Eq. 4.82.

There are two problems when minimizing Eq. 4.82 or equivalently Eq. 4.85 for the estimation of optimal generator G^*. First, during the minimization of Eq. 4.85, we simultaneously minimize the KL divergence $\mathcal{D}_{\text{KL}}(p_g(\mathbf{x})\|p_{\text{data}}(\mathbf{x}))$ and maximizing the JS divergence $\mathcal{D}_{\text{JS}}(p_{\text{data}}(\mathbf{x})\|p_g(\mathbf{x}))$. This circumstance is very strange and inconsistent and makes the calculation of gradients unstable. Second, KL divergence is *not* symmetric. In addition to the condition of global optimum $p_g(\mathbf{x}) = p_{\text{data}}(\mathbf{x})$, there are two possibilities to obtain the lowest divergence

$$\mathcal{D}_{\text{KL}}(p_g(\mathbf{x})\|p_{\text{data}}(\mathbf{x})) \to 0, \quad \text{if } p_g(\mathbf{x}) \to 0 \text{ and } p_{\text{data}}(\mathbf{x}) \to 1 \tag{4.86}$$

and

$$\mathcal{D}_{\mathrm{KL}}(p_{\mathrm{data}}(\mathbf{x})\|p_g(\mathbf{x})) \to 0, \quad \text{if } p_{\mathrm{data}}(\mathbf{x}) \to 0 \text{ and } p_g(\mathbf{x}) \to 1. \tag{4.87}$$

Eq. 4.86 corresponds to the case that the generator is too weak to produce the samples with sufficient variety. Eq. 4.87 is seen as the case that the generator is too arbitrary to generate realistic and meaningful samples. Because of these two different possibilities, the generative model based on GAN may repeatedly produce similar samples or weakly synthesized samples with lack of variety. Both penalties considerably cause the issue of mode collapse in the training procedure of the generative model.

In what next, the generative adversarial network is employed to conduct the probabilistic autoencoder as a new type of deep generative model. The variational inference in Section 2.3 is again introduced in the implementation.

4.6.5 Adversarial Autoencoder

An AAE [122] was proposed by incorporating the generative adversarial network (GAN), as addressed in Section 4.6, into the construction of a VAE, as described in Section 4.5. An AAE is seen as a variant of a VAE where the latent variable $\mathbf{z}$ is learned via adversarial learning through a discriminator. Alternatively, an AAE is also viewed as a variant of GAN driven by variational inference where the discriminator in adversarial learning is trained to tell if the latent variable $\mathbf{z}$ is either inferred from real data $\mathbf{x}$ with a variational distribution $q_\phi(\mathbf{z}|\mathbf{x})$ or artificially generated from a prior density $p(\mathbf{z})$. The neural network parameter ϕ is estimated to calculate the data-dependent mean and variance vectors $\{\mu_\phi(\mathbf{x}), \sigma^2_\phi(\mathbf{x})\}$ of a Gaussian distribution of $\mathbf{z}$ in the encoder output of a neural network where

$$\mathbf{z} \sim \mathcal{N}(\mathbf{z}|\mu_\phi(\mathbf{x}), \sigma^2_\phi(\mathbf{x})\mathbf{I}). \tag{4.88}$$

Figure 4.24 depicts how the adversarial learning is merged in an adversarial autoencoder. An AAE is constructed with three components, which are encoder, discriminator, and decoder. The building components of an AAE are different from those of a VAE consisting of an encoder and a decoder and those of GAN, containing a generator and a discriminator. Similar to a VAE the latent code $\mathbf{z}$ in an AAE is sampled from a variational distribution $q_\phi(\mathbf{z}|\mathbf{x})$ through an inference process. Different from a VAE the decoder in an AAE is learned as a deep generative model that maps the imposed prior $p(\mathbf{z})$ to the reconstructed data distribution. Such a mapping is estimated by adversarial learning through a discriminator that is optimally trained to judge $\mathbf{z}$ is drawn from a variational posterior $q_\phi(\mathbf{z}|\mathbf{x})$ or came from an imposed prior $p(\mathbf{z})$.

A VAE and AAE are further compared as follows. First of all, a VAE directly reconstructs an input data through an encoder and a decoder while an AAE indirectly generates data with an auxiliary discriminator that never sees the data $\mathbf{x}$ but classifies the latent variable $\mathbf{z}$. Second, a VAE optimizes the encoder and decoder by maximizing the evidence lower bound $\mathcal{L}(\theta, \phi)$, which consists of a log likelihood term for the lowest reconstruction error and a regularization term for the best variational posterior $q_\phi(\mathbf{z}|\mathbf{x})$. The derivative of the learning objective is calculated with respect to the decoder

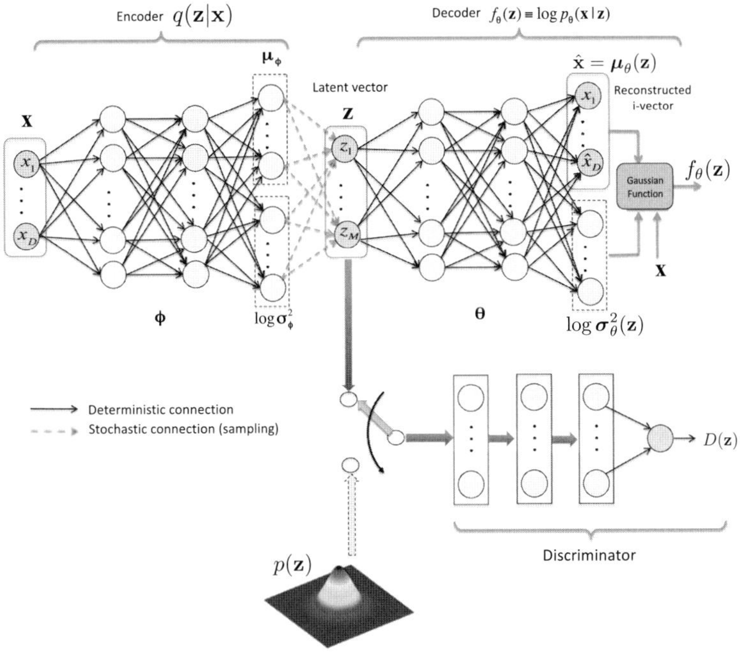

Figure 4.24 Adversarial autoencoder for data reconstruction, which consists of an encoder, a decoder, and a discriminator.

parameter $\boldsymbol{\theta}$ and encoder parameter $\boldsymbol{\phi}$ through the continuous hidden variable $\mathbf{z}$. Error backpropagation could not directly run with discrete latent variable $\mathbf{z}$. On the other hand, an AAE optimizes the encoder, decoder, and discriminator by maximizing the evidence lower bound $\mathcal{L}(\boldsymbol{\theta}, \boldsymbol{\phi})$ as well as minimizing the cross-entropy error function of binary classification of latent variable $\mathbf{z}$ either from the variational posterior $q(\mathbf{z}|\mathbf{x})$ or from the imposed prior $p(\mathbf{z})$. The derivative of the learning objective over three components via minimax procedure is likely unstable. Third, a VAE and an AAE may be both underfitting but caused by different reasons. The capacity of a VAE is degraded due to the Bayesian approximation in variational inference while the capacity of an AAE is considerably affected by the non-convergence in the learning procedure due to minimax optimization. From the perspective of data generation, a VAE basically synthesizes the data where the global composition is assured but with blurred details. On the contrary, an AAE is trained for data generation where the local features of training data are captured but the global structure of data manifold is disregarded.

To balance the tradeoff between a VAE and an AAE, a new combination of a VAE and GAN was proposed in [123] as a new type of autoencoder. The construction of a hybrid VAE and GAN is illustrated in Figure 4.25. Using this hybrid VAE and GAN, the feature representation learned in GAN discriminator is used as the basis for a VAE reconstruction objective. The hybrid model is trained to estimate the optimal discriminator with parameter $\boldsymbol{\theta}_d$, which almost could not tell the difference between true data $\mathbf{x}$ or fake data $\widehat{\mathbf{x}}$, which is reconstructed by a decoder or a generator with parameter $\boldsymbol{\theta}_g$ from

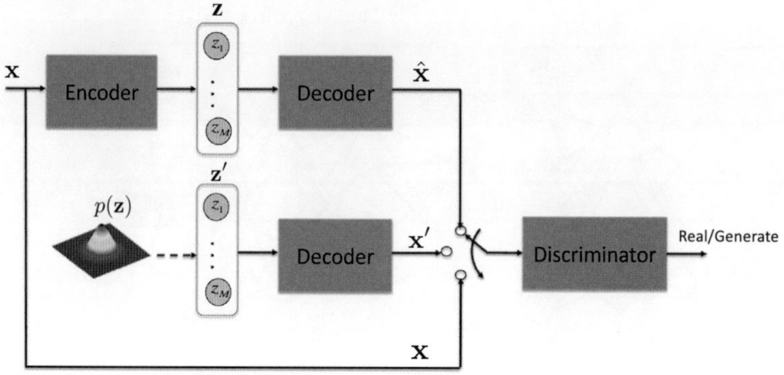

Figure 4.25 A hybrid VAE and GAN model where the decoder and generator are unified.

an inferred latent variable $\mathbf{z}$ through an encoder with parameter $\boldsymbol{\theta}_e$. The parameters of the encoder, generator, and discriminator $\{\boldsymbol{\theta}_e, \boldsymbol{\theta}_g, \boldsymbol{\theta}_d\}$ are jointly learned by a minimax optimization procedure over a learning objective consisting of an evidence lower bound and a cross-entropy error function. Different from an AAE, this hybrid model adopts the cross-entropy function measured in data domain $\mathbf{x}$ rather than in feature domain $\mathbf{z}$. Using a hybrid VAE and GAN, the evidence lower bound is maximized to pursue the learned similarity in feature space $\mathbf{z}$, which is different from the learned similarity of a VAE in data space $\mathbf{x}$. Abstract or high-level reconstruction error is minimized. Global structure is captured with the preserved local features.

4.7 Deep Transfer Learning

Traditional machine learning algorithms in speaker recognition work well under a common assumption that training and test data are in the same feature space and follow the same distribution. However, the real-world speech signals of different speakers may not follow this assumption due to the varying feature space or the mismatch between training and test conditions. In practical circumstances, we may train a speaker recognition system in a *target domain*, but there are only sufficient training data in the other *source domain*. The training data in different domains basically follow by different distributions or locate in different feature spaces. As shown in Figure 4.26(a), the traditional learning models are separately trained from scratch by using newly collected data from individual domains in presence of various distributions and feature spaces. Learning tasks in existing domains are assumed to be independent. However, the collection and labeling of training data in new domain are expensive. Solely relying on supervised learning is impractical. It is crucial to conduct the knowledge transfer with semi-supervised learning so as to reduce the labeling cost and improve the system utility.

Knowledge is transferred across different domains through learning a good feature representation or a desirable model adaptation method. More specifically, it is beneficial to cotrain the feature representation and the speaker recognition model to build a domain

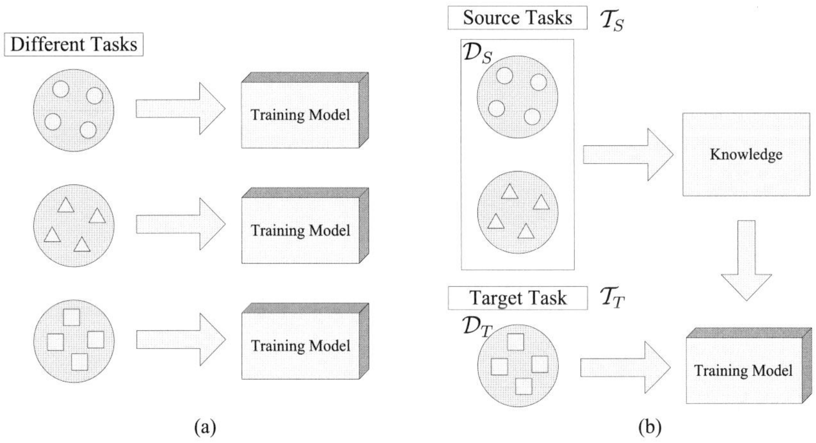

Figure 4.26 Learning processes based on (a) traditional machine learning and (b) transfer learning.

invariant system without the labeling in the target domain. Such an emerging research topic on transfer learning or in particular domain adaptation has been recently attracting the attention of many researchers in speaker recognition areas. Accordingly, this section will first address the general definition of transfer learning and discuss the relationships among different transfer learning settings such as multi-task learning and domain adaptation. Then, domain adaptation is emphasized and described in different types of solutions. Finally, a variety of deep learning solutions to transfer learning or domain adaptation are detailed for the development of modern speaker recognition systems.

4.7.1 Transfer Learning

Figure 4.26(b) conceptually depicts how knowledge transfer is performed for domains from different source tasks to a target task [124]. The domains and tasks are defined as follows. A domain $\mathcal{D} = \{\mathcal{X}, p(X)\}$ is composed of feature space $\mathcal{X}$ and a marginal probability distribution $p(X)$, where $X = \{\mathbf{x}_1, \ldots, \mathbf{x}_n\} \subset \mathcal{X}$. Here, $\mathbf{x}_i$ means the ith training sample and $\mathcal{X}$ is the space of all samples. Fundamentally, if two domains are different, their observation data may follow different marginal probability distributions in different feature spaces. In addition, the task is denoted as $\mathcal{T} = \{\mathcal{Y}, f(\cdot)\}$. Basically, a task is composed of a label space $\mathcal{Y}$ and an objective predictive function $f(\cdot)$, which is expressed as $p(Y|X)$ from a probabilistic perspective and can be learned from the training data. Let $\mathcal{D}_S = \{(\mathbf{x}_1^s, \mathbf{y}_1^s), \ldots, (\mathbf{x}_m^s, \mathbf{y}_m^s)\}$ denotes the training data in a source domain, where $\mathbf{x}_i^s \in \mathcal{X}_s$ means the observation input and $\mathbf{y}_i^s \in \mathcal{Y}_s$ corresponds to its label information. Similarly, the training data in the target domain are denoted as $\mathcal{D}_T = \{(\mathbf{x}_1^t, \mathbf{y}_1^t), \ldots, (\mathbf{x}_n^t, \mathbf{y}_n^t)\}$, where $\mathbf{x}_i^t \in \mathcal{X}_t$ and $\mathbf{y}_i^t \in \mathcal{Y}_t$.

Traditional machine learning methods strongly assume that the source domain and the target domain are the same, i.e., $\mathcal{D}_S = \mathcal{D}_T$, and that the source task and the target task are identical, i.e., $\mathcal{T}_S = \mathcal{T}_T$. When such an assumption does not exist in learning scenario,

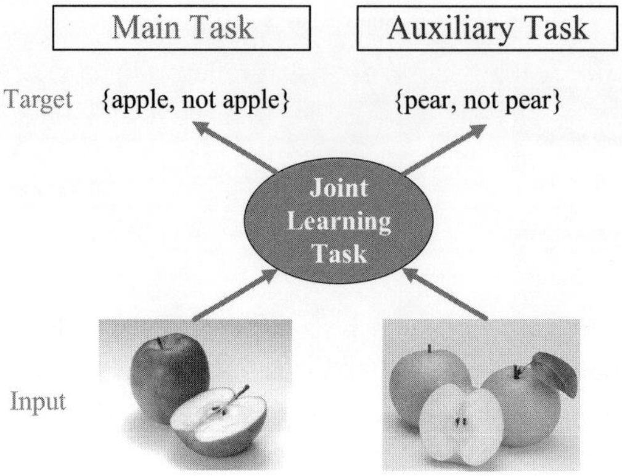

Figure 4.27 Illustration for multi-task learning with a main task and an auxiliary task.

the learning problem turns out to be a transfer learning problem. In general, transfer learning commonly assumes that the source and target domains are different $\mathcal{D}_S \neq \mathcal{D}_T$ or the source task and target are different $\mathcal{T}_S \neq \mathcal{T}_T$. Transfer learning aims to improve the target predictive function $f_T(\cdot)$ of task $\mathcal{T}_T$ in target domain $\mathcal{D}_T$ by transferring the knowledge from source domain $\mathcal{D}_S$ under source task $\mathcal{T}_S$. Here, the condition $\mathcal{D}_S \neq \mathcal{D}_T$ means that either $\mathcal{X}_S \neq \mathcal{X}_T$ or $p(X^s) \neq p(X^t)$, whereas the condition $\mathcal{T}_S \neq \mathcal{T}_T$ means that either $\mathcal{T}_S \neq \mathcal{T}_T$ or $p(Y^s|X^s) \neq p(Y^t|X^t)$.

There are two popular styles of transfer learning. One is multi-task learning, while the other is domain adaptation.

Multi-Task Learning

Multi-task learning is seen as an inductive transfer learning mechanism. Rather than only concerning about the target task, multi-task learning aims to learn all of the source and target tasks simultaneously. The goal of multi-task learning is to improve the generalization performance by using the domain information contained in the training data across different related tasks. For example, as shown in Figure 4.27, we may find that learning to recognize the pears in an auxiliary task might help to build a classifier to recognize the apples in a main task. More specifically, multi-task learning is seen as a learning strategy where the main task is performed together with the other related tasks at the same time so as to discover the shared feature representations that help in improving the learning fidelity of each individual task.

Previous works on multi-task learning were proposed to learn a common feature representation shared by different tasks [125–127]. To encourage a model to also work well on other tasks that are different but related is a better regularization than uninformed regularization such as weight decay. A standard learning objective with model regularization is expressed by

$$\min_{\boldsymbol{\theta}} \mathcal{L}(\mathcal{D}, \boldsymbol{\theta}) + \lambda \Omega(\boldsymbol{\theta}) \tag{4.89}$$

where $\boldsymbol{\theta}$ denotes the model parameter and λ denotes the regularization parameter. Multi-task learning adopts the regularization $\Omega(\cdot)$ that reflects the shared information among different tasks. Thus, we can obtain an improved model for the main task by learning an integrated model for multiple tasks jointly.

4.7.2 Domain Adaptation

Domain adaptation is another popular case of transfer learning that has been recently developed for speaker recognition [128, 129]; see also Chapter 6. Assume that the collected speech samples are available in the source domain and also in the target domain. Let $\{X^s, Y^s\} = \{(\mathbf{x}_1^s, \mathbf{y}_1^s), \ldots, (\mathbf{x}_m^s, \mathbf{y}_m^s)\}$, which denotes the labeled training samples in the source domain. Here, $\mathbf{x}_i^s$ denotes the training token and $\mathbf{y}_i^s$ denotes the corresponding label information. On the other hand, we have the unlabeled data from the target domain $\{X^t, Y^t\} = \{(\mathbf{x}_1^t, \mathbf{y}_1^t), \ldots, (\mathbf{x}_n^t, \mathbf{y}_n^t)\}$ where the label information $\mathbf{y}_i^t$ is unseen. One important assumption in domain adaptation is that two domains $\{\mathcal{D}_S, \mathcal{D}_T\}$ are related but different, namely the joint distributions of samples and labels in two domains $p(X^s, Y^s)$ and $p(X^t, Y^t)$ are different. Domain adaptation is a special category under transfer learning where the marginal distributions in two domains $p(X^s)$ and $p(X^t)$ are different while making the assumption that two conditional distributions $p(Y^s|X^s)$ and $p(Y^t|X^t)$ are identical. Typically, the condition of distinct marginal distributions in training data and test data is also known as the sample selection bias or the covariate shift [130, 131]. There are two types of domain adaptation methods that are instance-based domain adaptation and feature-based domain adaptation.

Instance-Based Domain Adaptation

Using the instance-based method, importance reweighting or importance sampling is performed to reweight the labeled instances from the source domain. The divergence between marginal distributions in two domains can be compensated. In general, we minimize the expected risk or the expectation of a loss function $\mathcal{L}(\mathbf{x}, \mathbf{y}, \boldsymbol{\theta})$ to learn the optimal model parameter $\boldsymbol{\theta}^*$ via empirical risk minimization:

$$\begin{aligned}\boldsymbol{\theta}^* &= \underset{\boldsymbol{\theta}}{\operatorname{argmin}}\, \mathbb{E}_{(\mathbf{x},\mathbf{y})\sim p(\mathbf{x},\mathbf{y})}[\mathcal{L}(\mathbf{x}, \mathbf{y}, \boldsymbol{\theta})] \\ &= \underset{\boldsymbol{\theta}}{\operatorname{argmin}}\, \frac{1}{n}\sum_{i=1}^{n} \mathcal{L}(\mathbf{x}_i, \mathbf{y}_i, \boldsymbol{\theta})\end{aligned} \tag{4.90}$$

where n is the number of training samples. Domain adaptation or covariate shift aims at learning the optimal model parameters $\boldsymbol{\theta}^*$ for the *target* domain by minimizing the expected risk

$$\boldsymbol{\theta}^* = \underset{\boldsymbol{\theta}}{\operatorname{argmin}} \sum_{(\mathbf{x},\mathbf{y})\in\mathcal{D}_T} p(\mathcal{D}_T)[\mathcal{L}(\mathbf{x}, \mathbf{y}, \boldsymbol{\theta})]. \tag{4.91}$$

In traditional machine learning, the optimal model parameter $\boldsymbol{\theta}^*$ is learned by assuming $p(\mathcal{D}_S) = p(\mathcal{D}_T)$ and accordingly using the source domain data $\mathcal{D}_s$. However, the distributions are different $p(\mathcal{D}_S) \neq p(\mathcal{D}_T)$ in the real-world speaker recognition where the domain adaptation is required. Furthermore, no labeled data in the target domain are available while a lot of labeled data in source domain are available in this situation. Accordingly, we modify the optimization problem to fit the target domain using a source domain data as follows [124]:

$$\begin{aligned}\boldsymbol{\theta}^* &= \underset{\boldsymbol{\theta}}{\operatorname{argmin}} \sum_{(\mathbf{x},\mathbf{y})\in\mathcal{D}_S} \frac{p(\mathcal{D}_T)}{p(\mathcal{D}_S)} p(\mathcal{D}_S)\mathcal{L}(\mathbf{x},\mathbf{y},\boldsymbol{\theta}) \\ &\approx \underset{\boldsymbol{\theta}}{\operatorname{argmin}} \sum_{i=1}^{m} \left(\frac{p_T(\mathbf{x}_i^t,\mathbf{y}_i^t)}{p_S(\mathbf{x}_i^s,\mathbf{y}_i^s)}\right) \mathcal{L}(\mathbf{x}_i^s,\mathbf{y}_i^s,\boldsymbol{\theta}),\end{aligned} \tag{4.92}$$

where m is the number of training samples in the source domain. The optimization problem in Eq. 4.92 is solved to learn a model for the target domain by giving different weight $\frac{p_T(\mathbf{x}_i^t,\mathbf{y}_i^t)}{p_S(\mathbf{x}_i^s,y_i^s)}$ to each instance $(\mathbf{x}_i^s,\mathbf{y}_i^s)$ in the source domain. Under the assumption that the conditional distributions $p(Y^s|X^s)$ and $p(Y^t|X^t)$ are the same, the difference between $p(D_S)$ and $p(D_T)$ is caused by $p(X^s)$ and $p(X^t)$. It is because that the assumption of domain adaptation yields

$$\frac{p_T(\mathbf{x}_i^t,\mathbf{y}_i^t)}{p_S(\mathbf{x}_i^s,\mathbf{y}_i^s)} = \frac{p(\mathbf{x}_i^t)}{p(\mathbf{x}_i^s)}, \tag{4.93}$$

where $\frac{p(\mathbf{x}_i^t)}{p(\mathbf{x}_i^s)}$ is a reweighting factor for the loss function of individual training sample. Or equivalently, the reweighting factor can be used for adjusting the difference between the source and target distributions. Finally, the domain adaptation problem is simplified as a problem of estimating the factor $\frac{p(\mathbf{x}_i^t)}{p(\mathbf{x}_i^s)}$ for calculating the learning objective using each sample $\mathbf{x}_i^s$ in the training procedure. Figure 4.28 illustrates a regression example before and after the instance-based domain adaptation where the data distributions and the learned models are shown.

Feature-Based Domain Adaptation

Feature-based domain adaptation is a common approach. Instead of reweighting the training instances, the feature-based method is developed with the assumption that there exists a domain-invariant feature space. The goal of feature-based approach is to span a feature space by minimizing the divergence between two domains and simultaneously preserving the discriminative information for classification in test session. In [132], the structural correspondence learning was developed to match the correspondence between the source domain and the target domain. This method provides an important technique that makes use of the unlabeled data in the target domain to discover relevant features that could reduce the divergence between the source and target domains. The key idea was to evaluate the feature correspondence between the source and target domains, which are based on the correlation governed by the pivot features. Pivot features behaved similarly for discriminative learning in both domains. Non-pivot features are correlated with some of the pivot features. A shared low-dimensional real-valued feature space was learned by using the pivot features. However, these previous studies do not minimize the divergence between different domains directly.

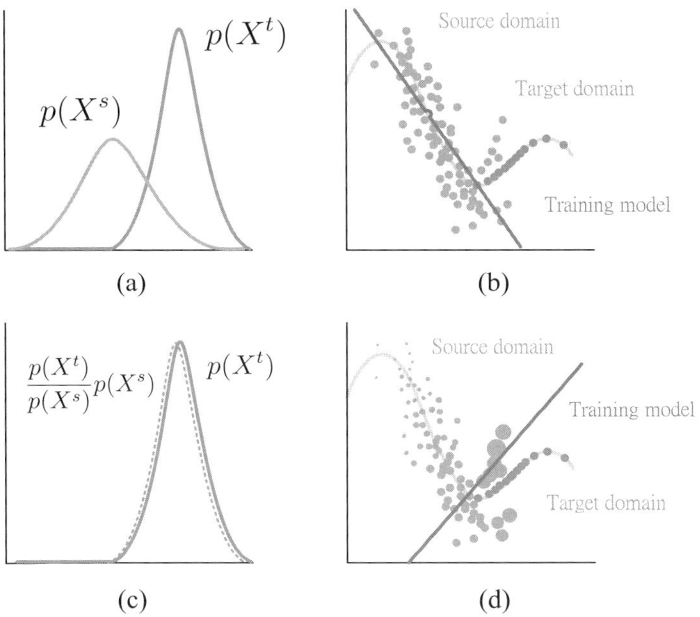

Figure 4.28 Illustration of instance-based domain adaptation. (a) Data distributions of source and target domains. (b) Straight line is a linear model learned from the source domain. (c) Distribution of the source domain is adjusted by the reweighting factor. (d) After reweighting the data in source domain, the learned model fits the target domain.

In [133], a stationary subspace analysis was proposed to match distributions in a low-dimensional space. In [134], a method based on dimensionality reduction was presented to simultaneously minimize the divergence between source distribution and target distribution and minimize the information loss in data. Divergence reduction can be also performed via the maximum mean discrepancy, which will be addressed in the next section.

4.7.3 Maximum Mean Discrepancy

Maximum mean discrepancy (MMD) [135, 136] provides an effective approach to domain adaptation based on the distribution matching, which is a major work in domain adaptation by compensating the mismatch between the marginal distributions in the source and target domains. There exist many criteria that can be used to estimate the distance between the source and target distributions. However, many of them, such as the Kullback–Leibler (KL) divergence, require estimating the densities before estimating the divergence. Traditionally, it was popular to carry out the distribution matching based on the estimation of parametric distributions.

Compared with the parametric distributions, it is more attractive to implement the nonparametric solution to the distance between distributions without the need of density estimation. Maximum mean discrepancy is a well-known nonparametric method that is referred as a divergence between distributions of two data sets in the reproducing kernel Hilbert space (RKHS) denoted by $\mathcal{H}$. MMD can directly evaluate whether two distributions are different on the basis of data samples from each of them. A key superiority of MMD to the other methods is that MMD does not require the density estimation.

The MMD criterion has been employed in many domain adaptation methods for reweighting data or building domain-invariant feature space. Let the kernel-induced feature map be denoted by ϕ. The training observations in the source domain and the target domain are denoted by $X^s = \{\mathbf{x}_1^s, \ldots, \mathbf{x}_m^s\}$ and $X^t = \{\mathbf{x}_1^t, \ldots, \mathbf{x}_n^t\}$, which are drawn independently and identically distributed (i.i.d.) from $p(X^s)$ and $p(X^t)$ (p_S and p_T in short), respectively. The MMD between $\{\mathbf{x}_1^s, \ldots, \mathbf{x}_m^s\}$ and $\{\mathbf{x}_1^t, \ldots, \mathbf{x}_n^t\}$ is defined as follows

$$\text{MMD}(p_S, p_T) = \sup_{\|f\|_{\mathcal{H}} \leqslant 1} (\mathbb{E}_{\mathbf{x}^s}[f(\mathbf{x}^s)] - \mathbb{E}_{\mathbf{x}^t}[f(\mathbf{x}^t)]), \tag{4.94}$$

where $f \in \mathcal{F}$ are functions belonging to the unit ball in a reproducing kernel Hilbert space $\mathcal{H}$. Then, we obtain a biased empirical estimate of MMD by replacing the population expectations with empirical expectations

$$\text{MMD}(X^s, X^t) = \sup_{\|f\|_{\mathcal{H}} \leqslant 1} \left(\frac{1}{m} \sum_{i=1}^{m} f(\mathbf{x}_i^s) - \frac{1}{n} \sum_{i=1}^{n} f(\mathbf{x}_i^t) \right), \tag{4.95}$$

where $\|\cdot\|_{\mathcal{H}}$ denotes the RKHS norm. Due to the property of RKHS, the function evaluation can be rewritten as $f(\mathbf{x}) = \langle \phi(\mathbf{x}), f \rangle$, where $\phi(\mathbf{x}) \colon \mathcal{X} \to \mathcal{H}$, the empirical estimate of MMD is expressed as

$$\text{MMD}(X^s, X^t) = \left\| \frac{1}{m} \sum_{i=1}^{m} \phi(\mathbf{x}_i^s) - \frac{1}{n} \sum_{i=1}^{n} \phi(\mathbf{x}_i^t) \right\|_{\mathcal{H}}. \tag{4.96}$$

Rewriting the norm in Eq. 4.96 as an inner product in RKHS and using the reproducing property, we have

$$\begin{aligned} &\text{MMD}(X^s, X^t) \\ &= \left[\frac{1}{m^2} \sum_{i,j=1}^{m} k(\mathbf{x}_i^s, \mathbf{x}_j^s) - \frac{2}{mn} \sum_{i,j=1}^{m,n} k(\mathbf{x}_i^s, \mathbf{x}_j^t) + \frac{1}{n^2} \sum_{i,j=1}^{n} k(\mathbf{x}_i^t, \mathbf{x}_j^t) \right]^{1/2}, \end{aligned} \tag{4.97}$$

where $k(\cdot, \cdot)$ denotes the characteristic function based on the positive semidefinite kernel. One commonly used kernel is the Gaussian kernel in the form

$$k(\mathbf{x}_i, \mathbf{x}_j) = \exp\left(-\|\mathbf{x}_i - \mathbf{x}_j\|^2 / 2\sigma^2 \right) \tag{4.98}$$

with a variance parameter σ^2.

In summary, the distance between two distributions is equivalent to the distance between the means of two samples in the RKHS. The value of MMD is nonnegative, and it vanishes if and only if two distributions are the same. In Chapter 6, the domain adaptation techniques are further specialized for the application in speaker recognition. We will also address how MMD is developed in a type of deep learning solution to speaker recognition in Section 6.4. In what follows, we introduce the realization of deep neural networks for transfer learning.

4.7.4 Neural Transfer Learning

In accordance with the fundamentals of transfer learning, hereafter, we address the implementation of deep neural networks for the fulfillment of generalizable learning as well as for the realization of classification learning with training data in source domain and test data in target domain. Neural transfer learning is carried out as a new type of machine learning in a form of transfer learning where deep neural networks are configured.

Generalizable Learning

Neural transfer learning is basically feasible to conduct the so-called generalizable learning with two types of learning strategies in presence of multiple learning tasks. First, as depicts in Figure 4.29(a), deep transfer learning can be realized for multi-task learning from raw data through learning the shared representation in intermediate layers of a deep neural network. Such a shared intermediate representation is learned in an unsupervised manner that conveys deep latent information generalizable across different tasks. As we can see, the outputs of the shared intermediate layer are forwarded to carry out different tasks. On the other hand, neural transfer learning can also be implemented as a kind of generalizable learning via the partial feature sharing as shown in Figure 4.29(b). In this case, the low-level features and high-level features are partially connected and shared between different layers toward the outputs $\{y_1, \ldots, y_N\}$ of N tasks. The mixed mode learning is performed as another type of generalizable learning where different compositions of functions are calculated in different nodes and different layers.

Typically, transfer learning is implemented as a learning algorithm that can discover the relations across different tasks as well as share and transfer knowledge across multiple domains. It is because that the representation of deep models has the capability of capturing the underlying factors from training data in different domains that are virtually associated with individual tasks. Correspondingly, the feature representation using deep models with multi-task learning is advantageous and meaningful for system improvement due to the shared factors across tasks.

In [137], a hierarchical feed-forward model was trained by leveraging the cross-domain knowledge via transfer learning from some pseudo tasks without supervision. In [138], the multi-task learning was applied to elevate the performance of deep models by transferring the shared domain-specific information that was contained in the related tasks. In particular, the multi-task learning problem can be formulated as a special realization based on the neural network architecture as displayed in Figure 4.30(a) where a main task and N auxiliary tasks are considered. Let $\mathcal{D}_m = \{\mathbf{x}_n, y_{mn}\}$ denote N input samples of main task, which is indexed by m. The optimization problem is formulated as

$$\min_{\theta} \mathcal{L}(\mathcal{D}_m, \boldsymbol{\theta}) + \lambda\Omega(\boldsymbol{\theta}) \tag{4.99}$$

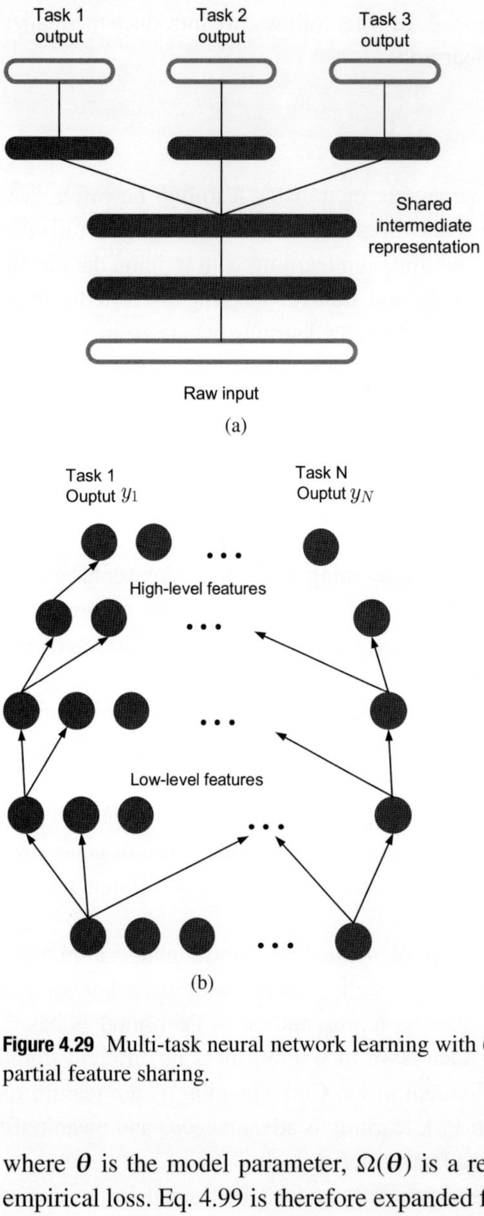

Figure 4.29 Multi-task neural network learning with (a) the shared representation and (b) the partial feature sharing.

where $\boldsymbol{\theta}$ is the model parameter, $\Omega(\boldsymbol{\theta})$ is a regularization term and $\mathcal{L}$ amounts to an empirical loss. Eq. 4.99 is therefore expanded for the main task in a form of

$$\min_{\mathbf{w}_m} \left[\sum_n \mathcal{L}\left(y_{mn}, \mathbf{w}_m^\top f(\mathbf{x}_n; \boldsymbol{\theta})\right) + \lambda \|\mathbf{w}_m^\top \mathbf{w}_m\| \right] \tag{4.100}$$

where $f(\mathbf{x}_n; \boldsymbol{\theta}) = \mathbf{z} = [z_1 \cdots z_K]^\top$ denotes the hidden units of a network, $\mathbf{w}_m$ denotes the model parameter and $\mathcal{L}(\cdot)$ denotes the error function for the main task or the target task. In multi-task learning, $\Omega(\boldsymbol{\theta})$ is implemented by introducing N auxiliary tasks. Each auxiliary task is learned by using $\mathcal{D}_k = \{\mathbf{x}_n, y_{kn}\}$, where $y_{kn} = g_k(\mathbf{x}_n)$ denotes the output of an auxiliary function. The regularization term $\Omega(\boldsymbol{\theta})$ is also incorporated in the learning objective:

$$\min_{\{\mathbf{w}_k\}} \sum_k \left[\sum_n \mathcal{L}\left(y_{kn}, \mathbf{w}_k^\top f(\mathbf{x}_n; \boldsymbol{\theta})\right) + \lambda \|\mathbf{w}_k^\top \mathbf{w}_k\| \right]. \tag{4.101}$$

As we can see, a model is encouraged to work well in the other tasks. This provides a better regularization than the noninformative regularization. Thus, we are able to achieve a better feature representation for the main task through the help of deep models jointly trained from multiple tasks. Next, neural transfer learning is further implemented as a general solution to build a classification system based on semi-supervised learning.

Semi-Supervised Domain Adaptation

Figure 4.30(b) depicts the learning strategy based on a deep semi-supervised model adaptation. The learning objective is to train a classification system for the target domain

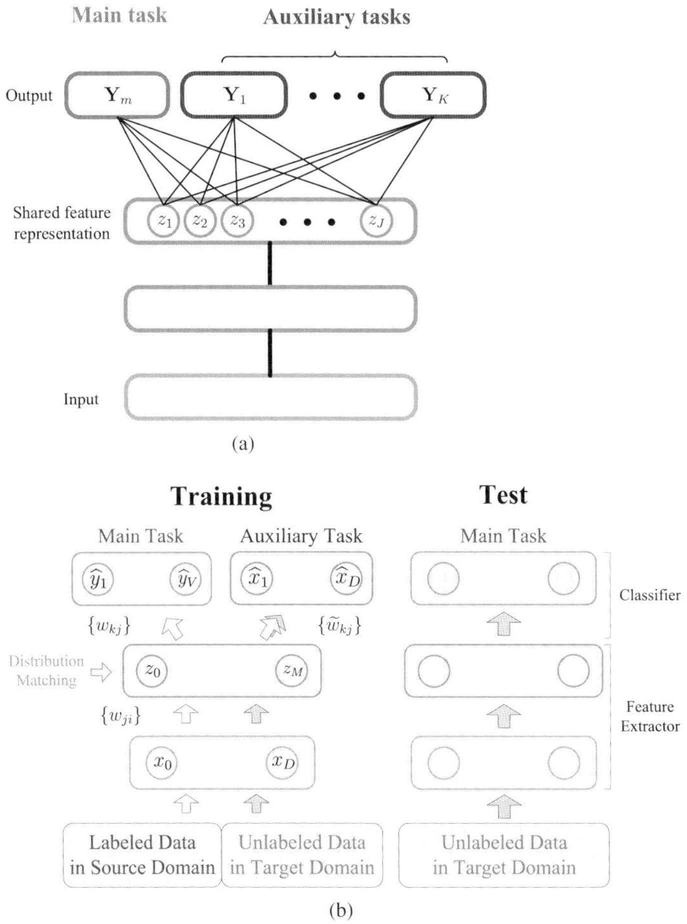

Figure 4.30 (a) Deep transfer learning via a main task and an auxiliary task. (b) A classification system is built with training data in source domain and test data in target domain. [Reprinted from *Deep Semi-Supervised Learning for Domain Adaptation (Figure 1), by H.Y. Chen and J.T. Chien, Proceedings IEEE Workshop on Machine Learning for Signal Processing, Boston, 2015, with permission of IEEE.*]

where a set of labeled samples in source domain and a set of unlabeled samples or test samples in target domain are available. This domain adaptation problem requires assigning a group of output neurons for the main classification task and a group of neurons for the auxiliary regression tasks, where training is jointly performed to acquire the shared information between two domains. Such information is conveyed through different hidden layers consisting of different levels of hidden features.

Basically, Figure 4.30(b) shows a hybrid model of deep neural network (DNN) for the application of semi-supervised domain adaptation. The DNN classifier is trained and applied for prediction of class labels for test samples. For the case of D dimensional input data $\mathbf{x} = [x_1 \dots x_D]^\top$, the auxiliary task for regression can be designed as the reconstruction of input data as $\widehat{\mathbf{x}} = [\widehat{x}_1 \dots \widehat{x}_D]^\top$. We therefore develop a semi-supervised domain adaptation by matching the distributions between two domains based on the shared information in hidden layers. There are two learning tasks in an integrated objective function, which is formulated to find the solution to semi-supervised domain adaptation. One is the cross-entropy error function, which is minimized for optimal classification while the other is the reconstruction error function, which is minimized for optimal regression. Based on this learning strategy, the semi-supervised learning is conducted under multiple objectives. Notably, the main and auxiliary tasks are separated in the output layer by using separate weights, $\{w_{kj}\}$ and $\{\widetilde{w}_{kj}\}$. The layer next to the output layer is known as the layer for distribution matching. Those low-level features were shared for both classification and regression using weight parameters $\{w_{ji}\}$.

In this chapter, we have addressed the fundamentals of deep learning ranging from the traditional models including restricted Boltzmann machines, deep neural networks, and the deep belief networks to the advanced models including variational autoencoders, generative adversarial networks, and the neural transfer learning machines. These models will be employed in two categories of speaker recognition systems. One is for robust speaker recognition while the other is for domain adaptive speaker recognition, which will be detailed in Chapter 5 and Chapter 6, respectively.

5 Robust Speaker Verification

5.1 DNN for Speaker Verification

Recently, deep learning and deep neural networks (DNNs) have changed the research landscape in speech processing [139–142]. This is mainly due to their superb performance. In speaker recognition, the most common strategy is to train a DNN with a bottleneck layer from which either frame-based features or utterance-based features can be extracted. The network can be trained to produce senone posteriors or speaker posteriors. For the latter, a pooling procedure is applied to convert variable-length utterances into a fixed-length feature vector. Then, standard back-end classifiers can be applied for scoring.

The relationship between i-vectors and background noise is not straightforward and may not be linear. Linear models such as PLDA and LDA cannot fully capture this complex relationship. Recently, a number of studies have demonstrated that DNNs are more capable of modeling this complex relationship. For instance, Isik et al. [143] extracted speaker vectors from i-vectors through disentangling the latent dependence between speaker and channel components, and [144, 145] used the weights of a stacked restricted Boltzmann machine (RBM) to replace the parameters of a PLDA model. In [146], noisy i-vectors were mapped to their clean counterparts by a discriminative denoising autoencoder (DDAE) using the speaker identities and the information in the clean i-vectors.

5.1.1 Bottleneck Features

DNNs have been used as frame-based feature extractors. This is achieved by reading the activations of the bottleneck layer of a DNN that is trained to produce the senone posteriors of a short segment of speech [147, 148]. The procedure of this approach is illustrated in Figure 5.1. The main idea is to replace the MFCC-UBM by the bottleneck-feature based UBM (BNF-UBM). As the DNN is trained to produce senone posteriors, the bottleneck features contain more phonetic information and the alignments with the BNF-UBM are more relevant and reliable for speaker verification. In addition to computing the zeroth-order statistics (alignment), the bottleneck features can also be used as acoustic features for i-vector extraction.

Alternatively, bottleneck features can be extracted at the conversation level by averaging the activation of the last hidden layer [149, 150]. This procedure leads to the

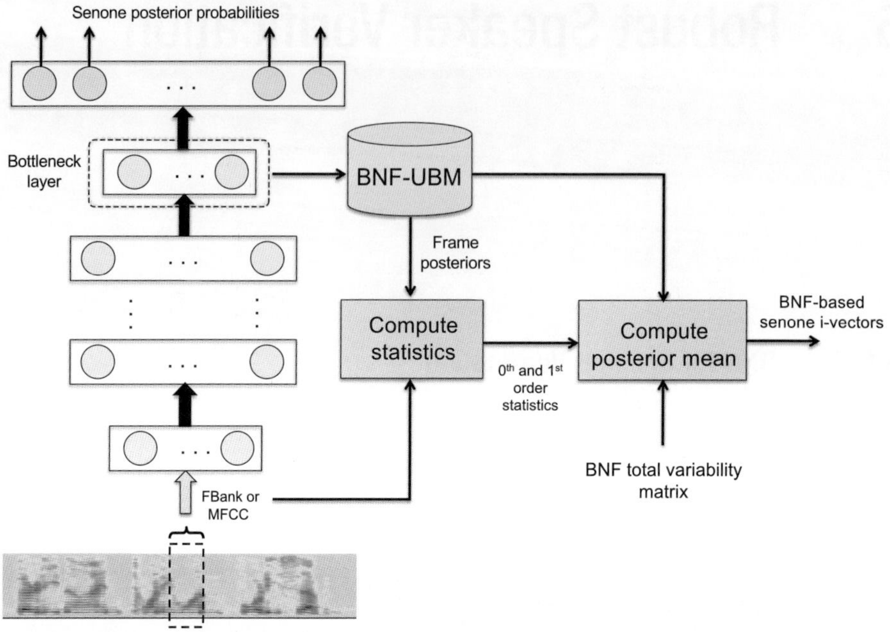

Figure 5.1 The procedure of extracting i-vectors from DNN-based bottleneck features. Instead of using an MFCC-based UBM for frame alignment, an UBM trained by DNN-based bottleneck features (BNF) is used.

d-vectors. The structure of the DNN is almost the same as Figure 5.1 except that the DNN is trained to output speaker posteriors instead of senone posteriors. This straightforward use of DNNs, however, can barely achieve significant performance gain, despite some success under reverberant environments [151] or with the help of a denoising autoencoder [2, 148]. The idea is illustrated in Figure 5.2.

5.1.2 DNN for I-Vector Extraction

One of the promising approaches is to use a DNN to compute the frame posteriors (senone posteriors) for i-vector extraction [3, 84, 85, 152–154] (see Section 3.6.10). In this method, the component posteriors of a GMM-based universal background model (UBM) are replaced by the output of a phonetically aware DNN. This means that an UBM is not required for frame alignment. Instead, the DNN estimates the posterior probabilities of thousands of senones given multiple contextual acoustic frames. However, as the i-vector extractor is still based on a GMM-based factor analyzer, the total variability matrix is still formed by stacking multiple loading matrices, one for each Gaussian. The resulting i-vectors are known as senone i-vectors or DNN i-vectors. Figure 5.3 shows the procedure of senone i-vector extraction.

The idea of replacing the UBM by a DNN is fascinating because the latter facilitates the comparison of speakers as if they were pronouncing the same context.

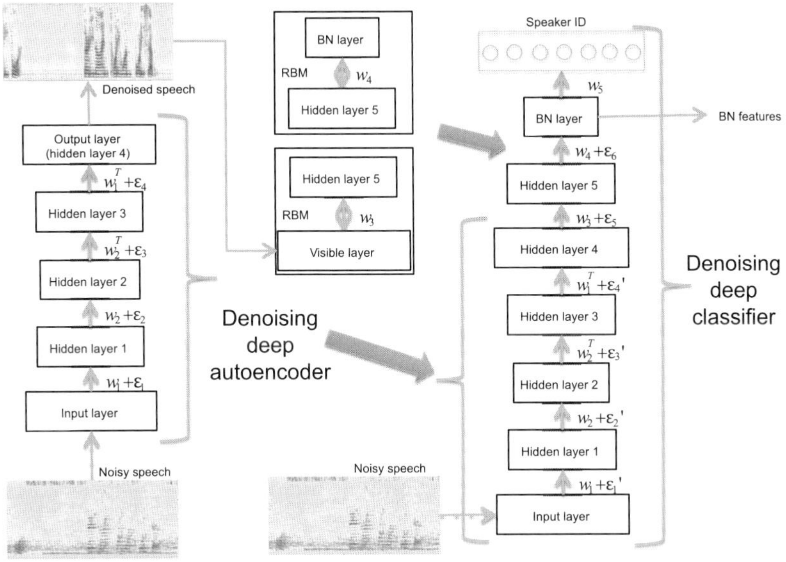

Figure 5.2 Procedure for creating a DNN classifier by stacking restricted Boltzmann machines and the extraction of bottleneck (BN) features from the BN layer. A softmax layer outputting speaker IDs is put on top of the denoising autoencoder. After backpropagation fine-tuning, BN features extracted from the BN layer can be used for speaker recognition [2].

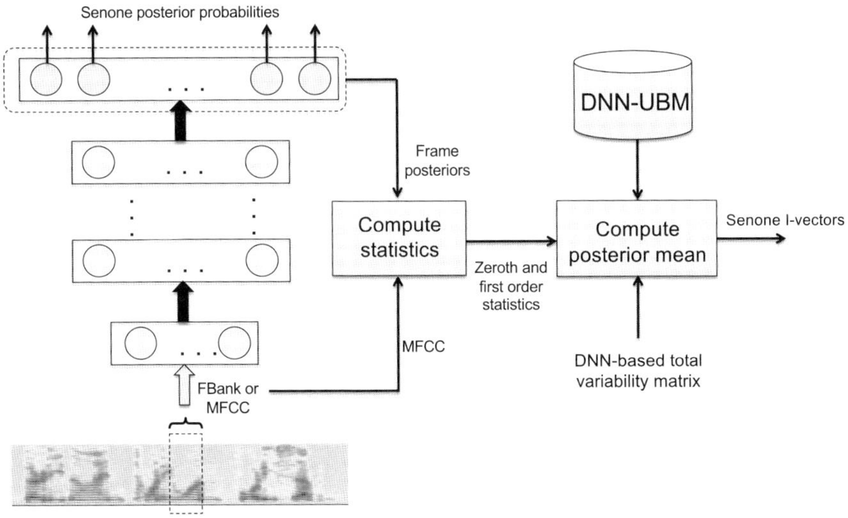

Figure 5.3 Extracting senone i-vectors. The DNN–UBM contains the mean vectors and covariance matrices computed from the frame posteriors and acoustic vectors (typically MFCCs) using Eq. 3.161 and Eq. 3.163, respectively. The DNN-based total variability matrix is obtained by Eq. 3.129 using the frame posteriors computed by the DNN. In some systems [3], the MFCCs are replaced by bottleneck features.

5.2 Speaker Embedding

The use of deep neural networks to create an embedded space that captures most of the speaker characteristics has attracted a lot of attention in the last few years. Strictly speaking, the d-vector [150] is also a kind of speaker embedding. However, the most promising approach today is the x-vectors [155, 156].

5.2.1 X-Vectors

Despite the outstanding performance of the senone i-vectors, training the DNN to output senone posteriors is computationally demanding and requires language-specific resources such as transcribed data. So far, senone i-vectors are limited to English only.

To remove the language constraint in senone i-vectors, Snyder et al. [155, 156] proposed to train a network to output speaker posteriors instead of senone posteriors. The concept is similar to the d-vector. However, unlike the d-vector, a time-delay neural network (TDNN) is used to capture the temporal information in the acoustic frames and statistical pooling is applied to aggregate the frame-level information into utterance-level information. Statistical pooling computes the mean and standard deviation of the TDNN's outputs across consecutive frames in the utterance. Speaker-dependent embedded features are then extracted at the layer(s) after the pooling. The authors name the resulting feature vectors as x-vectors. The TDNN and the pooling mechanism is shown to be very effective in converting acoustic frames into fixed-length feature vectors for PLDA scoring. Figure 5.4 shows the structure of the x-vector extractor.

In addition to the TDNN and the statistical pooling mechanism, another difference between the x-vector extractor and the d-vector extractor is that the former outputs the speaker posteriors at the utterance-level, whereas the latter outputs the speaker posterior at the frame level.

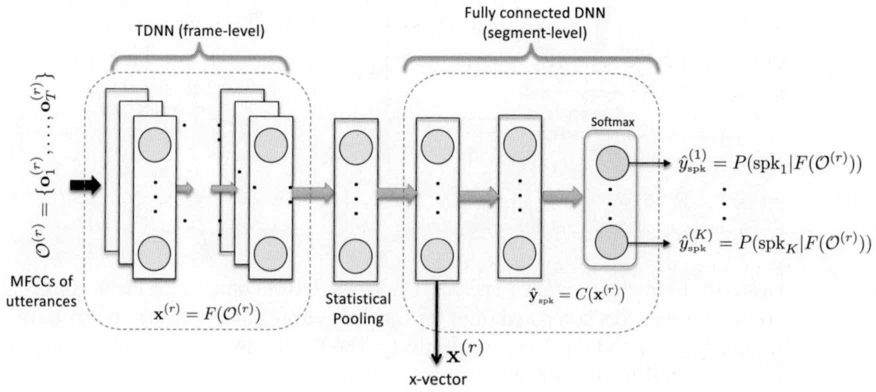

Figure 5.4 Structure of an x-vector extractor. The TDNN aims to capture the temporal information of contextual acoustic frames and the pooling layer aggregates the temporal information to form the utterance-level representation called the x-vector.

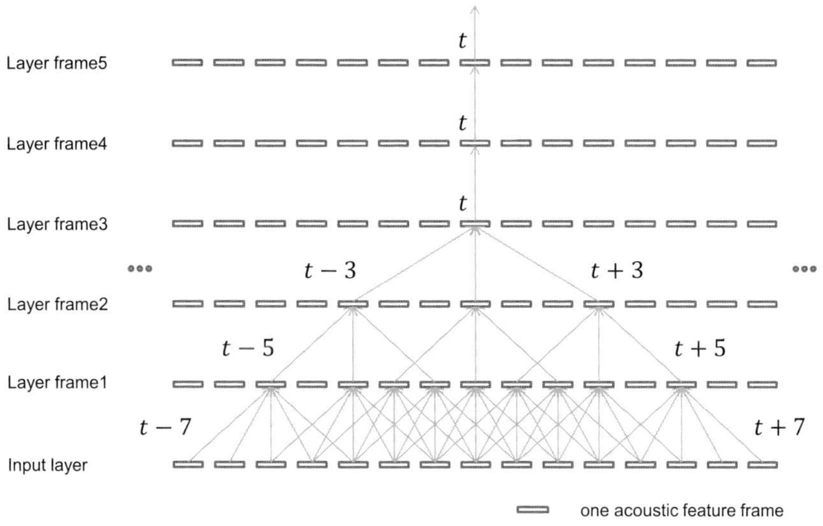

Figure 5.5 The frame context at different layers of an x-vector extraction DNN (courtesy of Youzhi Tu).

The TDNN structure allows neurons in the upper layers to receive signals that span across multiple frames. For example, in [155], at a particular frame t of an input utterance, the first hidden layer receives five frames between $t-2$ and $t+2$ from the input layer; the second hidden layer receives three frames at $t-2$, t, and $t+2$ from the first hidden layer; the third hidden layer receives three frames at $t-3$, t, and $t+3$ from the second hidden layer. As a result, each neuron in the third hidden layer will have a temporal span of 15 frames, as shown in Figures 5.5 and 5.6.

The authors in [156] shows that their x-vector system significantly outperforms their i-vector systems in the Cantonese part of NIST 2016 SRE. It was demonstrated that the performance of x-vectors can be further improved by increasing the diversity of training speech. This was done by adding noise, music, babble, and reverberation effect to the original speech samples. The combined speech samples are then used for training the DNN in Figure 5.4.

Very recently, the x-vector embedding has been enhanced by adding LSTM layers on top of the last TDNN layer, followed by applying statistical pooling on both the TDNN and LSTM layers [157, 158]. Another improvement is to add an attention mechanism to the statistical pooling process [159, 160], which are shown to be better than simply averaging the frame-level features.

5.2.2 Meta-Embedding

The idea of meta-embedding [62] is motivated by the difficulty of propagating the uncertainty of i-vectors and x-vectors to the PLDA models. Although it has been shown that the uncertainty of i-vectors (represented by the their posterior covariances) can

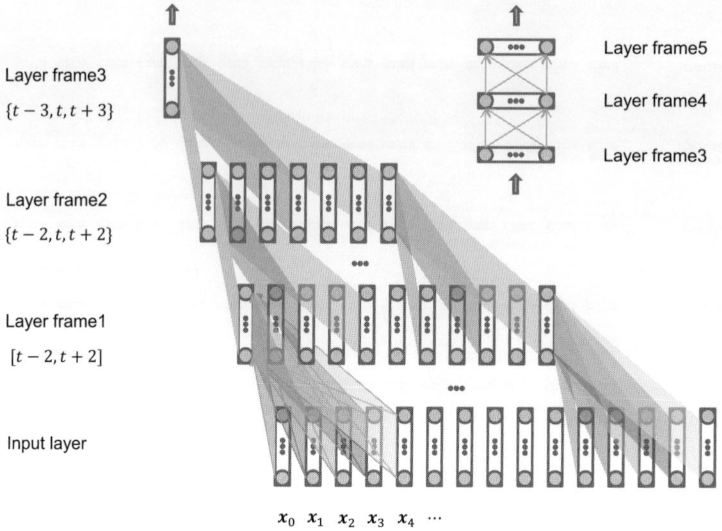

Figure 5.6 Architecture of the TDNN in the x-vector extractor. At the input layer, the vertical bars represent acoustic frames. In the hidden layers, the vertical bars comprises neurons with connections to contextual nodes in the layer below. (Courtesy of Youzhi Tu)

be propagated to the PLDA models [50, 80], the method increases the scoring time significantly [82, 83].

In the conventional i-vector/PLDA framework, speaker identity is represented by the a point estimate of the posterior density (Eq. 3.76) and the uncertainty of the point estimate is discarded. Meta-embedding overcomes this limitation by considering the likelihood function for the speaker identity variable:

$$f(\mathbf{z}) \propto P(r|\mathbf{z}), \tag{5.1}$$

where $\mathbf{z} \in \mathbb{R}^d$ is the speaker identity variable (which is hidden) and r is a representation of the audio recording, e.g., i-vector or MFCCs. Therefore, given the jth recording, instead of using the point estimate r_j to represent the speaker, we keep all of the information about the speaker in $f_j(\mathbf{z}) = k_j P(r_j|\mathbf{z})$ where k_j is an arbitrary constant. It is important to note that meta-embedding is the whole function f_j instead of some point estimates that live in $\mathbb{R}^d$.

Given the representation of two recordings, r_1 and r_2, we have two hypotheses:

$$H_1 : r_1 \text{ and } r_2 \text{ are obtained from the same speaker}$$
$$H_2 : r_1 \text{ and } r_2 \text{ are obtained from two different speakers}$$

To apply meta-embedding for speaker verification, we need to compute the likelihood ratio

$$\frac{p(r_1, r_2|H_1)}{p(r_1, r_2|H_2)} = \frac{\int_{\mathbb{R}^d} f_1(\mathbf{z}) f_2(\mathbf{z}) \pi(\mathbf{z}) d\mathbf{z}}{\left[\int_{\mathbb{R}^d} f_1(\mathbf{z}) \pi(\mathbf{z}) d\mathbf{z}\right] \left[\int_{\mathbb{R}^d} f_2(\mathbf{z}) \pi(\mathbf{z}) d\mathbf{z}\right]}, \tag{5.2}$$

where $\pi(\mathbf{z})$ is the prior density of $\mathbf{z}$. In [62], $\pi(\mathbf{z}) = \mathcal{N}(\mathbf{z}|\mathbf{0}, \mathbf{I})$. The integrals in Eq. 5.2 are the expectation of f_j:

$$\int_{\mathbb{R}^d} f_j(\mathbf{z})\pi(\mathbf{z})d\mathbf{z} = \mathbb{E}_{\mathbf{z}\sim\pi(\mathbf{z})}\{f_j(\mathbf{z})\}.$$

For Gaussian meta-embedding (GME), the expectation is given by [161]

$$\begin{aligned}
\mathbb{E}_{\mathbf{z}\sim\pi(\mathbf{z})}\{f(\mathbf{z})\} &= \int \mathcal{N}(\mathbf{z}|\mathbf{0}, \mathbf{I}) f(\mathbf{z})d\mathbf{z} \\
&= \int \frac{1}{\sqrt{(2\pi)^d}} \exp\left\{-\frac{1}{2}\mathbf{z}^{\mathsf{T}}\mathbf{z}\right\} \cdot \exp\left\{\mathbf{a}^{\mathsf{T}}\mathbf{z} - \frac{1}{2}\mathbf{z}^{\mathsf{T}}\mathbf{B}\mathbf{z}\right\} d\mathbf{z} \\
&= \int \frac{1}{\sqrt{(2\pi)^d}} \exp\left\{\mathbf{a}^{\mathsf{T}}\mathbf{z} - \frac{1}{2}\mathbf{z}^{\mathsf{T}}(\mathbf{B}+\mathbf{I})\mathbf{z}\right\} d\mathbf{z} \\
&= |\mathbf{B}+\mathbf{I}|^{-\frac{1}{2}} \exp\left\{\frac{1}{2}\mathbf{a}^{\mathsf{T}}(\mathbf{B}+\mathbf{I})^{-1}\mathbf{a}\right\},
\end{aligned} \tag{5.3}$$

where $\mathbf{a}$ and $\mathbf{B}$ are the natural parameters of the Gaussian likelihood function $f_j(\mathbf{z})$.

5.3 Robust PLDA

One school of thought to enhance the robustness of speaker verification systems against background noise is to improve the robustness of the PLDA model. To this end, both the SNR and duration variabilities can be incorporated into the PLDA model so that these variabilities can be marginalized during the scoring stage. The method is called SNR- and duration-invariant PLDA [55, 162, 163]. Table 5.1 summarizes the nomenclature of various types of PLDA models to be discussed in this section.

5.3.1 SNR-Invariant PLDA

An i-vector $\mathbf{x}_{ij}$ is generated from a linear model of the form [164, 165]:

$$\mathbf{x}_{ij} = \mathbf{m} + \mathbf{V}\mathbf{h}_i + \mathbf{G}\mathbf{r}_{ij} + \epsilon_{ij}, \tag{5.4}$$

Table 5.1 Abbreviations of various PLDA models.

Abbr.	Model Name	Formula
PLDA	Probabilistic LDA	$\mathbf{x}_{ij} = \mathbf{m} + \mathbf{V}\mathbf{h}_i + \epsilon_{ij}$ (Eq. 5.5)
SI-PLDA	SNR-invariant PLDA	$\mathbf{x}_{ij}^k = \mathbf{m} + \mathbf{V}\mathbf{h}_i + \mathbf{U}\mathbf{w}_k + \epsilon_{ij}^k$ (Eq. 5.6)
DI-PLDA	Duration-invariant PLDA	$\mathbf{x}_{ij}^p = \mathbf{m} + \mathbf{V}\mathbf{h}_i + \mathbf{R}\mathbf{y}_p + \epsilon_{ij}^p$ (Eq. 5.7)
SDI-PLDA	SNR- and duration-invariant PLDA	$\mathbf{x}_{ij}^{kp} = \mathbf{m} + \mathbf{V}\mathbf{h}_i + \mathbf{U}\mathbf{w}_k + \mathbf{R}\mathbf{y}_p + \epsilon_{ij}^{kp}$ (Eq. 5.29)

where $\mathbf{V}$ and $\mathbf{G}$ define the speaker subspace and channel subspace, respectively. $\mathbf{h}_i$ and $\mathbf{r}_{ij}$ are the speaker and channel factors, respectively, and ϵ_{ij} is the residue with a Gaussian distribution, $\mathcal{N}(\epsilon|\mathbf{0}, \mathbf{\Sigma})$. In most cases, $\mathbf{\Sigma}$ is a diagonal covariance matrix that represents the variation that cannot be described by $\mathbf{G}\mathbf{G}^{\mathsf{T}}$ and $\mathbf{V}\mathbf{V}^{\mathsf{T}}$.

The PLDA model in Eq. 5.4 has two components [61, 66], namely the speaker component ($\mathbf{m}+\mathbf{V}\mathbf{h}_i$) that represents the characteristics of speaker i and the channel component ($\mathbf{G}\mathbf{r}_{ij}+\epsilon_{ij}$) that characterizes the speaker as well as the channel. Because the dimension of i-vectors is sufficiently low, the covariance $\mathbf{G}\mathbf{G}^{\mathsf{T}}$ can be absorbed into $\mathbf{\Sigma}$ provided that it is a full covariance matrix. As a result, the PLDA model reduces to [166]:

$$\mathbf{x}_{ij} = \mathbf{m} + \mathbf{V}\mathbf{h}_i + \epsilon_{ij}, \tag{5.5}$$

where $\epsilon_{ij} \sim \mathcal{N}(\mathbf{0}, \mathbf{\Sigma})$ with $\mathbf{\Sigma}$ being a full covariance matrix.

SNR-invariant PLDA (SI-PLDA) was developed [55, 162] to address the limitation of PLDA in modeling SNR and duration variabilities in i-vectors. To train an SNR-invariant model, we partition the training set into K groups based on the SNR of the training utterances so that each i-vector is assigned to one SNR group. Let $\mathbf{x}_{ij}^{k}$ represents the jth i-vector from the ith speaker in the kth SNR group. Extending the classical PLDA to modeling SNR and duration variabilities, we may write $\mathbf{x}_{ij}^{k}$ as

$$\mathbf{x}_{ij}^{k} = \mathbf{m} + \mathbf{V}\mathbf{h}_i + \mathbf{U}\mathbf{w}_k + \epsilon_{ij}^{k}, \tag{5.6}$$

where $\mathbf{V}$ and $\mathbf{U}$ represent the speaker and SNR subspaces respectively, $\mathbf{h}_i$ and $\mathbf{w}_k$ are speaker and SNR factors with a standard normal prior, and ϵ_{ij}^{k} is a residue term following a Gaussian distribution $\mathcal{N}(\epsilon|\mathbf{0}, \mathbf{\Sigma})$. In [55, 162], the channel variability is modeled by the full covariance matrix $\mathbf{\Sigma}$.

The SI-PLDA (Eq. 5.6) and the classical PLDA (Eq. 5.4) are different. Specifically, in SI-PLDA, variability in i-vectors caused by noise level variation is modeled by $\mathbf{U}\mathbf{U}^{\mathsf{T}}$, whereas in classical PLDA, the channel variability is modeled by $\mathbf{G}\mathbf{G}^{\mathsf{T}}$. Therefore, the SNR factor ($\mathbf{w}_k$ in Eq. 5.6) depends on the SNR groups, whereas the channel factor ($\mathbf{r}_{ij}$ in Eq. 5.4) depends on the speaker and channel.

Figure 5.7 shows the use of speaker and SNR labels when training a PLDA model (a) and an SNR-invariant PLDA model (b). In PLDA, the SNR group labels are ignored during PLDA training, whereas, in SI-PLDA, the SNR labels are used to group the i-vectors according to the SNR of their respective utterances. These extra SNR labels enable the PLDA model to find an SNR subspace, which helps to reduce i-vector variability due to noise level variation.

5.3.2 Duration-Invariant PLDA

In some datasets, the duration of utterances varies widely. Figure 5.8 shows the histogram of utterance durations in NIST 2016 SRE. According to [167, 168], duration variability in the i-vectors can be modeled as additive noise in i-vector space. Adding a duration factor to the PLDA model results in a duration-invariant PLDA (DI-PLDA) [163].

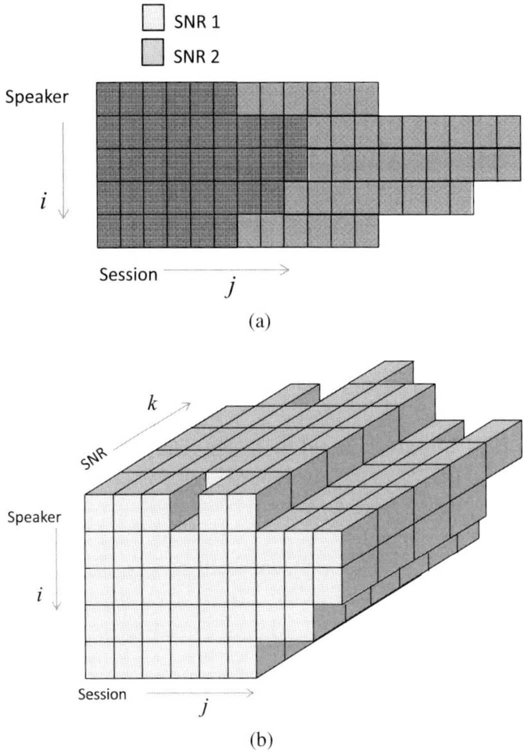

Figure 5.7 (a) Grouping of i-vectors under multi-condition training in conventional PLDA. The rectangles represents i-vectors. Although the training set contains two groups of i-vectors (with different SNRs), the training algorithm of PLDA will not consider this information and will sum over the statistics of both groups. (b) Grouping of i-vectors under SNR-invariant PLDA in which each cube represents an i-vector. There are $H_i(k)$ i-vectors from speaker i whose utterances fall into SNR group k. The training algorithm of SNR-invariant PLDA will take the SNR labels and speaker labels into consideration and will sum over the statistics within individual groups. [Reprinted from *Discriminative Subspace Modeling of SNR and Duration Variabilities for Robust Speaker Verification (Figure 1), N. Li, M.W. Mak, W.W. Lin and J.T. Chien, Computer Speech and Language, vol. 45, pp. 87–103, 2017, with permission of Elsevier*]

EM Formulations

Denote $\mathcal{X} = \{\mathbf{x}_{ij}^p | i = 1, \ldots, S; j = 1, \ldots, H_i(p); p = 1, \ldots, P\}$ as a set of i-vectors obtained from S speakers, where $\mathbf{x}_{ij}^p$ is utterance j from speaker i at duration group p. Speaker i posseses $H_i(p)$ i-vectors from duration group p. If the term associated with the SNR is replaced by a term related to duration, Eq. 5.6 becomes DI-PLDA, i.e.,

$$\mathbf{x}_{ij}^p = \mathbf{m} + \mathbf{V}\mathbf{h}_i + \mathbf{R}\mathbf{y}_p + \epsilon_{ij}^p, \tag{5.7}$$

where the duration subspace is defined by $\mathbf{R}$ and $\mathbf{y}_p$ denotes the duration factor whose prior is a standard Gaussian. The meanings of other terms remain the same as those in Eq. 5.6.

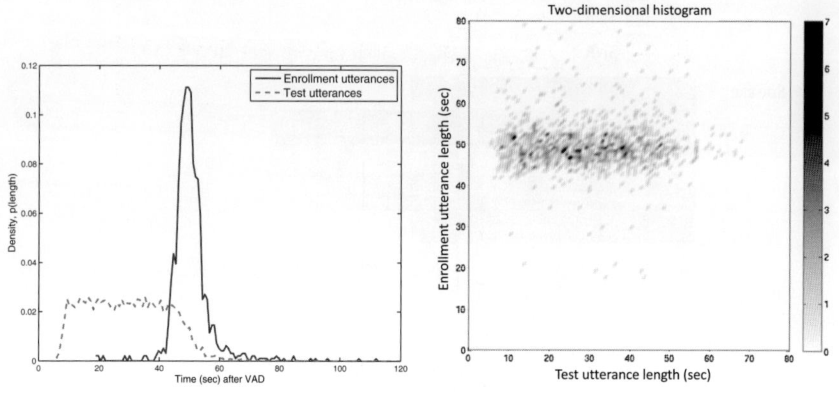

(a) (b)

Figure 5.8 (a) Duration distributions of enrollment and test utterances in NIST 2016 SRE. (b) A two-dimensional histogram illustrating the duration distribution of all possible target-test pairs.

To simplify computation, the latent factors $\mathbf{h}_i$ and $\mathbf{y}_p$ are assumed to be posteriorly independent in [55]. However, a better approach is to consider them posteriorly dependent, as in [163]. In the latter case, variational Bayes methods [1] can be used for deriving the EM algorithms for training the SI-PLDA and DI-PLDA models.

Suppose there are $N_i = \sum_{p=1}^{P} H_i(p)$ training utterances from speaker i and $B_p = \sum_{i=1}^{S} H_i(p)$ training utterances in duration group p. Denote $\boldsymbol{\theta} = \{\mathbf{m}, \mathbf{V}, \mathbf{R}, \boldsymbol{\Sigma}\}$ as the old estimate of the model parameters, a new estimate $\boldsymbol{\theta}'$ can be computed by maximizing the auxiliary function:

$$
\begin{aligned}
Q(\boldsymbol{\theta}'|\boldsymbol{\theta}) &= \mathbb{E}_{q(\underline{\mathbf{h}},\underline{\mathbf{y}})}\left\{\log p(\mathcal{X}, \underline{\mathbf{h}}, \underline{\mathbf{y}}|\boldsymbol{\theta}')\middle|\mathcal{X}, \boldsymbol{\theta}\right\} \\
&= \mathbb{E}_{q(\underline{\mathbf{h}},\underline{\mathbf{y}})}\left\{\sum\nolimits_{ijp} \log\left[p(\mathbf{x}_{ij}^{p}|\mathbf{h}_i, \mathbf{y}_p, \boldsymbol{\theta}')p(\mathbf{h}_i, \mathbf{y}_p)\right]\middle|\mathcal{X}, \boldsymbol{\theta}\right\},
\end{aligned}
\tag{5.8}
$$

where $\underline{\mathbf{h}} = \{\mathbf{h}_1, \ldots, \mathbf{h}_s\}$, $\underline{\mathbf{y}} = \{\mathbf{y}_1, \ldots, \mathbf{y}_P\}$, and $q(\underline{\mathbf{h}}, \underline{\mathbf{y}})$ is the variational posterior density of $\underline{\mathbf{h}}$ and $\underline{\mathbf{y}}$. $Q(\boldsymbol{\theta}'|\boldsymbol{\theta})$ can then be maximized by setting the derivative of $Q(\boldsymbol{\theta}'|\boldsymbol{\theta})$ with respect to the model parameters $\{\mathbf{m}, \mathbf{V}, \mathbf{R}, \boldsymbol{\Sigma}\}$ to $\mathbf{0}$, which results in

$$
\mathbf{m} = \frac{1}{N}\sum_{i=1}^{S}\sum_{p=1}^{P}\sum_{j=1}^{H_i(p)} \mathbf{x}_{ij}^{p}, \tag{5.9a}
$$

$$
\begin{aligned}
\mathbf{V}' = &\left\{\sum_{i=1}^{S}\sum_{p=1}^{P}\sum_{j=1}^{H_i(p)} \left[(\mathbf{x}_{ij}^{p} - \mathbf{m})\langle\mathbf{h}_i|\mathcal{X}\rangle - \mathbf{R}\langle\mathbf{y}_p\mathbf{h}_i^{\mathsf{T}}|\mathcal{X}\rangle\right]\right\} \\
&\left[\sum_{i=1}^{S}\sum_{p=1}^{P}\sum_{j=1}^{H_i(p)} \langle\mathbf{h}_i\mathbf{h}_i^{\mathsf{T}}|\mathcal{X}\rangle\right]^{-1},
\end{aligned}
\tag{5.9b}
$$

$$\mathbf{R}' = \left\{\sum_{i=1}^{S}\sum_{p=1}^{P}\sum_{j=1}^{H_i(p)}\left[(\mathbf{x}_{ij}^{p}-\mathbf{m})\langle\mathbf{y}_p|\mathcal{X}\rangle - \mathbf{V}\langle\mathbf{h}_i\mathbf{y}_p^{\mathsf{T}}|\mathcal{X}\rangle\right]\right\},$$

$$\left[\sum_{i=1}^{S}\sum_{p=1}^{P}\sum_{j=1}^{H_i(p)}\langle\mathbf{y}_p\mathbf{y}_p^{\mathsf{T}}|\mathcal{X}\rangle\right]^{-1} \tag{5.9c}$$

$$\boldsymbol{\Sigma}' = \frac{1}{N}\sum_{i=1}^{S}\sum_{p=1}^{P}\sum_{j=1}^{H_i(p)}\left[(\mathbf{x}_{ij}^{p}-\mathbf{m})(\mathbf{x}_{ij}^{p}-\mathbf{m})^{\mathsf{T}} - \mathbf{V}\langle\mathbf{h}_i|\mathcal{X}\rangle(\mathbf{x}_{ij}^{p}-\mathbf{m})^{\mathsf{T}}\right] -$$

$$\frac{1}{N}\sum_{i=1}^{S}\sum_{p=1}^{P}\sum_{j=1}^{H_i(p)}\mathbf{R}\langle\mathbf{y}_p|\mathcal{X}\rangle(\mathbf{x}_{ij}^{p}-\mathbf{m})^{\mathsf{T}}, \tag{5.9d}$$

where $N = \sum_{i=1}^{S} N_i = \sum_{p=1}^{P} B_p$.

The global mean, Eq. 5.9(a), can be computed in one iteration. However, the other model parameters, Eqs. 5.9(b)–(d), which make up the M-step of the EM algorithm, require iterative update because the posterior distribution of $\mathbf{h}_i$ and $\mathbf{y}_p$ are not known. Below, we will explain how variational Bayes method can be used to compute these posteriors.

In variational Bayes, the true posterior $p(\underline{\mathbf{h}},\underline{\mathbf{y}}|\mathcal{X})$ is approximated by a variational posterior $q(\underline{\mathbf{h}},\underline{\mathbf{y}})$, and the marginal likelihood of $\mathcal{X}$ is written as

$$\begin{aligned}
\log p(\mathcal{X}) &= \int\!\!\int q(\underline{\mathbf{h}},\underline{\mathbf{y}})\log p(\mathcal{X})d\underline{\mathbf{h}}d\underline{\mathbf{y}}\\
&= \int\!\!\int q(\underline{\mathbf{h}},\underline{\mathbf{y}})\log\left[\frac{p(\underline{\mathbf{h}},\underline{\mathbf{y}},\mathcal{X})}{p(\underline{\mathbf{h}},\underline{\mathbf{y}}|\mathcal{X})}\right]d\underline{\mathbf{h}}d\underline{\mathbf{y}}\\
&= \int\!\!\int q(\underline{\mathbf{h}},\underline{\mathbf{y}})\log\left[\frac{p(\underline{\mathbf{h}},\underline{\mathbf{y}},\mathcal{X})}{q(\underline{\mathbf{h}},\underline{\mathbf{y}})}\right]d\underline{\mathbf{h}}d\underline{\mathbf{y}} + \int\!\!\int q(\underline{\mathbf{h}},\underline{\mathbf{y}})\log\left[\frac{q(\underline{\mathbf{h}},\underline{\mathbf{y}})}{p(\underline{\mathbf{h}},\underline{\mathbf{y}}|\mathcal{X})}\right]d\underline{\mathbf{h}}d\underline{\mathbf{y}}\\
&= \mathcal{L}(q) + \mathcal{D}_{\mathrm{KL}}(q(\underline{\mathbf{h}},\underline{\mathbf{y}})\|p(\underline{\mathbf{h}},\underline{\mathbf{y}}|\mathcal{X})).
\end{aligned} \tag{5.10}$$

In Eq. 5.10, $\mathcal{D}_{\mathrm{KL}}(q||p)$ denotes the KL-divergence between distribution q and distribution p; also,

$$\mathcal{L}(q) = \int\!\!\int q(\underline{\mathbf{h}},\underline{\mathbf{y}})\log\left[\frac{p(\underline{\mathbf{h}},\underline{\mathbf{y}},\mathcal{X})}{q(\underline{\mathbf{h}},\underline{\mathbf{y}})}\right]d\underline{\mathbf{h}}d\underline{\mathbf{y}} \tag{5.11}$$

is the variational lower-bound of the marginal likelihood. Because of the non-negativity of KL-divergence, maximizing the lower bound with respect to $q(\underline{\mathbf{h}})$ will also maximize the marginal likelihood. The maximum is attained when $q(\underline{\mathbf{h}},\underline{\mathbf{y}})$ is the same as the true posterior $p(\underline{\mathbf{h}},\underline{\mathbf{y}}|\mathcal{X})$. Also, it is assumed that the approximated posterior $q(\underline{\mathbf{h}},\underline{\mathbf{y}})$ is factorizable, i.e.,

$$\log q(\underline{\mathbf{h}},\underline{\mathbf{y}}) = \log q(\underline{\mathbf{h}}) + \log q(\underline{\mathbf{y}}) = \sum_{i=1}^{S}\log q(\mathbf{h}_i) + \sum_{p=1}^{P}\log q(\mathbf{y}_p). \tag{5.12}$$

When the lower bound $\mathcal{L}(q)$ in Eq. 5.11 is maximized, we have [1, 30]

$$\begin{aligned}\log q(\underline{\mathbf{h}}) &= \mathbb{E}_{q(\underline{\mathbf{y}})}\{\log p(\underline{\mathbf{h}}, \underline{\mathbf{y}}, \mathcal{X})\} + \text{const} \\ \log q(\underline{\mathbf{y}}) &= \mathbb{E}_{q(\underline{\mathbf{h}})}\{\log p(\underline{\mathbf{h}}, \underline{\mathbf{y}}, \mathcal{X})\} + \text{const},\end{aligned} \tag{5.13}$$

where $\mathbb{E}_{q(\underline{\mathbf{y}})}\{Z\}$ is the expectation of Z using $q(\underline{\mathbf{y}})$ as the density.

Note that $\log q(\underline{\mathbf{h}})$ in Eq. 5.13 can be written as

$$\begin{aligned}&\log q(\underline{\mathbf{h}}) \\ &= \sum\nolimits_i \log q(\mathbf{h}_i) \\ &= \langle \log p(\underline{\mathbf{h}}, \underline{\mathbf{y}}, \mathcal{X})\rangle_{\underline{\mathbf{y}}} + \text{const} \\ &= \langle \log p(\mathcal{X}|\underline{\mathbf{h}}, \underline{\mathbf{y}})\rangle_{\underline{\mathbf{y}}} + \langle \log p(\underline{\mathbf{h}}, \underline{\mathbf{y}})\rangle_{\underline{\mathbf{y}}} + \text{const} \\ &= \sum\nolimits_{ijp} \Big\langle \log \mathcal{N}(\mathbf{x}_{ij}^p | \mathbf{m} + \mathbf{V}\mathbf{h}_i + \mathbf{R}\mathbf{y}_p, \boldsymbol{\Sigma})\Big\rangle_{\mathbf{y}_p} + \sum\nolimits_i \langle \log \mathcal{N}(\mathbf{h}_i|\mathbf{0}, \mathbf{I})\rangle_{\underline{\mathbf{y}}} \\ &\quad + \sum\nolimits_p \langle \log \mathcal{N}(\mathbf{y}_p|\mathbf{0}, \mathbf{I})\rangle_{\mathbf{y}_p} + \text{const} \\ &= -\frac{1}{2}\sum\nolimits_{ijp}(\mathbf{x}_{ij}^p - \mathbf{m} - \mathbf{V}\mathbf{h}_i - \mathbf{R}\mathbf{y}_p^*)^{\mathsf{T}}\boldsymbol{\Sigma}^{-1}(\mathbf{x}_{ij}^p - \mathbf{m} - \mathbf{V}\mathbf{h}_i - \mathbf{R}\mathbf{y}_p^*) \\ &\quad - \frac{1}{2}\sum\nolimits_i \mathbf{h}_i^{\mathsf{T}}\mathbf{h}_i + \text{const} \\ &= \sum\nolimits_i \left[\mathbf{h}_i^{\mathsf{T}}\mathbf{V}^{\mathsf{T}}\boldsymbol{\Sigma}^{-1}\sum\nolimits_{jp}(\mathbf{x}_{ij}^p - \mathbf{m} - \mathbf{R}\mathbf{y}_p^*) - \frac{1}{2}\mathbf{h}_i^{\mathsf{T}}\left(\mathbf{I} + \sum\nolimits_p H_i(p)\mathbf{V}^{\mathsf{T}}\boldsymbol{\Sigma}^{-1}\mathbf{V}\right)\mathbf{h}_i^{\mathsf{T}}\right] \\ &\quad + \text{const},\end{aligned} \tag{5.14}$$

where $\mathbf{y}_p^* \equiv \langle \mathbf{y}_p|\mathcal{X}\rangle_{\mathbf{y}_p}$ is the posterior mean of $\mathbf{y}_p$ in the previous iteration and $\langle . \rangle_{\mathbf{y}_p}$ denotes the expectation with respect to $\mathbf{y}_p$. Note that $\langle \log \mathcal{N}(\mathbf{y}_p|\mathbf{0}, \mathbf{I})\rangle_{\mathbf{y}_p}$ is the differential entropy of a normal distribution, which does not depend on $\mathbf{h}_i$ [169, Chapter 8].

By extracting $\mathbf{h}_i$ in Eq. 5.14 and comparing with $\sum_i \log q(\mathbf{h}_i)$, we can see that $q(\mathbf{h}_i)$ is a Gaussian with mean vector and precision matrix as follows:

$$\mathbb{E}_{q(\mathbf{h}_i)}\{\mathbf{h}_i|\mathcal{X}\} = \langle \mathbf{h}_i|\mathcal{X}\rangle = \left(\mathbf{L}_i^{(1)}\right)^{-1}\mathbf{V}^{\mathsf{T}}\boldsymbol{\Sigma}^{-1}\sum_{p=1}^{P}\sum_{j=1}^{H_i(p)}(\mathbf{x}_{ij}^p - \mathbf{m} - \mathbf{R}\mathbf{y}_p^*)$$

and

$$\mathbf{L}_i^{(1)} \equiv \mathbf{I} + \sum_{p=1}^{P} H_i(p)\mathbf{V}^{\mathsf{T}}\boldsymbol{\Sigma}^{-1}\mathbf{V}.$$

Using this precision matrix, we can compute the second-order moment (which will be needed in the M-step) as follows:

$$\langle \mathbf{h}_i\mathbf{h}_i^{\mathsf{T}}|\mathcal{X}\rangle = \left(\mathbf{L}_i^{(1)}\right)^{-1} + \langle \mathbf{h}_i|\mathcal{X}\rangle\langle \mathbf{h}_i|\mathcal{X}\rangle^{\mathsf{T}}.$$

In the same manner, we obtain the posterior mean and second-order moment of $\mathbf{y}_p$ by comparing the terms in $\log q(\mathbf{y}_p)$ with a Gaussian distribution.

The posterior moment $\langle \mathbf{h}_i \mathbf{y}_p^{\mathsf{T}} | \mathcal{X} \rangle$ is also needed in the M-step. We can use variational Bayes to approximate it as follows:

$$p(\mathbf{h}_i, \mathbf{y}_p | \mathcal{X}) \approx q(\mathbf{h}_i) q(\mathbf{y}_p), \tag{5.15}$$

where both $q(\mathbf{h}_i)$ and $q(\mathbf{y}_p)$ are Gaussians. According to the law of total expectation [170], we factorize Eq. 5.15 to get

$$\begin{aligned} \langle \mathbf{y}_p \mathbf{h}_i^{\mathsf{T}} | \mathcal{X} \rangle &\approx \langle \mathbf{y}_p | \mathcal{X} \rangle \langle \mathbf{h}_i | \mathcal{X} \rangle^{\mathsf{T}} \\ \langle \mathbf{h}_i \mathbf{y}_p^{\mathsf{T}} | \mathcal{X} \rangle &\approx \langle \mathbf{h}_i | \mathcal{X} \rangle \langle \mathbf{y}_p | \mathcal{X} \rangle^{\mathsf{T}}. \end{aligned}$$

Therefore, the equations for the variational E-step are as follows:

$$\mathbf{L}_i^{(1)} = \mathbf{I} + N_i \mathbf{V}^{\mathsf{T}} \boldsymbol{\Sigma}^{-1} \mathbf{V} \quad i = 1, \ldots, S \tag{5.16a}$$

$$\mathbf{L}_p^{(2)} = \mathbf{I} + B_p \mathbf{R}^{\mathsf{T}} \boldsymbol{\Sigma}^{-1} \mathbf{R} \quad p = 1, \ldots, P \tag{5.16b}$$

$$\langle \mathbf{h}_i | \mathcal{X} \rangle = (\mathbf{L}_i^{(1)})^{-1} \mathbf{V}^{\mathsf{T}} \boldsymbol{\Sigma}^{-1} \sum_{p=1}^{P} \sum_{j=1}^{H_i(p)} (\mathbf{x}_{ij}^p - \mathbf{m} - \mathbf{R} \mathbf{y}_p^*) \tag{5.16c}$$

$$\langle \mathbf{y}_p | \mathcal{X} \rangle = (\mathbf{L}_p^{(2)})^{-1} \mathbf{R}^{\mathsf{T}} \boldsymbol{\Sigma}^{-1} \sum_{i=1}^{S} \sum_{j=1}^{H_i(p)} (\mathbf{x}_{ij}^p - \mathbf{m} - \mathbf{V} \mathbf{h}_i^*) \tag{5.16d}$$

$$\langle \mathbf{h}_i \mathbf{h}_i^{\mathsf{T}} | \mathcal{X} \rangle = (\mathbf{L}_i^{(1)})^{-1} + \langle \mathbf{h}_i | \mathcal{X} \rangle \langle \mathbf{h}_i | \mathcal{X} \rangle^{\mathsf{T}} \tag{5.16e}$$

$$\langle \mathbf{y}_p \mathbf{y}_p^{\mathsf{T}} | \mathcal{X} \rangle = (\mathbf{L}_p^{(2)})^{-1} + \langle \mathbf{y}_p | \mathcal{X} \rangle \langle \mathbf{y}_p | \mathcal{X} \rangle^{\mathsf{T}} \tag{5.16f}$$

$$\langle \mathbf{y}_p \mathbf{h}_i^{\mathsf{T}} | \mathcal{X} \rangle \approx \langle \mathbf{y}_p | \mathcal{X} \rangle \langle \mathbf{h}_i | \mathcal{X} \rangle^{\mathsf{T}} \tag{5.16g}$$

$$\langle \mathbf{h}_i \mathbf{y}_p^{\mathsf{T}} | \mathcal{X} \rangle \approx \langle \mathbf{h}_i | \mathcal{X} \rangle \langle \mathbf{y}_p | \mathcal{X} \rangle^{\mathsf{T}} \tag{5.16h}$$

Algorithm 4 shows the procedure of training a duration-invariant PLDA model.

Derivation of VB Posteriors

Our goal is to approximate the true joint posteriors $p(\underline{\mathbf{h}}, \underline{\mathbf{w}} | \mathcal{X})$ by a distribution $q(\underline{\mathbf{h}}, \underline{\mathbf{w}})$ such that

$$q(\underline{\mathbf{h}}, \underline{\mathbf{w}}) = q(\underline{\mathbf{h}}) q(\underline{\mathbf{w}}).$$

The derivation of the VB posteriors begins with minimizing the KL-divergence

$$\begin{aligned} \mathcal{D}_{\mathrm{KL}}(q(\underline{\mathbf{h}}, \underline{\mathbf{w}}) || p(\underline{\mathbf{h}}, \underline{\mathbf{w}} | \mathcal{X})) &= \int \int q(\underline{\mathbf{h}}, \underline{\mathbf{w}}) \log \left[\frac{q(\underline{\mathbf{h}}, \underline{\mathbf{w}})}{p(\underline{\mathbf{h}}, \underline{\mathbf{w}} | \mathcal{X})} \right] \mathrm{d}\underline{\mathbf{h}} \mathrm{d}\underline{\mathbf{w}} \\ &= - \int \int q(\underline{\mathbf{h}}, \underline{\mathbf{w}}) \log \left[\frac{p(\underline{\mathbf{h}}, \underline{\mathbf{w}} | \mathcal{X})}{q(\underline{\mathbf{h}}, \underline{\mathbf{w}})} \right] \mathrm{d}\underline{\mathbf{h}} \mathrm{d}\underline{\mathbf{w}} \\ &= - \int \int q(\underline{\mathbf{h}}, \underline{\mathbf{w}}) \log \left[\frac{p(\underline{\mathbf{h}}, \underline{\mathbf{w}}, \mathcal{X})}{q(\underline{\mathbf{h}}, \underline{\mathbf{w}}) p(\mathcal{X})} \right] \mathrm{d}\underline{\mathbf{h}} \mathrm{d}\underline{\mathbf{w}} \\ &= -\mathcal{L}(q) + \log p(\mathcal{X}), \end{aligned} \tag{5.17}$$

Algorithm 4 Variational Bayes EM algorithm for duration-invariant PLDA

Input:

Development data set consisting of i-vectors $\mathcal{X} = \{\mathbf{x}_{ij}^{p} | i = 1, \ldots, S; j = 1, \ldots, H_i(p); p = 1, \ldots, P\}$, with identity labels and duration group labels.

Initialization:

$\mathbf{y}_p^* \leftarrow \mathbf{0}$;
$\boldsymbol{\Sigma} \leftarrow 0.01\mathbf{I}$;
$\mathbf{V}, \mathbf{R} \leftarrow$ eigenvectors of PCA projection matrix learned using data set $\mathcal{X}$;

Parameter Estimation:

Compute $\mathbf{m}$ via Eq. 5.9(a);
Compute $\mathbf{L}_i^{(1)}$ and $\mathbf{L}_p^{(2)}$ according to Eqs. 5.16(a) and (b), respectively;
Set $\mathbf{y}_p^*$ to the posterior mean of $\mathbf{y}_p$. Compute the posterior mean of $\mathbf{h}_i$ using Eq. 5.16(c);
Use the posterior mean of $\mathbf{h}_i$ computed in Step 3 to update the posterior mean of $\mathbf{y}_p$ according to Eq. 5.16(d);
Compute the other terms in the E-step, i.e., Eq. 5.16(e)–(h);
Update the model parameters using Eq. 5.9(a)–(c);
Go to Step 2 until convergence;

Return: the parameters of the duration-invariant PLDA model $\theta = \{\mathbf{m}, \mathbf{V}, \mathbf{R}, \boldsymbol{\Sigma}\}$.

where $\mathcal{L}(q)$ is the variational Bayes lower bound (VBLB). The lower bound is given by:

$$\begin{aligned}
\mathcal{L}(q) &= \int\int q(\underline{\mathbf{h}}, \underline{\mathbf{w}}) \log \left[\frac{p(\underline{\mathbf{h}}, \underline{\mathbf{w}}, \mathcal{X})}{q(\underline{\mathbf{h}}, \underline{\mathbf{w}})} \right] \mathrm{d}\underline{\mathbf{h}}\mathrm{d}\underline{\mathbf{w}} \\
&= \int\int q(\underline{\mathbf{h}}, \underline{\mathbf{w}}) \log p(\underline{\mathbf{h}}, \underline{\mathbf{w}}, \mathcal{X}) \mathrm{d}\underline{\mathbf{h}}\mathrm{d}\underline{\mathbf{w}} - \int\int q(\underline{\mathbf{h}}, \underline{\mathbf{w}}) \log q(\underline{\mathbf{h}}, \underline{\mathbf{w}}) \mathrm{d}\underline{\mathbf{h}}\mathrm{d}\underline{\mathbf{w}} \\
&= \int\int q(\underline{\mathbf{h}}) q(\underline{\mathbf{w}}) \log p(\underline{\mathbf{h}}, \underline{\mathbf{w}}, \mathcal{X}) \mathrm{d}\underline{\mathbf{h}}\mathrm{d}\underline{\mathbf{w}} - \int\int q(\underline{\mathbf{h}}) q(\underline{\mathbf{w}}) \log q(\underline{\mathbf{h}}) \mathrm{d}\underline{\mathbf{h}}\mathrm{d}\underline{\mathbf{w}} \\
&\quad - \int\int q(\underline{\mathbf{h}}) q(\underline{\mathbf{w}}) \log q(\underline{\mathbf{w}}) \mathrm{d}\underline{\mathbf{h}}\mathrm{d}\underline{\mathbf{w}} \\
&= \int\int q(\underline{\mathbf{h}}) q(\underline{\mathbf{w}}) \log p(\underline{\mathbf{h}}, \underline{\mathbf{w}}, \mathcal{X}) \mathrm{d}\underline{\mathbf{h}}\mathrm{d}\underline{\mathbf{w}} - \int q(\underline{\mathbf{h}}) \log q(\underline{\mathbf{h}}) \mathrm{d}\underline{\mathbf{h}} \\
&\quad - \int q(\underline{\mathbf{w}}) \log q(\underline{\mathbf{w}}) \mathrm{d}\underline{\mathbf{w}}.
\end{aligned} \tag{5.18}$$

Note that the first term in Eq. 6.39 can be written as

$$\begin{aligned}
\int\int q(\underline{\mathbf{h}}) q(\underline{\mathbf{w}}) \log p(\underline{\mathbf{h}}, \underline{\mathbf{w}}, \mathcal{X}) \mathrm{d}\underline{\mathbf{h}}\mathrm{d}\underline{\mathbf{w}} &= \int q(\underline{\mathbf{h}}) \left[\int q(\underline{\mathbf{w}}) \log p(\underline{\mathbf{h}}, \underline{\mathbf{w}}, \mathcal{X}) \mathrm{d}\underline{\mathbf{w}} \right] \mathrm{d}\underline{\mathbf{h}} \\
&= \int q(\underline{\mathbf{h}}) \mathbb{E}_{q(\underline{\mathbf{w}})} \{\log p(\underline{\mathbf{h}}, \underline{\mathbf{w}}, \mathcal{X})\} \mathrm{d}\underline{\mathbf{h}}.
\end{aligned} \tag{5.19}$$

Define a distribution of $\underline{\mathbf{h}}$ as

$$q^*(\underline{\mathbf{h}}) \equiv \frac{1}{Z} \exp \left\{ \mathbb{E}_{q(\underline{\mathbf{w}})} \{\log p(\underline{\mathbf{h}}, \underline{\mathbf{w}}, \mathcal{X})\} \right\}, \tag{5.20}$$

where Z is to normalize the distribution. Using Eqs. 5.19 and 5.20, Eq. 6.39 can be written as:

$$\begin{aligned}\mathcal{L}(q) &= \int q(\underline{\mathbf{h}}) \log q^*(\underline{\mathbf{h}}) d\underline{\mathbf{h}} - \int q(\underline{\mathbf{h}}) \log q(\underline{\mathbf{h}}) d\underline{\mathbf{h}} - \int q(\underline{\mathbf{w}}) \log q(\underline{\mathbf{w}}) d\underline{\mathbf{w}} + \log Z \\ &= -\int q(\underline{\mathbf{h}}) \log \left[\frac{q(\underline{\mathbf{h}})}{q^*(\underline{\mathbf{h}})} \right] d\underline{\mathbf{h}} + \mathcal{H}(q(\underline{\mathbf{w}})) + \log Z \\ &= -\mathcal{D}_{\mathrm{KL}}(q(\underline{\mathbf{h}}) || q^*(\underline{\mathbf{h}})) + \mathcal{H}(q(\underline{\mathbf{w}})) + \log Z.\end{aligned}$$

$\mathcal{L}(q)$ will attain its maximum when the KL-divergence vanishes, i.e.,

$$\log q(\underline{\mathbf{h}}) = \log q^*(\underline{\mathbf{h}}) = \mathbb{E}_{q(\underline{\mathbf{w}})}\{\log p(\underline{\mathbf{h}}, \underline{\mathbf{w}}, \mathcal{X})\} + \text{const.} \tag{5.21}$$

Now, we write Eq. 6.41 as

$$\log p(\mathcal{X}) = \mathcal{L}(q) + \mathcal{D}_{\mathrm{KL}}(q(\underline{\mathbf{h}}, \underline{\mathbf{w}}) || p(\underline{\mathbf{h}}, \underline{\mathbf{w}} | \mathcal{X})).$$

Because KL-divergence is nonnegative, we may maximize the data likelihood $p(\mathcal{X})$ by maximizing the VB lower bound $\mathcal{L}(q)$, which can be achieved by Eq. 5.21. A similar treatment can be applied to $q(\underline{\mathbf{w}})$.

Likelihood Ratio Scores

Suppose we do not know (or not use) the duration of the target and test utterances. Then, the scoring function of SI-PLDA [55] can be used for computing the likelihood ratio score. However, because we know the duration of the the target and test utterances (ℓ_s and ℓ_t), the likelihood-ratio score can be computed as follows:

$$S_{\mathrm{LR}}(\mathbf{x}_s, \mathbf{x}_t | \ell_s, \ell_t) = \log \frac{p(\mathbf{x}_s, \mathbf{x}_t | \text{same-speaker}, \ell_s, \ell_t)}{p(\mathbf{x}_s, \mathbf{x}_t | \text{different-speakers}, \ell_s, \ell_t)}, \tag{5.22}$$

where $\mathbf{x}_s$ and $\mathbf{x}_t$ are the i-vectors of the target speaker and test speaker, respectively.

There are two approaches to calculating the score in Eq. 5.22, depending on the sharpness of the posterior density of $\mathbf{y}_p$. Consider the case where the posterior density of $\mathbf{y}_p$ is sharp at its mean $\mathbf{y}_p^*$.[1] Assuming that ℓ belongs to duration group p, the marginal-likelihood of i-vector $\mathbf{x}$ can then be expressed as:

$$\begin{aligned}p(\mathbf{x}|\ell \in p\text{th duration group}) &= \int_{\mathbf{h}} p(\mathbf{x}|\mathbf{h}, \mathbf{y}_p^*) p(\mathbf{h}) d\mathbf{h} \\ &= \int_{\mathbf{h}} \mathcal{N}(\mathbf{x}|\mathbf{m} + \mathbf{V}\mathbf{h} + \mathbf{R}\mathbf{y}_p^*, \mathbf{\Sigma}) \mathcal{N}(\mathbf{h}|\mathbf{0}, \mathbf{I}) d\mathbf{h} \\ &= \mathcal{N}(\mathbf{x}|\mathbf{m} + \mathbf{R}\mathbf{y}_p^*, \mathbf{V}\mathbf{V}^{\mathsf{T}} + \mathbf{\Sigma}),\end{aligned} \tag{5.23}$$

where $\mathbf{y}_p^* \equiv \langle \mathbf{y}_p | \mathcal{X} \rangle$ is the posterior mean of $\mathbf{y}_p$. Given a test i-vector $\mathbf{x}_t$ and a target i-vector $\mathbf{x}_s$, we can use Eq. 5.23 to compute the likelihood ratio score:

[1] As suggested by Eq. 5.16(b), this occurs when the number of training i-vectors (B_p) in duration group p is large.

$$
\begin{aligned}
S_{\text{LR}}(\mathbf{x}_s, \mathbf{x}_t | \ell_s, \ell_t) &= \log \frac{p(\mathbf{x}_s, \mathbf{x}_t | \text{same-speaker}, \ell_s, \ell_t)}{p(\mathbf{x}_s, \mathbf{x}_t | \text{different-speakers}, \ell_s, \ell_t)} \\
&= \log \frac{\mathcal{N}\left(\begin{bmatrix} \mathbf{x}_s \\ \mathbf{x}_t \end{bmatrix} \middle| \begin{bmatrix} \mathbf{m} + \mathbf{R}\mathbf{y}^*_{p_s} \\ \mathbf{m} + \mathbf{R}\mathbf{y}^*_{p_t} \end{bmatrix}, \begin{bmatrix} \boldsymbol{\Psi} & \boldsymbol{\Sigma}_{ac} \\ \boldsymbol{\Sigma}_{ac} & \boldsymbol{\Psi} \end{bmatrix} \right)}{\mathcal{N}\left(\begin{bmatrix} \mathbf{x}_s \\ \mathbf{x}_t \end{bmatrix} \middle| \begin{bmatrix} \mathbf{m} + \mathbf{R}\mathbf{y}^*_{p_s} \\ \mathbf{m} + \mathbf{R}\mathbf{y}^*_{p_t} \end{bmatrix}, \begin{bmatrix} \boldsymbol{\Psi} & \mathbf{0} \\ \mathbf{0} & \boldsymbol{\Psi} \end{bmatrix} \right)} \\
&= \frac{1}{2}[\bar{\mathbf{x}}_s^{\mathsf{T}} \mathbf{Q} \bar{\mathbf{x}}_s + 2\bar{\mathbf{x}}_s^{\mathsf{T}} \mathbf{P} \bar{\mathbf{x}}_t + \bar{\mathbf{x}}_t^{\mathsf{T}} \mathbf{Q} \bar{\mathbf{x}}_t] + \text{const}
\end{aligned} \tag{5.24}
$$

where

$$
\begin{aligned}
\bar{\mathbf{x}}_s &= \mathbf{x}_s - \mathbf{m} - \mathbf{R}\mathbf{y}^*_{p_s} \\
\bar{\mathbf{x}}_t &= \mathbf{x}_t - \mathbf{m} - \mathbf{R}\mathbf{y}^*_{p_t} \\
\mathbf{Q} &= \boldsymbol{\Psi}^{-1} - (\boldsymbol{\Psi} - \boldsymbol{\Sigma}_{ac}\boldsymbol{\Psi}^{-1}\boldsymbol{\Sigma}_{ac})^{-1} \\
\mathbf{P} &= \boldsymbol{\Psi}^{-1}\boldsymbol{\Sigma}_{ac}(\boldsymbol{\Psi} - \boldsymbol{\Sigma}_{ac}\boldsymbol{\Psi}^{-1}\boldsymbol{\Sigma}_{ac})^{-1} \\
\boldsymbol{\Psi} &= \mathbf{V}\mathbf{V}^{\mathsf{T}} + \boldsymbol{\Sigma}; \quad \boldsymbol{\Sigma}_{ac} = \mathbf{V}\mathbf{V}^{\mathsf{T}}.
\end{aligned}
$$

Now, let's consider the case where the posterior density of $\mathbf{y}_p$ is *moderately* sharp and is a Gaussian $\mathcal{N}(\mathbf{y}_p | \boldsymbol{\mu}^*_p, \boldsymbol{\Sigma}^*_p)$.[2] If ℓ falls on duration group p, the marginal-likelihood of i-vector $\mathbf{x}$ is:

$$
\begin{aligned}
p(\mathbf{x}|\ell \in p\text{th duration group}) &= \int_{\mathbf{h}} \int_{\mathbf{y}_p} p(\mathbf{x}|\mathbf{h}, \mathbf{y}_p) p(\mathbf{h}) p(\mathbf{y}_p) \mathrm{d}\mathbf{h} \mathrm{d}\mathbf{y}_p \\
&= \int_{\mathbf{h}} \int_{\mathbf{y}_p} \mathcal{N}(\mathbf{x}|\mathbf{m} + \mathbf{V}\mathbf{h} + \mathbf{R}\mathbf{y}_p, \boldsymbol{\Sigma}) \mathcal{N}(\mathbf{h}|\mathbf{0}, \mathbf{I}) \mathcal{N}(\mathbf{y}_p|\mathbf{0}, \mathbf{I}) \mathrm{d}\mathbf{h} \mathrm{d}\mathbf{y}_p \\
&= \int_{\mathbf{y}_p} \mathcal{N}(\mathbf{x}|\mathbf{m} + \mathbf{R}\mathbf{y}_p, \mathbf{V}\mathbf{V}^{\mathsf{T}} + \boldsymbol{\Sigma}) \mathcal{N}(\mathbf{y}_p|\mathbf{0}, \mathbf{I}) \mathrm{d}\mathbf{y}_p \\
&= \mathcal{N}(\mathbf{x}|\mathbf{m} + \mathbf{R}\boldsymbol{\mu}^*_p, \mathbf{V}\mathbf{V}^{\mathsf{T}} + \mathbf{R}\boldsymbol{\Sigma}^*_p\mathbf{R}^{\mathsf{T}} + \boldsymbol{\Sigma}),
\end{aligned} \tag{5.25}
$$

where we may use Eq. 5.16(d) to compute $\boldsymbol{\mu}^*_p$ and Eq. 5.16(b) to estimate $\boldsymbol{\Sigma}^*_p$. Given a test i-vector $\mathbf{x}_t$ and a target i-vector $\mathbf{x}_s$, the likelihood ratio score can be computed as:

$$
S_{\text{LR}}(\mathbf{x}_s, \mathbf{x}_t | \ell_s, \ell_t) = \log \frac{p(\mathbf{x}_s, \mathbf{x}_t | \text{same-speaker}, \ell_s, \ell_t)}{p(\mathbf{x}_s, \mathbf{x}_t | \text{different-speakers}, \ell_s, \ell_t)} \tag{5.26}
$$

$$
= \log \frac{\mathcal{N}\left(\begin{bmatrix} \mathbf{x}_s \\ \mathbf{x}_t \end{bmatrix} \middle| \begin{bmatrix} \mathbf{m} + \mathbf{R}\boldsymbol{\mu}^*_{p_s} \\ \mathbf{m} + \mathbf{R}\boldsymbol{\mu}^*_{p_t} \end{bmatrix}, \begin{bmatrix} \boldsymbol{\Sigma}_s & \boldsymbol{\Sigma}_{ac} \\ \boldsymbol{\Sigma}_{ac} & \boldsymbol{\Sigma}_t \end{bmatrix} \right)}{\mathcal{N}\left(\begin{bmatrix} \mathbf{x}_s \\ \mathbf{x}_t \end{bmatrix} \middle| \begin{bmatrix} \mathbf{m} + \mathbf{R}\boldsymbol{\mu}^*_{p_s} \\ \mathbf{m} + \mathbf{R}\boldsymbol{\mu}^*_{p_t} \end{bmatrix}, \begin{bmatrix} \boldsymbol{\Sigma}_s & \mathbf{0} \\ \mathbf{0} & \boldsymbol{\Sigma}_t \end{bmatrix} \right)} \tag{5.27}
$$

$$
= \frac{1}{2}[\bar{\mathbf{x}}_s^{\mathsf{T}} \mathbf{A}_{s,t} \bar{\mathbf{x}}_s + 2\bar{\mathbf{x}}_s^{\mathsf{T}} \mathbf{B}_{s,t} \bar{\mathbf{x}}_t + \bar{\mathbf{x}}_t^{\mathsf{T}} \mathbf{C}_{s,t} \bar{\mathbf{x}}_t] + \text{const} \tag{5.28}
$$

[2] This happens when there are a moderate number of training i-vectors (B_p) in duration group p, as evident by Eq. 5.16(b).

where

$$\begin{aligned}
\bar{\mathbf{x}}_s &= \mathbf{x}_s - \mathbf{m} - \mathbf{R}\mu^*_{p_s} \\
\bar{\mathbf{x}}_t &= \mathbf{x}_t - \mathbf{m} - \mathbf{R}\mu^*_{p_t} \\
\mathbf{A}_{s,t} &= \boldsymbol{\Sigma}_s^{-1} - (\boldsymbol{\Sigma}_s - \boldsymbol{\Sigma}_{ac}\boldsymbol{\Sigma}_t^{-1}\boldsymbol{\Sigma}_{ac})^{-1} \\
\mathbf{B}_{s,t} &= \boldsymbol{\Sigma}_s^{-1}\boldsymbol{\Sigma}_{ac}(\boldsymbol{\Sigma}_t - \boldsymbol{\Sigma}_{ac}\boldsymbol{\Sigma}_s^{-1}\boldsymbol{\Sigma}_{ac})^{-1} \\
\mathbf{C}_{s,t} &= \boldsymbol{\Sigma}_t^{-1} - (\boldsymbol{\Sigma}_t - \boldsymbol{\Sigma}_{ac}\boldsymbol{\Sigma}_s^{-1}\boldsymbol{\Sigma}_{ac})^{-1} \\
\boldsymbol{\Sigma}_s &= \mathbf{V}\mathbf{V}^{\mathsf{T}} + \mathbf{R}\boldsymbol{\Sigma}^*_{p_s}\mathbf{R}^{\mathsf{T}} + \boldsymbol{\Sigma} \\
\boldsymbol{\Sigma}_t &= \mathbf{V}\mathbf{V}^{\mathsf{T}} + \mathbf{R}\boldsymbol{\Sigma}^*_{p_t}\mathbf{R}^{\mathsf{T}} + \boldsymbol{\Sigma} \\
\boldsymbol{\Sigma}_{ac} &= \mathbf{V}\mathbf{V}^{\mathsf{T}}.
\end{aligned}$$

5.3.3 SNR- and Duration-Invariant PLDA

The SNR-invariant PLDA in Section 5.3.1 and the duration-invariant PLDA in Section 5.3.2 can be combined to address both SNR and duration variabilities in utterances. It is called SNR- and duration-invariant PLDA (SDI-PLDA) in [163]. The model has three latent factors, and they represent speaker, SNR, and duration information, respectively.

Generative Model

In classical PLDA, i-vectors from the same speaker share the same speaker factor. Inspired by this notion, SDI-PLDA assumes (1) that utterances with similar SNR will lead to i-vectors that possess similar SNR information and (2) that utterances of approximately the same length should lead to i-vectors that own similar duration information.

Figure 5.9 demonstrates that the assumptions above are reasonable. In Figure 5.9(a), we project the i-vectors to the first three principal components and denote three groups of i-vectors by three different colors. Each group corresponds to the same set of speakers whose utterances' SNR belong to the same SNR group. Grouping the i-vectors in this way is to ensure that any displacement in the clusters is not caused by speaker variability. Figure 5.9(a) clearly shows three clusters, and each cluster corresponds to one SNR group. To show that the second assumption is valid, we divided the clean i-vectors according to their utterance duration and plotted them on the first three principal components. Here, the word "clean" means that the i-vectors were obtained from clean telephone utterances. Again, each duration group in the legend of Figure 5.9(b) corresponds to the same set of speakers. Results clearly show that i-vectors are duration dependent and that the cluster locations depend on the utterance duration.

With the above assumptions, i-vectors can be generated from a linear model that comprises five components, namely speaker, SNR, duration, channel, and residue. Suppose we have a set of D-dimensional i-vectors

$$\mathcal{X} = \{\hat{\mathbf{x}}_{ij}^{kp} | i = 1, \ldots, S; k = 1, \ldots, K; p = 1, \ldots, P; j = 1, \ldots, H_{ik}(p)\}$$

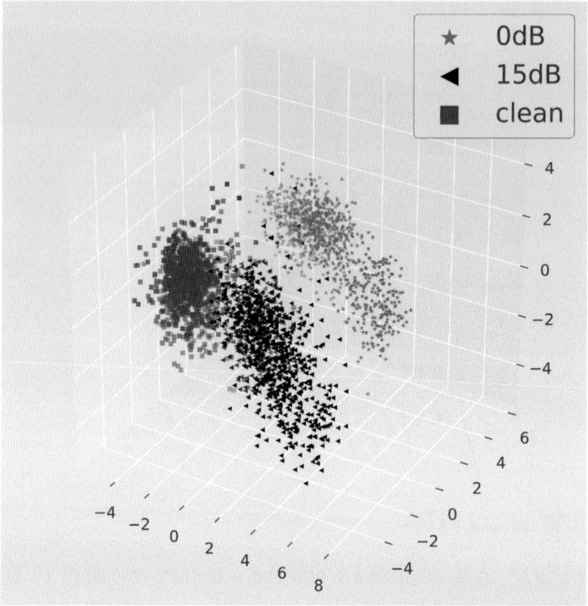

(a)

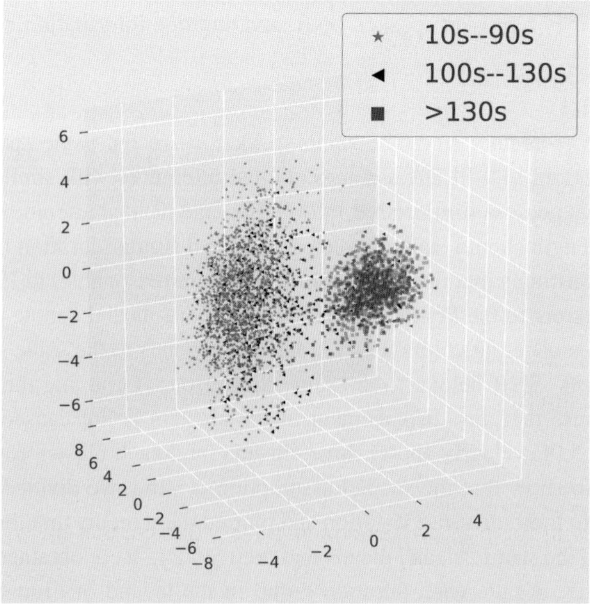

(b)

Figure 5.9 Projection of i-vectors on the first three principal components. (a) Utterances of different SNRs. (b) Utterances of different durations. *It may be better to view the color version of this figure, which is available at https://github.com/enmwmak/ML-for-Spkrec.*

Table 5.2 The definitions of the variables in Eq. 5.29.

Symbol	Dimension	Definition
$\mathbf{m}$	D	Global mean vector
$\mathbf{h}_i$	Q_1	Speaker factor with Gaussian prior $\mathcal{N}(\mathbf{0},\mathbf{I})$
$\mathbf{w}_k$	Q_2	SNR factor with Gaussian prior $\mathcal{N}(\mathbf{0},\mathbf{I})$
$\mathbf{y}_p$	Q_3	Duration factor with Gaussian prior $\mathcal{N}(\mathbf{0},\mathbf{I})$
ϵ_{ij}^{kp}	D	Residual vector with Gaussian prior $\mathcal{N}(\mathbf{0},\mathbf{\Sigma})$
$\mathbf{V}$	$D \times Q_1$	Speaker loading matrix
$\mathbf{U}$	$D \times Q_2$	SNR loading matrix
$\mathbf{R}$	$D \times Q_3$	Duration loading matrix

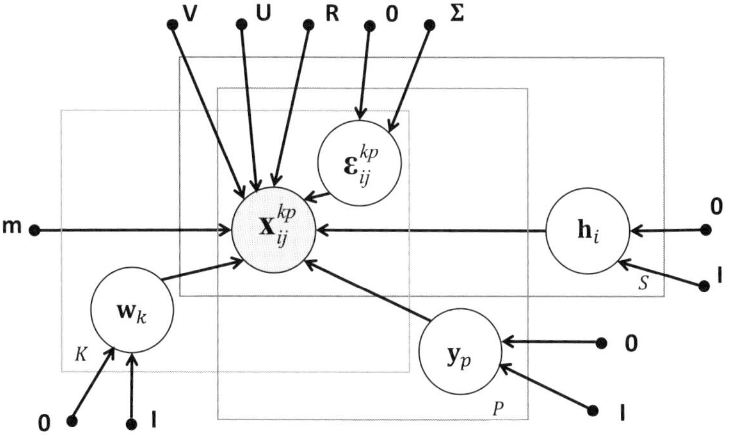

Figure 5.10 Probabilistic graphical model of SDI-PLDA.

obtained from S speakers, where $\hat{\mathbf{x}}_{ij}^{kp}$ is the jth i-vector from speaker i and k and p denote the SNR and duration groups, respectively. In SDI-PLDA, $\hat{\mathbf{x}}_{ij}^{kp}$ can be considered generated from the model:

$$\hat{\mathbf{x}}_{ij}^{kp} = \mathbf{m} + \mathbf{V}\mathbf{h}_i + \mathbf{U}\mathbf{w}_k + \mathbf{R}\mathbf{y}_p + \epsilon_{ij}^{kp}, \tag{5.29}$$

where the variables in Eq. 5.29 are defined in Table 5.2. Note that $\mathbf{h}_i$, $\mathbf{w}_k$, and $\mathbf{y}_p$ are assumed to be independent in their prior. Figure 5.10 shows the graphical model of SDI-PLDA.

The SNR- and duration-invariant PLDA extends the conventional PLDA in that the former takes the advantage of having multiple labels (speaker IDs, SNR levels, and duration ranges) in the training set to train the loading matrices, whereas the latter only uses the speaker IDs. SDI-PLDA uses the additional duration subspace to leverage the duration information in the utterances. Different from the session- and speaker-dependent term $\mathbf{G}\mathbf{r}_{ij}$ in Eq. 5.4, the duration term $\mathbf{R}\mathbf{y}_p$ and the SNR term $\mathbf{U}\mathbf{w}_k$ in Eq. 5.29 depend on the duration and SNR groups, respectively.

Variational Bayes EM Algorithm

The parameters $\boldsymbol{\theta} = \{\mathbf{m}, \mathbf{V}, \mathbf{U}, \mathbf{R}, \boldsymbol{\Sigma}\}$ of the SNR- and duration-invariant PLDA model can be obtained by using the maximum likelihood principle. Specifically, denote $\boldsymbol{\theta}$ as the old parameters. We estimate the new parameters $\boldsymbol{\theta}'$ by maximizing the auxiliary function:

$$\begin{aligned} Q(\boldsymbol{\theta}'|\boldsymbol{\theta}) &= \mathbb{E}_{q(\underline{\mathbf{h}},\underline{\mathbf{w}},\underline{\mathbf{y}})}\Big[\log p(\mathcal{X},\underline{\mathbf{h}},\underline{\mathbf{w}},\underline{\mathbf{y}}|\boldsymbol{\theta}')\Big|\mathcal{X},\boldsymbol{\theta}\Big] \\ &= \mathbb{E}_{q(\underline{\mathbf{h}},\underline{\mathbf{w}},\underline{\mathbf{y}})}\Big[\sum\nolimits_{ikpj}\log\Big(p(\hat{\mathbf{x}}_{ij}^{kp}|\mathbf{h}_i,\mathbf{w}_k,\mathbf{y}_p,\boldsymbol{\theta}')p(\mathbf{h}_i,\mathbf{w}_k,\mathbf{y}_p)\Big)\Big|\mathcal{X},\boldsymbol{\theta}\Big]. \end{aligned} \tag{5.30}$$

Taking expectation with respect to $q(\underline{\mathbf{h}},\underline{\mathbf{w}},\underline{\mathbf{y}})$ suggests that it is necessary to compute the posterior distributions of the latent variables given the model parameters $\boldsymbol{\theta}$. Let $N_i = \sum_{kp} H_{ik}(p)$ be the number of training i-vectors from speaker i, $M_k = \sum_{ip} H_{ik}(p)$ be the number of training i-vectors from SNR group k, and $B_p = \sum_{ik} H_{ik}(p)$ be the number of training i-vectors from duration group p. Similar to duration-invariant PLDA, we may use variational Bayes to compute these posteriors:

$$\mathbf{L}_i^{(1)} = \mathbf{I} + N_i\mathbf{V}^{\mathsf{T}}\boldsymbol{\Sigma}^{-1}\mathbf{V} \quad i = 1,\ldots,S \tag{5.31a}$$
$$\mathbf{L}_k^{(2)} = \mathbf{I} + M_k\mathbf{U}^{\mathsf{T}}\boldsymbol{\Sigma}^{-1}\mathbf{U} \quad k = 1,\ldots,K \tag{5.31b}$$
$$\mathbf{L}_p^{(3)} = \mathbf{I} + B_p\mathbf{R}^{\mathsf{T}}\boldsymbol{\Sigma}^{-1}\mathbf{R} \quad p = 1,\ldots,P \tag{5.31c}$$
$$\langle\mathbf{h}_i|\mathcal{X}\rangle = (\mathbf{L}_i^{(1)})^{-1}\mathbf{V}^{\mathsf{T}}\boldsymbol{\Sigma}^{-1}\sum\nolimits_{kpj}(\hat{\mathbf{x}}_{ij}^{kp} - \mathbf{m} - \mathbf{U}\mathbf{w}_k^* - \mathbf{R}\mathbf{y}_p^*) \tag{5.31d}$$
$$\langle\mathbf{w}_k|\mathcal{X}\rangle = (\mathbf{L}_k^{(2)})^{-1}\mathbf{U}^{\mathsf{T}}\boldsymbol{\Sigma}^{-1}\sum\nolimits_{ipj}(\hat{\mathbf{x}}_{ij}^{kp} - \mathbf{m} - \mathbf{V}\mathbf{h}_i^* - \mathbf{R}\mathbf{y}_p^*) \tag{5.31e}$$
$$\langle\mathbf{y}_p|\mathcal{X}\rangle = (\mathbf{L}_p^{(3)})^{-1}\mathbf{R}^{\mathsf{T}}\boldsymbol{\Sigma}^{-1}\sum\nolimits_{ikj}(\hat{\mathbf{x}}_{ij}^{kp} - \mathbf{m} - \mathbf{V}\mathbf{h}_i^* - \mathbf{U}\mathbf{w}_k^*) \tag{5.31f}$$
$$\langle\mathbf{h}_i\mathbf{h}_i^{\mathsf{T}}|\mathcal{X}\rangle = (\mathbf{L}_i^{(1)})^{-1} + \langle\mathbf{h}_i|\mathcal{X}\rangle\langle\mathbf{h}_i|\mathcal{X}\rangle^{\mathsf{T}} \tag{5.31g}$$
$$\langle\mathbf{w}_k\mathbf{w}_k^{\mathsf{T}}|\mathcal{X}\rangle = (\mathbf{L}_k^{(2)})^{-1} + \langle\mathbf{w}_k|\mathcal{X}\rangle\langle\mathbf{w}_k|\mathcal{X}\rangle^{\mathsf{T}} \tag{5.31h}$$
$$\langle\mathbf{y}_p\mathbf{y}_p^{\mathsf{T}}|\mathcal{X}\rangle = (\mathbf{L}_p^{(3)})^{-1} + \langle\mathbf{y}_p|\mathcal{X}\rangle\langle\mathbf{y}_p|\mathcal{X}\rangle^{\mathsf{T}} \tag{5.31i}$$
$$\langle\mathbf{w}_k\mathbf{h}_i^{\mathsf{T}}|\mathcal{X}\rangle \approx \langle\mathbf{w}_k|\mathcal{X}\rangle\langle\mathbf{h}_i|\mathcal{X}\rangle^{\mathsf{T}} \tag{5.31j}$$
$$\langle\mathbf{h}_i\mathbf{w}_k^{\mathsf{T}}|\mathcal{X}\rangle \approx \langle\mathbf{h}_i|\mathcal{X}\rangle\langle\mathbf{w}_k|\mathcal{X}\rangle^{\mathsf{T}} \tag{5.31k}$$
$$\langle\mathbf{w}_k\mathbf{y}_p^{\mathsf{T}}|\mathcal{X}\rangle \approx \langle\mathbf{w}_k|\mathcal{X}\rangle\langle\mathbf{y}_p|\mathcal{X}\rangle^{\mathsf{T}} \tag{5.31l}$$
$$\langle\mathbf{y}_p\mathbf{w}_k^{\mathsf{T}}|\mathcal{X}\rangle \approx \langle\mathbf{y}_p|\mathcal{X}\rangle\langle\mathbf{w}_k|\mathcal{X}\rangle^{\mathsf{T}} \tag{5.31m}$$
$$\langle\mathbf{h}_i\mathbf{y}_p^{\mathsf{T}}|\mathcal{X}\rangle \approx \langle\mathbf{h}_i|\mathcal{X}\rangle\langle\mathbf{y}_p|\mathcal{X}\rangle^{\mathsf{T}} \tag{5.31n}$$
$$\langle\mathbf{y}_p\mathbf{h}_i^{\mathsf{T}}|\mathcal{X}\rangle \approx \langle\mathbf{y}_p|\mathcal{X}\rangle\langle\mathbf{h}_i|\mathcal{X}\rangle^{\mathsf{T}}, \tag{5.31o}$$

where $\mathbf{w}_k^*$, $\mathbf{y}_p^*$, and $\mathbf{h}_i^*$ denote the posterior mean of $\mathbf{w}_k$, $\mathbf{y}_p$, and $\mathbf{h}_i$ in the previous iteration, respectively.

Using Eq. 5.31(a)–(o), the model parameters $\boldsymbol{\theta}'$ can be estimated by maximizing the auxiliary function in Eq. 5.30, which gives

$$\mathbf{m} = \frac{1}{N}\sum\nolimits_{ikpj}\hat{\mathbf{x}}_{ij}^{kp} \tag{5.32a}$$

$$\mathbf{V}' = \left\{\sum_{ikpj}\left[(\hat{\mathbf{x}}_{ij}^{kp} - \mathbf{m})\langle\mathbf{h}_i|\mathcal{X}\rangle^{\mathsf{T}} - \mathbf{U}\langle\mathbf{w}_k\mathbf{h}_i^{\mathsf{T}}|\mathcal{X}\rangle - \mathbf{R}\langle\mathbf{y}_p\mathbf{h}_i^{\mathsf{T}}|\mathcal{X}\rangle\right]\right\}$$
$$\left\{\sum_{ikpj}\langle\mathbf{h}_i\mathbf{h}_i^{\mathsf{T}}|\mathcal{X}\rangle\right\}^{-1} \tag{5.32b}$$
$$\mathbf{U}' = \left\{\sum_{ikpj}\left[(\hat{\mathbf{x}}_{ij}^{kp} - \mathbf{m})\langle\mathbf{w}_k|\mathcal{X}\rangle^{\mathsf{T}} - \mathbf{V}\langle\mathbf{h}_i\mathbf{w}_k^{\mathsf{T}}|\mathcal{X}\rangle - \mathbf{R}\langle\mathbf{y}_p\mathbf{w}_k^{\mathsf{T}}|\mathcal{X}\rangle\right]\right\}$$
$$\left\{\sum_{ikpj}\langle\mathbf{w}_k\mathbf{w}_k^{\mathsf{T}}|\mathcal{X}\rangle\right\}^{-1} \tag{5.32c}$$
$$\mathbf{R}' = \left\{\sum_{ikpj}\left[(\hat{\mathbf{x}}_{ij}^{kp} - \mathbf{m})\langle\mathbf{y}_p|\mathcal{X}\rangle^{\mathsf{T}} - \mathbf{V}\langle\mathbf{h}_i\mathbf{y}_p^{\mathsf{T}}|\mathcal{X}\rangle - \mathbf{U}\langle\mathbf{w}_k\mathbf{y}_p^{\mathsf{T}}|\mathcal{X}\rangle\right]\right\}$$
$$\left\{\sum_{ikpj}\langle\mathbf{y}_p\mathbf{y}_p^{\mathsf{T}}|\mathcal{X}\rangle\right\}^{-1} \tag{5.32d}$$
$$\boldsymbol{\Sigma}' = \frac{1}{N}\sum_{ikpj}\left[(\hat{\mathbf{x}}_{ij}^{kp} - \mathbf{m})(\hat{\mathbf{x}}_{ij}^{kp} - \mathbf{m})^{\mathsf{T}} - \mathbf{V}\langle\mathbf{h}_i|\mathcal{X}\rangle(\hat{\mathbf{x}}_{ij}^{kp} - \mathbf{m})^{\mathsf{T}}\right.$$
$$\left. - \mathbf{U}\langle\mathbf{w}_k|\mathcal{X}\rangle(\hat{\mathbf{x}}_{ij}^{kp} - \mathbf{m})^{\mathsf{T}} - \mathbf{R}\langle\mathbf{y}_p|\mathcal{X}\rangle(\hat{\mathbf{x}}_{ij}^{kp} - \mathbf{m})^{\mathsf{T}}\right], \tag{5.32e}$$

where $N = \sum_{i=1}^{S} N_i = \sum_{k=1}^{K} M_k$. Algorithm 5 shows the procedure of applying the variational EM algorithm.

Algorithm 5 Variational Bayes EM algorithm for SNR- and duration-invariant PLDA

Input:

Development data set comprising i-vectors or LDA-projected i-vectors $\mathcal{X} = \{\hat{\mathbf{x}}_{ij}^{kp} | i = 1,\ldots,S; k = 1,\ldots,K; p = 1,\ldots,P; j = 1,\ldots,H_{ik}(p)\}$, with speaker labels, SNR group labels, and duration labels.

Initialization:

$\mathbf{y}_p^* \leftarrow \mathbf{0}$, $\mathbf{w}_k^* \leftarrow \mathbf{0}$;
$\boldsymbol{\Sigma} \leftarrow 0.01\mathbf{I}$;
$\mathbf{V}, \mathbf{U}, \mathbf{R} \leftarrow$ eigenvectors obtained from the PCA of $\mathcal{X}$;

Parameter Estimation:

Compute $\mathbf{m}$ via Eq. 5.32(a);
Compute $\mathbf{L}_i^{(1)}$, $\mathbf{L}_k^{(2)}$, and $\mathbf{L}_p^{(3)}$ according to Eq. 5.31(a) to Eq. 5.31(c), respectively;
Compute the posterior mean of $\mathbf{h}_i$ using Eq. 5.31(d);
Use the posterior mean of $\mathbf{h}_i$ computed in Step 3 to update the posterior means of $\mathbf{w}_k$ and $\mathbf{y}_p$ using Eq. 5.31(e)–(f);
Compute the other terms in the E-step, i.e., Eq. 5.31(g)–(o);
Update the model parameters using Eq. 5.32(b)–(e);
Set $\mathbf{y}_p^* = \langle\mathbf{y}_p|\mathcal{X}\rangle$, $\mathbf{w}_k^* = \langle\mathbf{w}_k|\mathcal{X}\rangle$, and $\mathbf{h}_i^* = \langle\mathbf{h}_i|\mathcal{X}\rangle$;
Go to Step 2 until convergence;

Return: the parameters of the SNR- and duration-invariant PLDA model $\boldsymbol{\theta} = \{\mathbf{m}, \mathbf{V}, \mathbf{U}, \mathbf{R}, \boldsymbol{\Sigma}\}$.

Likelihood Ratio Scores

Consider the general case where both SNR and duration of the target speaker's utterance and the test speaker's utterance are unknown. Let $\mathbf{x}_s$ and $\mathbf{x}_t$ be the i-vectors of the target speaker and test speaker respectively, the likelihood ratio score is

$$
\begin{aligned}
S_{\mathrm{LR}}(\mathbf{x}_s,\mathbf{x}_t) &= \log \frac{P\left(\mathbf{x}_s,\mathbf{x}_t|\text{same-speaker}\right)}{P(\mathbf{x}_s,\mathbf{x}_t|\text{different-speakers})} \\
&= \text{const} + \frac{1}{2}\bar{\mathbf{x}}_s^{\mathsf{T}}\mathbf{Q}\bar{\mathbf{x}}_s + \frac{1}{2}\bar{\mathbf{x}}_t^{\mathsf{T}}\mathbf{Q}\bar{\mathbf{x}}_t + \bar{\mathbf{x}}_s^{\mathsf{T}}\mathbf{P}\bar{\mathbf{x}}_t,
\end{aligned}
\tag{5.33}
$$

where

$$
\begin{aligned}
\bar{\mathbf{x}}_s &= \mathbf{x}_s - \mathbf{m}, \qquad \bar{\mathbf{x}}_t = \mathbf{x}_t - \mathbf{m}, \\
\mathbf{P} &= \boldsymbol{\Sigma}_{tot}^{-1}\boldsymbol{\Sigma}_{ac}(\boldsymbol{\Sigma}_{tot} - \boldsymbol{\Sigma}_{ac}\boldsymbol{\Sigma}_{tot}^{-1}\boldsymbol{\Sigma}_{ac})^{-1}, \\
\mathbf{Q} &= \boldsymbol{\Sigma}_{tot}^{-1} - (\boldsymbol{\Sigma}_{tot} - \boldsymbol{\Sigma}_{ac}\boldsymbol{\Sigma}_{tot}^{-1}\boldsymbol{\Sigma}_{ac})^{-1}, \\
\boldsymbol{\Sigma}_{ac} &= \mathbf{V}\mathbf{V}^{\mathsf{T}}, \text{ and } \boldsymbol{\Sigma}_{tot} = \mathbf{V}\mathbf{V}^{\mathsf{T}} + \mathbf{U}\mathbf{U}^{\mathsf{T}} + \mathbf{R}\mathbf{R}^{\mathsf{T}} + \boldsymbol{\Sigma}.
\end{aligned}
$$

See Section 3.4.3 for the derivation of Eq. 5.33. If we know the utterance duration and SNR, we may use the principle in Eq. 5.26 to derive the scoring function. SDI-PLDA scoring is computationally light because both **P** and **Q** can be computed during training.

5.4 Mixture of PLDA

In duration- and SNR-invariant PLDA, the duration and SNR variabilities are modeled by the duration and SNR subspace in the PLDA model. In case the variabilities are very large, e.g., some utterances are very clean but some are very noisy, then the variabilities may be better modeled by a mixture of PLDA models in which each model is responsible for a small range of variability. In this section, we will discuss the mixture models that are good at modeling utterances with a wide range of SNR. Much of the materials in this section are based on the authors' previous work on PLDA mixture models [47, 95, 171–173].

5.4.1 SNR-Independent Mixture of PLDA

In Eq. 3.70, the PLDA model assumes that the length-normalized i-vectors are Gaussian distributed. The modeling capability of a single Gaussian, however, is rather limited, causing deficiency of the Gaussian PLDA model in tackling noise and reverberation with varying levels. In such scenarios, it is more appropriate to model the i-vectors by a mixture of K factor analysis (FA) models. Specifically, the parameters of the mixture FA model are given by $\underline{\omega} = \{\varphi_k, \mathbf{m}_k, \mathbf{V}_k, \boldsymbol{\Sigma}_k\}_{k=1}^{K}$, where φ_k, $\mathbf{m}_k$, $\mathbf{V}_k$, and $\boldsymbol{\Sigma}_k$ are the mixture, coefficient, mean vector, speaker loading matrix and covariance matrix of the k-th mixture, respectively. It is assumed that the full covariance matrices $\boldsymbol{\Sigma}_k$'s are capable of modeling the channel effect. Because there is no direct relationship between the mixture components and the SNR of utterances, we refer to this type of mixture models as SNR-independent mixture of PLDA (SI-mPLDA).

The following subsection will provide the derivation of the EM formulations for the scenario in which an i-vector is connected with K latent factors. After that, the formulations will be extended to the scenario in which each speaker has only one latent factor across all mixtures.

EM Formulation

Denote $\mathcal{Y} = \{y_{ijk}\}_{k=1}^{K}$ as the latent indicator variables identifying which of the K factor analysis models $\{\varphi_k, \mathbf{m}_k, \mathbf{V}_k, \boldsymbol{\Sigma}_k\}_{k=1}^{K}$ produces $\mathbf{x}_{ij}$. Specifically, $y_{ijk} = 1$ if the FA model k produces $\mathbf{x}_{ij}$, and $y_{ijk} = 0$ otherwise. Also denote $\mathcal{Z} = \{\mathbf{z}_{ik}\}_{k=1}^{K}$ as the latent factors for the K mixtures. Then, the auxiliary function in Eq. 3.71 can be written as

$$
\begin{aligned}
Q(\underline{\omega}'|\underline{\omega}) &= \mathbb{E}_{\mathcal{Y},\mathcal{Z}}\{\log p(\mathcal{X},\mathcal{Y},\mathcal{Z}|\underline{\omega}')|\mathcal{X},\underline{\omega}\} \\
&= \mathbb{E}_{\mathcal{Y},\mathcal{Z}}\left\{\sum\nolimits_{ijk} y_{ijk}\log\left[p(y_{ijk}|\underline{\omega}')p(\mathbf{x}_{ij}|\mathbf{z}_{ik}, y_{ijk}=1, \omega'_k)p(\mathbf{z}_{ik}|\underline{\omega}')\right]\middle|\mathcal{X},\underline{\omega}\right\} \\
&= \sum_{i=1}^{N}\sum_{j=1}^{H_i}\sum_{k=1}^{K}\mathbb{E}_{\mathcal{Y},\mathcal{Z}}\left\{y_{ijk}\log\left[\varphi'_k\mathcal{N}(\mathbf{x}_{ij}|\mathbf{m}'_k + \mathbf{V}'_k\mathbf{z}_{ik}, \boldsymbol{\Sigma}'_k)\mathcal{N}(\mathbf{z}_{ik}|\mathbf{0},\mathbf{I})\right]\middle|\mathcal{X},\underline{\omega}\right\}.
\end{aligned}
\tag{5.34}
$$

Although the ith training speaker have multiple (H_i) sessions, these sessions share the same set of latent variables $\{\mathbf{z}_{ik}\}_{k=1}^{K}$. To simplify notations, we will drop the $'$ symbol in Eq. 5.34:

$$
\begin{aligned}
Q(\underline{\omega}) &= \sum\nolimits_{ijk}\mathbb{E}_{\mathcal{Y},\mathcal{Z}}\left\{y_{ijk}\log\left[\varphi_k\mathcal{N}(\mathbf{x}_{ij}|\mathbf{m}_k + \mathbf{V}_k\mathbf{z}_{ik}, \boldsymbol{\Sigma}_k)\mathcal{N}(\mathbf{z}_{ik}|\mathbf{0},\mathbf{I})\right]\middle|\mathcal{X}\right\} \\
&= \sum_{ijk}\left\langle y_{ijk}\left\{\log\varphi_k - \frac{1}{2}\log|\boldsymbol{\Sigma}_k|\right\}\middle|\mathcal{X}\right\rangle \\
&\quad - \frac{1}{2}\sum_{ijk}\left\langle y_{ijk}\left\{(\mathbf{x}_{ij} - \mathbf{m}_k - \mathbf{V}_k\mathbf{z}_{ik})^{\mathsf{T}}\boldsymbol{\Sigma}_k^{-1}(\mathbf{x}_{ij} - \mathbf{m}_k - \mathbf{V}_k\mathbf{z}_{ik})\right\}\middle|\mathcal{X}\right\rangle \\
&\quad - \frac{1}{2}\sum_{ijk}\left\langle y_{ijk}\mathbf{z}_{ik}{}^{\mathsf{T}}\mathbf{z}_{ik}\middle|\mathcal{X}\right\rangle \\
&= \sum_{ijk}\langle y_{ijk}|\mathcal{X}\rangle\left[\log\varphi_k - \frac{1}{2}\log|\boldsymbol{\Sigma}_k| - \frac{1}{2}(\mathbf{x}_{ij} - \mathbf{m}_k)^{\mathsf{T}}\boldsymbol{\Sigma}_k^{-1}(\mathbf{x}_{ij} - \mathbf{m}_k)\right] \\
&\quad + \sum_{ijk}(\mathbf{x}_{ij} - \mathbf{m}_k)^{\mathsf{T}}\boldsymbol{\Sigma}_k^{-1}\mathbf{V}_k\left\langle y_{ijk}\mathbf{z}_{ik}|\mathcal{X}\right\rangle \\
&\quad - \frac{1}{2}\left[\sum_{ijk}\mathrm{tr}\left\{\left(\mathbf{V}_k^{\mathsf{T}}\boldsymbol{\Sigma}_k^{-1}\mathbf{V}_k + 1\right)\left\langle y_{ijk}\mathbf{z}_{ik}\mathbf{z}_{ik}^{\mathsf{T}}|\mathcal{X}\right\rangle\right\}\right].
\end{aligned}
\tag{5.35}
$$

To compute $\mathbf{m}_k$, we follow the two-step EM as suggested in [174]. Specifically, we drop $\mathbf{z}_{ik}$ at this stage of EM and setting[3]

$$
\frac{\partial Q}{\partial \mathbf{m}_k} = -\sum_{ij}\langle y_{ijk}|\mathcal{X}\rangle\left(\boldsymbol{\Sigma}_k^{-1}\mathbf{m}_k - \boldsymbol{\Sigma}_k^{-1}\mathbf{x}_{ij}\right) = 0,
$$

[3] In [174], the first step in the two-step EM does not consider $\mathbf{z}_{ik}$. As a result, the 6th row of Eq. 5.35 is not included in the derivative.

which results in

$$\mathbf{m}_k = \frac{\sum_{i=1}^{N}\sum_{j=1}^{H_i}\langle y_{ijk}|\mathcal{X}\rangle \mathbf{x}_{ij}}{\sum_{i=1}^{N}\sum_{j=1}^{H_i}\langle y_{ijk}|\mathcal{X}\rangle}.$$

To find $\mathbf{V}_k$, we compute

$$\frac{\partial Q}{\partial \mathbf{V}_k} = \sum_{ij}\mathbf{\Sigma}_k^{-1}(\mathbf{x}_{ij}-\mathbf{m}_k)\langle y_{ijk}\mathbf{z}_{ik}^{\mathsf{T}}|\mathcal{X}\rangle - \sum_{ij}\mathbf{\Sigma}_k^{-1}\mathbf{V}_k\left\langle y_{ijk}\mathbf{z}_{ik}\mathbf{z}_{ik}^{\mathsf{T}}|\mathcal{X}\right\rangle. \tag{5.36}$$

Setting $\frac{\partial Q}{\partial \mathbf{V}_k} = 0$, we obtain

$$\mathbf{V}_k = \left[\sum_{ij}(\mathbf{x}_{ij}-\mathbf{m}_k)\langle y_{ijk}\mathbf{z}_{ik}|\mathcal{X}\rangle^{\mathsf{T}}\right]\left[\sum_{ij}\left\langle y_{ijk}\mathbf{z}_{ik}\mathbf{z}_{ik}^{\mathsf{T}}|\mathcal{X}\right\rangle\right]^{-1}. \tag{5.37}$$

To find $\mathbf{\Sigma}_k$, we evaluate

$$\begin{aligned}\frac{\partial Q}{\partial \mathbf{\Sigma}_k^{-1}} &= \frac{1}{2}\sum_{ij}\langle y_{ij}|\mathcal{X}\rangle\left[\mathbf{\Sigma}_k - (\mathbf{x}_{ij}-\mathbf{m}_k)(\mathbf{x}_{ij}-\mathbf{m}_k)^{\mathsf{T}}\right]\\ &\quad + \sum_{ij}(\mathbf{x}_{ij}-\mathbf{m}_k)\langle y_{ijk}\mathbf{z}_{ik}^{\mathsf{T}}|\mathcal{X}\rangle\mathbf{V}_k^{\mathsf{T}} - \frac{1}{2}\mathbf{V}_k\left[\sum_{ij}\langle y_{ijk}\mathbf{z}_{ik}\mathbf{z}_{ik}^{\mathsf{T}}|\mathcal{X}\rangle\right]\mathbf{V}_k^{\mathsf{T}}.\end{aligned} \tag{5.38}$$

Substituting Eq. 5.37 into Eq. 5.38 and setting $\frac{\partial Q}{\partial \mathbf{\Sigma}_k^{-1}} = 0$, we have

$$\begin{aligned}&\sum_{ij}\langle y_{ijk}|\mathcal{X}\rangle\mathbf{\Sigma}_k\\ &= \sum_{ij}\left[\langle y_{ijk}|\mathcal{X}\rangle(\mathbf{x}_{ij}-\mathbf{m}_k)(\mathbf{x}_{ij}-\mathbf{m}_k)^{\mathsf{T}} - (\mathbf{x}_{ij}-\mathbf{m}_k)\langle y_{ijk}\mathbf{z}_{ik}^{\mathsf{T}}|\mathcal{X}\rangle\mathbf{V}_k^{\mathsf{T}}\right].\end{aligned}$$

Rearranging, we have

$$\mathbf{\Sigma}_k = \frac{\sum_{ij}\left[\langle y_{ijk}|\mathcal{X}\rangle(\mathbf{x}_{ij}-\mathbf{m}_k)(\mathbf{x}_{ij}-\mathbf{m}_k)^{\mathsf{T}} - \mathbf{V}_k\langle y_{ijk}\mathbf{z}_{ik}|\mathcal{X}\rangle(\mathbf{x}_{ij}-\mathbf{m}_k)^{\mathsf{T}}\right]}{\sum_{ij}\langle y_{ijk}|\mathcal{X}\rangle}.$$

To compute φ_k, we optimize $Q(\underline{\omega})$ subject to the constraint $\sum_k \varphi_k = 1$. This can be achieved by introducing a Lagrange multiplier λ such that $Q'(\underline{\omega}) = Q(\underline{\omega}) + \lambda(\sum_k \varphi_k - 1)$. λ can be found by setting $\frac{\partial Q'}{\partial \varphi_k} = 0$, which results in

$$-\varphi_k\lambda = \sum_{ij}\langle y_{ijk}|\mathcal{X}\rangle. \tag{5.39}$$

Summing both side of Eq. 5.39 from 1 to K, we obtain

$$\lambda = -\sum_{ijk}\langle y_{ijk}|\mathcal{X}\rangle. \tag{5.40}$$

Substituting Eq. 5.40 into Eq. 5.39 and rearranging, we obtain

$$\varphi_k = \frac{\sum_{ij} \langle y_{ijk} | \mathcal{X} \rangle}{\sum_{ijl} \langle y_{ijl} | \mathcal{X} \rangle}.$$

E-Step: Computing Posterior Expectations

Because the mixture model comprises more than one latent variable, formally these latent variables should be inferred through the variational Bayes (VB) method in which the variational distributions over the latent variables $\mathbf{z}_{ik}$ and y_{ijk} are assumed to be factorizable. Specifically, we have $q(\mathbf{z}_{ik}, y_{ijk}) = q(\mathbf{z}_{ik})q(y_{ijk})$. The VB method is able to find the closest variational distribution to the true joint posterior distribution of two dependent latent variables $p(\mathbf{z}_{ik}, y_{ijk}|\mathcal{X})$. In the VB-E step, the optimal variational distribution or variational parameters with the largest lower-bound of the likelihood are estimated. Then, in the VB-M step, given the updated variational distribution, we update the model parameters $\{\varphi_k, \mathbf{m}_k, \mathbf{V}_k, \boldsymbol{\Sigma}_k\}_{k=1}^K$ by maximizing the variational lower bound.

Rather than using the more complex VB method, we assume that the latent variable $\mathbf{z}_{ik}$ is posteriorly independent of y_{ijk}, that is,

$$p(\mathbf{z}_{ik}, y_{ijk}|\mathcal{X}_i) = p(\mathbf{z}_{ik}|\mathcal{X}_i)p(y_{ijk}|\mathbf{x}_{ij}).$$

This assumption is similar to that of traditional continuous density HMM in which the HMM states and Gaussian mixtures are assumed independent. Therefore, in the E-step, we compute the posterior means $\langle y_{ijk}\mathbf{z}_{ik}|\mathcal{X}\rangle$ and posterior moment $\langle y_{ijk}\mathbf{z}_{ik}\mathbf{z}_{ik}^{\mathsf{T}}|\mathcal{X}\rangle$, where $j = 1, \ldots, H_i$. The joint posterior expectations are given by:

$$\begin{aligned} \langle y_{ijk}\mathbf{z}_{ik}|\mathcal{X}_i\rangle &= \langle y_{ijk}|\mathbf{x}_{ij}\rangle\langle \mathbf{z}_{ik}|\mathcal{X}_i\rangle \\ \langle y_{ijk}\mathbf{z}_{ik}\mathbf{z}_{ik}^{\mathsf{T}}|\mathcal{X}_i\rangle &= \langle y_{ijk}|\mathbf{x}_{ij}\rangle\langle \mathbf{z}_{ik}\mathbf{z}_{ik}^{\mathsf{T}}|\mathcal{X}_i\rangle. \end{aligned} \tag{5.41}$$

Figure 5.11 shows the relation among $\mathbf{z}_{ik}$, $\mathbf{x}_{ij}$ and y_{ijk} [171]. In the figure, $\underline{\omega} = \{\varphi_k, \mathbf{m}_k, \boldsymbol{\Sigma}_k, \mathbf{V}_k\}_{k=1}^K$. In the diagram, $\underline{\mathbf{m}} = \{\mathbf{m}_k\}_{k=1}^K$, $\underline{\mathbf{V}} = \{\mathbf{V}_k\}_{k=1}^K$, $\underline{\boldsymbol{\Sigma}} = \{\boldsymbol{\Sigma}_k\}_{k=1}^K$, and $\underline{\varphi} = \{\varphi_k\}_{k=1}^K$.

Let $y_{i\cdot k}$ be the indicator variables of the H_i i-vectors from the ith speaker for the kth mixture. We use the Bayes rule to estimate the joint posterior expectations as follows:

$$\begin{aligned} p(\mathbf{z}_{ik}, y_{i.k}|\mathcal{X}_i) &\propto p(\mathcal{X}_i|\mathbf{z}_{ik}, y_{i.k} = 1)p(\mathbf{z}_{ik}, y_{i.k}) \\ &= p(\mathcal{X}_i|\mathbf{z}_{ik}, y_{i.k} = 1)p(y_{i.k})p(\mathbf{z}_{ik}) \qquad \because \mathbf{z}_{ik} \text{ and } y_{i.k} \text{ are independent} \\ &= \prod_{j=1}^{H_i} \left[\varphi_k p(\mathbf{x}_{ij}|y_{ijk} = 1, \mathbf{z}_{ik})\right]^{y_{ijk}} p(\mathbf{z}_{ik}) \qquad [1, \text{Eq. } 9.38] \\ &= \underbrace{p(\mathbf{z}_{ik}) \prod_{j=1}^{H_i} \left[\mathcal{N}(\mathbf{x}_{ij}|\mathbf{m}_k + \mathbf{V}_k\mathbf{z}_{ik}, \boldsymbol{\Sigma}_k)\right]^{y_{ijk}}}_{\propto\, p(\mathbf{z}_{ik}|\mathcal{X}_i)} \varphi_k^{y_{ijk}}. \end{aligned} \tag{5.42}$$

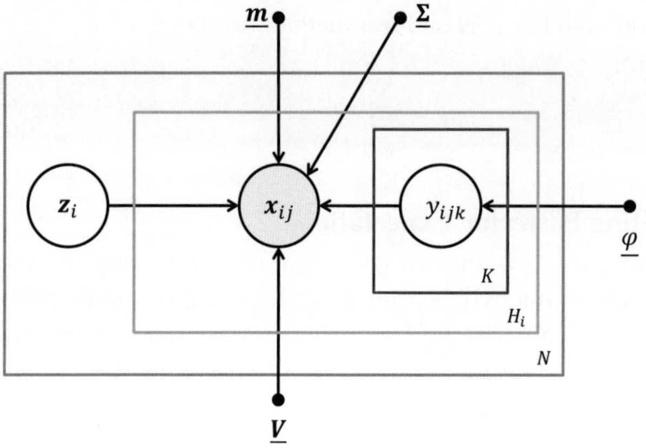

Figure 5.11 Probabilistic graphical model representing SNR-independent mixture of PLDA. [Reprinted from *Mixture of PLDA for Noise Robust I-Vector Speaker Verification (Figure 2), M.W. Mak, X.M Pang and J.T. Chien., IEEE/ACM Trans. on Audio Speech and Language Processing, vol. 24, No. 1, pp. 130–142, Jan. 2016, with permission of IEEE*]

To find the posterior of $\mathbf{z}_{ik}$, we extract terms dependent on $\mathbf{z}_{ik}$ from Eq. 5.42 as follows:

$$\begin{aligned}
& p(\mathbf{z}_{ik}|\mathcal{X}_i) \\
& \propto \exp\left\{-\frac{1}{2}\sum_{j=1}^{H_i} y_{ijk}(\mathbf{x}_{ij}-\mathbf{m}_k-\mathbf{V}_k\mathbf{z}_{ik})^{\mathsf{T}}\boldsymbol{\Sigma}_k^{-1}(\mathbf{x}_{ij}-\mathbf{m}_k-\mathbf{V}_k\mathbf{z}_{ik})-\frac{1}{2}\mathbf{z}_{ik}^{\mathsf{T}}\mathbf{z}_{ik}\right\} \\
& = \exp\left\{\mathbf{z}_{ik}^{\mathsf{T}}\mathbf{V}_k^{\mathsf{T}}\sum_{j\in\mathcal{H}_{ik}}\boldsymbol{\Sigma}_k^{-1}(\mathbf{x}_{ij}-\mathbf{m}_k)-\frac{1}{2}\mathbf{z}_{ik}^{\mathsf{T}}\left(\mathbf{I}+\sum_{j\in\mathcal{H}_{ik}}\mathbf{V}_k^{\mathsf{T}}\boldsymbol{\Sigma}_k^{-1}\mathbf{V}_k\right)\mathbf{z}_{ik}\right\}, \qquad (5.43)
\end{aligned}$$

where $\mathcal{H}_{ik}$ comprises the indexes of speaker i's i-vectors that aligned to mixture k. Comparing Eq. 3.77 with Eq. 5.43, we have

$$\begin{aligned}
\langle\mathbf{z}_{ik}|\mathcal{X}_i\rangle &= \mathbf{L}_{ik}^{-1}\mathbf{V}_k^{\mathsf{T}}\boldsymbol{\Sigma}_k^{-1}\sum_{j\in\mathcal{H}_{ik}}(\mathbf{x}_{ij}-\mathbf{m}_k) \\
\langle\mathbf{z}_{ik}\mathbf{z}_{ik}^{\mathsf{T}}|\mathcal{X}_i\rangle &= \mathbf{L}_{ik}^{-1}+\langle\mathbf{z}_{ik}|\mathcal{X}_i\rangle\langle\mathbf{z}_{ik}^{\mathsf{T}}|\mathcal{X}_i\rangle,
\end{aligned} \qquad (5.44)$$

where

$$\mathbf{L}_{ik}=\mathbf{I}+H_{ik}\mathbf{V}_k^{\mathsf{T}}\boldsymbol{\Sigma}_k^{-1}\mathbf{V}_k, \qquad (5.45)$$

where H_{ik} is the number (occupancy count) of i-vectors from speaker i aligned to mixture j.

Eq. 5.44 suggests that only H_{ik} out of H_i i-vectors from speaker i are generated by mixture k. This characteristic is crucial for proper modeling of i-vectors when they are obtained from utterances with large difference in SNR. This is because these i-vectors tend to fall on different areas of the i-vector space (see Hypothesis I in Section II of [171]). For example, the training i-vectors in [171] were obtained from utterances having three noise levels: 6dB, 15dB, and clean. Therefore, when $K = 3$, Mixtures 1,

2, and 3 will be responsible for generating i-vectors whose utterances have noise level of 6dB, 15dB, and clean, respectively. This characteristic also causes the PLDA mixture model different from that of [175]. In particular, in [175], a mixture component is first chosen for the i-vector of a particular speaker; then the selected component generates all of the i-vectors from that speaker. Clearly, this strategy is constrained to the case in which the i-vectors from a speaker are obtained from a similar acoustic environment.

To estimate the posterior expectation of y_{ijk}, we may use the Bayes rule:

$$\begin{aligned}\langle y_{ijk}|\mathcal{X}_i\rangle &= P(y_{ijk}=1|\mathbf{x}_{ij},\underline{\omega})\\ &= \frac{P(y_{ijk}=1)p(\mathbf{x}_{ij}|y_{ijk}=1,\underline{\omega})}{\sum_{r=1}^{K}P(y_{ijr}=1)p(\mathbf{x}_{ij}|y_{ijr}=1,\underline{\omega})}\\ &= \frac{\varphi_k\mathcal{N}(\mathbf{x}_{ij}|\mathbf{m}_k,\mathbf{V}_k\mathbf{V}_k^{\mathsf{T}}+\boldsymbol{\Sigma}_k)}{\sum_{r=1}^{K}\varphi_r\mathcal{N}(\mathbf{x}_{ij}|\mathbf{m}_r,\mathbf{V}_r\mathbf{V}_r^{\mathsf{T}}+\boldsymbol{\Sigma}_r)}.\end{aligned} \tag{5.46}$$

Note that in Eq. 5.46,

$$\begin{aligned}p(\mathbf{x}_{ij}|y_{ijk}=1,\underline{\omega}) &= \int p(\mathbf{x}_{ij}|\mathbf{z},y_{ijk}=1,\underline{\omega})p(\mathbf{z})d\mathbf{z}\\ &= \int \mathcal{N}(\mathbf{x}_{ij}|\mathbf{m}_k+\mathbf{V}_k\mathbf{z},\boldsymbol{\Sigma}_k)\mathcal{N}(\mathbf{z}|\mathbf{0},\mathbf{I})d\mathbf{z}\\ &= \mathcal{N}(\mathbf{x}_{ij}|\mathbf{m}_k,\mathbf{V}_k\mathbf{V}_k^{\mathsf{T}}+\boldsymbol{\Sigma}_k).\end{aligned}$$

Sharing Latent Factors

Note that in Eq. 5.44, each speaker needs K latent factors $\mathbf{z}_{ik}$. Another way is to let the speaker to share the same latent factor, i.e., $\mathbf{z}_{ik}=\mathbf{z}_i\ \forall i$. Therefore, Eq. 5.34 changes into

$$Q(\underline{\omega}'|\underline{\omega}) = \sum_{ijk}\mathbb{E}_{\mathcal{Y},\mathcal{Z}}\left\{y_{ijk}\log\left[\varphi_k'\mathcal{N}(\mathbf{x}_{ij}|\mathbf{m}_k'+\mathbf{V}_k'\mathbf{z}_i,\boldsymbol{\Sigma}_k')\mathcal{N}(\mathbf{z}_i|\mathbf{0},\mathbf{I})\right]\middle|\mathcal{X},\underline{\omega}\right\}. \tag{5.47}$$

Let $y_{i\cdot\cdot}$ be the indicator variables for all possible sessions and mixture components for speaker i. The joint posterior density in Eq. 5.42 turns into

$$\begin{aligned}p(\mathbf{z}_i,y_{i\cdot\cdot}|\mathcal{X}_i) &\propto p(\mathcal{X}_i|\mathbf{z}_i,y_{i\cdot\cdot}=1)p(\mathbf{z}_i,y_{i\cdot\cdot})\\ &= p(\mathcal{X}_i|\mathbf{z}_i,y_{i\cdot\cdot}=1)p(y_{i\cdot\cdot})p(\mathbf{z}_i) && \because \mathbf{z}_i \text{ and } y_{i\cdot\cdot} \text{ are independent}\\ &= \prod_{j=1}^{H_i}\prod_{k=1}^{K}\left[\varphi_k p(\mathbf{x}_{ij}|y_{ijk}=1,\mathbf{z}_i)\right]^{y_{ijk}}p(\mathbf{z}_i) && [1, \text{Eq. } 9.38]\\ &= \underbrace{p(\mathbf{z}_i)\prod_{j=1}^{H_i}\prod_{k=1}^{K}\left[\mathcal{N}(\mathbf{x}_{ij}|\mathbf{m}_k+\mathbf{V}_k\mathbf{z}_i,\boldsymbol{\Sigma}_k)\right]^{y_{ijk}}}_{\propto p(\mathbf{z}_i|\mathcal{X}_i)}\varphi_k^{y_{ijk}},\end{aligned} \tag{5.48}$$

where we have used the property that for each i and j, only one of y_{ijk}'s equals to 1, the rest are 0. Taking out terms depending on $\mathbf{z}_i$ from Eq. 5.48 and contrasting with Eq. 3.77, we have the posterior expectations as follows:

$$
\begin{aligned}
\langle \mathbf{z}_i|\mathcal{X}_i\rangle &= \mathbf{L}_i^{-1}\sum_{k=1}^{K}\sum_{j\in\mathcal{H}_{ik}}\mathbf{V}_k^{\mathsf{T}}\boldsymbol{\Sigma}_k^{-1}(\mathbf{x}_{ij}-\mathbf{m}_k)\\
\langle \mathbf{z}_i\mathbf{z}_i^{\mathsf{T}}|\mathcal{X}_i\rangle &= \mathbf{L}_i^{-1}+\langle \mathbf{z}_i|\mathcal{X}_i\rangle\langle \mathbf{z}_i^{\mathsf{T}}|\mathcal{X}_i\rangle.
\end{aligned} \tag{5.49}
$$

We also have the posterior precision as follows:

$$
\mathbf{L}_i = \mathbf{I}+\sum_{k=1}^{K}\sum_{j\in\mathcal{H}_{ik}}\mathbf{V}_k^{\mathsf{T}}\boldsymbol{\Sigma}_k^{-1}\mathbf{V}_k, \tag{5.50}
$$

where $\mathcal{H}_{ik}$ contains the indexes of the i-vectors from speaker i that aligned to mixture k.

In summary, we have the following EM formulations:

E-Step:

$$
\langle y_{ijk}|\mathcal{X}\rangle = \frac{\varphi_k\mathcal{N}(\mathbf{x}_{ij}|\mathbf{m}_k,\mathbf{V}_k\mathbf{V}_k^{\mathsf{T}}+\boldsymbol{\Sigma}_k)}{\sum_{l=1}^{K}\varphi_l\mathcal{N}(\mathbf{x}_{ij}|\mathbf{m}_l,\mathbf{V}_l\mathbf{V}_l^{\mathsf{T}}+\boldsymbol{\Sigma}_l)} \tag{5.51a}
$$

$$
\mathbf{L}_i = \mathbf{I}+\sum_{k=1}^{K}H_{ik}\mathbf{V}_k^{\mathsf{T}}\boldsymbol{\Sigma}_k^{-1}\mathbf{V}_k \tag{5.51b}
$$

$$
\langle y_{ijk}\mathbf{z}_i|\mathcal{X}\rangle = \langle y_{ijk}|\mathcal{X}\rangle\langle \mathbf{z}_i|\mathcal{X}\rangle \tag{5.51c}
$$

$$
\langle \mathbf{z}_i|\mathcal{X}\rangle = \mathbf{L}_i^{-1}\sum_{k=1}^{K}\mathbf{V}_k^{\mathsf{T}}\boldsymbol{\Sigma}_k^{-1}\sum_{j\in\mathcal{H}_{ik}}(\mathbf{x}_{ij}-\mathbf{m}_k) \tag{5.51d}
$$

$$
\langle y_{ijk}\mathbf{z}_i\mathbf{z}_i^{\mathsf{T}}|\mathcal{X}\rangle = \langle y_{ijk}|\mathcal{X}\rangle\langle \mathbf{z}_i\mathbf{z}_i^{\mathsf{T}}|\mathcal{X}\rangle \tag{5.51e}
$$

$$
\langle \mathbf{z}_i\mathbf{z}_i^{\mathsf{T}}|\mathcal{X}\rangle = \mathbf{L}_i^{-1}+\langle \mathbf{z}_i|\mathcal{X}\rangle\langle \mathbf{z}_i|\mathcal{X}\rangle^{\mathsf{T}} \tag{5.51f}
$$

M-Step:

$$
\mathbf{m}'_k = \frac{\sum_{i=1}^{N}\sum_{j=1}^{H_i}\langle y_{ijk}|\mathcal{X}\rangle\mathbf{x}_{ij}}{\sum_{i=1}^{N}\sum_{j=1}^{H_i}\langle y_{ijk}|\mathcal{X}\rangle} \tag{5.52a}
$$

$$
\varphi'_k = \frac{\sum_{ij}\langle y_{ijk}|\mathcal{X}\rangle}{\sum_{ijl}\langle y_{ijl}|\mathcal{X}\rangle} \tag{5.52b}
$$

$$
\mathbf{V}'_k = \left[\sum_{ij}(\mathbf{x}_{ij}-\mathbf{m}'_k)\langle y_{ijk}\mathbf{z}_i|\mathcal{X}\rangle^{\mathsf{T}}\right]\left[\sum_{ij}\left\langle y_{ijk}\mathbf{z}_i\mathbf{z}_i^{\mathsf{T}}|\mathcal{X}\right\rangle\right]^{-1} \tag{5.52c}
$$

$$
\boldsymbol{\Sigma}'_k = \frac{\sum_{ij}\left[\langle y_{ijk}|\mathcal{X}\rangle(\mathbf{x}_{ij}-\mathbf{m}'_k)(\mathbf{x}_{ij}-\mathbf{m}'_k)^{\mathsf{T}}-\mathbf{V}'\langle y_{ijk}\mathbf{z}_i|\mathcal{X}\rangle(\mathbf{x}_{ij}-\mathbf{m}'_k)^{\mathsf{T}}\right]}{\sum_{ij}\langle y_{ijk}|\mathcal{X}\rangle} \tag{5.52d}
$$

where $\langle y_{ijk}|\mathcal{X}\rangle \equiv \mathbb{E}_{\mathcal{Y}}\left\{y_{ijk}|\mathcal{X},\underline{\omega}\right\}$, $\langle y_{ijk}\mathbf{z}_i|\mathcal{X}\rangle \equiv \mathbb{E}_{\mathcal{Y},\mathcal{Z}}\left\{y_{ijk}\mathbf{z}_i|\mathcal{X},\underline{\omega}\right\}$, and $\langle y_{ijk}\mathbf{z}_i\mathbf{z}_i^{\mathsf{T}}|\mathcal{X}\rangle \equiv \mathbb{E}_{\mathcal{Y},\mathcal{Z}}\left\{y_{ijk}\mathbf{z}_i\mathbf{z}_i^{\mathsf{T}}|\mathcal{X},\underline{\omega}\right\}$.

Mixture PLDA Scoring

Given the target speaker's i-vector $\mathbf{x}_s$ and a test i-vector $\mathbf{x}_t$, the same-speaker likelihood (numerator) can be written as:

$$
\begin{aligned}
&p(\mathbf{x}_s, \mathbf{x}_t | \text{same-speaker}) \\
&= \sum_{k_s=1}^{K} \sum_{k_t=1}^{K} \int p(\mathbf{x}_s, \mathbf{x}_t, y_{k_s} = 1, y_{k_t} = 1, \mathbf{z} | \underline{\omega}) d\mathbf{z} \\
&= \sum_{k_s=1}^{K} \sum_{k_t=1}^{K} P(y_{k_s} = 1, y_{k_t} = 1 | \underline{\omega}) \int p(\mathbf{x}_s, \mathbf{x}_t | y_{k_s} = 1, y_{k_t} = 1, \mathbf{z}, \underline{\omega}) p(\mathbf{z}) d\mathbf{z} \\
&= \sum_{k_s=1}^{K} \sum_{k_t=1}^{K} \varphi_{k_s} \varphi_{k_t} \int p(\mathbf{x}_s, \mathbf{x}_t | y_{k_s} = 1, y_{k_t} = 1, \mathbf{z}, \underline{\omega}) p(\mathbf{z}) d\mathbf{z} \\
&= \sum_{k_s=1}^{K} \sum_{k_t=1}^{K} \varphi_{k_s} \varphi_{k_t} \mathcal{N}\left(\left[\mathbf{x}_s^{\mathsf{T}} \ \mathbf{x}_t^{\mathsf{T}} \right]^{\mathsf{T}} \middle| \left[\mathbf{m}_{k_s}^{\mathsf{T}} \ \mathbf{m}_{k_t}^{\mathsf{T}} \right]^{\mathsf{T}}, \hat{\mathbf{V}}_{k_s k_t} \hat{\mathbf{V}}_{k_s k_t}^{\mathsf{T}} + \hat{\mathbf{\Sigma}}_{k_s k_t} \right).
\end{aligned}
\tag{5.53}
$$

A similar procedure is applied to compute the different-speaker likelihood. Therefore, the likelihood ratio score for the mixture of PLDA becomes is

$$
\begin{aligned}
&S_{\text{mLR}}(\mathbf{x}_s, \mathbf{x}_t) \\
&= \frac{\sum_{k_s=1}^{K} \sum_{k_t=1}^{K} \varphi_{k_s} \varphi_{k_t} \mathcal{N}\left(\left[\mathbf{x}_s^{\mathsf{T}} \ \mathbf{x}_t^{\mathsf{T}} \right]^{\mathsf{T}} | \left[\mathbf{m}_{k_s}^{\mathsf{T}} \ \mathbf{m}_{k_t}^{\mathsf{T}} \right]^{\mathsf{T}}, \hat{\mathbf{V}}_{k_s k_t} \hat{\mathbf{V}}_{k_s k_t}^{\mathsf{T}} + \hat{\mathbf{\Sigma}}_{k_s k_t} \right)}{\left[\sum_{k_s=1}^{K} \varphi_{k_s} \mathcal{N}\left(\mathbf{x}_s | \mathbf{m}_{k_s}, \mathbf{V}_{k_s} \mathbf{V}_{k_s}^{\mathsf{T}} + \mathbf{\Sigma}_{k_s} \right) \right] \left[\sum_{k_t=1}^{K} \varphi_{k_t} \mathcal{N}\left(\mathbf{x}_t | \mathbf{m}_{k_t}, \mathbf{V}_{k_t} \mathbf{V}_{k_t}^{\mathsf{T}} + \mathbf{\Sigma}_{k_t} \right) \right]}
\end{aligned}
\tag{5.54}
$$

where $\hat{\mathbf{\Sigma}}_{k_s k_t} = \text{diag}\{\mathbf{\Sigma}_{k_s}, \mathbf{\Sigma}_{k_t}\}$ and $\hat{\mathbf{V}}_{k_s k_t} = [\mathbf{V}_{k_s}^{\mathsf{T}} \ \ \mathbf{V}_{k_t}^{\mathsf{T}}]^{\mathsf{T}}$.

5.4.2 SNR-Dependent Mixture of PLDA

The SNR-dependent mixture of PLDA (SD-mPLDA) uses the SNR of utterances to guide the clustering of i-vectors. As a result, the alignment of i-vectors is based on the posterior probabilities of SNR rather than the posterior probabilities of i-vectors, as in the SI-PLDA in Section 5.4.1.

In SD-mPLDA, i-vectors are modeled by a mixture of SNR-dependent factor analyzers with parameters:

$$
\underline{\theta} = \{\underline{\Lambda}, \underline{\omega}\} = \{\Lambda_k, \omega_k\}_{k=1}^{K} = \{\pi_k, \mu_k, \sigma_k, \mathbf{m}_k, \mathbf{V}_k, \mathbf{\Sigma}_k\}_{k=1}^{K}, \tag{5.55}
$$

where $\Lambda_k = \{\pi_k, \mu_k, \sigma_k\}$ contains the mixture coefficient, mean and standard deviation of the kth SNR group, and $\omega_k = \{\mathbf{m}_k, \mathbf{V}_k, \mathbf{\Sigma}_k\}$ comprises the mean i-vector, factor loading matrix, and residual covariance matrix of the FA model associated with SNR group k.

M-Step: Maximizing Expectation of Complete Likelihood

Let $\mathcal{Y} = \{y_{ijk}\}_{k=1}^{K}$ be the latent indicator variables identifying which of the K FA models should be selected based on the SNR of the training utterances. More precisely, if the kth FA model produces $\mathbf{x}_{ij}$, then $y_{ijk} = 1$; otherwise $y_{ijk} = 0$. We also let $\mathcal{L} = \{\ell_{ij}; i = 1, \ldots, N; j = 1, \ldots, H_i\}$ be the SNR of the training utterances. Then, the auxiliary function for deriving the EM formulation can be written as

$$
\begin{aligned}
Q(\underline{\theta}'|\underline{\theta}) &= \mathbb{E}_{\mathcal{Y},\mathcal{Z}}\{\log p(\mathcal{X},\mathcal{L},\mathcal{Y},\mathcal{Z}|\underline{\theta}')|\mathcal{X},\mathcal{L},\underline{\theta}\} \\
&= \mathbb{E}_{\mathcal{Y},\mathcal{Z}}\left\{\sum_{ijk} y_{ijk}\log\left[p(\ell_{ij}|y_{ijk}=1)p(y_{ijk})p(\mathbf{x}_{ij}|\mathbf{z}_i,y_{ijk}=1,\omega'_k)p(\mathbf{z}_i)\right]\middle|\mathcal{X},\mathcal{L},\underline{\theta}\right\} \\
&= \sum_{i=1}^{N}\sum_{k=1}^{K}\sum_{j=1}^{H_i}\mathbb{E}_{\mathcal{Y},\mathcal{Z}}\Big\{y_{ijk}\log\big[\mathcal{N}(\ell_{ij}|\mu'_k,\sigma'_k)\pi'_k\mathcal{N}(\mathbf{x}_{ij}|\mathbf{m}'_k+\mathbf{V}'_k\mathbf{z}_i,\boldsymbol{\Sigma}'_k) \\
&\qquad \mathcal{N}(\mathbf{z}_i|\mathbf{0},\mathbf{I})\big]\Big|\mathcal{X},\mathcal{L},\underline{\theta}\Big\}.
\end{aligned}
\tag{5.56}
$$

For notation simplicity, we will remove the symbol ($'$) in Eq. 5.56 and ignore the constants that do not depend on the model parameters. Then, Eq. 5.56 is written as

$$
\begin{aligned}
Q(\underline{\theta}) &= \sum_{ijk}\langle y_{ijk}|\mathcal{L}\rangle\left[-\log\sigma_k-\frac{1}{2}\sigma_k^{-2}(\ell_{ij}-\mu_k)^2+\log\pi_k\right] \\
&\quad+\sum_{ijk}\Big\langle y_{ijk}\Big[-\frac{1}{2}\log|\boldsymbol{\Sigma}_k| \\
&\quad-\frac{1}{2}(\mathbf{x}_{ij}-\mathbf{m}_k-\mathbf{V}_k\langle\mathbf{z}_i|\mathcal{X}\rangle)^{\mathsf{T}}\boldsymbol{\Sigma}_k^{-1}(\mathbf{x}_{ij}-\mathbf{m}_k-\mathbf{V}_k\langle\mathbf{z}_i|\mathcal{X}\rangle)\Big]\Big\rangle \\
&\quad-\frac{1}{2}\sum_{ijk}\left\langle y_{ijk}\mathbf{z}_i{}^{\mathsf{T}}\mathbf{z}_i|\mathcal{X},\mathcal{L}\right\rangle \\
&= \sum_{ijk}\langle y_{ijk}|\mathcal{L}\rangle\left[-\log\sigma_k-\frac{1}{2}\sigma_k^{-2}(\ell_{ij}-\mu_k)^2+\log\pi_k\right] \\
&\quad+\sum_{ijk}\langle y_{ijk}|\mathcal{L}\rangle\left[-\frac{1}{2}\log|\boldsymbol{\Sigma}_k|-\frac{1}{2}(\mathbf{x}_{ij}-\mathbf{m}_k)^{\mathsf{T}}\boldsymbol{\Sigma}_k^{-1}(\mathbf{x}_{ij}-\mathbf{m}_k)\right] \\
&\quad+\sum_{ijk}(\mathbf{x}_{ij}-\mathbf{m}_k)^{\mathsf{T}}\boldsymbol{\Sigma}_k^{-1}\mathbf{V}_k\langle y_{ijk}\mathbf{z}_i|\mathcal{X},\mathcal{L}\rangle \\
&\quad-\frac{1}{2}\left[\sum_{ijk}\left\langle y_{ijk}\mathbf{z}_i^{\mathsf{T}}\mathbf{V}_k^{\mathsf{T}}\boldsymbol{\Sigma}_k^{-1}\mathbf{V}_k\mathbf{z}_i|\mathcal{X},\mathcal{L}\right\rangle+\left\langle y_{ijk}\mathbf{z}_i^{\mathsf{T}}\mathbf{z}_i|\mathcal{X},\mathcal{L}\right\rangle\right] \\
&= \sum_{ijk}\langle y_{ijk}|\mathcal{L}\rangle\left[-\log\sigma_k-\frac{1}{2}\sigma_k^{-2}(\ell_{ij}-\mu_k)^2+\log\pi_k\right] \\
&\quad+\sum_{ijk}\langle y_{ijk}|\mathcal{L}\rangle\left[-\frac{1}{2}\log|\boldsymbol{\Sigma}_k|-\frac{1}{2}(\mathbf{x}_{ij}-\mathbf{m}_k)^{\mathsf{T}}\boldsymbol{\Sigma}_k^{-1}(\mathbf{x}_{ij}-\mathbf{m}_k)\right] \\
&\quad+\sum_{ijk}(\mathbf{x}_{ij}-\mathbf{m}_k)^{\mathsf{T}}\boldsymbol{\Sigma}_k^{-1}\mathbf{V}_k\langle y_{ijk}\mathbf{z}_i|\mathcal{X},\mathcal{L}\rangle \\
&\quad-\frac{1}{2}\left[\sum_{ijk}\mathrm{tr}\left\{\left(\mathbf{V}_k^{\mathsf{T}}\boldsymbol{\Sigma}_k^{-1}\mathbf{V}_k+\mathbf{I}\right)\left\langle y_{ijk}\mathbf{z}_i\mathbf{z}_i^{\mathsf{T}}|\mathcal{X},\mathcal{L}\right\rangle\right\}\right],
\end{aligned}
\tag{5.57}
$$

where we have used the identity

$$\left\langle \mathbf{x}^{\mathsf{T}}\mathbf{A}\mathbf{x}\right\rangle = \operatorname{tr}\left\{\mathbf{A}\left[\left\langle \mathbf{x}\mathbf{x}^{T}\right\rangle - \langle \mathbf{x}\rangle\langle \mathbf{x}^{\mathsf{T}}\rangle\right]\right\} + \langle \mathbf{x}^{\mathsf{T}}\rangle\mathbf{A}\langle \mathbf{x}\rangle.$$

The derivation of SD-mPLDA parameters is almost the same as that of SI-mPLDA in Section 5.4.1, except for the parameters of the SNR model. Specifically, to compute μ_k and σ_k, we set $\frac{\partial Q}{\partial \mu_k} = 0$ and $\frac{\partial Q}{\partial \sigma_k} = 0$, which result in

$$\mu_k = \frac{\sum_{ij}\langle y_{ijk}|\mathcal{L}\rangle \ell_{ij}}{\sum_{ij}\langle y_{ij}|\mathcal{L}\rangle} \quad \text{and} \quad \sigma_k^2 = \frac{\sum_{ij}\langle y_{ijk}|\mathcal{L}\rangle(\ell_{ij}-\mu_k)^2}{\sum_{ij}\langle y_{ijk}|\mathcal{L}\rangle}.$$

E-Step: Computing Posterior Expectations

The E-step is almost the same as the "Sharing Latent Factors" in Section 5.4.1, except for the additional observations $\mathcal{L}$ (the SNR of utterances). Let $\mathcal{X}_i$ and $\mathcal{L}_i$ be the i-vectors and SNR of utterances from the ith speaker, respectively. We begin with the joint posterior density:

$$\begin{aligned}
p(\mathbf{z}_i, y_{i..}|\mathcal{X}_i,\mathcal{L}_i) &\propto p(\mathcal{X}_i,\mathcal{L}_i|\mathbf{z}_i, y_{i..}=1)p(\mathbf{z}_i, y_{i..}) \\
&= p(\mathcal{X}_i|\mathbf{z}_i, y_{i..}=1)p(\mathcal{L}_i|y_{i..}=1)p(y_{i..})p(\mathbf{z}_i) \quad \because \mathbf{z}_i \text{ and } y_{i..} \text{ are independent} \\
&= \prod_{j=1}^{H_i}\prod_{k=1}^{K}\left[\pi_k p(\mathbf{x}_{ij}|y_{ijk}=1,\mathbf{z}_i)p(\ell_{ij}|y_{ijk}=1)\right]^{y_{ijk}} p(\mathbf{z}_i) \quad [1, \text{Eq. } 9.38] \qquad (5.58) \\
&= \underbrace{p(\mathbf{z}_i)\left\{\prod_{j=1}^{H_i}\prod_{k=1}^{K}\left[\mathcal{N}(\mathbf{x}_{ij}|\mathbf{m}_k+\mathbf{V}_k\mathbf{z}_i,\mathbf{\Sigma}_k)\right]^{y_{ijk}}\right\}}_{\propto\, p(\mathbf{z}_i|\mathcal{X}_i)}\left\{\prod_{j=1}^{H_i}\prod_{k=1}^{K}\left[\pi_k\mathcal{N}(\ell_{ij}|\mu_k,\sigma_k^2)\right]^{y_{ijk}}\right\}, \qquad (5.59)
\end{aligned}$$

where we have used the property that y_{ijk} is governed by ℓ_{ij}. Notice that the posterior density of $\mathbf{z}_i$ has the identical form as Eq. 5.48. The only dissimilarity is the posterior of y_{ijk}, which can be estimated by the Bayes rule:

$$\begin{aligned}
\langle y_{ijk}|\mathcal{L}\rangle &= \langle y_{ijk}|\ell_{ij}\rangle \\
&= P(y_{ijk}=1|\ell_{ij},\underline{\Lambda}) \\
&= \frac{P(y_{ijk}=1)p(\ell_{ij}|y_{ijk}=1,\underline{\Lambda})}{\sum_{r=1}^{K}P(y_{ijr}=1)p(\ell_{ij}|y_{ijr}=1,\underline{\Lambda})} \\
&= \frac{\pi_k\mathcal{N}(\ell_{ij}|\mu_k,\sigma_k^2)}{\sum_{r=1}^{K}\pi_r\mathcal{N}(\ell_{ij}|\mu_r,\sigma_r^2)}. \qquad (5.60)
\end{aligned}$$

Then, the joint posterior expectations are given by

$$\begin{aligned}
\langle y_{ijk}\mathbf{z}_i|\mathcal{X}_i,\mathcal{L}_i\rangle &= \langle y_{ijk}|\ell_{ij}\rangle\langle \mathbf{z}_i|\mathcal{X}_i\rangle \\
\langle y_{ijk}\mathbf{z}_i\mathbf{z}_i^{\mathsf{T}}|\mathcal{X}_i,\mathcal{L}_i\rangle &= \langle y_{ijk}|\ell_{ij}\rangle\langle \mathbf{z}_i\mathbf{z}_i^{\mathsf{T}}|\mathcal{X}_i\rangle.
\end{aligned} \qquad (5.61)$$

To make sure that the clustering process is guided by the SNR of utterances instead of their i-vectors, we assume that y_{ijk} is posteriorly independent of $\mathbf{x}_{ij}$, i.e., $\langle y_{ijk}|\ell_{ij},\mathbf{x}_{ij}\rangle =$

$\langle y_{ijk}|\ell_{ij}\rangle$. This assumption causes the alignments of i-vectors depend on the SNR of utterances only.

In summary, the EM steps of SI-mPLDA are given below:

E-Step:

$$\mathbb{E}_{\mathcal{Y}}\left\{y_{ijk}=1|\mathcal{L},\underline{\Lambda}\right\} \equiv \langle y_{ijk}|\mathcal{L}\rangle = \frac{\pi_k \mathcal{N}(\ell_{ij}|\mu_k,\sigma_k^2)}{\sum_{r=1}^K \pi_r \mathcal{N}(\ell_{ij}|\mu_r,\sigma_r^2)} \tag{5.62a}$$

$$\mathbf{L}_i = \mathbf{I} + \sum_{k=1}^K H_{ik}\mathbf{V}_k^{\mathsf{T}}\boldsymbol{\Sigma}_k^{-1}\mathbf{V}_k \tag{5.62b}$$

$$\langle y_{ijk}\mathbf{z}_i|\mathcal{X},\mathcal{L}\rangle = \langle y_{ijk}|\mathcal{L}\rangle\langle \mathbf{z}_i|\mathcal{X}\rangle \tag{5.62c}$$

$$\langle \mathbf{z}_i|\mathcal{X}\rangle = \mathbf{L}_i^{-1}\sum_{k=1}^K\sum_{j\in\mathcal{H}_{ik}}\mathbf{V}_k^{\mathsf{T}}\boldsymbol{\Sigma}_k^{-1}(\mathbf{x}_{ij}-\mathbf{m}_k) \tag{5.62d}$$

$$\langle y_{ijk}\mathbf{z}_i\mathbf{z}_i^{\mathsf{T}}|\mathcal{X},\mathcal{L}\rangle = \langle y_{ijk}|\mathcal{L}\rangle\langle \mathbf{z}_i\mathbf{z}_i^{\mathsf{T}}|\mathcal{X}\rangle \tag{5.62e}$$

$$\langle \mathbf{z}_i\mathbf{z}_i^{\mathsf{T}}|\mathcal{X}\rangle = \mathbf{L}_i^{-1} + \langle \mathbf{z}_i|\mathcal{X}\rangle\langle \mathbf{z}_i|\mathcal{X}\rangle^{\mathsf{T}} \tag{5.62f}$$

M-Step:

$$\pi_k' = \frac{\sum_{i=1}^N\sum_{j=1}^{H_i}\langle y_{ijk}|\mathcal{L}\rangle}{\sum_{i=1}^N\sum_{j=1}^{H_i}\sum_{l=1}^K\langle y_{ijl}|\mathcal{L}\rangle} \tag{5.63a}$$

$$\mu_k' = \frac{\sum_{i=1}^N\sum_{j=1}^{H_i}\langle y_{ijk}|\mathcal{L}\rangle\ell_{ij}}{\sum_{i=1}^N\sum_{j=1}^{H_i}\langle y_{ijk}|\mathcal{L}\rangle} \tag{5.63b}$$

$$\sigma_k^{2\prime} = \frac{\sum_{i=1}^N\sum_{j=1}^{H_i}\langle y_{ijk}|\mathcal{L}\rangle(\ell_{ij}-\mu_k')^2}{\sum_{i=1}^N\sum_{j=1}^{H_i}\langle y_{ijk}|\mathcal{L}\rangle} \tag{5.63c}$$

$$\mathbf{m}_k' = \frac{\sum_{i=1}^N\sum_{j=1}^{H_i}\langle y_{ijk}|\mathcal{L}\rangle\mathbf{x}_{ij}}{\sum_{i=1}^N\sum_{j=1}^{H_i}\langle y_{ijk}|\mathcal{L}\rangle} \tag{5.63d}$$

$$\mathbf{V}_k' = \left[\sum_{i=1}^N\sum_{j=1}^{H_i}\mathbf{f}_{ijk}'\langle y_{ijk}\mathbf{z}_i|\mathcal{X},\mathcal{L}\rangle^{\mathsf{T}}\right]\left[\sum_{i=1}^N\sum_{j=1}^{H_i}\left\langle y_{ijk}\mathbf{z}_i\mathbf{z}_i^{\mathsf{T}}|\mathcal{X},\mathcal{L}\right\rangle\right]^{-1}$$

$$\boldsymbol{\Sigma}_k' = \frac{\sum_{i=1}^N\sum_{j=1}^{H_i}\left[\langle y_{ijk}|\mathcal{L}\rangle\mathbf{f}_{ijk}'\mathbf{f}_{ijk}'^{\mathsf{T}} - \mathbf{V}_k'\langle y_{ijk}\mathbf{z}_i|\mathcal{X},\mathcal{L}\rangle\mathbf{f}_{ijk}'^{\mathsf{T}}\right]}{\sum_{i=1}^N\sum_{j=1}^{H_i}\langle y_{ijk}|\mathcal{L}\rangle} \tag{5.63e}$$

$$\mathbf{f}_{ijk}' = \mathbf{x}_{ij} - \mathbf{m}_k' \tag{5.63f}$$

Likelihood Ratio Scores

Denote $\mathbf{x}_s$ and $\mathbf{x}_t$ as the i-vectors of the target speaker and the test speaker, respectively. Also denote ℓ_s and ℓ_t as the SNR (in dB) of the corresponding utterances. Then, the same-speaker marginal likelihood is

$$
\begin{aligned}
& p(\mathbf{x}_s, \mathbf{x}_t, \ell_s, \ell_t | \text{same-speaker}) \\
&= p(\ell_s) p(\ell_t) p(\mathbf{x}_s, \mathbf{x}_t | \ell_s, \ell_t, \text{same-speaker}) \\
&= p_{st} \sum_{k_s=1}^{K} \sum_{k_t=1}^{K} \int p(\mathbf{x}_s, \mathbf{x}_t, y_{k_s} = 1, y_{k_t} = 1, \mathbf{z} | \underline{\boldsymbol{\theta}}, \ell_s, \ell_t) \mathrm{d}\mathbf{z} \\
&= p_{st} \sum_{k_s=1}^{K} \sum_{k_t=1}^{K} \gamma_{\ell_s, \ell_t}(y_{k_s}, y_{k_t}) \int p(\mathbf{x}_s, \mathbf{x}_t | y_{k_s} = 1, y_{k_t} = 1, \mathbf{z}, \underline{\omega}) p(\mathbf{z}) \mathrm{d}\mathbf{z} \\
&= p_{st} \sum_{k_s=1}^{K} \sum_{k_t=1}^{K} \gamma_{\ell_s, \ell_t}(y_{k_s}, y_{k_t}) \mathcal{N}\left(\left[\mathbf{x}_s^{\mathsf{T}} \ \mathbf{x}_t^{\mathsf{T}} \right]^{\mathsf{T}} \middle| \left[\mathbf{m}_{k_s}^{\mathsf{T}} \ \mathbf{m}_{k_t}^{\mathsf{T}} \right]^{\mathsf{T}}, \hat{\mathbf{V}}_{k_s k_t} \hat{\mathbf{V}}_{k_s k_t}^{\mathsf{T}} + \hat{\boldsymbol{\Sigma}}_k \right)
\end{aligned}
$$

where $p_{st} = p(\ell_s)p(\ell_t)$, $\hat{\mathbf{V}}_{k_s k_t} = [\mathbf{V}_{k_s}^{\mathsf{T}} \ \ \mathbf{V}_{k_t}^{\mathsf{T}}]^{\mathsf{T}}$, $\hat{\boldsymbol{\Sigma}}_k = \mathrm{diag}\{\boldsymbol{\Sigma}_{k_s}, \boldsymbol{\Sigma}_{k_t}\}$ and

$$
\begin{aligned}
\gamma_{\ell_s, \ell_t}(y_{k_s}, y_{k_t}) &\equiv P(y_{k_s} = 1, y_{k_t} = 1 | \ell_s, \ell_t, \underline{\boldsymbol{\lambda}}) \\
&= \frac{\pi_{k_s} \pi_{k_t} \mathcal{N}([\ell_s \ \ \ell_t]^{\mathsf{T}} | [\mu_{k_s} \ \ \mu_{k_t}]^{\mathsf{T}}, \mathrm{diag}\{\sigma_{k_s}^2, \sigma_{k_t}^2\})}{\sum_{k_s'=1}^{K} \sum_{k_t'=1}^{K} \pi_{k_s'} \pi_{k_t'} \mathcal{N}([\ell_s \ \ \ell_t]^{\mathsf{T}} | [\mu_{k_s'} \ \ \mu_{k_t'}]^{\mathsf{T}}, \mathrm{diag}\{\sigma_{k_s'}^2, \sigma_{k_t'}^2\})}.
\end{aligned}
$$

Likewise, the different-speaker marginal likelihood is

$$
p(\mathbf{x}_s, \mathbf{x}_t, \ell_s, \ell_t | \text{different-speaker}) = p(\mathbf{x}_s, \ell_s | \text{Spk } s) p(\mathbf{x}_t, \ell_t | \text{Spk } t), \qquad \text{Spk } s \neq \text{Spk } t,
$$

where

$$
\begin{aligned}
p(\mathbf{x}_s, \ell_s | \text{Spk } s) &= p(\ell_s) \sum\nolimits_{k_s=1}^{K} \int p(\mathbf{x}_s, y_{k_s} = 1, \mathbf{z} | \underline{\boldsymbol{\theta}}, \ell_s) \mathrm{d}\mathbf{z} \\
&= p(\ell_s) \sum\nolimits_{k_s=1}^{K} \gamma_{\ell_s}(y_{k_s}) \mathcal{N}\left(\mathbf{x}_s \middle| \mathbf{m}_{k_s}, \mathbf{V}_{k_s} \mathbf{V}_{k_s}^{\mathsf{T}} + \boldsymbol{\Sigma}_{k_s} \right),
\end{aligned}
$$

and similarly for $p(\mathbf{x}_t, \ell_t | \text{Spk } t)$. Therefore, the likelihood ratio $S_{\text{mLR-SNR}}$ is given by:

$$
S_{\text{mLR-SNR}}(\mathbf{x}_s, \mathbf{x}_t) = \frac{\sum_{k_s=1}^{K} \sum_{k_t=1}^{K} \gamma_{\ell_s, \ell_t}(y_{k_s}, y_{k_t}) \mathcal{N}\left(\left[\mathbf{x}_s^{\mathsf{T}} \ \mathbf{x}_t^{\mathsf{T}} \right]^{\mathsf{T}} \middle| \left[\mathbf{m}_{k_s}^{\mathsf{T}} \ \mathbf{m}_{k_t}^{\mathsf{T}} \right]^{\mathsf{T}}, \hat{\mathbf{V}}_{k_s k_t} \hat{\mathbf{V}}_{k_s k_t}^{\mathsf{T}} + \hat{\boldsymbol{\Sigma}}_{k_s k_t} \right)}{\left[\sum_{k_s=1}^{K} \gamma_{\ell_s}(y_{k_s}) \mathcal{N}\left(\mathbf{x}_s | \mathbf{m}_{k_s}, \mathbf{V}_{k_s} \mathbf{V}_{k_s}^{\mathsf{T}} + \boldsymbol{\Sigma}_{k_s} \right) \right] \left[\sum_{k_t=1}^{K} \gamma_{\ell_t}(y_{k_t}) \mathcal{N}\left(\mathbf{x}_t | \mathbf{m}_{k_t}, \mathbf{V}_{k_t} \mathbf{V}_{k_t}^{\mathsf{T}} + \boldsymbol{\Sigma}_{k_t} \right) \right]} \tag{5.64}
$$

Because the determinant of $\hat{\mathbf{V}}_{k_s} \hat{\mathbf{V}}_{k_s}^{\mathsf{T}} + \hat{\boldsymbol{\Sigma}}_{k_s}$ could exceed the double-precision representation, direct evaluations of Eq. 5.64 could cause numerical problems. However, the problems can be overcome by noting the identity: $|\alpha \mathbf{A}| = \alpha^D |\mathbf{A}|$ where α is a scalar and $\mathbf{A}$ is a $D \times D$ matrix. Thus, we can rewrite Eq. 5.64 as

$$
S_{\text{mLR-SNR}}(\mathbf{x}_s, \mathbf{x}_t) =
$$
$$
= \frac{\sum_{k_s=1}^{K}\sum_{k_t=1}^{K} \gamma_{\ell_s,\ell_t}(y_{k_s}, y_{k_t}) e^{\left\{-\frac{1}{2}\log|\alpha\hat{\mathbf{\Lambda}}_{k_s k_t}| - \frac{1}{2}\mathcal{D}\left(\left[\mathbf{x}_s^{\mathsf{T}}\ \mathbf{x}_t^{\mathsf{T}}\right]^{\mathsf{T}} \| \left[\mathbf{m}_{k_s}^{\mathsf{T}}\ \mathbf{m}_{k_t}^{\mathsf{T}}\right]^{\mathsf{T}}\right)\right\}}}{\left[\sum_{k_s=1}^{K} \gamma_{\ell_s}(y_{k_s}) e^{\left\{-\frac{1}{2}\log|\alpha\mathbf{\Lambda}_{k_s}| - \frac{1}{2}\mathcal{D}(\mathbf{x}_s\|\mathbf{m}_{k_s})\right\}}\right]\left[\sum_{k_t=1}^{K} \gamma_{\ell_t}(y_{k_t}) e^{\left\{-\frac{1}{2}\log|\alpha\mathbf{\Lambda}_{k_t}| - \frac{1}{2}\mathcal{D}(\mathbf{x}_t\|\mathbf{m}_{k_t})\right\}}\right]}, \quad (5.65)
$$

where $\hat{\mathbf{\Lambda}}_{k_s k_t} = \hat{\mathbf{V}}_{k_s}\hat{\mathbf{V}}_{k_t}^{\mathsf{T}} + \hat{\mathbf{\Sigma}}_{k_s k_t}$, $\mathbf{\Lambda}_{k_s} = \mathbf{V}_{k_s}\mathbf{V}_{k_s}^{\mathsf{T}} + \mathbf{\Sigma}_{k_s}$, $\hat{\mathbf{\Sigma}}_{k_s k_t} = \text{diag}\{\mathbf{\Sigma}_{k_s}, \mathbf{\Sigma}_{k_t}\}$, and $\mathcal{D}(\mathbf{x}\|\mathbf{y})$ is the Mahalanobis distance between $\mathbf{x}$ and $\mathbf{y}$, $\mathcal{D}(\mathbf{x}\|\mathbf{y}) = (\mathbf{x}-\mathbf{y})^{\mathsf{T}}\mathbf{S}^{-1}(\mathbf{x}-\mathbf{y})$, where $\mathbf{S} = \text{cov}(\mathbf{x}, \mathbf{x})$. In [171], α was set to 5.

5.4.3 DNN-Driven Mixture of PLDA

The DNN-driven mixture of PLDA [95, 176] is an extension of SNR-dependent mixture of PLDA in Section 5.4.2. The main idea is to compute the posteriors of mixture components by using an SNR-aware DNN instead of using a one-D SNR-GMM model. Figure 5.12 shows the difference between the scoring process of (a) SNR-dependent mixture of PLDA and (b) DNN-driven mixture of PLDA. In the former, an SNR estimator is applied to estimate the SNR of the target utterance and the test utterance. The posterior probabilities $\gamma_\ell(y_k)$ of mixture components depend purely on the SNR ℓ:

$$
\gamma_\ell(y_k) \equiv P(y_k = 1|\ell, \breve{}) = \frac{\pi_k \mathcal{N}(\ell|\mu_k, \sigma_k^2)}{\sum_{k'=1}^{K} \pi_{k'} \mathcal{N}(\ell|\mu_{k'}, \sigma_{k'}^2)}, \quad (5.66)
$$

where $\{\pi_k, \mu_k, \sigma_k^2\}_{k=1}^K$ are the parameters of the one-D SNR-GMM model. For the latter, an SNR-aware DNN is trained to produce the posteriors of SNR $\gamma_{\mathbf{x}}(y_k)$ based on the input i-vector $\mathbf{x}$:

$$
\gamma_{\mathbf{x}}(y_k) \equiv P(y_k = 1|\mathbf{x}, \mathbf{w}), \quad (5.67)
$$

where $\mathbf{w}$ denotes the weights of the SNR-aware DNN. Plugging these mixture posteriors into the mixture of PLDA scoring function in Eq. 5.65, we have

$$
S_{\text{DNN-mPLDA}}(\mathbf{x}_s, \mathbf{x}_t) =
$$
$$
= \frac{\sum_{k_s=1}^{K}\sum_{k_t=1}^{K} \gamma_{\mathbf{x}_s}(y_{k_s})\gamma_{\mathbf{x}_t}(y_{k_t}) e^{\left\{-\frac{1}{2}\log|\alpha\hat{\mathbf{\Lambda}}_{k_s k_t}| - \frac{1}{2}\mathcal{D}\left(\left[\mathbf{x}_s^{\mathsf{T}}\ \mathbf{x}_t^{\mathsf{T}}\right]^{\mathsf{T}} \| \left[\mathbf{m}_{k_s}^{\mathsf{T}}\ \mathbf{m}_{k_t}^{\mathsf{T}}\right]^{\mathsf{T}}\right)\right\}}}{\left[\sum_{k_s=1}^{K} \gamma_{\mathbf{x}_s}(y_{k_s}) e^{\left\{-\frac{1}{2}\log|\alpha\mathbf{\Lambda}_{k_s}| - \frac{1}{2}\mathcal{D}(\mathbf{x}_s\|\mathbf{m}_{k_s})\right\}}\right]\left[\sum_{k_t=1}^{K} \gamma_{\mathbf{x}_t}(y_{k_t}) e^{\left\{-\frac{1}{2}\log|\alpha\mathbf{\Lambda}_{k_t}| - \frac{1}{2}\mathcal{D}(\mathbf{x}_t\|\mathbf{m}_{k_t})\right\}}\right]}, \quad (5.68)
$$

Figure 5.13(a) shows the graphical model of SNR-dependent mixture of PLDA with parameters $\{\pi_k, \mu_k, \sigma_k, \mathbf{m}_k, \mathbf{V}_k, \mathbf{\Sigma}_k\}_{k=1}^K$. In the diagram, $\underline{\pi} = \{\pi_k\}_{k=1}^K$, $\underline{\mu} = \{\mu_k\}_{k=1}^K$, $\underline{\sigma} = \{\sigma_k\}_{k=1}^K$, $\underline{\mathbf{m}} = \{\mathbf{m}_k\}_{k=1}^K$, $\underline{\mathbf{V}} = \{\mathbf{V}_k\}_{k=1}^K$, $\underline{\mathbf{\Sigma}} = \{\mathbf{\Sigma}_k\}_{k=1}^K$. Figure 5.13(b) shows the graphical model of DNN-driven mixture of PLDA with parameters $\{\mathbf{m}_k, \mathbf{V}_k, \mathbf{\Sigma}_k\}_{k=1}^K$.

The key advantage of the DNN-driven mixture of PLDA is that it uses a classifier to guide the PLDA training, such that each i-vector cluster is modeled by one mixture component. During testing, we combine the PLDA scores using dynamic weights that depend on the classifier's output. This approach is flexible because there is no restriction

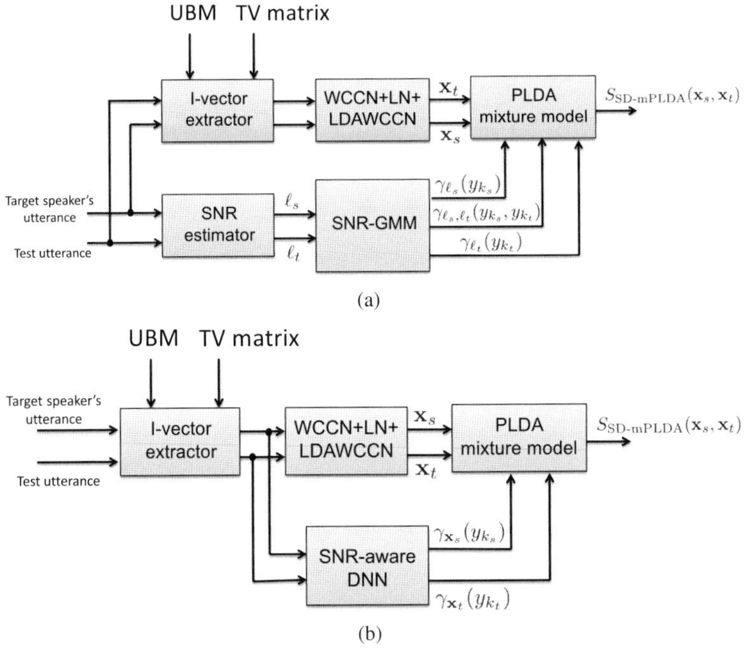

Figure 5.12 (a) Procedure of computing the score of two utterances with SNR ℓ_s and ℓ_t in SNR-dependent mixture of PLDA. (b) Same as (a) but replacing the SNR-dependent mixture of PLDA by DNN-driven mixture of PLDA.

on the type of classifiers to be used. In fact, any classifier, such as DNN, SVM, and logistic regression, can be used as long as they can leverage the cluster property in the training data. However, experimental results suggest that DNN classifiers give the best performance [95].

5.5 Multi-Task DNN for Score Calibration

Because adverse acoustic conditions and duration variability in utterances could change the distribution of PLDA scores, a number of back ends have been investigated to replace the PLDA models. These back ends include support vector machines (SVMs) [177] and end-to-end models [178].

Rather than enhancing the PLDA back end, score calibration methods have been proposed to compensate for the harmful effect on the PLDA scores due to background noise and duration variability. For example, the score shifts in in [167, 179, 180] are deterministic and the shifts are assumed to be linearly related to the utterances' SNR and/or to the logarithm of utterance duration. In [181], the authors assume that the shift follows a Gaussian distribution with quality-dependent mean and variance. A Bayesian network was used to infer the posterior distributions of the target and nontarget hypotheses; a calibrated likelihood-ratio is then computed from the posterior distributions. On

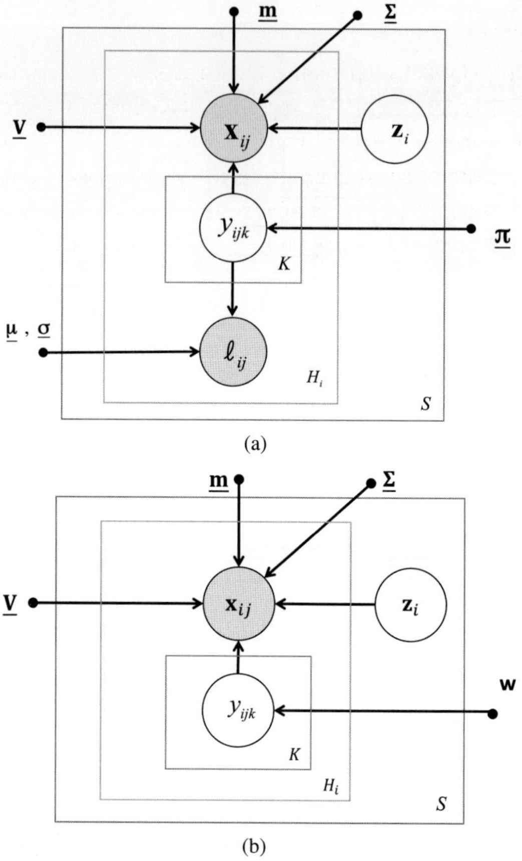

Figure 5.13 The graphical model of (a) SNR-dependent mixture of PLDA and (b) DNN-driven mixture of PLDA. [Reprinted with permission from *DNN-driven Mixture of PLDA for Robust Speaker Verification (Figure 2), N. Li, M.W. Mak, and J.T. Chien., IEEE/ACM Trans. on Audio Speech and Language Processing, vol. 25, no. 6, pp. 1371–1383, June 2017, with permission of IEEE*]

the other hand, the score shift in [182, 183] was assumed to be bilinear or considered as a cosine-distance function of the two quality vectors derived from the target-speaker and test i-vectors.

There are methods that compensate for the duration mismatch only [167, 184, 185] or both duration and SNR mismatch [179–182]. A common property of these methods is that all of them assume that the detrimental effect can be compensated by shifting the PLDA scores. These methods aim to estimate the shift based on some meta data, including duration and SNR [179, 180]), or from the i-vectors [182] to counteract the effect. Despite these methods use linear models, they have demonstrated improvement in performance in a number of systems. However, score shift may not be linearly related to SNR and log-duration. As a result, cosine-distance scores and bilinear transformation may not be able to capture the true relationship accurately.

5.5.1 Quality Measure Functions

A quality measure function (QMF) maps some measurable quantities, such as SNR and duration of utterances, to score shifts that represent the detrimental effect of background noise and duration variability on the PLDA scores [167, 179, 180]. Denote S as the *uncalibrated* PLDA score of a test utterance and a target speaker utterance. Also denote λ_{tgt} and λ_{tst} as the quality measures of the target speaker and test utterances, respectively. The *calibrated* score S' can then be evaluated according to:

$$S' = w_0 + w_1 S + Q(\lambda_{tgt}, \lambda_{tst}), \tag{5.69}$$

where $Q(\lambda_{tgt}, \lambda_{tst})$ is a QMF. In [179, 180], the QMFs were based on the duration (d_{tst}) and SNR (SNR_{tst}) of the test utterance:

$$\begin{aligned} Q_{\text{SNR}}(\text{SNR}_{tst}) &= w_2 \text{SNR}_{tst} \\ Q_{\text{Dur}}(d_{tst}) &= w_2 \log(d_{tst}) \\ Q_{\text{SNR+Dur}}(\text{SNR}_{tst}, d_{tst}) &= w_2 \text{SNR}_{tst} + w_3 \log(d_{tst}), \end{aligned} \tag{5.70}$$

where w_2 and w_3 are the weights of the respective meta information. In case the effect of noise in the target utterance is also considered, Q_{SNR} becomes:

$$Q_{\text{SNR2}}(\text{SNR}_{tgt}, \text{SNR}_{tst}) = w_2 \text{SNR}_{tgt} + w_3 \text{SNR}_{tst}, \tag{5.71}$$

where SNR_{tgt} is the SNR of the target speaker utterance. In Eqs. 5.69–5.71, the weights $w_i, i = 0, \ldots, 3$, can be estimated by logistic regression [186].

Ferrer et al. [183] and Nautsch et al. [182] derived a quality vector $\mathbf{q}$ based on the posterior probabilities of various acoustic conditions given an i-vector based on the assumption that i-vectors are acoustic-condition dependent. In the method, every i-vector has its corresponding quality vector. Given the i-vectors of a verification trial, the score shift is a function of the quality vectors in that trial.

Nautsch et al. [182] name the function the "function of quality estimate (FQE)." Specifically, i-vectors derived from utterances of 55 combinations of different durations and SNRs were used to train 55 Gaussian models $\Lambda_j = \{\boldsymbol{\mu}_j, \boldsymbol{\Sigma}\}_{j=1}^{55}$. These Gaussians have their own mean $\boldsymbol{\mu}_j$ estimated from the i-vectors of the respective conditions; however, they share the same global covariance matrix $\boldsymbol{\Sigma}$. For an i-vector $\mathbf{x}$, the jth component of $\mathbf{q}$ is the posterior of condition j:

$$q_j = \frac{\mathcal{N}(\mathbf{x}|\boldsymbol{\mu}_j, \boldsymbol{\Sigma})}{\sum_{j'} \mathcal{N}(\mathbf{x}|\boldsymbol{\mu}_{j'}, \boldsymbol{\Sigma})}, \quad j = 1, \ldots, 55. \tag{5.72}$$

Given the i-vectors $\mathbf{x}_{tgt}$ and $\mathbf{x}_{tst}$ from a target speaker and a test speaker, respectively, the corresponding quality vectors $\mathbf{q}_{tgt}$ and $\mathbf{q}_{tst}$ are obtained from Eq. 5.72. Then, we can obtain the score shift as follows:

$$\begin{aligned} Q_{\text{UAC}}(\mathbf{q}_{tgt}, \mathbf{q}_{tst}) &= w_2 \mathbf{q}_{tgt}^{\mathsf{T}} \mathbf{W} \mathbf{q}_{tst} \\ Q_{\text{qvec}}(\mathbf{q}_{tgt}, \mathbf{q}_{tst}) &= w_2 \cos(\mathbf{q}_{tgt}, \mathbf{q}_{tst}), \end{aligned} \tag{5.73}$$

where $\mathbf{W}$ is a symmetric bilinear matrix and $\cos(\mathbf{a}, \mathbf{b})$ means cosine-distance score between vectors $\mathbf{a}$ and $\mathbf{b}$:

$$\cos(\mathbf{a}, \mathbf{b}) = \frac{\mathbf{a}^{\mathsf{T}}\mathbf{b}}{\|\mathbf{a}\|\|\mathbf{b}\|}.$$

In Eq. 5.73, the components of a quality vector are the posteriors of the respective SNR/duration groups. The cosine-distance and bilinear transformation of the two quality vectors reflect the similarity between the two vectors in terms of SNR and duration. They are similar to the QMFs in Eq. 5.70 and Eq. 5.71 in that the score shift is linearly related to the SNR of the utterances and/or to the logarithm of utterance duration. As will be discussed later, the relationship among score shift, SNR, and duration is rather complex. As a result, the information in SNR and duration is not enough to estimate the ideal score shift. It turns out that i-vectors can provide essential information for estimating the ideal score shift.

Theoretically, the FQE in Eq. 5.73 is better than the QMF in Eq. 5.69 because the former does not use SNR information directly but instead uses it implicitly through the i-vectors and the Gaussian models. Nevertheless, at low SNR, it is unclear whether the cosine distance and bilinear transformation can accurately estimate the score shift. As Figure 5.14 shows, the score shifts and the SNR of utterances have a nonlinear and complex relationship. Because of the noise-level dependence of i-vectors [171] (also see Figure 5.9), it is reasonable to explicitly predict the score shifts from i-vectors instead of implicitly predict them through the Gaussian models of the i-vectors such as the FQE method. In the next section, we explain how the score shifts can be accurately estimated through multi-task deep learning.

5.5.2 DNN-Based Score Calibration

In [3], a DNN-based score calibration algorithm was proposed to mitigate the limitations of the score calibration algorithms described in Section 5.5.1. The key idea is to apply a DNN to determine a suitable score shift for each pair of i-vectors or to estimate the clean PLDA score given a pair of noisy i-vector and a noisy PLDA score. The DNN is trained to carry out *score compensation*, and it plays the same role as function Q in Eq. 5.69. Nevertheless, when the DNN is trained to produce clean PLDA scores, it essentially carries out *score transformation*. For whatever purposes and roles, a further *calibration* process is necessary because there is no guarantee that the DNN can produce true log-likelihood ratios in its output. The authors in [3] collectively refer to the score compensation, transformation, and calibration processes as DNN-based score calibration. Figure 5.15 shows the full process.

In some reports [187, 188], the term *calibration* referred strictly to the process of score adjustment and the adjustment will not change the equal error rate (EER). Here, the terminology in [179, 180, 182] was adopted, i.e., we relax the definition of calibration so that it also includes the processes that could reduce the EER.

The most basic form of DNN-based score calibration is to use a DNN as shown in Figure 5.16 to estimate the appropriate score shift given the target and test i-vector pairs

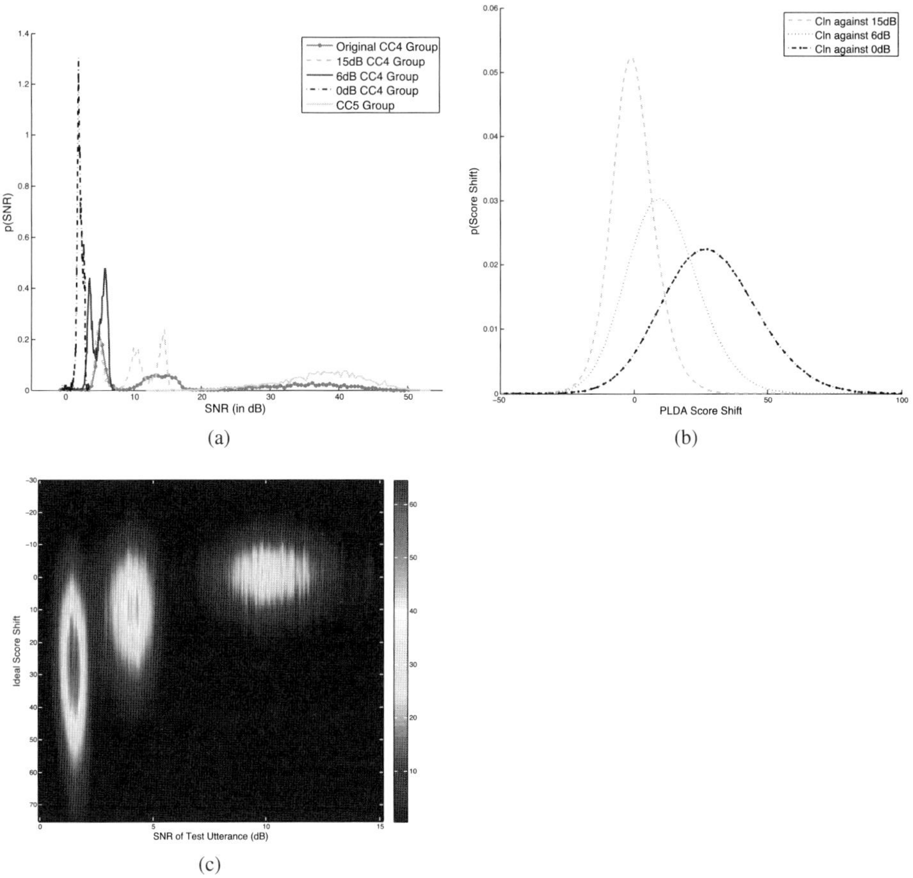

Figure 5.14 Nonlinear relationship between SNRs and socre shifts, and large score shift at low SNR. (a) Distribution of the SNR of the noise-contaminated test utterances and the original utterances in CC4 and CC5 (male) of NIST 2012 SRE. Babble noise was added to to original utterances at SNR of 15dB, 6DB, and 0dB. (b) Distributions of ideal score shifts ($S - S_{cln}$) under three SNR conditions in the test utterances and clean condition in the target speakers' utterances. S_{cln} (clean scores) were obtained by scoring clean test i-vectors against clean i-vectors from target speakers. (c) Distributions of score shifts with respect to test utterances' SNR under clean utterances from target speakers. [Reprinted from *Denoised Senone I-Vectors for Robust Speaker Verification (Figures 6, 8, and 9), Z.L. Tan, M.W. Mak, et al., IEEE/ACM Trans. on Audio Speech and Language Processing, vol. 26, no. 4, pp. 820–830, April 2018, with permission of IEEE*]

($\mathbf{x}_{tgt}$ and $\mathbf{x}_{tst}$) and the uncalibrated PLDA score S. Given an uncalibrated PLDA score S of a verification trial, the compensated score is given by:

$$S'_{\text{st}} = S + \text{DNN}_{\text{st}}(\mathbf{x}_s, \mathbf{x}_t, S), \tag{5.74}$$

where the subscript "st" denotes the output of a *single-task* DNN. With the i-vector pair and the uncalibrated score as input, the DNN outputs the shift of the PLDA score due to the deviation of the acoustic condition from the clean one:

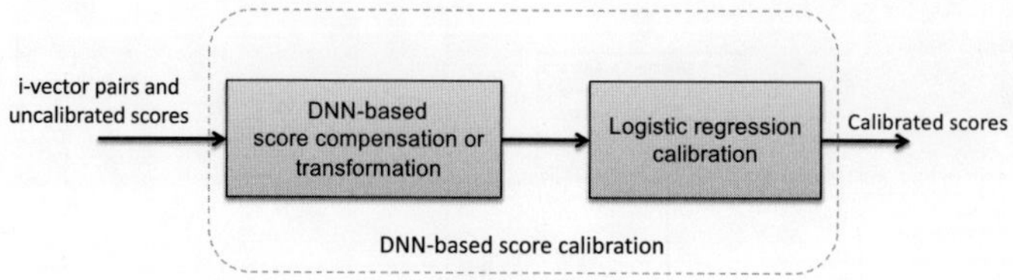

Figure 5.15 DNN-based score calibration.

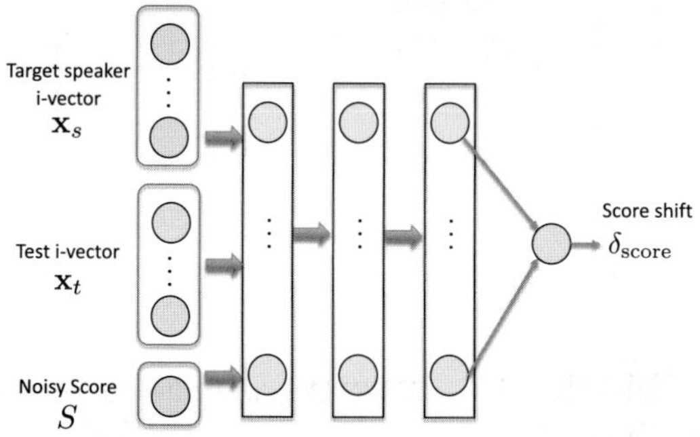

Figure 5.16 The most basic form of DNN-based score calibration in which a DNN predicts the score shift given the target speaker i-vector, the test i-vector, and noisy PLDA score as inputs.

$$\text{DNN}_{\text{st}}(\mathbf{x}_s, \mathbf{x}_t, S) \approx \delta_{\text{score}} = S_{\text{cln}} - S, \tag{5.75}$$

where S_{cln} is the PLDA score if both $\mathbf{x}_s$ and $\mathbf{x}_t$ were derived from clean utterances. Substituting Eq. 5.75 to Eq. 5.74, we have:

$$S'_{\text{st}} \approx S + (S_{\text{cln}} - S) = S_{\text{cln}}. \tag{5.76}$$

Eq. 5.76 suggests that the clean score can be recovered.

The PLDA score of i-vector pair ($\mathbf{x}_s$, $\mathbf{x}_t$) is the log-likelihood ratio (Eq. 3.85):

$$\begin{aligned} S &= \text{LLR}(\mathbf{x}_s, \mathbf{x}_t) \\ &= \frac{1}{2}\mathbf{x}_s^{\mathsf{T}} \boldsymbol{Q} \mathbf{x}_t + \frac{1}{2}\mathbf{x}_t^{\mathsf{T}} \boldsymbol{Q} \mathbf{x}_t + \mathbf{x}_s^{\mathsf{T}} \boldsymbol{P} \mathbf{x}_t + \text{const}, \end{aligned} \tag{5.77}$$

where $\boldsymbol{P}$ and $\boldsymbol{Q}$ and are derived from the across-speaker covariances and total covariances of i-vectors. Using Eq. 5.77, the general form of score shift is:

$$\begin{aligned} \delta_{\text{score}} &= \text{LLR}(\mathbf{x}_{s\_\text{cln}}, \mathbf{x}_{t\_\text{cln}}) - \text{LLR}(\mathbf{x}_s, \mathbf{x}_t) \\ &= \frac{1}{2}\mathbf{x}_{s\_\text{cln}}^{\mathsf{T}} \boldsymbol{Q} \mathbf{x}_{s\_\text{cln}} - \frac{1}{2}\mathbf{x}_s^{\mathsf{T}} \boldsymbol{Q} \mathbf{x}_s + \frac{1}{2}\mathbf{x}_{t\_\text{cln}}^{\mathsf{T}} \boldsymbol{Q} \mathbf{x}_{t\_\text{cln}} \\ &\quad - \frac{1}{2}\mathbf{x}_t^{\mathsf{T}} \boldsymbol{Q} \mathbf{x}_t + \mathbf{x}_{s\_\text{cln}}^{\mathsf{T}} \boldsymbol{P} \mathbf{x}_{t\_\text{cln}} - \mathbf{x}_s^{\mathsf{T}} \boldsymbol{P} \mathbf{x}_t. \end{aligned} \tag{5.78}$$

Take notice of the differences between Eq. 5.78 and the bilinear transformation in Eq. 5.73. Specifically, the score shift in Eq. 5.78 involves both of the bilinear transformation of clean and noisy test i-vectors and the bilinear transformation between the target

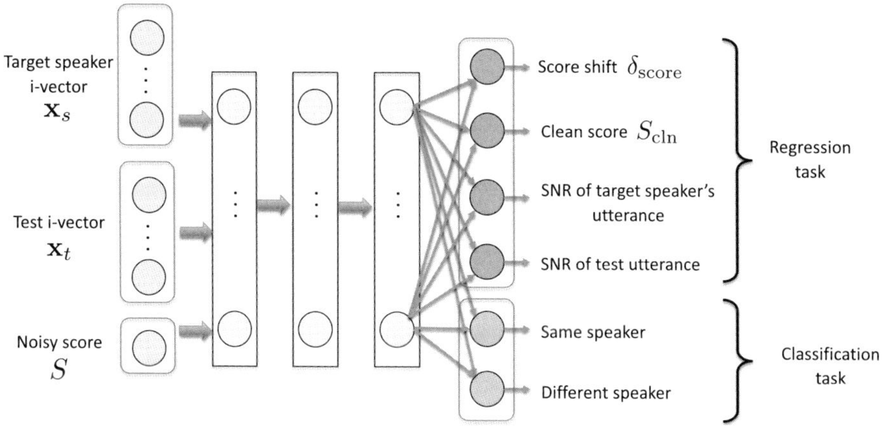

Figure 5.17 Multi-task DNN for score calibration.

speaker and test i-vectors. If the clean test i-vector ($\mathbf{x}_{t\_\text{cln}}$) for every noisy test i-vector ($\mathbf{x}_t$) were known, then Eq. 5.78 can be easily computed without a DNN. However, $\mathbf{x}_{t\_\text{cln}}$ is unknown; therefore, it is necessary for the DNN to learn the complex relationship between the score shifts and the i-vector pairs.

To train the DNN in Figure 5.16, it is necessary to propagate the output node's errors to hundreds of hidden nodes as well as to the input layers. Because there is only one output node, the error propagation is very inefficient. The problem could be solved by introducing some auxiliary tasks for the network to learn, i.e., multi-task learning [189, 190]. The idea is based on the notion that the auxiliary information in the output layer of a multi-task DNN may help to improve the learning efficiency.

In Figure 5.17, a DNN uses multi-task learning to learn two tasks: main and auxiliary. In the main task, the network is trained to produce score shift δ_{score} and clean score S_{cln}, whereas in the auxiliary tasks, the network produces the SNR of target speaker's utterance and the test utterance and outputs the same-speaker and different-speaker posteriors. Both the clean score S_{cln} and ideal score shift δ_{score} are the target outputs of the multi-task DNN.

The regression task uses four output nodes, two of which aim to predict the SNRs of the target speaker's utterance (SNR_s) and the test utterance (SNR_t). They are part of the auxiliary task that helps the network to estimate the score shift. The DNN's regression part uses minimum mean squared error as the optimization criterion and linear activation functions in its output nodes. Another auxiliary task is to enable the network to verify speakers. To this end, we add two classification nodes at the output to indicate whether the input i-vectors pair are from the same speaker or not. The classification part of the network uses softmax outputs and cross-entropy as the optimization criterion.

The multi-task DNN with two classification nodes and four regression nodes comprises two sets of vectors that are concatenated together:

$$\text{DNN}_{\text{mt}}(\mathbf{x}_s, \mathbf{x}_t, S) \approx \Big[\underbrace{[\delta_{\text{score}}, S_{\text{cln}}, \text{SNR}_s, \text{SNR}_t]}_{\text{Regression}},\ \underbrace{[p^+, p^-]}_{\text{Classification}} \Big]^{\mathsf{T}}, \tag{5.79}$$

where DNN_{mt} denotes the *multi-task* DNN outputs, and p^+ and p^- are the posterior probabilities of same-speaker and different-speaker hypotheses, respectively. Similar to Eq. 5.75, Eq. 5.79 means that the DNN uses the i-vector pair $(\mathbf{x}_s, \mathbf{x}_t)$ and the original score S as input. The multi-task learning strategy enables the network to output the score shift δ_{score}, the clean score S_{cln}, the SNR of target-speaker speech, the SNR of test speech, and the posterior probabilities (p^+ and p^-).

After training, the network can be used for score transformation and calibration. The calibration process will only use the score shift and clean score produced by the multi-task DNN:

$$\text{DNN}_{\text{mt,shift}}(\mathbf{x}_s, \mathbf{x}_t, S) \approx \delta_{\text{score}}, \tag{5.80}$$

and

$$\text{DNN}_{\text{mt,cln}}(\mathbf{x}_s, \mathbf{x}_t, S) \approx S_{\text{cln}}, \tag{5.81}$$

where the subscripts denote the output nodes corresponding to the score shift and clean score in Figure 5.17, respectively. Therefore, we have

$$S'_{\text{mt}} = S + \text{DNN}_{\text{mt,shift}}(\mathbf{x}_s, \mathbf{x}_t, S) \approx S_{cln}, \tag{5.82}$$

and

$$S'_{\text{mt}} = \text{DNN}_{\text{mt,cln}}(\mathbf{x}_s, \mathbf{x}_t, S) \approx S_{cln}. \tag{5.83}$$

5.6 SNR-Invariant Multi-Task DNN

In [191], Yao and Mak analyze how noisy speech affects the distribution of i-vectors (Figures 5.18 and 5.19). The analysis shows that background noise has detrimental effects on intra-speaker variability and that the variability depends on the SNR levels of the utterances. Yao and Mak argue that this SNR-dependent noise effect on the i-vectors is caused by SNR variability. This effect causes difficulty in the back end to separate the speaker and channel variabilities. To address this problem, Yao and Mak proposed using DNNs to suppress both SNR and channel variabilities in the i-vector space directly. To this end, they proposed two models. The first model is called hierarchical regression DNN (H-RDNN). It contains two denoising regression DNNs hierarchically stacked. The second model is called multi-task DNN (MT-DNN). It is trained to carry out speaker classification as well as i-vector denoising (regression).

Yao and Mak observed that i-vectors derived from noise-contaminated speech with similar SNRs tend to form SNR-dependent clusters. This SNR-grouping phenomenon inspires the use of PLDA mixture models in Section 5.4 (see also [95, 171]) so that each SNR group can be handled by a PLDA model. Yao and Mak [191] argued that rather than tackling the i-vectors belonging to the SNR-dependent groups, as in [171], it is better to compensate for the variability directly in the i-vector space.

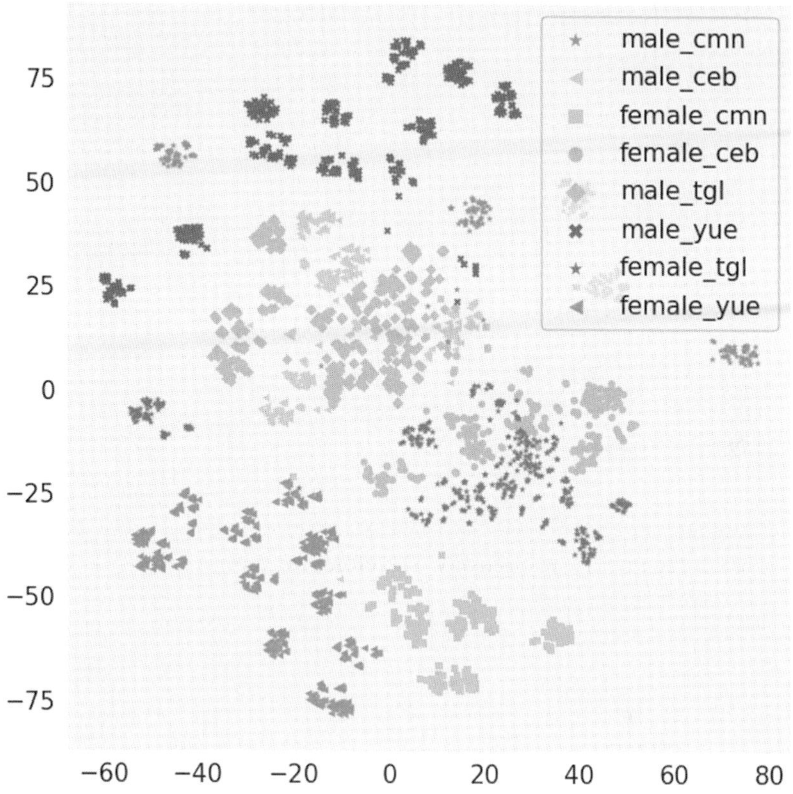

Figure 5.18 Distribution of gender- and language-dependent i-vectors in SRE16 on a two-dimensional t-SNE space. "cmn": Mandarin; "ceb": Cebuano; "yue": Cantonese; "tgl": Tagalog. *It may be better to view the color version of this figure, which is available at https://github.com/enmwmak/ML-for-Spkrec.*

5.6.1 Hierarchical Regression DNN

Figure 5.20 shows the structure of a hierarchical regression DNN (H-RDNN). Denote $\{\mathbf{x}_n\}$ as the i-vectors preprocessed by within-class covariance normalization (WCCN) and length normalization (LN), and $\mathbf{t}_n$ as target i-vectors obtained by averaging speaker-dependent i-vectors derived from clean utterances. Also denote $\mathcal{S} = \{\mathbf{x}_n, \mathbf{t}_n; n = 1, \ldots, N\}$ as a training set comprising N i-vector pairs. Training is divided into two stages. In the first stage, the regression network $f_{\mathbf{\Theta}}^{reg}(\cdot)$ works toward minimizing the MSE and the Frobenius norm of weight paramters:[4]

$$\min_{\mathbf{\Theta}} \frac{1}{N} \sum_{n=1}^{N} \frac{1}{2} || f_{\mathbf{\Theta}}^{reg}(\mathbf{x}_n) - \mathbf{t}_n ||_2^2 + \frac{\beta_{reg_1}}{2} ||\mathbf{\Theta}||_2^2, \tag{5.84}$$

[4] If $\mathbf{t}_n$ and $\mathbf{x}_n$ are derived from a clean utterance and its corresponding noise-contaminated version, Eq. 5.84 leads to the denoising autoencoder (DAE) [3].

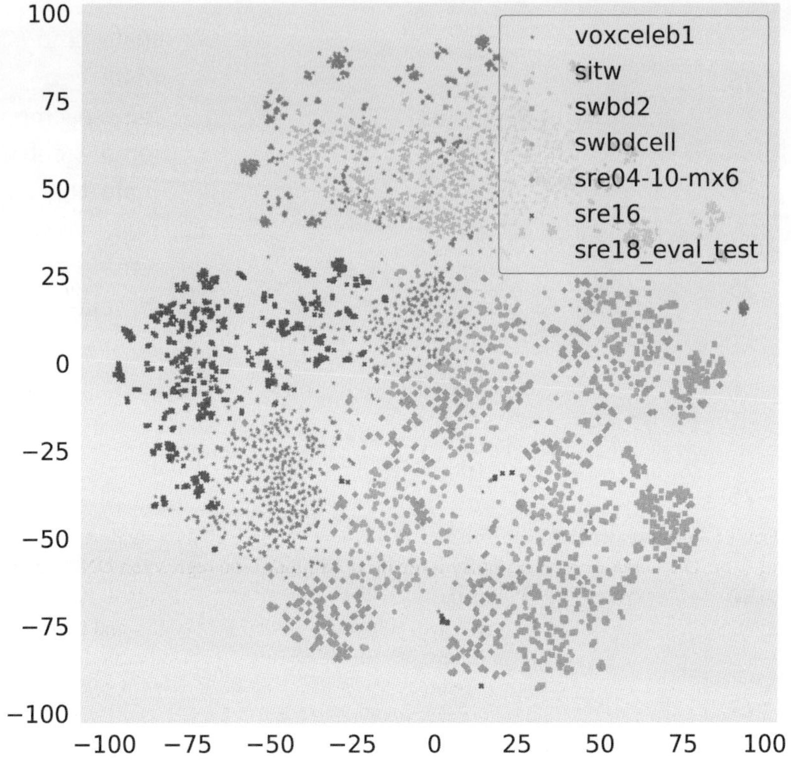

Figure 5.19 Distribution of x-vectors of various datasets on a two-dimensional t-SNE space. *It may be better to view the color version of this figure, which is available at https://github.com/enmwmak/ML-for-Spkrec.*

where $f_{\mathbf{\Theta}}^{reg}(\mathbf{x}_n)$ is the output of the top regression layer of the first (left) DNN in Figure 5.20; $\mathbf{\Theta}$ denotes the weights in the first regression network and β_{reg_1} controls the degree of regularization. The first regression DNN is trained to suppress channel and SNR variations simultaneously within each speaker cluster.

In the second stage, another regression DNN (the DNN on the right of Figure 5.20) is trained to constrain the outliers that the first DNN cannot properly denoise. Assume that all i-vectors have been processed by the first DNN followed by WCCN whitening and LN. Given a training set comprising N i-vector pairs: $\mathcal{S}' = \{\mathbf{x}'_n, \mathbf{t}'_n; n = 1, \ldots, N\}$, a regularization term and the MSE of the regression network $g_{\mathbf{\Phi}}^{reg}(\cdot)$ are minimized jointly:

$$\min_{\mathbf{\Phi}} \frac{1}{N} \sum_{n=1}^{N} \frac{1}{2} ||g_{\mathbf{\Phi}}^{reg}(\mathbf{x}'_n) - \mathbf{t}'_n||_2^2 + \frac{\beta_{reg2}}{2} ||\mathbf{\Phi}||_2^2, \tag{5.85}$$

where $\mathbf{x}'_n$ is the nth i-vector denoised by the first DNN, i.e., $\mathbf{x}'_n = f_{\mathbf{\Theta}}^{reg}(\mathbf{x}_n)$; $\mathbf{t}'_n$ is the associated i-vector obtained from the original i-vector set (without noise contamination) and then denoised by the first DNN, i.e., $\mathbf{t}'_n = f_{\mathbf{\Theta}}^{reg}(\mathbf{x}_n^{org})$; $\mathbf{x}''_n = g_{\mathbf{\Phi}}^{reg}(\mathbf{x}'_n)$ is the output of

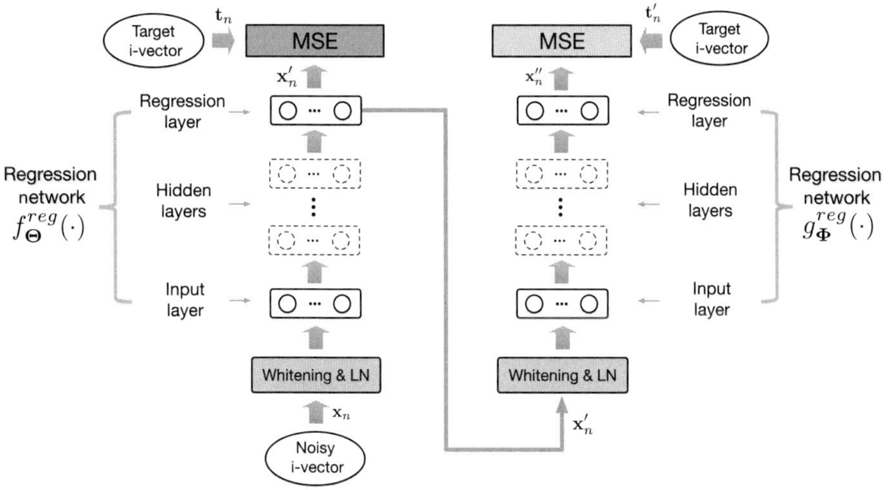

Figure 5.20 Structure of the hierarchical regression DNN (H-RDNN) with arrows denote the flow of data. The network receives noisy i-vector $\mathbf{x}_n$ as input. The regression DNN on the right receives $\mathbf{x}'_n$ as input and produces the final denoised i-vector $\mathbf{x}''_n$ as output. $f^{reg}_{\boldsymbol{\Theta}}(\cdot)$ and $g^{reg}_{\boldsymbol{\Phi}}(\cdot)$ are the mapping functions of the two regression networks, respectively. $\mathbf{t}_n$ and $\mathbf{t}'_n$ are the target i-vectors in Eq. 5.84 and Eq. 5.85, respectively. MSE: mean squared error. [Reprinted from *SNR-Invariant Multi-Task Deep Neural Networks for Robust Speaker Verification (Figure 1), Q. Yao and M.W. Mak, IEEE Signal Processing Letters, vol. 25, no. 11, pp. 1670–1674, Nov. 2018, with permission of IEEE*]

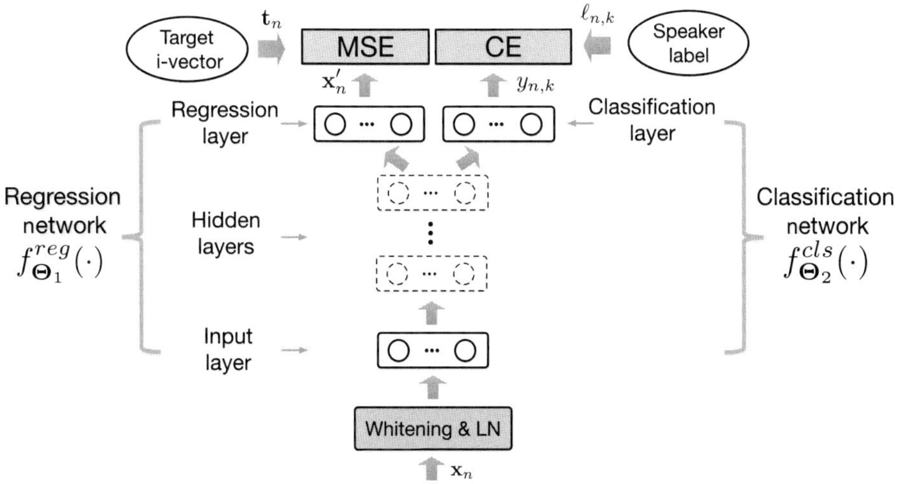

Figure 5.21 The structure of the multi-task DNN (MT-DNN). It receives noisy i-vector $\mathbf{x}_n$ as input and produces $\mathbf{x}'_n$ and $y_{n,k}$ as output. $\mathbf{t}_n$ in Eq. 5.84 and the target label for the classification task $\ell_{n,k}$ in Eq. 5.86 are the targets for the regression and classification tasks, respectively. MSE: mean squared error; CE: cross entropy. [Reprinted from *SNR-Invariant Multi-Task Deep Neural Networks for Robust Speaker Verification (Figure 2), Q. Yao and M.W. Mak, IEEE Signal Processing Letters, vol. 25, no. 11, pp. 1670–1674, Nov. 2018, with permission of IEEE*]

the second regression DNN; $\mathbf{\Phi}$ denotes the weights of the second regression DNN and β_{reg_2} controls the degree of regularization.

5.6.2 Multi-Task DNN

The regression task should avoid losing speaker information. To this end, we need the DNN-transformed i-vectors to form a small within-speaker scatter and a large between-speaker scatter. This is realized by a multi-task DNN (MT-DNN) shown in Figure 5.21.

Denote $\boldsymbol{\ell}_n$ as the speaker label in one-hot format of the nth utterance. Also denote $\mathbf{x}_n$ and $\mathbf{t}_n$ as the preprocessed i-vector and the target i-vector, respectively. Assume that we have a training set $\mathcal{S}' = \{\mathbf{x}_n, \mathbf{t}_n, \boldsymbol{\ell}_n; n = 1, \ldots, N\}$ derived from N utterances. To train the regression network $f^{reg}_{\mathbf{\Theta}_1}(\cdot)$ in Figure 5.21, the MSE is minimized in the identical manner as Eq. 5.84. The top regression layer's output is the denoised i-vector $\mathbf{x}'_n$, i.e., $\mathbf{x}'_n = f^{reg}_{\mathbf{\Theta}_1}(\mathbf{x}_n)$. To train the classification network $f^{cls}_{\mathbf{\Theta}_2}(\cdot)$, the cross-entropy (CE) cost together with the Frobenius norm of weights in the classification network are jointly minimized:

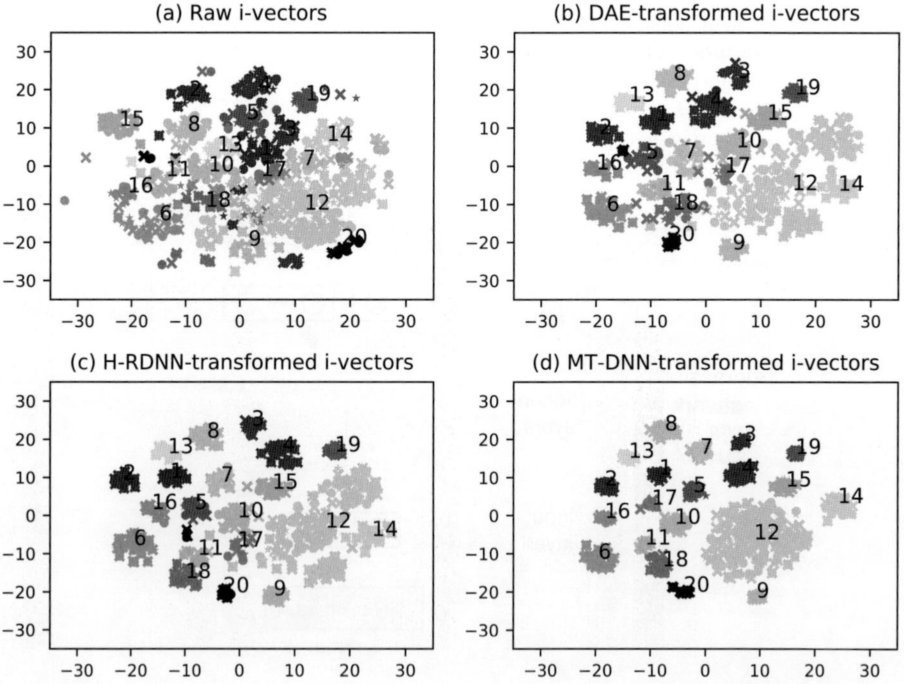

Figure 5.22 T-SNE visualization of 20 speakers from three SNR groups (org+15dB+6dB, telephone speech, babble noise). The colors of the markers represent speakers and the markers with different shapes (○, ×, and ∗) correspond to different SNR groups. *It may be better to view the color version of this figure, which is available at https://github.com/enmwmak/ML-for-Spkrec.* [Reprinted from *SNR-Invariant Multi-Task Deep Neural Networks for Robust Speaker Verification (Figure 3), Q. Yao and M.W. Mak, IEEE Signal Processing Letters, vol. 25, no. 11, pp. 1670–1674, Nov. 2018, with permission of IEEE*]

$$\min_{\mathbf{\Theta}_2} -\frac{1}{N}\sum_{n=1}^{N}\sum_{k=1}^{K} \ell_{n,k} \log y_{n,k} + \frac{\beta_{cls}}{2}||\mathbf{\Theta}_2||_2^2. \tag{5.86}$$

In Eq. 5.86, $\ell_{n,k}$ is the kth element of $\boldsymbol{\ell}_n$; if the utterance of $\mathbf{x}_n$ is spoken by speaker k, then $\ell_{n,k} = 1$, otherwise it is equal to 0; $y_{n,k}$ is the posterior probability of the kth speaker (totally K speakers), which is a component of $\mathbf{y}_n$. Note that $\mathbf{y}_n$ is the classification network's output, i.e., $\mathbf{y}_n = f^{cls}_{\mathbf{\Theta}_2}(\mathbf{x}_n)$; $\mathbf{\Theta}_2$ denotes the weights in the classification network and β_{cls} controls the degree of regularization.

The classification and regression tasks are trained in an alternating manner. Specifically, at iteration t, weights are updated according to the gradient of regression loss, and at iteration $t+1$, weights are updated according to the gradient of classification loss. Then, the cycle repeats.

The distributions of 20 speaker clusters are shown in Figure 5.22. These clusters are formed by the raw i-vectors (original, 15 dB, and 6 dB) and the i-vectors transformed by different DNN models. Obviously, the original i-vectors (top-left) could not form

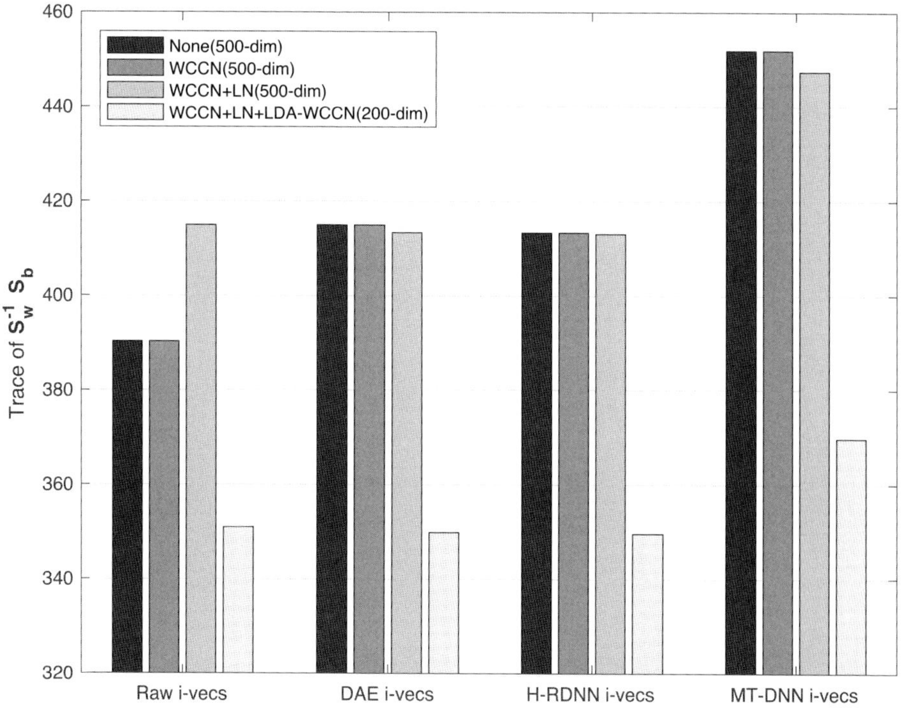

Figure 5.23 Degree of separation among the clusters of 20 speaker from three SNR groups: org+15dB+6dB, telephone speech, and babble noise. The x-axis labels correspond to different types of DNN transformation methods. The vertical axis shows the values of $\text{Tr}(\mathbf{S}_w^{-1} S_b)$. The gray levels in the legend correspond to different i-vector post-processing methods applied to the DNN-transformed i-vectors. [Reprinted from *SNR-Invariant Multi-Task Deep Neural Networks for Robust Speaker Verification (Figure. 4), Q. Yao and M.W. Mak, IEEE Signal Processing Letters, vol. 25, no. 11, pp. 1670–1674, Nov. 2018, with permission of IEEE*]

distinguishable clusters; on the other hand, the clusters formed by the DNN-transformed i-vectors are very apparent. Consider speaker cluster 6 (darkest shade) in the top-left figure, the left-most • of this speaker drifts from its center significantly. As this i-vector was extracted from an uncontaminated utterance, channel effect is the main reason for the deviation. The 15dB and 6dB i-vectors (marked with crosses and asterisks) have large speaker clusters. After denoising, i-vectors produced by the MT-DNN have the most compact clusters. This suggests that the i-vectors transformed by the MT-DNN are less dependent on the channel and background noise but more dependent on speakers, which is a favorable property for PLDA modeling.

Figure 5.23 shows the trace of $\mathbf{S}_w^{-1}\mathbf{S}_b$, where $\mathbf{S}_b$ and $\mathbf{S}_w$ are the between- and within-speaker scatter matrices obtained from different types of i-vectors (see the labels in the x-axis). These matrices were obtained from the same set of i-vectors used for producing Figure 5.22. The traces measure the dispersion of speaker clusters. The bars indicated by different gray level represent different i-vector post-processing methods. Because $\text{Tr}\{\mathbf{S}_w^{-1}\mathbf{S}_b\}$ will not be affected by WCCN, the value of "None" and "WCCN" are the same. The figure suggest that MT-DNN can produce speaker clusters with the largest degree of separation.

6 Domain Adaptation

To make speaker verification commercially viable, it is important to adapt a system from a resource-rich domain to a resource-limited domain – a topic known as domain adaptation. Alternatively, we may build systems that are domain-invariant. This chapter covers supervised and unsupervised domain adaptation (DA) techniques. Adaptation in feature space and model space will be discussed.

6.1 Overview of Domain Adaptation

As the effect of domain mismatch on PLDA parameters is more pronounced than the UBM and the total variability matrix [192], previous research mainly focused on adapting the PLDA parameters. Earlier approaches to reducing the domain mismatch require speaker labels in the in-domain data. For instance, Garcial et al. [192] used the speaker labels in the in-domain data to compute the in-domain within-speaker and across-speaker covariance matrices by MAP estimation. Villaba et al. [193] treated these matrices as latent variables and used variational Bayes to factorize the joint posterior distribution. Then, the point estimates were used for in-domain scoring.

There are attempts to remove the need of speaker labels. For example, the Bayesian adaptation in [193] is extended by treating the *unknown* speaker labels in the in-domain data as latent variables [194]. In [195], the requirement of speaker-label is removed by assuming independence in the in-domain data. *Hypothesized* speaker labels via unsupervised clustering is also a popular approach [196, 197]. With the hypothesized speaker labels, the within speaker covariance matrices of in-domain data can be computed. They can be interpolated with the covariance matrices of the out-of-domain data to get the adapted covariance matrix from which the adapted PLDA model can be obtained. Yet another approach is to use the cosine distances among the in-domain data to regularize the discriminative training of PLDA models [198]. In [199], a network combining an autoencoder and a denosing autoencoder is trained to adapt data from a resource-rich domain to a resource-limited domain.

Instead of adapting the PLDA parameters, it is possible to transform the i-vectors to reduce the domain mismatch. For example, Aronowitz [200, 201] proposed a method called inter-dataset variability compensation (IDVC) to compensate for domain mismatch in the i-vector space. The idea is to remove the subspace corresponding to domain variability via nuisance attribute projection (NAP). Unlike the IDVC, Rahman et al.

[202] and Kanagasundaram et al. [203] attempted to map the i-vectors to a domain-invariant space through a transformation similar to the within-class covariance normalization. In [204], Glembek et al. *corrected* the within-speaker covariance matrix by adding a between-dataset covariance matrix to it. The correction will cause the LDA to find a direction with minimum dataset shift. Singer et al. [205] divided the out-of-domain data into a number of subgroups and computed a whitening transform for each group; during verification, the transform that gives the maximum likelihood of the test i-vector is selected for transforming the i-vectors. The advantage of the method is that in-domain data are not required for training. In [206], domain-adapted vectors derived from a convolutional neural network (CNN) is appended to i-vectors for domain normalization. The CNN is trained to minimize the domain-adapted triplet loss in which positive samples comprise in-domain data, whereas negative samples comprise out-of-domain data.

A more recent direction is to apply adversarial learning. For example, in [207, 208], an encoder network and a discriminator network are adversarially trained so that the encoder network can produce bottleneck features that will be considered as clean by the discriminator network even if the input to the encoder network is noisy. In [209], a generative adversarial network (GAN) is applied to generate long-duration i-vectors from short-duration i-vectors. The method is also applicable to domain adaptation if the training objective changes from duration mismatch compensation to domain mismatch compensation. The speaker-invariant training in [210] can also be applied to domain-invariant training by replacing the speaker loss with domain loss. In [211], a domain adversarial network is adversarially trained to transform i-vectors to a domain-invariant space. The results show that the transformed i-vectors are more robust to domain mismatch than those found by conventional domain adaptation methods.

In summary, the above methods either adapt PLDA parameters [192, 194, 196–198, 212], transform the i-vectors to a domain-invariant space [200–205, 211], or find the common properties of both source and target domains [199, 213].

6.2 Feature-Domain Adaptation/Compensation

This type of approaches attempt to (1) transform the i/x-vectors so that they become domain invariant, (2) adapt the i/x-vectors to fit the target domain better, and (3) incorporate the in-domain information into the i-vector extraction process.

6.2.1 Inter-Dataset Variability Compensation

Inter-dataset variability compensation (IDVC) was proposed in [200] to reduce the variability of i-vectors due to domain mismatches. The method uses the idea of nuisance attributed projection (see Section 3.2.4) to determine the subspace with the highest inter-dataset variability and remove the variability in the subspace. To achieve this goal, the i-vectors $\mathbf{x}$'s are transformed as follows:

$$\hat{\mathbf{x}} = (\mathbf{I} - \mathbf{W}\mathbf{W}^{\mathsf{T}})\mathbf{x}, \tag{6.1}$$

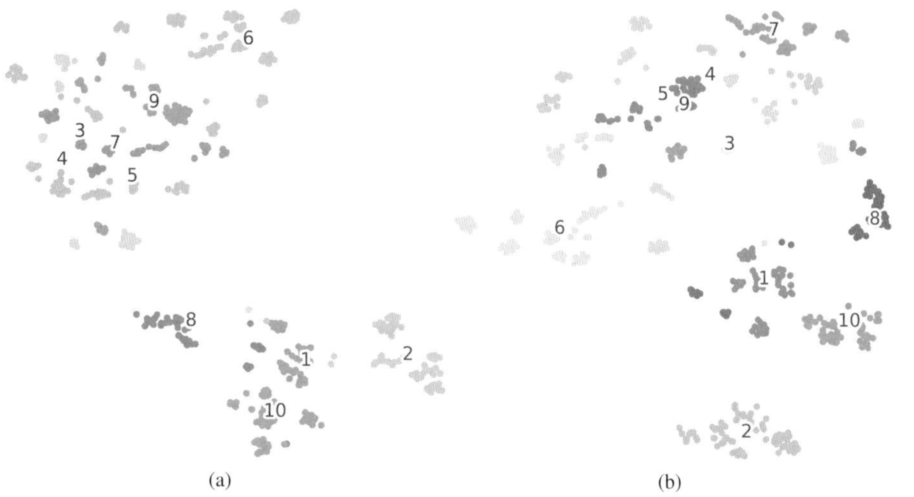

Figure 6.1 (a) X-vectors and (b) IDVC-transformed x-vectors of the development data of NIST 2018 SRE on a two-dimensional t-SNE space. Each speaker is represented by one color. Ideally, points (x-vectors) of the same speaker should be close to each other. *It may be better to view the color version of this figure, which is available at https://github.com/enmwmak/ML-for-Spkrec.*

where the columns of $\mathbf{W}$ represent the subspace of unwanted variability. The matrix $\mathbf{W}$ is composed of the eigenvectors of subset means' covariance matrix. Aronowitz [214] extended IDVC to the suppression of the variability in the between- and within-speaker covariance matrices in the PLDA model caused by domain mismatch. Recently, Aronowitz [201] extended IDVC to reduce the detrimental effect of domain mismatch on PLDA scores.

Figure 6.1 shows the x-vectors and IDVC-transformed x-vectors of SRE18-dev data on a two-dimensional t-SNE space. The figures show the x-vectors of 10 speakers before and after IDVC transformation. Evidently, after the transformation, the speakers are more distinguishable.

IDVC uses the covariances of datasets' means to represent the domain mismatch. In practical situations, however, the mismatch of datasets may not only reveal in the means of the datasets, but also in their higher-order statistics. Totally relying on the covariance of the dataset means is the major limitation of IDVC. We may appreciate this by assuming that each dataset is modeled by one Gaussian distribution and these Gaussian distributions have identical mean but very different covariance matrices. Although there are severe mismatches among these Gaussians, IDVC will treat them equally and will not remove any subspace ($\mathbf{W}$ in Eq. 6.1 is a null matrix) to reduce the mismatches.

6.2.2 Dataset-Invariant Covariance Normalization

A variant of IDVC is the dataset-invariant covariance normalization (DICN) [202]. In this approach, a global mean i-vector μ is first estimated by pooling the in-domain and out-domain i-vectors:

$$\mu = \frac{1}{N_{\text{in}} + N_{\text{out}}} \left[\sum_{\mathbf{x} \in \mathcal{X}_{\text{in}}} \mathbf{x} + \sum_{\mathbf{x} \in \mathcal{X}_{\text{out}}} \mathbf{x} \right], \tag{6.2}$$

where N_{in} and N_{out} are the numbers of in-domain and out-domain i-vectors, respectively. Then, domain variability is represented by the covariance matrix

$$\boldsymbol{\Sigma}_{\text{DICN}} = \frac{1}{N_{\text{in}} + N_{\text{out}}} \left[\sum_{\mathbf{x} \in \mathcal{X}_{\text{in}}} (\mathbf{x} - \mu)(\mathbf{x} - \mu)^{\mathsf{T}} + \sum_{\mathbf{x} \in \mathcal{X}_{\text{out}}} (\mathbf{x} - \mu)(\mathbf{x} - \mu)^{\mathsf{T}} \right] \tag{6.3}$$

The domain dependency of i-vectors can be reduced by performing covariance normalization as follows:

$$\mathbf{x} \leftarrow \mathbf{A}^{\mathsf{T}}\mathbf{x}, \qquad \mathbf{x} \in \mathcal{X}_{\text{out}}, \tag{6.4}$$

where the normalization matrix is the square root of $\boldsymbol{\Sigma}_{\text{DICN}}^{-1}$, i.e., $\mathbf{A}\mathbf{A}^{\mathsf{T}} = \boldsymbol{\Sigma}_{\text{DICN}}^{-1}$. The normalized $\mathbf{x}$'s with speaker labels can then be used for training a PLDA model.

The authors of DICN also proposed a variant of DICN by replacing the mean in Eq. 6.2 with the mean of in-domain i-vectors so that inter-dataset variability is represented by the covariance of out-of-domain i-vectors with respect to the in-domain mean [203]. Specifically, Eq. 6.3 is modified to

$$\boldsymbol{\Sigma}_{\text{IDV}} = \frac{1}{N_{\text{out}}} \sum_{\mathbf{x} \in \mathcal{X}_{\text{out}}} (\mathbf{x} - \mu_{\text{in}})(\mathbf{x} - \mu_{\text{in}})^{\mathsf{T}},$$

where the subscript "IDV" stands for inter-dataset variability and

$$\mu_{\text{in}} = \frac{1}{N_{\text{in}}} \sum_{\mathbf{x} \in \mathcal{X}_{\text{in}}} \mathbf{x}.$$

I-vectors are then transformed using Eq. 6.4 with $\mathbf{A}$ being the square root of $\boldsymbol{\Sigma}_{\text{IDV}}^{-1}$.

6.2.3 Within-Class Covariance Correction

In many i-vector/PLDA systems, i-vectors are transformed by linear discriminant analysis (LDA) or nonparametric discriminant analysis (NDA) before being scored by the PLDA back end (see Section 3.6.6). This preprocessing step aims to reduce the dimension of i-vectors and to enhance speaker discrimination. The use of LDA means that at a global perspective, the distribution of i-vectors are unimodal (there is only one global means), and at a local perspective, the i-vectors of each speaker form a cluster. After LDA projection, the clusters become further apart, which results in better speaker discrimination. The LDA projection works well when all training i-vectors are derived from the same domain (or same dataset). However, various studies [54, 55, 200, 215] have shown that the distribution of i-vectors are multimodal when they are derived from multiple domains (or multiple datasets), with each mode corresponds to a data subset. The existence of multiple modes can lead to incorrect LDA projection because the within-speaker scatter matrices in LDA cannot differentiate between-dataset variability from between-speaker variability. This may cause misinterpretation of the speaker and channel information in the i-vectors.

To address the above issues, Glembek et al. [204] proposed to adapt the within-speaker scatter matrix in LDA and called their method within-class covariance correction (WCC). In their method, i-vectors are labeled with datasets and speakers. Specifically, denote

$$\mathcal{X} = \{\mathbf{x}_{n,d,s}; d = 1,\ldots,D; s = 1,\ldots,S_d; n = 1,\ldots,N_{d,s}\}$$

as a collection of training i-vectors from D datasets (domains), where in each domain d, there are S_d speakers, each has $N_{d,s}$ utterances. The dataset variability is described by a between-dataset scatter matrix:[1]

$$\mathbf{\Sigma}_{\text{BD}} = \frac{1}{D}\sum_{d=1}^{D}(\boldsymbol{\mu}_d - \boldsymbol{\mu})(\boldsymbol{\mu}_d - \boldsymbol{\mu})^{\mathsf{T}},$$

where $\boldsymbol{\mu}_d$ and $\boldsymbol{\mu}$ are the mean of the dth dataset and the global mean, respectively, i.e.,

$$\boldsymbol{\mu}_d = \frac{1}{N_d}\sum_{s=1}^{S_d}\sum_{n=1}^{N_{d,s}}\mathbf{x}_{n,d,s} \quad \text{and} \quad \boldsymbol{\mu} = \frac{1}{N}\sum_{d=1}^{D}\sum_{s=1}^{S_d}\sum_{n=1}^{N_{d,s}}\mathbf{x}_{n,d,s},$$

where $N_d = \sum_{s=1}^{S_d} N_{d,s}$ is the number of i-vectors in the dth dataset and $N = \sum_{d=1}^{D}\sum_{s=1}^{S_d} N_{d,s}$ is the total number of training i-vectors.

The inter-session variability that describes the average speaker variability within a dataset is given by

$$\mathbf{\Sigma}_{\text{IS}} = \frac{1}{N}\sum_{d=1}^{D}\sum_{s=1}^{S_d}\sum_{n=1}^{N_{d,s}} N_{d,s}(\mathbf{x}_{n,d,s} - \boldsymbol{\mu}_{d,s})(\mathbf{x}_{n,d,s} - \boldsymbol{\mu}_{d,s})^{\mathsf{T}},$$

where

$$\boldsymbol{\mu}_{d,s} = \frac{1}{N_{d,s}}\sum_{n=1}^{N_{d,s}}\mathbf{x}_{n,d,s} \tag{6.5}$$

is the mean i-vector of speaker s in dataset d. Note that $\mathbf{\Sigma}_{\text{IS}}$ is shared across all datasets. In case speakers do not overlap across datasets, $\mathbf{\Sigma}_{\text{IS}}$ will also represent within-speaker variability. However, there might be case in which speakers appear in multiple datasets. In such scenario, the within-speaker variability is the sum of between-dataset and inter-session variabilities:

$$\mathbf{\Sigma}_{\text{WS}} = \mathbf{\Sigma}_{\text{BD}} + \mathbf{\Sigma}_{\text{IS}}. \tag{6.6}$$

The total variability is the sum of the between-speaker variability and within-speaker variability:

$$\begin{aligned}\mathbf{\Sigma}_{\text{Total}} &= \mathbf{\Sigma}_{\text{BS}} + \mathbf{\Sigma}_{\text{WS}}\\ &= \mathbf{\Sigma}_{\text{BS}} + \mathbf{\Sigma}_{\text{BD}} + \mathbf{\Sigma}_{\text{IS}},\end{aligned}$$

[1] To avoid clutter with notations, in the sequel, we denote $\boldsymbol{\mu}_d$ and $\boldsymbol{\mu}_s$ as the dataset means and speaker means, respectively.

where $\mathbf{\Sigma}_{\text{BS}}$ is the between-speaker scatter matrix. Because speakers may overlap across datasets, to compute $\mathbf{\Sigma}_{\text{BS}}$, we need to group the i-vectors into speaker subsets. Specifically, denote $\mathcal{X}_i$ as the set of i-vectors belonging to speaker i. Then, we have

$$\mathbf{\Sigma}_{\text{BS}} = \frac{1}{N}\sum_{i=1}^{S} |\mathcal{X}_i|(\boldsymbol{\mu}_i - \boldsymbol{\mu})(\boldsymbol{\mu}_i - \boldsymbol{\mu})^{\mathsf{T}},$$

where $|\mathcal{X}_i|$ is the number of i-vectors from speaker i, S is the number of speakers across all datasets and

$$\boldsymbol{\mu}_i = \frac{1}{|\mathcal{X}_i|}\sum_{\mathbf{x}\in\mathcal{X}_i} \mathbf{x}$$

is the mean i-vector of speaker i.

Note that if speakers do not overlap across datasets, $\mathbf{\Sigma}_{\text{WS}} \approx \mathbf{\Sigma}_{\text{IS}}$ and $\mathbf{\Sigma}_{\text{BD}}$ in Eq. 6.6 will diminish. This means that dataset variability can only be reflected in $\mathbf{\Sigma}_{\text{BS}}$. But unfortunately, the objective function of LDA aims to maximize the between-speaker variability. While it is desirable to maximize between-speaker variability, the maximization will also increase between-dataset variability, which contradicts to our goal. To minimize the between-dataset variability after LDA projection, the within-speaker scatter matrix can be *corrected* by the between-dataset variability as follows:

$$\mathbf{\Sigma}_{\text{WS}}^{\text{new}} = \mathbf{\Sigma}_{\text{WS}} + \alpha\mathbf{\Sigma}_{\text{BD}}^{\text{WCC}}, \tag{6.7}$$

where α is a scaling factor ($\alpha = 10$ in [204]) and

$$\mathbf{\Sigma}_{\text{BD}}^{\text{WCC}} = \frac{1}{D}\sum_{d=1}^{D}(\boldsymbol{\mu}_d - \hat{\boldsymbol{\mu}})(\boldsymbol{\mu}_d - \hat{\boldsymbol{\mu}})^{\mathsf{T}},$$

where $\hat{\boldsymbol{\mu}} = \frac{1}{D}\sum_{d=1}^{D}\boldsymbol{\mu}_d$. By using $\mathbf{\Sigma}_{\text{BS}}$ and $\mathbf{\Sigma}_{\text{WS}}^{\text{new}}$ in the objective function of LDA (Eq. 3.140), the resulting LDA projection matrix will minimize dataset variability in the LDA-projected space.

6.2.4 Source-Normalized LDA

The source-normalized LDA [215] is based on the notion that if training speakers do not overlap across multiple sources (datasets), the within-speaker scatter matrix will underestimate the truth speaker variability during verification. This is because during verification, the same speaker (target or nontarget) could provide speech from multiple sources (domains). Also, having disjoint speakers in different datasets will lead to over estimation of the between-speaker scatter.

To avoid inaccurate estimation of the scatter matrices, McLaren and Van Leeuwen [215] computed the between-speaker matrices of individual datasets separately. The overall between-speaker scatter is the weighted sum of the dataset-dependent between-speaker scatter. The within-speaker scatter is the difference between the total scatter and the between-speaker scatter.

Following the notations in Section 6.2.3, the between-speaker scatter matrix in [215] is given by

$$\boldsymbol{\Sigma}_{\text{BS}} = \sum_{d=1}^{D}\sum_{s=1}^{S_d} N_{d,s}(\boldsymbol{\mu}_{d,s} - \boldsymbol{\mu}_d)(\boldsymbol{\mu}_{d,s} - \boldsymbol{\mu}_d)^{\mathsf{T}}, \tag{6.8}$$

where $\boldsymbol{\mu}_{d,s}$ is computed as in Eq. 6.5. The total scatter matrix is given by

$$\boldsymbol{\Sigma}_{\text{T}} = \sum_{d=1}^{D}\sum_{s=1}^{S_d}\sum_{n=1}^{N_{d,s}} (\mathbf{x}_{n,d,s} - \boldsymbol{\mu})(\mathbf{x}_{n,d,s} - \boldsymbol{\mu})^{\mathsf{T}}.$$

Then, the within-speaker scatter matrix is

$$\boldsymbol{\Sigma}_{\text{WS}} = \boldsymbol{\Sigma}_{\text{T}} - \boldsymbol{\Sigma}_{\text{BS}}.$$

The LDA projection matrix is obtained from the eigenvectors of $\boldsymbol{\Sigma}_{\text{WS}}^{-1}\boldsymbol{\Sigma}_{\text{BS}}$.

6.2.5 Nonstandard Total-Factor Prior

Assume that we have a set of in-domain i-vectors $\mathcal{X}^{\text{in}} = \{\mathbf{x}_1^{\text{in}}, \ldots, \boldsymbol{x}_N^{\text{in}}\}$ and an i-vector extractor with total variability matrix $\mathbf{T}$ and super-covariance matrix (typically from the UBM) $\boldsymbol{\Sigma}^{(b)}$. Instead of using the standard prior $\mathcal{N}(\mathbf{0}, \mathbf{I})$ for the total factor $\mathbf{w}$ in Eq. 3.119, Rahman et al. [216] proposed using a nonstandard Gaussian prior [217] with mean and covariance matrix given by

$$\begin{aligned}\boldsymbol{\mu}_w &= \frac{1}{N}\sum_{j=1}^{N}\mathbf{x}_j^{\text{in}} \\ \boldsymbol{\Sigma}_w &= \frac{1}{N}\sum_{j=1}^{N}(\mathbf{x}_j^{\text{in}} - \boldsymbol{\mu}_w)(\mathbf{x}_j^{\text{in}} - \boldsymbol{\mu}_w)^{\mathsf{T}} + \frac{1}{N}\sum_{j=1}^{N}\left(\mathbf{I} + \mathbf{T}^{\mathsf{T}}(\boldsymbol{\Sigma}^{(b)})^{-1}\mathbf{N}_j^{\text{in}}\mathbf{T}\right)^{-1},\end{aligned} \tag{6.9}$$

where the diagonal of $\mathbf{N}_j^{\text{in}}$ contains the zeroth order sufficient statistics of the jth in-domain utterance. Therefore, we have $\mathbf{w} \sim \mathcal{N}(\boldsymbol{\mu}_w, \boldsymbol{\Sigma}_w)$. Replacing the term $\frac{1}{2}\mathbf{w}_i^{\mathsf{T}}\mathbf{w}_i$ in Eq. 3.122 by $\frac{1}{2}(\mathbf{w}_i - \boldsymbol{\mu}_w)^{\mathsf{T}}\boldsymbol{\Sigma}_w^{-1}(\mathbf{w}_i - \boldsymbol{\mu}_w)$ and performing the same derivations as in Eqs. 3.123–3.132, we obtain the new i-vectors as follows:

$$\mathbf{x}_i = \left(\boldsymbol{\Sigma}_w^{-1} + \mathbf{T}^{\mathsf{T}}(\boldsymbol{\Sigma}^{(b)})^{-1}\mathbf{N}_i^{\text{out}}\mathbf{T}\right)^{-1}\left(\mathbf{T}^{\mathsf{T}}(\boldsymbol{\Sigma}^{(b)})^{-1}\tilde{\mathbf{f}}_i^{\text{out}} + \boldsymbol{\Sigma}_w^{-1}\boldsymbol{\mu}_w\right). \tag{6.10}$$

These i-vectors incorporate the prior information of the in-domain i-vectors through $\boldsymbol{\mu}_w$ and $\boldsymbol{\Sigma}_w$ in Eq. 6.9.

To compensate for the domain mismatch, it is necessary to reduce the mismatch in the first-order statistics between the in-domain and out-of-domain i-vectors. To achieve this, the out-domain first-order statistics for utterance i are re-centered:

$$\begin{aligned}\hat{\mathbf{f}}_{ic}^{\text{out}} &= \sum_t \gamma(\ell_{i,t,c})\left(\mathbf{o}_{it}^{\text{out}} - \boldsymbol{\mu}_c^{(b)} - \mathbf{T}\boldsymbol{\mu}_w'\right) \\ &= \tilde{\mathbf{f}}_{ic}^{\text{out}} - \mathbf{N}_{ic}^{\text{out}}\mathbf{T}\boldsymbol{\mu}_w',\end{aligned}$$

where $\gamma(\ell_{i,t,c})$ is the posterior probability of the cth Gaussian given acoustic vector $\mathbf{o}_{it}$ at frame t, $\boldsymbol{\mu}_c^{(b)}$ is the cth UBM mean, and $\boldsymbol{\mu}'_w$ is the mean of the domain mismatch given by

$$\boldsymbol{\mu}'_w = \frac{1}{M}\sum_{i=1}^{M}\left(\mathbf{x}_i^{\text{out}} - \boldsymbol{\mu}_w\right),$$

where $\mathbf{x}_i^{\text{out}}$ is the ith out-domain i-vector and M is the number of out-domain i-vectors. Finally, the domain-mismatch compensated i-vectors are given by

$$\mathbf{x}^{\text{comp}} = \left(\boldsymbol{\Sigma}_w^{-1} + \mathbf{T}^{\mathsf{T}}(\boldsymbol{\Sigma}^{(b)})^{-1}\mathbf{N}^{\text{out}}\mathbf{T}\right)^{-1}\cdot \left(\mathbf{T}^{\mathsf{T}}(\boldsymbol{\Sigma}^{(b)})^{-1}(\tilde{\mathbf{f}}^{\text{out}} - \mathbf{N}^{\text{out}}\mathbf{T}\boldsymbol{\mu}'_w) + \boldsymbol{\Sigma}_w^{-1}\boldsymbol{\mu}_w\right).$$

6.2.6 Aligning Second-Order Statistics

The mismatch between the source and target domains is considered as domain shift. Alam et al. [218] proposed using an unsupervised domain adaptation method called CORelation ALignment (CORAL) [219] to compensate for the domain shift. The major advantage of CORAL is its simplicity and not requiring any speaker labels.

The goal is to align the second-order statistics of the in-domain (e.g., SRE16 data) and the out-domain (e.g., pre-SRE16 data). Denote the covariance matrices of the i-vectors/x-vectors in the in- and out-domains as $\boldsymbol{\Sigma}_{\text{in}}$ and $\boldsymbol{\Sigma}_{\text{out}}$, respectively. Then, CORAL aims to find a transformation matrix $\mathbf{A}$ to minimize the matrix distance:

$$\mathbf{A}^* = \underset{\mathbf{A}}{\text{argmin}}\, \|\mathbf{A}^{\mathsf{T}}\boldsymbol{\Sigma}_{\text{out}}\mathbf{A} - \boldsymbol{\Sigma}_{\text{in}}\|_F^2,$$

where $\|\cdot\|_F$ denotes the Frobenius norm. Sun et al. [219] shows that this minimization can be easily achieved by computing the covariance matrices of the out-of-domain and in-domain datasets, followed by whitening the out-of-domain data using the out-of-domain covariance matrix. Then, all data are recolored using the in-domain data covariance. Specifically, denote $\mathbf{X}_{\text{out}}$ as the row-wise data matrix of the out-domain data source. Data from the out-domain can be transformed by

$$\mathbf{X}_{\text{comp}} = \mathbf{X}_{\text{out}}\mathbf{C}_{\text{out}}^{-\frac{1}{2}}\mathbf{C}_{\text{in}}^{\frac{1}{2}},$$

where

$$\mathbf{C}_{\text{out}} = \boldsymbol{\Sigma}_{\text{out}} + \mathbf{I} \quad \text{and} \quad \mathbf{C}_{\text{in}} = \boldsymbol{\Sigma}_{\text{in}} + \mathbf{I}.$$

Then, $\mathbf{X}_{\text{comp}}$ can be used for training PLDA models and any out-domain test vectors can be transformed by

$$\mathbf{x}_{\text{comp}} = \mathbf{C}_{\text{in}}^{\frac{1}{2}}\mathbf{C}_{\text{out}}^{-\frac{1}{2}}\mathbf{x}_{\text{out}}.$$

Because of the simplicity and good performance of CORAL, it has been used by a number of sites in the NIST 2018 SRE.

6.2.7 Adaptation of I-Vector Extractor

The variability in i-vectors due to domain mismatch can be tackled by adapting an i-vector extractor to fit the in-domain data [220]. More precisely, after training the i-vector extractor using a large amount of out-of-domain data, a few iterations of minimum divergence training (Eq. 3.130) [72] are performed using the sufficient statistics obtained from unlabelled in-domain data.

6.2.8 Appending Auxiliary Features to I-Vectors

Instead of adapting or transforming the i-vectors, Bahmaninezhad and Hansen [206] proposed appending a speaker-independent auxiliary vector to each of the i-vectors. The auxiliary feature vectors do not contain any speaker information but contain the information of the in-domain environment. The concatenated vectors, called the a-vectors, contain both speaker information (from the i-vector part) and the in-domain information (from the auxiliary part), which results in better performance in the in-domain environment.

The auxiliary feature vectors were obtained from a simplified inception-v4 network [221] that takes i-vectors as the input and produces auxiliary feature vectors as the output. The inception-v4 network was trained to minimize the domain-adapted triplet loss. In the triplet set, positive samples are in-domain unlabeled vectors (unlabeled data in SRE16) and the negative samples are out-of-domain vectors selected from the NIST SRE of past years.

Denote $\mathbf{x}^a$, $\mathbf{x}^p$, and x^n as the anchor, positive, and negative i-vectors, respectively. Denote also the output of the simplified inception-v4 network as $f(\mathbf{x})$ (i.e., the auxiliary feature vector), where $\mathbf{x}$ is an input i-vector. The network aims to make $f(\mathbf{x})$ of the anchors close to the positive samples and far away from the negative samples in the auxiliary feature space. This is achieved by training the network to meet the following criterion:

$$\|f(\mathbf{x}_i^a) - f(\mathbf{x}_i^p)\|_2^2 + \alpha < \|f(\mathbf{x}_i^a) - f(\mathbf{x}_i^n)\|_2^2, \quad \forall\, (\mathbf{x}_i^a, \mathbf{x}_i^p, \mathbf{x}_i^n) \in \mathcal{T},$$

where $\mathcal{T}$ contains all possible triplets and α is a margin enforcing a relatively small distance between the anchor and the positive samples in the auxiliary space. To ensure the network meeting this criterion, we may set the loss function of the inception-v4 network to

$$\mathcal{L}_{\text{triplet}} = \sum_{i \in \mathcal{T}} \max(0, \Delta_i),$$

where

$$\Delta_i = \|f(\mathbf{x}_i^a) - f(\mathbf{x}_i^p)\|_2^2 - \|f(\mathbf{x}_i^a) - f(\mathbf{x}_i^n)\|_2^2 + \alpha.$$

The triplet loss leads to the minimization of the distance between the anchor and in-domain samples and the maximization of the distance between the anchor and out-domain samples. At the end-of-training, the network maps the i-vectors to the in-domain feature space. Because no speaker labels are used in the triplet loss (only domain labels are needed), the auxiliary vectors will not contain speaker information. When they are appended to the i-vectors, the concatenated vectors will have both speaker and in-domain information, which will perform better in the in-domain environment than using the i-vectors alone.

6.2.9 Nonlinear Transformation of I-Vectors

To reduce the effect of channel and domain variability, Bhattacharya et al. [222] trained a DNN to transform i-vectors to their speaker means and called the method nonlinear within-class normalization (NWCN). Specifically, given an i-vector $\mathbf{x}_i$, the NWCN-transformed vector is given by

$$\mathbf{x}_i^{\text{nwcn}} = \mathbf{W}_L f(\mathbf{W}_{L-1} f(\mathbf{W}_{L-2} f(\cdots f(\mathbf{W}_1 \mathbf{x}_i)))),$$

where f is a nonlinear function (sigmoid in [222]) and L is the number of layers in the DNN. For example, for a two-layer network, the transformed vector is $\mathbf{x}_i^{\text{nwcn}} = \mathbf{W}_2 f(\mathbf{W}_1 \boldsymbol{x}_i)$. Note that the network does not change the dimension of the i-vectors. To make the network transforming the i-vectors to their speaker mean, the angle loss is employed:

$$\mathcal{L}_{\text{nwcn}} = \sum_i \frac{\bar{\mathbf{x}}^{s_i} \cdot \mathbf{x}_i^{\text{nwcn}}}{\|\bar{\mathbf{x}}^{s_i}\| \|\mathbf{x}_i^{\text{nwcn}}\|},$$

where s_i is the speaker label of $\mathbf{x}_i$, $\bar{\boldsymbol{x}}^{s_i}$ is the mean i-vector of speaker s_i, and the loss is sum over a minibatch. Because each speaker mean is computed from multiple channels (or domains), the channel effect can be averaged out. As a result, minimizing $\mathcal{L}_{\text{nwcn}}$ will force the transformed vectors $\mathbf{x}^{\text{nwcn}}$ less channel-dependent and more speaker-dependent.

Domain information in i-vectors can be suppressed by transforming the i-vectors to a high-dimensional speaker-label space using a speaker classification network (SCN) [222]. Given an i-vector $\mathbf{x}_i$, the network out is

$$y(\mathbf{x}_i) = \text{softmax}(\mathbf{W}_L f(\mathbf{W}_{L-1} f(\mathbf{W}_{L-2} f(\cdots f(\mathbf{W}_1 \mathbf{x}_i))))),$$

where L is the number of layers. Because the network is trained to classify speaker, cross-entropy loss is used:

$$\mathcal{L}_{\text{scn}} = -\sum_i \sum_{c=1}^{C} t_{i,c} \log y(\mathbf{x}_i),$$

where C is the number of speakers in the training set, and

$$t_{i,c} = \begin{cases} 1 & \text{if } s_i = c \\ 0 & \text{otherwise.} \end{cases}$$

After training, SCN-transformed vectors can be extracted from the $(L-1)$th layer. Because the number of nodes in this layer is much larger than the dimension of the i-vectors, Bhattacharya et al. applied PCA to reduce the dimension of the SCN-transformed vectors to that of the i-vectors.

6.2.10 Domain-Dependent I-Vector Whitening

In i-vector/PLDA systems, applying length-normalization [66] on i-vectors is a simple but effective way to Gaussianize i-vectors so that they become amendable to Gaussian PLDA modeling. A necessary step before length-normalization is whitening:

$$\mathbf{x}^{\text{wht}} = \mathbf{W}^{\mathsf{T}}(\mathbf{x} - \bar{\mathbf{x}}), \tag{6.11}$$

where $\bar{\mathbf{x}}$ is for centering the i-vectors and $\mathbf{W}$ is a whitening matrix.

Ideally, $\mathbf{W}$ and $\bar{\mathbf{x}}$ in Eq. 6.11 are found from training data that match the distribution of the test data. However, in practice, domain mismatch between the training and test data causes error in the whitening process if $\mathbf{W}$ and $\bar{\mathbf{x}}$ are directly applied to whiten the test data. Also, the training data for estimating $\mathbf{W}$ and $\bar{\mathbf{x}}$ may come from multiple domains (e.g., telephone conversations, interview sessions, different languages), treating all domains by a single-mode distribution as in Eq. 6.11 may not be appropriate.

To address the multimodal issue, Singer and Reynolds [205] proposed using a library of whitening matrices and centering vectors to perform whitening. Specifically, given the training i-vectors of the dth domain, their mean $\boldsymbol{\mu}^{(d)}$ and covariance matrices $\boldsymbol{\Sigma}^{(d)}$ are estimated. Then, the i-vectors for individual domains are assumed to follow a Gaussian distribution $\mathcal{N}(\mathbf{x}|\boldsymbol{\mu}^{(d)}, \boldsymbol{\Sigma}^{(d)})$. During testing, given a test i-vector $\mathbf{x}_t$, the parameters of the Gaussian model that leads to the maximum likelihood are used for whiten the test i-vector, i.e.,

$$\mathbf{x}_t^{\text{wht}} = \left(\boldsymbol{\Sigma}^{(j)}\right)^{-1/2}\left(\mathbf{x}_t - \boldsymbol{\mu}^{(j)}\right),$$

where

$$j = \operatorname*{argmax}_{d\in\{1,\ldots,D\}} \mathcal{N}(\mathbf{x}_t|\boldsymbol{\mu}^{(d)}, \boldsymbol{\Sigma}^{(d)}).$$

By dividing the Switchboard data into six domains, Singer and Reynolds [205] show that this domain-dependent whitening approach can effectively reduce the domain mismatch between Switchboard data and NIST SRE data in the domain adaptation challenge (DAC13) [223]. Using a library of whiteners $\{\boldsymbol{\mu}^{(d)}, \boldsymbol{\Sigma}^{(d)}\}_{d=1}^{D}$ obtained from Switchboard, the performance is comparable to using the ideal whiteners obtained from NIST SRE data. The performance is also comparable to IDVC [200].

6.3 Adaptation of PLDA Models

The difference between the source domain and the target domain causes covariate shift [224, 225] in the i-vectors and x-vectors. Instead of compensating the shift in the feature space as in Section 6.2, we may adapt the PLDA model to fit the i-vectors or x-vectors in the target domain. Adaptation can be divided into supervised [192] and unsupervised [194].

For supervised adaptation, either the speaker labels in the in-domain data are known [192] or hypothesized by clustering [196, 197]. With the speaker labels in the in-domain data, the EM algorithm in Eq. 3.80 can be applied as usual to estimate the loading matrix $\mathbf{V}_{\text{in}}$ and within-speaker covariance matrix $\mathbf{\Sigma}_{\text{in}}$ from the in-domain data. Similarly, Eq. 3.80 is applied to out-of-domain data to estimate the loading matrix $\mathbf{V}_{\text{out}}$ and within-speaker covariance matrix $\mathbf{\Sigma}_{\text{out}}$. Define $\mathbf{\Gamma}_{\text{in}} = \mathbf{V}_{\text{in}}\mathbf{V}_{\text{in}}^{\mathsf{T}}$ and $\mathbf{\Gamma}_{\text{out}} = \mathbf{V}_{\text{out}}\mathbf{V}_{\text{out}}^{\mathsf{T}}$ as the in-domain and out-of-domain between-speaker covariance matrices, respectively. Then, the adapted between-speaker and within-speaker covariance matrices that fit the target domain can be obtained by interpolation:

$$\begin{aligned}\mathbf{\Gamma}_{\text{adt}} &= \alpha\mathbf{\Gamma}_{\text{in}} + (1-\alpha)\mathbf{\Gamma}_{\text{out}} \\ \mathbf{\Sigma}_{\text{adt}} &= \alpha\mathbf{\Sigma}_{\text{in}} + (1-\alpha)\mathbf{\Sigma}_{\text{out}},\end{aligned} \tag{6.12}$$

where $0 \le \alpha \le 1$ is an interpolation factor. The adapted covariance matrices can then be plugged into Eqs. 3.84 and 3.86 to compute the verification scores. Specifically, we replaced $\mathbf{\Sigma}_{tot}$ and $\mathbf{\Sigma}_{ac}$ in Eqs. 3.84 and 3.86 by

$$\begin{aligned}\mathbf{\Sigma}_{tot} &= \mathbf{\Gamma}_{\text{adt}} + \mathbf{\Sigma}_{\text{adt}} \\ \mathbf{\Sigma}_{ac} &= \mathbf{\Gamma}_{\text{adt}}.\end{aligned}$$

The PLDA interpolation in Eq. 6.12 is also applicable to senone i-vectors (using DNN to collect sufficient statistics for i-vector extraction) [197].

The PLDA interpolation in Eq. 6.12 will have difficulty in estimating the in-domain covariance matrices accurately when the number of samples per speaker is limited or when the channel diversity in the in-domain data is low, e.g., all utterances are obtained from the same microphone types. To address this issue, Borgström et al. [195] proposed using Bayesian adaptation of the PLDA parameters. They assume an additive noise model for the i-vectors $\mathbf{x}_{ij}$:

$$\mathbf{x}_{ij} = \mathbf{m} + \mathbf{y}_i + \boldsymbol{\epsilon}_{ij}, \tag{6.13}$$

where i and j index the speakers and sessions, respectively, $\mathbf{y}_i$ is the latent speaker component, and $\boldsymbol{\epsilon}_{ij}$ is the channel component. Note that the speaker component $\mathbf{y}_i$ in Eq. 6.13 is the term $\mathbf{V}\mathbf{z}_i$ in the simplified PLDA model in Eq. 3.69. The speaker and channel components follow Gaussian distributions:

$$\begin{aligned}\mathbf{y}_i &\sim \mathcal{N}(\mathbf{m}, \mathbf{\Sigma}_b) \\ \boldsymbol{\epsilon}_{ij} &\sim \mathcal{N}(\mathbf{0}, \mathbf{\Sigma}_w)\end{aligned}$$

where $\mathbf{\Sigma}_b$ and $\mathbf{\Sigma}_w$ are the between-speaker and within-speaker covariance matrices, respectively.

In [195], the joint posterior distribution of the model parameters $\mathbf{m}$, $\mathbf{\Sigma}_b$, and $\mathbf{\Sigma}_w$ and speaker components $\mathcal{Y} = \{\mathbf{y}_1, \ldots, \mathbf{y}_N\}$ is approximated by a factorized form using variational Bayes, i.e.,

$$p(\mathbf{m}, \mathbf{\Sigma}_b, \mathbf{\Sigma}_w, \mathcal{Y}|\mathcal{X}) \approx q(\mathcal{Y})q(\mathbf{m}, \mathbf{\Sigma}_b)q(\mathbf{\Sigma}_w).$$

To address the issue of low channel-diversity in the in-domain data $\mathcal{X}$, the authors assume that the samples in $\mathcal{X}$ are independent, i.e., each speaker contribute a single sample. Following the derivations in [193], given the between-speaker and within-speaker covariance matrices ($\boldsymbol{\Sigma}_b^{\text{out}}$ and $\boldsymbol{\Sigma}_w^{\text{out}}$) from the out-of-domain PLDA model and a set of in-domian i-vectors $\mathcal{X}$, the adapted between-speaker and within-speaker covariance matrices are

$$\boldsymbol{\Sigma}_b^{\text{adt}} = \alpha\left(\mathbf{H}_b\boldsymbol{\Sigma}_t\mathbf{H}_b^{\mathsf{T}} + (\boldsymbol{\Sigma}_b^{-1} + \boldsymbol{\Sigma}_w^{-1})^{-1}\right) + (1-\alpha)\boldsymbol{\Sigma}_b^{\text{out}}$$
$$\boldsymbol{\Sigma}_w^{\text{adt}} = \alpha\left(\mathbf{H}_w\boldsymbol{\Sigma}_t\mathbf{H}_w^{\mathsf{T}} + (\boldsymbol{\Sigma}_b^{-1} + \boldsymbol{\Sigma}_w^{-1})^{-1}\right) + (1-\alpha)\boldsymbol{\Sigma}_w^{\text{out}},$$

where $\boldsymbol{\Sigma}_t$ is the total covariance of $\mathcal{X}$, $\mathbf{H}_b = \boldsymbol{\Sigma}_b(\boldsymbol{\Sigma}_b + \boldsymbol{\Sigma}_w)^{-1}$ and $\mathbf{H}_w = \boldsymbol{\Sigma}_w(\boldsymbol{\Sigma}_b + \boldsymbol{\Sigma}_w)^{-1}$.

6.4 Maximum Mean Discrepancy Based DNN

Previous studies [226–229] have shown that it is important to minimize the divergence between the in-domain and out-domain distributions to obtain good representations for domain adaptation. Earlier approaches that only use the means vectors of multiple domains (e.g., IDVC) are not sufficient for finding good representations. This is because distributions with equal means could still have significant mismatch if their variances are very different. Therefore, to reduce the mismatch between datasets, it is important to go beyond the means.

To make good use of the statistics of the training data from different sources, maximum mean discrepancy (MMD) has been applied for domain adaptation [129]. MMD is a nonparametric approach to measuring the distance between two distributions [230–232]. With an appropriate kernel, MMD can measure the discrepancy up to any order of statistical moments in the two distributions. In [128], the MMD was generalized to measure the discrepancies among multiple domains. By incorporating MMD into the loss function of an autoencoder, we can make the autoencoder to find a latent space that contains minimal domain-specific information without losing task-specific information.

6.4.1 Maximum Mean Discrepancy

The idea of maximum mean discrepancy is to measure the distance between probability distributions in the reproducing kernel Hilbert space [230]. The method is a subset of the more general machine learning technique called kernel embedding of distributions [233]. Given two sets of samples $\mathcal{X} = \{\mathbf{x}_i\}_{i=1}^N$ and $\mathcal{Y} = \{\mathbf{y}_j\}_{j=1}^M$, we compute the mean squared difference of the statistics of the two datasets and call the difference MMD distance:

$$\mathcal{D}_{\text{MMD}} = \left\| \frac{1}{N}\sum_{i=1}^{N}\phi(\mathbf{x}_i) - \frac{1}{M}\sum_{j=1}^{M}\phi(\mathbf{y}_j)\right\|^2, \tag{6.14}$$

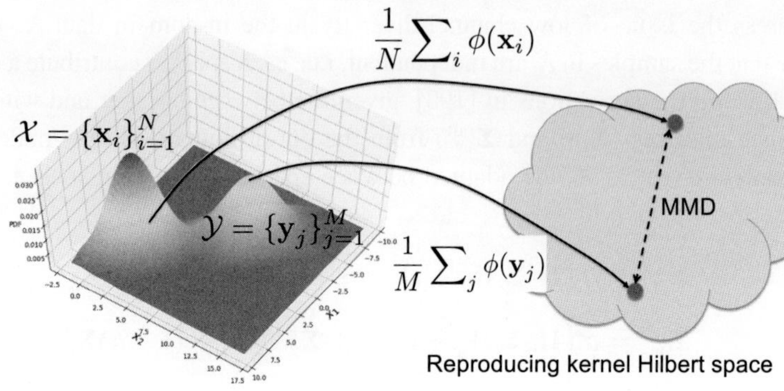

Figure 6.2 The concept of maximum mean discrepancy.

where ϕ is a feature map. Figure 6.2 illustrates the concept of MMD distance. When ϕ is the identity function, the MMD reduces to the distance between the sample means.

Eq. 6.14 can be expanded as:

$$\mathcal{D}_{\text{MMD}} = \frac{1}{N^2}\sum_{i=1}^{N}\sum_{i'=1}^{N}\phi(\mathbf{x}_i)^{\mathsf{T}}\phi(\mathbf{x}_{i'}) - \frac{2}{NM}\sum_{i=1}^{N}\sum_{j=1}^{M}\phi(\mathbf{x}_i)^{\mathsf{T}}\phi(\mathbf{y}_j)$$
$$+ \frac{1}{M^2}\sum_{j=1}^{M}\sum_{j'=1}^{M}\phi(\mathbf{y}_j)^{\mathsf{T}}\phi(\mathbf{y}_{j'}). \tag{6.15}$$

Because the terms in Eq. 6.15 involve dot products only, we may apply the kernel trick to Eq. 6.15 as follows:

$$\mathcal{D}_{\text{MMD}} = \frac{1}{N^2}\sum_{i=1}^{N}\sum_{i'=1}^{N}k(\mathbf{x}_i,\mathbf{x}_{i'}) - \frac{2}{NM}\sum_{i=1}^{N}\sum_{j=1}^{M}k(\mathbf{x}_i,\mathbf{y}_j) + \frac{1}{M^2}\sum_{j=1}^{M}\sum_{j'=1}^{M}k(\mathbf{y}_j,\mathbf{y}_{j'}), \tag{6.16}$$

where $k(\cdot,\cdot)$ is a kernel function. For quadratic kernels (Quad), we have:

$$k(\mathbf{x},\mathbf{y}) = \phi(\mathbf{x})^{\mathsf{T}}\phi(\mathbf{y}) = (\mathbf{x}^{\mathsf{T}}\mathbf{y} + c)^2. \tag{6.17}$$

To see the order of statistics that MMD can match, it is easier if we consider a simpler case where the feature dimension is 2 and the kernel is quadratic. Also, we consider one sample from $\mathcal{X}$ and one sample from $\mathcal{Y}$ and denote them as $\mathbf{x} = [x_1 \ \ x_2]^{\mathsf{T}}$ and $\mathbf{y} = [y_1 \ \ y_2]^{\mathsf{T}}$, respectively. Using Eq. 6.17, we have

$$\begin{aligned}\phi(\mathbf{x})^{\mathsf{T}}\phi(\mathbf{y}) &= \left([x_1 \ \ x_2]\begin{bmatrix} y_1 \\ y_2 \end{bmatrix} + c\right)^2 \\ &= (x_1y_1 + x_2y_2 + c)^2 \\ &= x_1^2y_1^2 + x_2^2y_2^2 + 2x_1y_1x_2y_2 + 2cx_1y_1 + 2cx_2y_2 + c^2. \end{aligned} \tag{6.18}$$

Eq. 6.18 suggests that

$$\phi(\mathbf{x}) = [x_1^2 \; x_2^2 \; \sqrt{2}x_1x_2 \; \sqrt{2c}x_1 \; \sqrt{2c}x_2 \; c]^\mathsf{T}$$
$$\phi(\mathbf{y}) = [y_1^2 \; y_2^2 \; \sqrt{2}y_1y_2 \; \sqrt{2c}y_1 \; \sqrt{2c}y_2 \; c]^\mathsf{T}.$$

Then, the MMD between $\mathbf{x}$ and $\mathbf{y}$ is

$$\begin{aligned} \|\phi(\mathbf{x}) - \phi(\mathbf{y})\|^2 &= \left\| \begin{matrix} x_1^2 - y_1^2 \\ x_2^2 - y_2^2 \\ \sqrt{2}(x_1x_2 - y_1y_2) \\ \sqrt{2c}(x_1 - y_1) \\ \sqrt{2c}(x_2 - y_2) \\ 0 \end{matrix} \right\|^2 \\ &= (x_1^2 - y_1^2)^2 + (x_2^2 - y_2^2)^2 + 2(x_1x_2 - y_1y_2)^2 + \\ &\quad + 2c(x_1 - y_1)^2 + 2c(x_2 - y_2)^2. \end{aligned} \tag{6.19}$$

The sum of the first three terms in Eq. 6.19 can be computed from the Frobenius norm:

$$\begin{aligned} &(x_1^2 - y_1^2)^2 + (x_2^2 - y_2^2)^2 + 2(x_1x_2 - y_1y_2)^2 \\ &= \left\| \begin{bmatrix} x_1^2 - y_1^2 & x_1x_2 - y_1y_2 \\ x_1x_2 - y_1y_2 & x_2^2 - y_2^2 \end{bmatrix} \right\|_F^2 \\ &= \left\| \begin{bmatrix} x_1^2 & x_1x_2 \\ x_1x_2 & x_2^2 \end{bmatrix} - \begin{bmatrix} y_1^2 & y_1y_2 \\ y_1y_2 & y_2^2 \end{bmatrix} \right\|_F^2 \\ &= \left\| \mathbf{x}\mathbf{x}^\mathsf{T} - \mathbf{y}\mathbf{y}^\mathsf{T} \right\|_F^2. \end{aligned} \tag{6.20}$$

The sum of the fourth and fifth terms in Eq. 6.19 is the square norm:

$$2c(x_1 - y_1)^2 + 2c(x_2 - y_2)^2 = 2c\|\mathbf{x} - \mathbf{y}\|^2. \tag{6.21}$$

Combining Eqs. 6.19–6.21, we have

$$\|\phi(\mathbf{x}) - \phi(\mathbf{y})\|^2 = 2c\|\mathbf{x} - \mathbf{y}\|^2 + \left\| \mathbf{x}\mathbf{x}^\mathsf{T} - \mathbf{y}\mathbf{y}^\mathsf{T} \right\|_F^2. \tag{6.22}$$

Now extending Eq. 6.19 to include the sums over all $\mathbf{x}_i$'s and $\mathbf{y}_j$'s in Eq. 6.14, the MMD becomes

$$\mathcal{D}_{\text{MMD}} = 2c \left\| \frac{1}{N}\sum_{i=1}^{N} \mathbf{x}_i - \frac{1}{M}\sum_{j=1}^{M} \mathbf{y}_j \right\|^2 + \left\| \frac{1}{N}\sum_{i=1}^{N} \mathbf{x}_i\mathbf{x}_i^\mathsf{T} - \frac{1}{M}\sum_{j=1}^{M} \mathbf{y}_j\mathbf{y}_j^\mathsf{T} \right\|_F^2, \tag{6.23}$$

where c controls the degree of match between the first- and second-order moments. The above analysis suggests that MMD with a quadratic kernel is able to match up to the second-order statistics.

The radial basis function (RBF) kernel

$$k(\mathbf{x}, \mathbf{y}) = \exp\left(-\frac{1}{2\sigma^2} \|\mathbf{x} - \mathbf{y}\|^2 \right), \tag{6.24}$$

is also commonly used in MMD. In this kernel, the hyperparameter σ is the kernel width that controls the influence of samples far away from the mean. The larger the width, the larger the influence. The RBF kernel induces a feature space of infinite dimensions. Minimizing the RBF-MMD of two distributions is equivalent to matching all of the moments of the two distributions [231].

Another possibility is to use a mixture of RBF kernels [232]:

$$k(\mathbf{x}, \mathbf{y}) = \sum_{q=1}^{K} \exp\left(-\frac{1}{2\sigma_q^2} \|\mathbf{x} - \mathbf{y}\|^2\right), \tag{6.25}$$

where σ_q is the width parameter of the qth RBF kernel.

6.4.2 Domain-Invariant Autoencoder

Now we apply the MMD in Section 6.4.1 to domain-invariant speaker verification and extend the discrepancy between two sources to multiple sources. Assume that we have in-domain data $\mathcal{X}^{\text{in}} = \{\mathbf{x}_i^{\text{in}}; i = 1, \ldots, N_{\text{in}}\}$ and out-domain data $\mathcal{X}^{\text{out}} = \{\mathbf{x}_j^{\text{out}}; j = 1, \ldots, N_{\text{out}}\}$. We want to learn a transform $\mathbf{h} = f(\mathbf{x})$ such that the transformed data $\{\mathbf{h}_i^{\text{in}}\}_{i=1}^{N_{\text{in}}}$ and $\{\mathbf{h}_j^{\text{out}}\}_{j=1}^{N_{\text{out}}}$ are as similar as possible. The mismatch between the transformed data is quantified by the MMD between $\mathcal{X}^{\text{in}}$ and $\mathcal{X}^{\text{out}}$:

$$\begin{aligned}\mathcal{D}_{\text{MMD}}(\mathcal{X}^{\text{in}}, \mathcal{X}^{\text{out}}) = &\frac{1}{N_{\text{in}}^2} \sum_{i=1}^{N_{\text{in}}} \sum_{i'=1}^{N_{\text{in}}} k(\mathbf{h}_i^{\text{in}}, \mathbf{h}_{i'}^{\text{in}}) - \frac{2}{N_{\text{in}} N_{\text{out}}} \sum_{i=1}^{N_{\text{in}}} \sum_{j=1}^{N_{\text{out}}} k(\mathbf{h}_i^{\text{in}}, \mathbf{h}_j^{\text{out}}) \\ &+ \frac{1}{N_{\text{out}}^2} \sum_{j=1}^{N_{\text{out}}} \sum_{j'=1}^{N_{\text{out}}} k(\mathbf{h}_j^{\text{out}}, \mathbf{h}_{j'}^{\text{out}}).\end{aligned} \tag{6.26}$$

For data obtained from multiple sources, we can apply MMD to create a feature space such that data in that space are invariant to the sources. To this end, we define a domain-wise MMD measure. Specifically, given D sets of data $\mathcal{X}^d = \{\mathbf{x}_i^d; i = 1, \ldots, N_d\}$, where $d = 1, \ldots, D$, we want the transformed data $\{\mathbf{h}_i^d\}_{i=1}^{N_d}$ to have small loss as defined by the following equation:

$$\begin{aligned}\mathcal{L}_{\text{mmd}} = \sum_{d=1}^{D} \sum_{\substack{d'=1 \\ d' \neq d}}^{D} \Bigg(&\frac{1}{N_d^2} \sum_{i=1}^{N_d} \sum_{i'=1}^{N_d} k(\mathbf{h}_i^d, \mathbf{h}_{i'}^d) - \frac{2}{N_d N_{d'}} \sum_{i=1}^{N_d} \sum_{j=1}^{N_{d'}} k(\mathbf{h}_i^d, \mathbf{h}_j^{d'}) \\ &+ \frac{1}{N_{d'}^2} \sum_{j=1}^{N_{d'}} \sum_{j'=1}^{N_{d'}} k(\mathbf{h}_j^{d'}, \mathbf{h}_{j'}^{d'}) \Bigg).\end{aligned} \tag{6.27}$$

We define another transform to reconstruct the input from $\mathbf{h}$:

$$\tilde{\mathbf{x}} = g(\mathbf{h}), \tag{6.28}$$

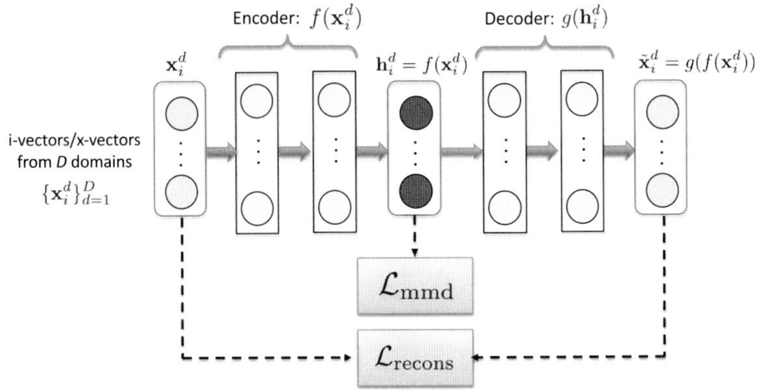

Figure 6.3 Structure of a DAE and the data sources from which the losses are minimized.

where $\tilde{\mathbf{x}}$ is the reconstruction of the input $\mathbf{x}$. It is necessary to keep non-domain related information in $\mathbf{h}$. To achieve this, we make $\tilde{\mathbf{x}}$ as close to $\mathbf{x}$ as possible by minimizing:

$$\mathcal{L}_{\text{recons}} = \frac{1}{2}\sum_{d=1}^{D}\sum_{i=1}^{N_d}\left\|\mathbf{x}_i^d - \tilde{\mathbf{x}}_i^d\right\|^2. \tag{6.29}$$

The objectives of small MMD and small reconstruction error can be achieved by an antoencoder using a total loss function:

$$\mathcal{L}_{\text{total}} = \mathcal{L}_{\text{mmd}} + \lambda\mathcal{L}_{\text{recons}}, \tag{6.30}$$

where λ is to control the contribution of the reconstruction loss. Note that both the encoder and decoder can be multilayer neural networks. Because the autoencoder aims to encode domain-invariant information, it was referred to as **d**omain-invariant **a**utoencoder (DAE) [128].[2]

Figure 6.3 shows the architecture of a DAE and the data sources from which the MMD loss and reconstruction loss are applied. Note that minimizing the MMD loss will only change the weights of the encoder. But minimizing the reconstruction loss will change the weights of both the encoder and the decoder. Figure 6.4 displays the hidden activations ($\mathbf{h}_i$ in Figure 6.3) of a DAE on a two-dimensional t-SNE space. Evidently, the domain-dependent clusters are less apparent in Figure 6.4(b) than in Figure 6.4(a). This suggestions that the DAE learns a domain-invariant representation.

6.4.3 Nuisance-Attribute Autoencoder

Rather than directly learning domain-invariant features, we may use the idea of IDVC to train an autoencoder to suppress the domain-specific information. The idea is inspired by rewriting Eq. 6.1 as

$$\hat{\mathbf{x}} = \mathbf{x} - \mathbf{W}\mathbf{W}^{\mathsf{T}}\mathbf{x}, \tag{6.31}$$

[2] Note that this is not the same as the denoising autoencoder proposed by Vincent et al [234].

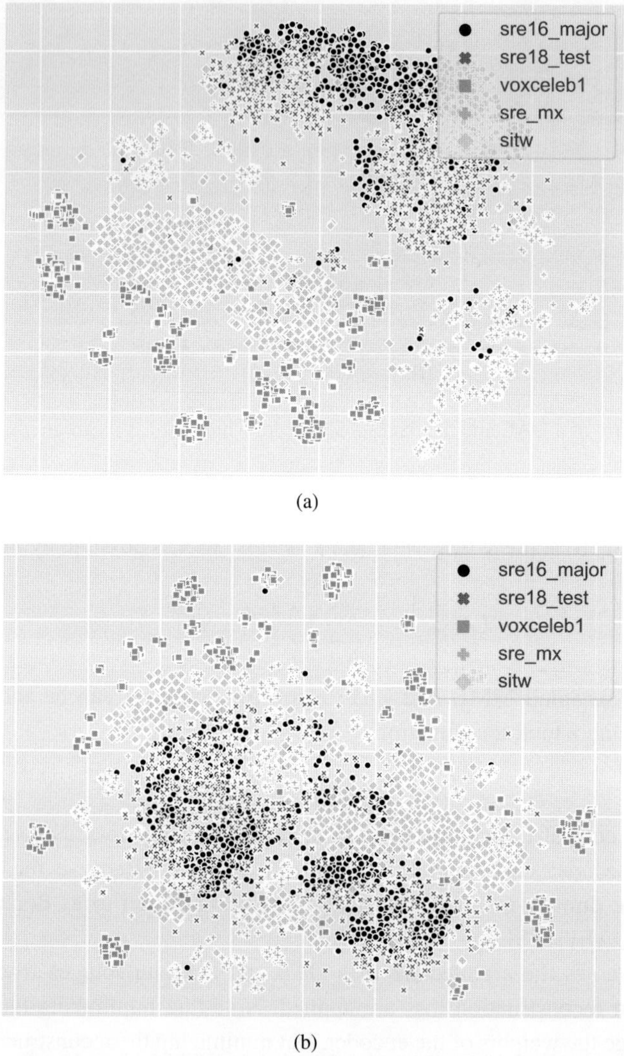

Figure 6.4 (a) X-vectors and (b) MMD-transformed x-vectors of different datasets. Each dataset is represented by one marker (color). In the legend, "sre_mx" means x-vectors from SRE04–10, Switchboard 2 Phase I–III, and augmented data (according to Kaldi's SRE16 recipe). *It may be better to view the color version of this figure, which is available at https://github.com/enmwmak/ML-for-Spkrec.*

where $\mathbf{W}\mathbf{W}^{\mathsf{T}}\mathbf{x}$ can be interpreted as the nuisance components. However, unlike IDVC, an autoencoder can be applied to estimate the nuisance components. Specifically, Eq. 6.31 can be replaced by:

$$\begin{aligned}\hat{\mathbf{x}} &= \mathbf{x} - \tilde{\mathbf{x}} \\ &= \mathbf{x} - g(f(\mathbf{x})),\end{aligned} \tag{6.32}$$

where $g(f(\mathbf{x}))$ is a special form of autoencoders. Similar to the DAE, the autoencoder comprises an encoder $\mathbf{h} = f(\mathbf{x})$ and a decoder $\tilde{\mathbf{x}} = g(\mathbf{h})$. However, unlike the DAE, the autoenconder is trained to make $\hat{\mathbf{x}}$ and $\mathbf{x}$ almost the same. According to Eq. 6.32, if $\hat{\mathbf{x}} \approx \mathbf{x}$, the autoencoder $g(f(\mathbf{x}))$ will be forced to represent the remaining loss, which corresponds to the domain mismatch. As a result, subtracting the nuisance component from the original i-vector in Eq. 6.32 will make $\hat{\mathbf{x}}$ less domain-dependent.

To attain the above goal, MMD can be used to measure the discrepancy between the distribution of $\hat{\mathbf{x}}$ across multiple datasets:

$$\mathcal{L}'_{\text{mmd}} = \sum_{d=1}^{D} \sum_{\substack{d'=1 \\ d' \neq d}}^{D} \left(\frac{1}{N_d^2} \sum_{i=1}^{N_d} \sum_{i'=1}^{N_d} k(\hat{\mathbf{x}}_i^d, \hat{\mathbf{x}}_{i'}^d) - \frac{2}{N_d N_{d'}} \sum_{i=1}^{N_d} \sum_{j=1}^{N_{d'}} k(\hat{\mathbf{x}}_i^d, \hat{\mathbf{x}}_j^{d'}) + \frac{1}{N_{d'}^2} \sum_{j=1}^{N_{d'}} \sum_{j'=1}^{N_{d'}} k(\hat{\mathbf{x}}_j^{d'}, \hat{\mathbf{x}}_{j'}^{d'}) \right). \tag{6.33}$$

We can also add the reconstruction loss to the MMD loss to obtain the total loss:

$$\mathcal{L}_{\text{total}} = \mathcal{L}'_{\text{mmd}} + \lambda \mathcal{L}_{\text{recons}}. \tag{6.34}$$

Beause this network encodes the unwanted domain-specific information, it is referred to as the **n**uisance-attribute **autoe**ncoder (NAE). Figure 6.5 shows the architecture of NAE.

NAE is able to make use of the speaker labels in the out-of-domain data via the triplet loss [235] or the center loss [236]. Center loss is inspired by the inability of the softmax loss (also known as cross-entropy loss) to take intra-class distances into account. By minimizing the distance between the class centers and data points for each class, center loss can help the network to learn more discriminative features. To incorporate the center loss into the objective function for training NAE, we consider the network outputs $\hat{\mathbf{x}}_i$'s in Eq. 6.32 as feature vectors in which training samples are drawn from a minibatch.

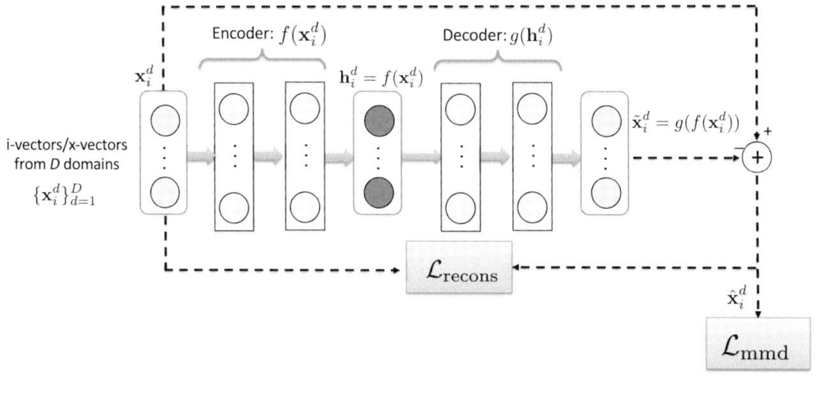

Figure 6.5 Structure of a nuisance-attribute autoencoder (NAE) and the data sources from which the losses are minimized.

Also denote $y_i \in \{1, \ldots, K\}$ as the class label for the ith training sample, where K is the number of classes. Then, center loss is defined as:

$$\mathcal{L}_{\text{center}} = \frac{1}{2} \sum_{i=1}^{B} \left\| \hat{\mathbf{x}}_i - \mathbf{c}_{y_i} \right\|^2, \tag{6.35}$$

where B is the number of samples in a minibatch and $\mathbf{c}_{y_i}$ is the center of the y_ith class. Note that $\hat{\mathbf{x}}_i$'s in the minibatch should be used to update $\mathbf{c}_{y_i}$.

Triplet loss aims to minimize intra-class distance and maximize the inter-class distance. Its operation relies on a set of triplets, where each triplet comprises an anchor $\hat{\mathbf{x}}_a$, a positive sample $\hat{\mathbf{x}}_p$, and a negative sample $\hat{\mathbf{x}}_n$. The positive sample has the same class as the anchor, whereas the negative sample belongs to a different class. Learning discriminative features amounts to maximizing inter-class distance while keeping intra-class distance small, which is achieved by the minimizing the triplet loss function:

$$\mathcal{L}_{\text{triplet}} = \max\left\{ \left\| \hat{\mathbf{x}}_a - \hat{\mathbf{x}}_p \right\|^2 - \left\| \hat{\mathbf{x}}_a - \hat{\mathbf{x}}_n \right\|^2 + m,\ 0 \right\}, \tag{6.36}$$

where m is a margin term.

We can incorporate center loss and triplet loss into the loss in Eq. 6.34 to obtain:

$$\mathcal{L}_{\text{total}} = \mathcal{L}_{\text{mismatch}} + \alpha\mathcal{L}_{\text{recons}} + \beta\mathcal{L}_{\text{center}} + \gamma\mathcal{L}_{\text{triplet}}, \tag{6.37}$$

where α, β, and γ are parameters controlling the importance of the respective losses. Eq. 6.37 enables the NAE to leverage both labeled and unlabeled data. For labeled data, all of the four terms will be involved in the minimization, whereas for unlabeled data, the center loss and triplet loss will be disabled by setting β and γ to 0. Therefore, we may make use of both unlabeled in-domain data and labeled out-domain data. Figure 6.6 shows the architecture of an NAE that leverages the center and triplet losses.

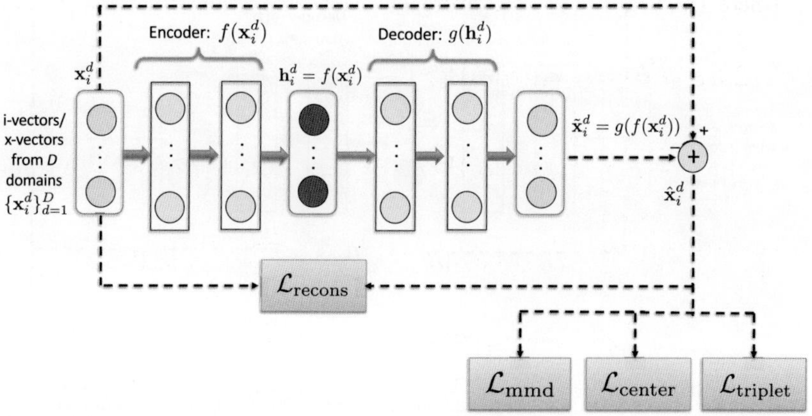

Figure 6.6 Structure of a nuisance-attribute autoencoder (NAE) with center loss and triplet loss.

6.5 Variational Autoencoders (VAE)

The PLDA and mixture of PLDA can be replaced by variational autoencoders [237] (see Section 4.5). Moreover, semi-supervised VAE [238] can be leveraged for domain adaptation, and the latent layer of an VAE can be used for feature extraction.

6.5.1 VAE Scoring

The variational autoencoder (VAE) in Section 4.5 considers the general case where the samples in the input space are independent. It also does not make use of the labels in the samples during training. In this section, we describe how the speaker labels can be leveraged during training to make the VAE suitable for scoring i-vectors or x-vectors.

Denote $\mathcal{X} = \{\mathbf{x}_{ij}; i = 1, \ldots, N; j = 1, \ldots, H_i\}$ as a set of length-normalized i-vectors, where i indexes speakers and j indexes the i-vectors of individual speakers. Define $\mathcal{X}_i = \{\mathbf{x}_{ij}; j = 1, \ldots, H_i\}$ as the set of i-vectors from speaker i. Because i-vectors in $\mathcal{X}_i$ are from the same speaker, they share the same latent variables $\mathbf{z}_i$. Due to the intractability of the true posterior $p_\theta(\mathbf{z}_i|\mathcal{X}_i)$, we approximate it by a variational posterior $q_\phi(\mathbf{z}_i|\mathcal{X}_i)$ parameterized by ϕ [237, 239]. Then, The marginal likelihood of $\mathcal{X}_i$ is

$$\begin{aligned}
\log p_\theta(\mathcal{X}_i) &= \int q_\phi(\mathbf{z}_i|\mathcal{X}_i) \log p_\theta(\mathcal{X}_i) d\mathbf{z}_i \\
&= \int q_\phi(\mathbf{z}_i|\mathcal{X}_i) \log \left[\frac{p_\theta(\mathbf{z}_i, \mathcal{X}_i)}{p_\theta(\mathbf{z}_i|\mathcal{X}_i)} \right] d\mathbf{z}_i \\
&= \int q_\phi(\mathbf{z}_i|\mathcal{X}_i) \log \left[\frac{p_\theta(\mathbf{z}_i, \mathcal{X}_i)}{q_\phi(\mathbf{z}_i|\mathcal{X}_i)} \cdot \frac{q_\phi(\mathbf{z}_i|\mathcal{X}_i)}{p_\theta(\mathbf{z}_i|\mathcal{X}_i)} \right] d\mathbf{z}_i \\
&= \int q_\phi(\mathbf{z}_i|\mathcal{X}_i) \log \left[\frac{p_\theta(\mathbf{z}_i, \mathcal{X}_i)}{q_\phi(\mathbf{z}_i|\mathcal{X}_i)} \right] d\mathbf{z}_i + \int q_\phi(\mathbf{z}_i|\mathcal{X}_i) \log \left[\frac{q_\phi(\mathbf{z}_i|\mathcal{X}_i)}{p_\theta(\mathbf{z}_i|\mathcal{X}_i)} \right] d\mathbf{z}_i \\
&= \mathcal{L}_1(\mathcal{X}_i, \phi, \theta) + \mathcal{D}_{\mathrm{KL}}(q_\phi(\mathbf{z}_i|\mathcal{X}_i) \| p_\theta(\mathbf{z}_i|\mathcal{X}_i)) \qquad (6.38) \\
&\geq \mathcal{L}_1(\mathcal{X}_i, \phi, \theta),
\end{aligned}$$

where $\mathcal{D}_{\mathrm{KL}}(q||p)$ is the KL-divergence between distributions q and p and

$$\begin{aligned}
\mathcal{L}_1(\mathcal{X}_i, \phi, \theta) &= \int q_\phi(\mathbf{z}_i|\mathcal{X}_i) \log \left[\frac{p_\theta(\mathcal{X}_i|\mathbf{z}_i) p_\theta(\mathbf{z}_i)}{q_\phi(\mathbf{z}_i|\mathcal{X}_i)} \right] d\mathbf{z}_i \\
&= \int q_\phi(\mathbf{z}_i|\mathcal{X}_i) \log \left[\frac{p_\theta(\mathbf{z}_i)}{q_\phi(\mathbf{z}_i|\mathcal{X}_i)} \right] d\mathbf{z}_i + \int q_\phi(\mathbf{z}_i|\mathcal{X}_i) \log p_\theta(\mathcal{X}_i|\mathbf{z}_i) d\mathbf{z}_i \\
&= -\mathcal{D}_{\mathrm{KL}}(q_\phi(\mathbf{z}_i|\mathcal{X}_i)||p_\theta(\mathbf{z}_i)) + \sum_{j=1}^{H_i} \mathbb{E}_{q_\phi(\mathbf{z}_i|\mathcal{X}_i)}\{\log p_\theta(\mathbf{x}_{ij}|\mathbf{z}_i)\} \\
&= -\mathcal{D}_{\mathrm{KL}}(q_\phi(\mathbf{z}_i|\mathcal{X}_i)||p_\theta(\mathbf{z}_i)) + \mathbb{E}_{q_\phi(\mathbf{z}_i|\mathcal{X}_i)}\{\log p_\theta(\mathcal{X}_i|\mathbf{z}_i)\} \qquad (6.39)
\end{aligned}$$

is the variational lower bound (VLB) of the marginal likelihood for speaker i. Assuming that speakers are independent, the total VBL can be used as a proxy for the total marginal likelihood, i.e., $\log p_\theta(\mathcal{X}) \geq \sum_{i=1}^{N} \mathcal{L}_1(\mathcal{X}_i, \phi, \theta)$. Note the similarity between Eq. 6.39

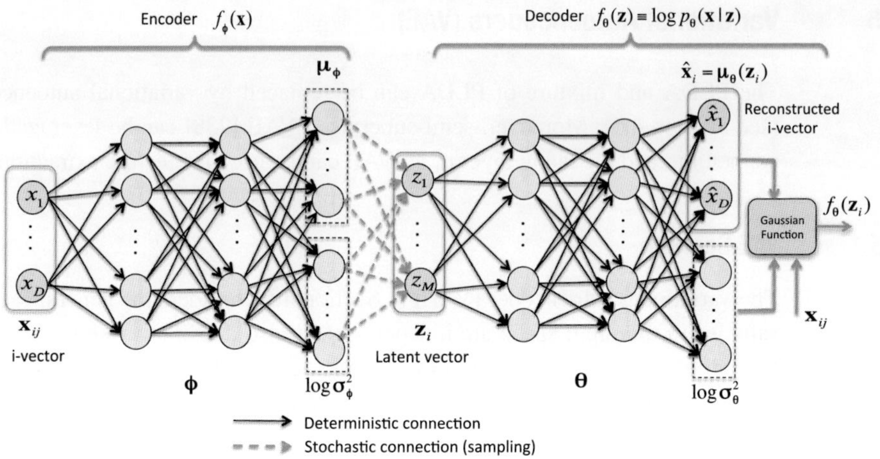

Figure 6.7 Architecture of the variational autoencoder for iVector–VAE scoring. The Gaussian function computes the log-likelihood of $\mathbf{x}$ using the reconstructed i-vector $\hat{\mathbf{x}}$ as the mean and $\sigma_\theta^2\mathbf{I}$ as the covariance matrix.

and the last line of Eq. 4.46. The only difference is that in Eq. 6.39, all of the samples from the same speaker are associated with one latent vector $\mathbf{z}$, whereas in Eq. 4.46, each sample $\mathbf{x}$ is associated with one latent vector $\mathbf{z}$. This means that during training, the lower bound in Eq. 6.39 is maximized in a speaker-wise manner.

We set the prior $p_\theta(\mathbf{z}_i) = \mathcal{N}(\mathbf{z}_i|\mathbf{0}, \mathbf{I})$ and posterior[3]

$$q_\phi(\mathbf{z}_i|\mathcal{X}_i) = \mathcal{N}(\mathbf{z}_i|\boldsymbol{\mu}_\phi(\mathcal{X}_i), \sigma_\phi^2(\mathcal{X}_i)\mathbf{I}),$$

where the mean $\boldsymbol{\mu}_\phi(\mathcal{X}_i)$ and variance $\sigma_\phi^2(\mathcal{X}_i)$ are determined by the hidden-layer outputs of a DNN, averaged over the H_i i-vectors in $\mathcal{X}_i$. Specifically, we have

$$\boldsymbol{\mu}_\phi(\mathcal{X}_i) = \frac{1}{H_i}\sum_{j=1}^{H_i} f_{\phi,\mu}(\mathbf{x}_{ij}) \tag{6.40a}$$

$$\log \sigma_\phi^2(\mathcal{X}_i) = \frac{1}{H_i}\sum_{j=1}^{H_i} f_{\phi,\sigma}(\mathbf{x}_{ij}), \tag{6.40b}$$

where $f_{\phi,\mu}$ and $f_{\phi,\sigma}$ correspond to the mean and variance nodes of the encoder parameterized by ϕ as shown in Figure 6.7. Having $q_\phi(\mathbf{z}_i|\mathcal{X}_i)$ and $p_\theta(\mathbf{z}_i)$ follow Gaussian distributions, the KL divergence can be evaluated analytically:

$$\begin{aligned}&\mathcal{D}_{\text{KL}}(q_\phi(\mathbf{z}_i|\mathcal{X}_i)||p_\theta(\mathbf{z}_i))\\ &= -\frac{1}{2}\sum_{m=1}^{M}\left[1 + \log(\sigma_{\phi,m}^2(\mathcal{X}_i)) - \mu_{\phi,m}^2(\mathcal{X}_i) - \sigma_{\phi,m}^2(\mathcal{X}_i)\right],\end{aligned} \tag{6.41}$$

where M is the dimension of $\mathbf{z}_i$.

[3] For clarity of presentation, we have slightly abused of notations here. To be more precise, the covariance matrix should be written as $\text{diag}\{\sigma_{\phi,1}^2, \ldots, \sigma_{\phi,D}^2\}$.

The expected negative reconstruction error (second term in Eq. 6.39) can be obtained by sampling the decoder outputs shown in Figure 6.7. More specifically, we have

$$\begin{aligned}\mathbb{E}_{q_\phi(\mathbf{z}_i|\mathcal{X}_i)}\{\log p_\theta(\mathcal{X}_i|\mathbf{z}_i)\} &= \mathbb{E}_{q_\phi(\mathbf{z}_i|\mathcal{X}_i)}\left\{\sum_{j=1}^{H_i}\log p_\theta(\mathbf{x}_{ij}|\mathbf{z}_i)\right\}\\ &\approx \frac{1}{L}\sum_{l=1}^{L}\sum_{j=1}^{H_i}\log p_\theta(\mathbf{x}_{ij}|\mathbf{z}_i^{(l)})\\ &= \frac{1}{L}\sum_{l=1}^{L}\sum_{j=1}^{H_i} f_\theta(\mathbf{z}_i^{(l)}),\end{aligned} \tag{6.42}$$

where L is the number of random samples and $f_\theta(\mathbf{z}) \equiv \log p_\theta(\mathbf{x}|\mathbf{z})$ denotes the output of the decoding network shown in Figure 6.7 using $\mathbf{z}$ as input. The expectation in Eq. 6.42 is not amenable to backpropagation because $\mathbf{z}_i^{(l)}$ is *stochastically* dependent on the model parameters ϕ through the sampling process. This problem can be solved by the re-parameterization trick [237, 240]:

$$\mathbf{z}_i^{(l)} = \boldsymbol{\mu}_\phi(\mathcal{X}_i) + \boldsymbol{\sigma}_\phi(\mathcal{X}_i) \odot \boldsymbol{\epsilon}^{(l)} \quad \text{with} \quad \boldsymbol{\epsilon}^{(l)} \sim \mathcal{N}(\mathbf{0}, \mathbf{I}), \tag{6.43}$$

where $\odot$ denotes element-wise products. This trick essentially moves the sampling process out of the hidden layer so that $\mathbf{z}_i$ becomes *deterministically* dependent on ϕ and $\mathcal{X}_i$. As a result, the error gradient can be written as

$$\nabla_\phi \mathbb{E}_{q_\phi(\mathbf{z}_i|\mathcal{X}_i)}\{\log p_\theta(\mathcal{X}_i|\mathbf{z}_i)\} = \mathbb{E}_{p(\epsilon)}\left\{\nabla_\phi f_\theta\left(\boldsymbol{\mu}_\phi(\mathcal{X}_i) + \boldsymbol{\sigma}_\phi(\mathcal{X}_i) \odot \boldsymbol{\epsilon}\right)\right\}. \tag{6.44}$$

During training, we group the i-vectors of the same speaker and make them to share the same $\boldsymbol{\mu}_\phi$ and $\boldsymbol{\sigma}_\phi^2$ while performing gradient ascent to maximize the sum of the VBL (Eq. 6.39) with respect to $\boldsymbol{\theta}$ and $\boldsymbol{\phi}$. This procedure effectively incorporates the speaker labels in the training process, which is in analogy to summing over the i-vectors of the same speaker during the computation of posterior means of speaker factors in PLDA training (see Eq. 3.78).

During scoring, given the target-speaker's i-vector $\mathbf{x}_s$ and test-speaker's i-vector $\mathbf{x}_t$, we use the VBLs as proxies to compute the likelihood ratio score:

$$\begin{aligned}S_{\text{LR}}(\mathbf{x}_s, \mathbf{x}_t) &= \log\left[\frac{p(\mathbf{x}_s, \mathbf{x}_t|\text{Same-spk})}{p(\mathbf{x}_s, \mathbf{x}_t|\text{Diff-spk})}\right]\\ &\approx \mathcal{L}_1(\mathbf{x}_s, \mathbf{x}_t, \boldsymbol{\theta}, \boldsymbol{\phi}) - \mathcal{L}_1(\mathbf{x}_s, \boldsymbol{\theta}, \boldsymbol{\phi}) - \mathcal{L}_1(\mathbf{x}_t, \boldsymbol{\theta}, \boldsymbol{\phi})\end{aligned} \tag{6.45}$$

where the same-speaker likelihood is

$$\mathcal{L}_1(\mathbf{x}_s, \mathbf{x}_t, \boldsymbol{\theta}, \boldsymbol{\phi}) = -\mathcal{D}_{\text{KL}}(q_\phi(\mathbf{z}|\mathbf{x}_s, \mathbf{x}_t)||p_\theta(\mathbf{z})) + \sum_{j\in\{s,t\}} \mathbb{E}_{q_\phi(\mathbf{z}|\mathbf{x}_j)} \log p_\theta(\mathbf{x}_j|\mathbf{z}). \tag{6.46}$$

The first term in Eq. 6.46 can be computed by replacing $\mathcal{X}_i$ in Eqs. 6.40 and 6.41 by $\{\mathbf{x}_s, \mathbf{x}_t\}$. The second term in Eq. 6.46 can be computed by replacing $\boldsymbol{x}_{ij}$ in Eq. 6.42 by $\mathbf{x}_s$ and $\boldsymbol{x}_t$, where $\mathbf{z}$ is stochastically generated by the DNN using $\mathbf{x}_s$ and $\mathbf{x}_t$ as inputs. The VB lower bounds $\mathcal{L}_1(\mathbf{x}_s, \boldsymbol{\theta}, \boldsymbol{\phi})$ and $\mathcal{L}_1(\boldsymbol{x}_t, \boldsymbol{\theta}, \boldsymbol{\phi})$ are computed similarly by replacing $\mathcal{X}_i$ with either $\mathbf{x}_s$ or $\mathbf{x}_t$.

6.5.2 Semi-Supervised VAE for Domain Adaptation

The VAE in Section 6.5.1 requires speaker labels for training. In many practical situations, we may need to adapt a well trained VAE to an adverse environment in which no speaker labels are available: a situation known as domain adaptation [192, 200, 202]. For example, in NIST 2016 SRE, participants are given a small subset of the CallMyNet collection for development. As the acoustic characteristics of CallMyNet are very different from previous years' SREs and Switchboard corpora, it is necessary to perform domain adaptation. Unfortunately, no speaker labels are available for the adaptation. Here, we extend the VAE in Section 6.5.1 and the semi-supervised VAE in [238] to overcome this difficulty.

Assume that there are K speakers in the CallMyNet subset and that the subset comprises N unlabeled conversations whose i-vectors are denoted as $\mathcal{X} = \{\mathbf{x}_n; n = 1,\ldots,N\}$. Also, denote $\mathcal{Y} = \{\mathbf{y}_n; n = 1,\ldots,N\}$ as a set of K-dimensional vectors in 1-of-K coding format indicating the *hypothesized* speaker IDs of the conversation. More precisely, $y_{n,k} = 1$ and $y_{n,r \neq k} = 0$ if the nth conversation is spoken by speaker k. Because we actually do not know the speaker labels, $\mathbf{y}_n$'s are treated as latent variables in the VAE [238], which leads to the following marginal likelihood:

$$
\begin{aligned}
\log p_\theta(\mathbf{x}_n) &= \sum_{\mathbf{y}_n} \int q_\phi(\mathbf{z}_n, \mathbf{y}_n|\mathbf{x}_n) \log p_\theta(\mathbf{x}_n) d\mathbf{z}_n \\
&= \sum_{\mathbf{y}_n} q_\phi(\mathbf{y}_n|\mathbf{x}_n) \int q_\phi(\mathbf{z}_n|\mathbf{x}_n, \mathbf{y}_n) \log \left[\frac{p_\theta(\mathbf{z}_n, \mathbf{x}_n, \mathbf{y}_n)}{p_\theta(\mathbf{z}_n, \mathbf{y}_n|\mathbf{x}_n)} \right] d\mathbf{z}_n \\
&= \sum_{\mathbf{y}_n} q_\phi(\mathbf{y}_n|\mathbf{x}_n) \int q_\phi(\mathbf{z}_n|\mathbf{x}_n, \mathbf{y}_n) \log \left[\frac{p_\theta(\mathbf{z}_n, \mathbf{x}_n, \mathbf{y}_n)}{q_\phi(\mathbf{z}_n, \mathbf{y}_n|\mathbf{x}_n)} \cdot \frac{q_\phi(\mathbf{z}_n, \mathbf{y}_n|\mathbf{x}_n)}{p_\theta(\mathbf{z}_n, \mathbf{y}_n|\mathbf{x}_n)} \right] d\mathbf{z}_n \\
&= \mathcal{L}_2(\mathbf{x}_n, \phi, \theta) + \mathcal{D}_{\text{KL}}(q_\phi(\mathbf{z}_n, \mathbf{y}_n|\mathbf{x}_n) \| p_\theta(\mathbf{z}_n, \mathbf{y}_n|\mathbf{x}_n)) \geq \mathcal{L}_2(\mathbf{x}_n, \phi, \theta),
\end{aligned} \tag{6.47}
$$

where the VB lower bound is

$$
\begin{aligned}
\mathcal{L}_2(\mathbf{x}_n, \phi, \theta) &= \sum_{\mathbf{y}_n} q_\phi(\mathbf{y}_n|\mathbf{x}_n) \int q_\phi(\mathbf{z}_n|\mathbf{x}_n, \mathbf{y}_n) \log \left[\frac{p_\theta(\mathbf{z}_n, \mathbf{x}_n, \mathbf{y}_n)}{q_\phi(\mathbf{z}_n, \mathbf{y}_n|\mathbf{x}_n)} \right] d\mathbf{z}_n \\
&= \sum_{\mathbf{y}_n} q_\phi(\mathbf{y}_n|\mathbf{x}_n) \int q_\phi(\mathbf{z}_n|\mathbf{x}_n, \mathbf{y}_n) \log \left[\frac{p_\theta(\mathbf{z}_n, \mathbf{y}_n)}{q_\phi(\mathbf{z}_n, \mathbf{y}_n|\mathbf{x}_n)} \right] d\mathbf{z}_n \\
&\quad + \sum_{\mathbf{y}_n} q_\phi(\mathbf{y}_n|\mathbf{x}_n) \int q_\phi(\mathbf{z}_n|\mathbf{x}_n, \mathbf{y}_n) \log p_\theta(\mathbf{x}_n|\mathbf{y}_n, \mathbf{z}_n) d\mathbf{z}_n \\
&= -\mathcal{D}_{\text{KL}}\left(q_\phi(\mathbf{z}_n, \mathbf{y}_n|\mathbf{x}_n) \| p_\theta(\mathbf{z}_n, \mathbf{y}_n)\right) + \mathbb{E}_{q_\phi(\mathbf{z}_n, \mathbf{y}_n|\mathbf{x}_n)}\{\log p_\theta(\mathbf{x}_n|\mathbf{y}_n, \mathbf{z}_n)\}.
\end{aligned} \tag{6.48}
$$

We assume that the joint posterior and joint prior have factorized forms:

$$
q_\phi(\mathbf{z}_n, \mathbf{y}_n|\mathbf{x}_n) = q_\phi(\mathbf{z}_n|\mathbf{x}_n) q_\phi(\mathbf{y}_n|\mathbf{x}_n)
$$

and

$$
p_\theta(\mathbf{z}_n, \mathbf{y}_n) = p_\theta(\mathbf{z}_n) p_\theta(\mathbf{y}_n).
$$

Then, the KL divergence in Eq. 6.48 becomes

$$\begin{aligned}&\mathcal{D}_{\text{KL}}\left(q_\phi(\mathbf{z}_n,\mathbf{y}_n|\mathbf{x}_n)\|p_\theta(\mathbf{z}_n,\mathbf{y}_n)\right)\\&\quad=\mathcal{D}_{\text{KL}}\left(q_\phi(\mathbf{z}_n|\mathbf{x}_n)\|p_\theta(\mathbf{z}_n)\right)+\mathcal{D}_{\text{KL}}\left(q_\phi(\mathbf{y}_n|\mathbf{x}_n)\|p_\theta(\mathbf{y}_n)\right),\end{aligned}\tag{6.49}$$

where the first term can be evaluated analytically as in Eq. 6.41. The second term in Eq. 6.49 is the KL-divergence between two categorical distributions:

$$\begin{aligned}\mathcal{D}_{\text{KL}}\left(q_\phi(\mathbf{y}_n|\mathbf{x}_n)\|p_\theta(\mathbf{y}_n)\right)&=\mathcal{D}_{\text{KL}}\left(\text{Cat}(\mathbf{y}_n|\pi_\phi(\mathbf{x}_n))\|\text{Cat}\left(\mathbf{y}_n|[1/K,\ldots,1/K]^{\mathsf{T}}\right)\right)\\&=\log K+\sum_{k=1}^{K}\left[\pi_\phi(\mathbf{x}_n)\right]_k\log\left[\pi_\phi(\mathbf{x}_n)\right]_k,\end{aligned}\tag{6.50}$$

where the probability vector $\pi_\phi(\mathbf{x}_n)$ can be obtained from the upper part of the encoding network in Figure 6.8 and $\left[\pi_\phi(\mathbf{x}_n)\right]_k$ denotes its kth element. Note that the softmax layer in Figure 6.8 is to ensure that $\pi_\phi(\mathbf{x}_n)$ is a probability vector such that $\sum_{k=1}^{K}\left[\pi_\phi(\mathbf{x}_n)\right]_k=1$.

The second term in Eq. 6.48 can be obtained from the decoding network $f_\theta(\mathbf{z},\mathbf{y})$ in Figure 6.8 through sampling:

$$\mathbb{E}_{q_\phi(\mathbf{z}_n,\mathbf{y}_n|\mathbf{x}_n)}\{\log p_\theta(\mathbf{x}_n|\mathbf{z}_n,\mathbf{y}_n)\}=\frac{1}{L}\sum_{l=1}^{L}f_\theta(\mathbf{z}_n^{(l)},\mathbf{y}_n^{(l)}).\tag{6.51}$$

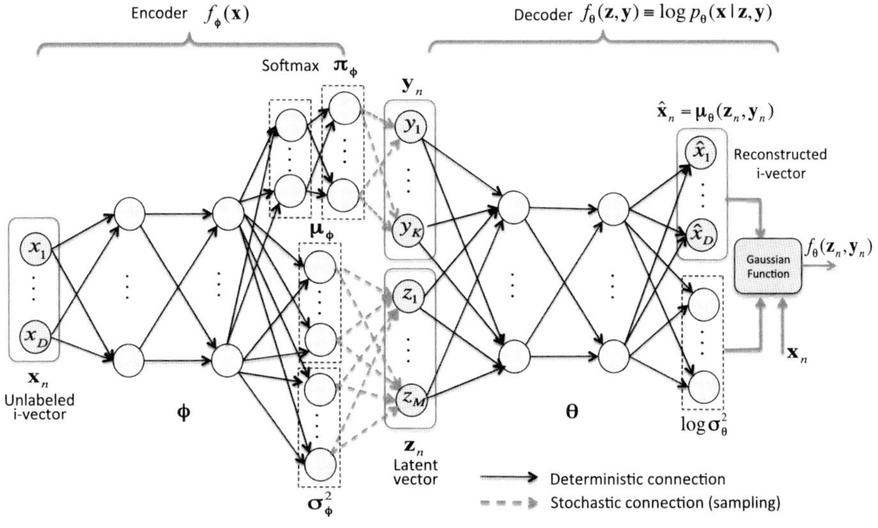

Figure 6.8 Architecture of the semi-supervised variational autoencoder for domain adaptation. Weights connected to μ_ϕ and σ_ϕ^2 are initialized from the VAE in Figure 6.7. Then, unsupervised adaptation of the whole network is performed using unlabeled i-vectors $\mathbf{x}_n$'s. Refer to the caption of Figure 6.7 for the operation of the Gaussian function.

Eq. 6.43 can be applied to reparameterize the Gaussian that generates $\mathbf{z}_n$, i.e., $\mathbf{z}_n^{(l)} = \boldsymbol{\mu}_\phi(\mathbf{x}_n) + \boldsymbol{\sigma}_\phi(\mathbf{x}_n) \odot \boldsymbol{\epsilon}_1^{(l)}$. Note that re-parameterization for $\text{Cat}(\mathbf{y}_n|\boldsymbol{\pi}_\phi(\mathbf{x}_n))$ is not necessary because $\mathbf{y}_n$ is deterministically dependent on $\boldsymbol{\phi}$. This can be observed through the sampling process:

$$c_n = \sum_{k=1}^{K} u\left[\epsilon_2 - F_k(\boldsymbol{\pi}_\phi(\mathbf{x}_n))\right] + 1 \quad \epsilon_2 \sim \mathcal{U}(0,1) \quad c_n \in \{1,\ldots,K\}, \tag{6.52}$$

where $u[\cdot]$ is the unit step function and $F_k(\boldsymbol{\pi}_\phi(\mathbf{x}_n)) = \sum_{j=1}^{k}[\boldsymbol{\pi}_\phi(\mathbf{x}_n)]_j$ is the cumulative distribution function. Given the random sample c_n, we set $y_{n,c} = 1$ and $y_{n,k} = 0$ $\forall k \neq c$. During backpropagation training, we need to ensure that $y_{n,k}$ is differentiable with respect to c_n. As both $y_{n,k}$ and c_n are discrete, we need to approximate the step function $u[\cdot]$ in Eq. 6.52 by a sigmoid function and $y_{n,k}$ by a softmax function:

$$c_n = \sum_{k=1}^{K} \frac{1}{1 + e^{-\alpha(\epsilon_2 - F_k(\boldsymbol{\pi}_\phi(\mathbf{x}_n)))}} + 1 \tag{6.53a}$$

$$y_{n,k} = \frac{e^{-\beta(c_n - k)^2}}{\sum_{j=1}^{K} e^{-\beta(c_n - j)^2}}, \quad \alpha, \beta > 0. \tag{6.53b}$$

The idea is similar to replacing the unit step function by the sigmoid function in [241].

Scoring in semi-supervised VAE is almost identical to that of VAE; the only difference is that we replace $\mathcal{L}_1$ in Eq. 6.45 by $\mathcal{L}_2$ in Eq. 6.48.

6.5.3 Variational Representation of Utterances

By adding class labels to Figure 6.7 and Eq. 6.38, we can apply VAE for i-vector extraction. Given an utterance with acoustic vectors $O = \{\boldsymbol{o}_t; t = 1,\ldots,T\}$ and a phonetically aware DNN [84, 148], we estimate the the posterior probabilities of the C senones for each frame by presenting its contextual frames to the DNN as shown in Figure 6.9. Denote $\mathcal{M} = \{m_{t,c}; t = 1,\ldots,T; c = 1,\ldots,C\}$ as the set of posterior probabilities. Also, denote $\mathbf{m}_t$ as a C-dimensional vector comprising the posteriors of C senones given $\mathbf{o}_t$. Then, Eq. 6.38 can be extended to

$$\begin{aligned}
\log p_\theta(O,\mathcal{M}) &= \int q_\phi(\mathbf{v}|O,\mathcal{M}) \log p_\theta(O,\mathcal{M}) d\mathbf{v} \\
&= \int q_\phi(\mathbf{v}|O,\mathcal{M}) \log \left[\frac{p_\theta(\mathbf{v},O,\mathcal{M})}{q_\phi(\mathbf{v}|O,\mathcal{M})}\right] d\mathbf{v} \\
&\quad + \int q_\phi(\mathbf{v}|O,\mathcal{M}) \log \left[\frac{q_\phi(\mathbf{v}|O,\mathcal{M})}{p_\theta(\mathbf{v}|O,\mathcal{M})}\right] d\mathbf{v} \\
&= \mathcal{L}_3(O,\mathcal{M},\phi,\theta) + \mathcal{D}_{\text{KL}}(q_\phi(\mathbf{v}|O,\mathcal{M}) \| p_\theta(\mathbf{v}|O,\mathcal{M})),
\end{aligned} \tag{6.54}$$

where $\mathbf{v}$ is the latent vector of a DNN shown in Figure 6.9.

Similar to the VAE in Section 6.5.1, we estimate the variational parameters ϕ and θ by maximizing the lower bound

$$\mathcal{L}_3(O, \mathcal{M}, \phi, \theta) = \int q_\phi(\mathbf{v}|O, \mathcal{M}) \log \left[\frac{p_\theta(O, \mathcal{M}|\mathbf{v}) p_\theta(\mathbf{v})}{q_\phi(\mathbf{v}|O, \mathcal{M})} \right] d\mathbf{v}$$
$$= \int q_\phi(\mathbf{v}|O, \mathcal{M}) \log \left[\frac{p_\theta(\mathbf{v})}{q_\phi(\mathbf{v}|O, \mathcal{M})} \right] d\mathbf{v} + \int q_\phi(\mathbf{v}|O, \mathcal{M}) \log p_\theta(O, \mathcal{M}|\mathbf{v}) d\mathbf{v}$$
$$= -\mathcal{D}_{\text{KL}}(q_\phi(\mathbf{v}|O, \mathcal{M})||p_\theta(\mathbf{v})) + \sum_{t=1}^{T} \mathbb{E}_{q_\phi(\mathbf{v}|O, \mathcal{M})}\{\log p_\theta(\mathbf{o}_t, \mathbf{m}_t|\mathbf{v})\}$$
$$= -\mathcal{D}_{\text{KL}}(q_\phi(\mathbf{v}|O, \mathcal{M})||p_\theta(\mathbf{v})) + \sum_{t=1}^{T} \mathbb{E}_{q_\phi(\mathbf{v}|O, \mathcal{M})}\{\log p_\theta(\mathbf{o}_t|\mathbf{m}_t, \mathbf{v}) + \log p(\mathbf{m}_t)\}. \tag{6.55}$$

Define the prior $p_\theta(\mathbf{v}) = \mathcal{N}(\mathbf{v}|\mathbf{0}, \mathbf{I})$ and posterior

$$q_\phi(\mathbf{v}|O, \mathcal{M}) = \mathcal{N}(\mathbf{v}|\boldsymbol{\mu}_\phi(O, \mathcal{M}), \sigma^2_\phi(O, \mathcal{M})\mathbf{I}),$$

where

$$\boldsymbol{\mu}_\phi(O, \mathcal{M}) = \frac{1}{T} \sum_{t=1}^{T} f_{\phi,\mu}(\mathbf{o}_t, \mathbf{m}_t) \tag{6.56a}$$

$$\log \boldsymbol{\sigma}^2_\phi(O, \mathcal{M}) = \frac{1}{T} \sum_{t=1}^{T} f_{\phi,\sigma}(\mathbf{x}_t, \mathbf{m}_t) \tag{6.56b}$$

are obtained from the encoder network as shown in Figure 6.9. The KL-divergence in Eq. 6.55 can be evaluated analytically:

$$\mathcal{D}_{\text{KL}}(q_\phi(\mathbf{v}|O, \mathcal{M})||p_\theta(\mathbf{v})) =$$
$$-\frac{1}{2} \sum_{d=1}^{D} \left[1 + \log(\sigma^2_{\phi,d}(O, \mathcal{M})) - \mu^2_{\phi,d}(O, \mathcal{M}) - \sigma^2_{\phi,d}(O, \mathcal{M}) \right], \tag{6.57}$$

where D is the dimension of $\mathbf{v}$. The expected negative reconstruction error can be obtained by stochastically sampling the decoder outputs shown in Figure 6.9:

$$\mathbb{E}_{q_\phi(\mathbf{v}|O, \mathcal{M})} \log p_\theta(\mathbf{o}_t|\mathbf{v}, \mathbf{m}_t) = \frac{1}{L} \sum_{l=1}^{L} \log p_\theta(\mathbf{o}_t|\mathbf{v}^{(l)}, \mathbf{m}_t)$$
$$= \frac{1}{L} \sum_{l=1}^{L} f_\theta(\mathbf{v}^{(l)}, \mathbf{m}_t) \tag{6.58}$$
$$= \frac{1}{L} \sum\nolimits_{l=1}^{L} f_\theta \left(\boldsymbol{\mu}_\phi(O, \mathcal{M}) + \sigma^2_\phi(O, \mathcal{M}) \odot \epsilon^{(l)}, \mathbf{m}_t \right).$$

Model training amounts to performing minibatch gradient ascent on the lower bound:

$$(\boldsymbol{\phi}, \boldsymbol{\theta}) \leftarrow (\boldsymbol{\phi}, \boldsymbol{\theta}) + \eta \nabla_{\phi,\theta} \sum_{i \in \text{minibatch}} \mathcal{L}_3(O_i, \mathcal{M}_i, \boldsymbol{\phi}, \boldsymbol{\theta}). \tag{6.59}$$

Given an utterance with acoustic vectors O_i, we present the acoustic vectors to the phonetically aware DNN to obtain the labels $\mathcal{M}_i$. Then, we compute the variational

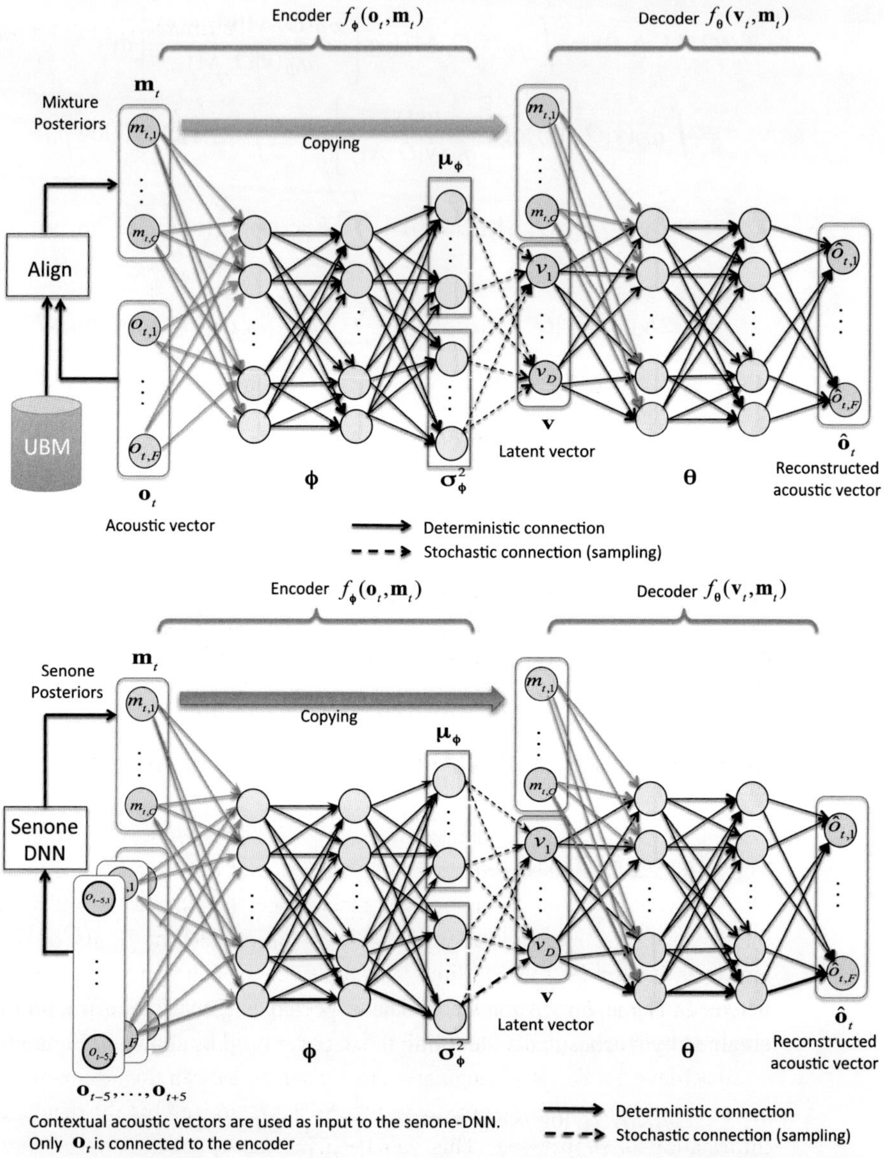

Figure 6.9 Architecture of the variational representation autoencoder using (a) a UBM and (b) a senone-DNN (phonetically aware DNN) to estimate senone posteriors. For simplicity, the network component that computes $p_\theta(\mathbf{x}|\mathbf{v}, \mathbf{m})$ is not shown.

representation vector $\mathbf{v}_i$ (referred to as v-vector) by sampling the latent-vector layer of the VAE:

$$\mathbf{v}_i \sim \mathcal{N}\left(\mu_\phi(\mathcal{O}_i, \mathcal{M}_i), \sigma^2_\phi(\mathcal{O}_i, \mathcal{M}_i)\right), \tag{6.60}$$

where the mean and variance are obtained by Eq. 6.56.

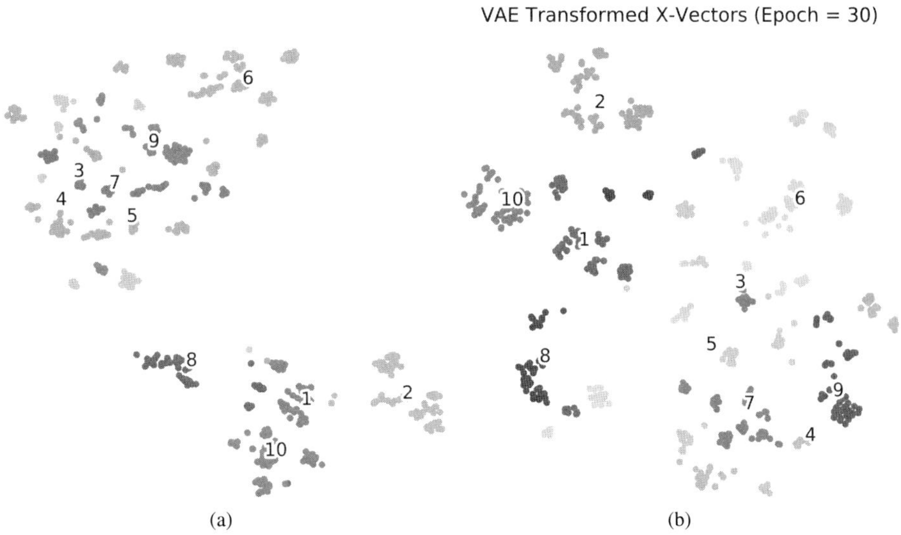

Figure 6.10 (a) X-vectors and (b) VAE-transformed x-vectors of the development data of NIST 2018 SRE on a two-dimensional t-SNE space. Each speaker is represented by one color. Ideally, points (x-vectors) of the same speaker should be close to each other. *It may be better to view the color version of this figure, which is available at https://github.com/enmwmak/ML-for-Spkrec.*

6.6 Generative Adversarial Networks for Domain Adaptation

Current state-of-the-art speaker verification systems are gender- and language-dependent. This is primarily because the i-vectors comprise not only speaker information but also other unwanted information such as genders and languages. To reduce the effect of this unwanted information, a system is trained to handle one gender and one language. However, it is undesirable to limit the languages spoken by the speakers and to switch between male and female systems during operation. Therefore, it is important to build a single system that can handle multiple languages and both genders (Figure 6.10).

To achieve gender- and language-independence, we can transform the i-vectors to an encoded space in which speaker information is maximized but the gender and language information is suppressed. This can be achieved by an adversarial structure shown in Figure 6.11,[4] which we refer to as Adversarial I-vector Transformer (AIT). The network has a feature encoder G, a speaker classifier C, and a gender discriminator D. During forward propagation, $\hat{\mathbf{z}}$ will be directly passed to the input of D; but during backpropagation, the gradient with respect to the weights of G will be reversed, causing the backpropagation on G to perform gradient ascend rather than gradient descend.[5] This reversal of gradient causes the encoder to produce encoded i-vectors $\hat{\mathbf{z}}$ that is confusable to the gender discriminator (suppression of gender information) but at the

[4] To simplify the figure, the language discriminator is omitted.

[5] This effect can be implemented by minimizing a *negative* cross-entropy or by a gradient reversal layer $R(\hat{\mathbf{z}}) = \hat{\mathbf{z}}$ and $\nabla_{\hat{\mathbf{z}}} R(\hat{\mathbf{z}}) = -\mathbf{I}$. Programmatically, the former implementation is simpler.

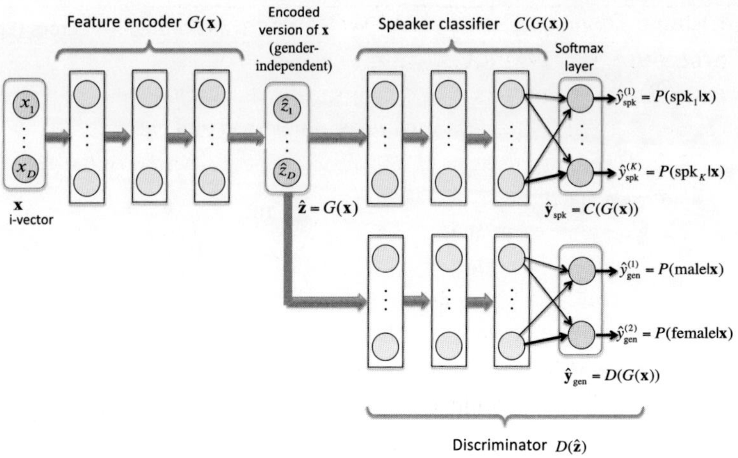

Figure 6.11 Adversarial i-vector transformer (AIT). For clarity, only the gender discriminator is shown. Extension to include the language discriminator is trivial.

same time discriminative for speaker classification (enhancing speaker information). The optimization criteria for G, C, and D in Figure 6.11 are as follows:

$$\text{Train } D : \max_D \left\{ \mathbb{E}_{\mathbf{x}\sim p_{\text{data}}(\mathbf{x})} \left[\sum_{k=1}^{2} y_{\text{gen}}^{(k)} \log D(G(\mathbf{x}))_k \right] \right\} \tag{6.61a}$$

$$\text{Train } G : \min_G \left\{ \mathbb{E}_{\mathbf{x}\sim p_{\text{data}}(\mathbf{x})} \left[\alpha \sum_{k=1}^{2} y_{\text{gen}}^{(k)} \log D(G(\mathbf{x}))_k \right] - \beta \sum_{k=1}^{K} y_{\text{spk}}^{(k)} \log C(G(\mathbf{x}))_k \right\} \tag{6.61b}$$

$$\text{Train } C : \min_C \left\{ -\mathbb{E}_{\mathbf{x}\sim p_{\text{data}}(\mathbf{x})} \sum_{k=1}^{K} y_{\text{spk}}^{(k)} \log C(G(\mathbf{x}))_k \right\}, \tag{6.61c}$$

where $D(\cdot)_k$ and $C(\cdot)_k$ denote the kth output of the discriminator and the classifier, respectively, and α and β control the tradeoff between the adversarial objectives. Note that the first term of Eq. 6.61(b) is the negative cross-entropy. Therefore, Eq. 6.61(b) and Eq. 6.61(a) impose min-max adversarial learning on the encoder G. More specifically, the adversarial learning forces G to produce encoded i-vectors $\hat{\mathbf{z}}$'s that are confusable to D. This means that the encoded i-vectors become gender independent. Algorithm 6 shows the training procedure.

As a proof of concept, we used a 1,024-mixture GMM and a 500-factor total variability matrix to compute the i-vectors of 8790 utterances from 126 male speakers and 167 female speakers in NIST 2006–2010 SREs. Each of these speakers has at least 30 utterances (i-vectors). These i-vectors, together with their speaker and gender labels, were used for training an adversarial i-vector transformer (Figure 6.11) with structure 500–128–128–500 for the feature encoder G, 500–128–128–293 for the speaker classifier C, and 500–128–128–2 for the gender discriminator D. We used Keras and TensorFlow

Algorithm 6 Training of adversarial I-vector transformer (Figure 6.11)

procedure AIT_TRAIN($\mathcal{X}, \mathcal{Y}_{\text{spk}}, \mathcal{Y}_{\text{gen}}$)
Input: Training i-vectors $\mathcal{X}$ and their speaker labels $\mathcal{Y}_{\text{spk}}$ and gender labels $\mathcal{Y}_{\text{gen}}$
Output: Weights of generator G, discriminator D and classifier C in Figure 6.11
    Initialize the weights of G, C, and D using the Xavier initializer [242]
    foreach epoch **do**
        Create N minibatch $\{\mathcal{X}_i, \mathcal{Y}_{\text{spk},i}, \mathcal{Y}_{\text{gen},i}\}_{i=1}^{N}$ of size B from $\{\mathcal{X}, \mathcal{Y}_{\text{spk}}, \mathcal{Y}_{\text{gen}}\}$
        for $i = 1$ to N **do**
            for $j = 1$ to B **do**
                Compute $\hat{\mathbf{z}}_{ij} = G(\mathbf{x}_{ij})$, where $\mathbf{x}_{ij} \in \mathcal{X}_i$
            end for
            Train discriminator D using $\{\hat{\mathbf{z}}_{ij}\}_{j=1}^{B}$ as input and $\{\mathbf{y}_{\text{gen},ij}\}_{j=1}^{B}$ as target outputs, where $\mathbf{y}_{\text{gen},ij} = [y_{\text{gen},ij}^{(1)} \quad y_{\text{gen},ij}^{(2)}]^{\mathsf{T}} \in \mathcal{Y}_{\text{gen},i}$, i.e., Eq. 6.61(a)
            Train adversarially generator G using $\{\mathbf{x}_{ij}\}_{j=1}^{B}$ as input, $\{\mathbf{y}_{\text{spk},ij}\}_{j=1}^{B}$ as target outputs of classifier C, and $\{\mathbf{y}_{\text{gen},ij}\}_{j=1}^{B}$ as target outputs of discriminator D, where $\mathbf{y}_{\text{spk},ij} = [y_{\text{spk},ij}^{(1)} \ \cdots \ y_{\text{spk},ij}^{(K)}]^{\mathsf{T}} \in \mathcal{Y}_{\text{spk},i}$, i.e., Eq. 6.61(b)
            Train classifier C using $\{\mathbf{x}_{ij}\}_{j=1}^{B}$ as input and $\{\mathbf{y}_{\text{spk},ij}\}_{j=1}^{B}$ as target output, i.e., Eq. 6.61(c)
        end for
    end foreach
end procedure

for the implementation. Each hidden node comprises a leaky ReLU (`alpha=0.2`) and each hidden layer is subject to batch normalization (`momentum=0.8`). The 500 output nodes of G are linear and the outputs of C and D use the softmax function to compute the posterior probabilities of speakers and genders, respectively. The Adam optimizer [243] (with `lr=0.0002`, `beta_1=0.5`, and the rest being the default values in Keras) was used for optimizing the objective functions in Eqs. 6.61(a)–(c). We set $\alpha = 0.1$ and $\beta = 0.9$ in Eq. 6.61(b) and trained the network for 125 epochs with a minibatch size of 128.

After training, we applied 422 i-vectors derived from 10 male and 342 i-vectors from 10 female speakers to the network and extracted the outputs from its feature encoder G. The 500-dimensional encoded vectors were than transformed by t-SNE [4] to a two-dimensional embedded space for visualization. The same set of i-vectors (without the AIT transformation) were also transformed by t-SNE to a two-dimensional space. Figure 6.12 shows the results. The left panel of the figure shows that without the adversarial transformation, the male and female i-vectors are clearly separated and they also occupy different regions in the embedded space. This is a clear evidence of gender-dependence in the i-vectors. On the other hand, the right panel clearly shows that with

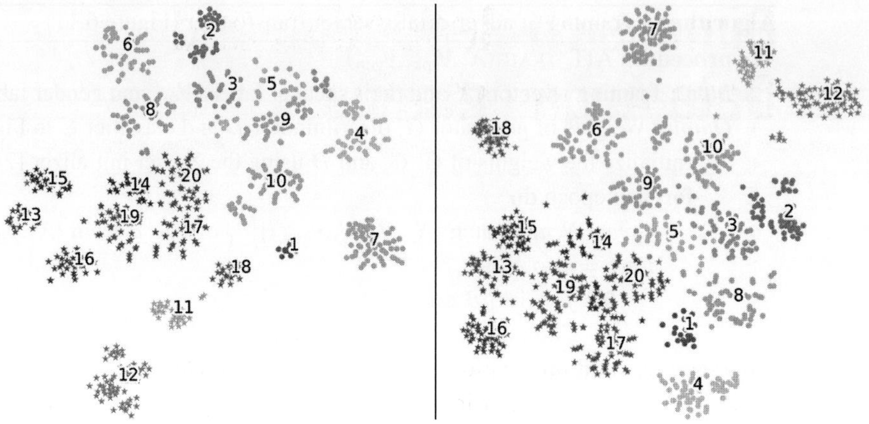

Figure 6.12 Visualization of i-vectors by t-SNE [4]. The i-vectors were derived from the utterances of 10 male speakers (•) and 10 female speakers (⋆). The numbers on top of each cluster are the speaker numbers (Speakers 1–10 are male and Speakers 11–20 are female) and each speaker is represented by one grey scale tone. *Left*: Projection of raw i-vectors onto the t-SNE space. *Right*: Transformation of i-vectors by the feature encoder $G(\mathbf{x})$ in Figure 6.11 to a 500-dimensional encoded space, followed by projection to a two-dimensional space using t-SNE. *It may be better to view the color version of this figure, which is available at https://github.com/enmwmak/ML-for-Spkrec.*

adversarial transformation, the transformed i-vectors are much less gender-dependent as there is no clear separation between male and female vectors. Although the results are still preliminary, they are very encouraging and strongly suggest that the adversarial transformation can reduce the gender-dependence in the i-vectors.

7 Dimension Reduction and Data Augmentation

In general, speaker recognition systems using the i-vectors [16] or x-vectors [156] as speaker features (addressed in Section 3.6) and the probabilistic linear discriminant analysis (PLDA) [17] (addressed in Section 3.5) as a scoring function can achieve good performance. PLDA is known as a linear latent variable model where the same speaker is assumed to be represented by a common factor and the i-vectors or x-vectors of different speakers are characterized by the corresponding latent variables or common factors. This chapter concerns several constraints that affect the performance of i-vectors or x-vectors in PLDA-based speaker recognition. The first constraint is about the insufficiency of speaker identity by using i-vectors, which is caused by the mixing of different factors, e.g., noise, channel, language, gender. In addition, the redundancy with high dimension in i-vectors likely happens. The constraint may be also caused by the insufficient number of training utterances or the imbalanced number of i-vectors among various speakers. The training performance is likely degraded due to the varying lengths of the enrolled sentences. The training of the speaker model may be downgraded by the sparse and heterogeneous utterances in data collection. It is beneficial to deal with the constraints of PLDA training by reducing the feature dimensions, especially in the case of insufficient training sentences [244].

This chapter deals with different constraints in the PLDA model and addresses a series of learning machines or neural networks that are specially designed for dimension reduction as well as data augmentation in speaker recognition systems based on the i-vector combined with PLDA. In particular, we describe how the expanding and emerging researches on a deep learning machine [95] based on the generative adversarial network (GAN) [118], as mentioned in Section 4.6, are developed to construct a subspace model with data augmentation.

Section 7.1 addresses how the objective of neighbor embedding is maximized to derive the low-dimensional representation for high-dimensional observation data based on i-vectors or x-vectors [245, 246]. Subspace mapping is performed in accordance with neighbor embedding. Supervised learning is implemented to assure a low-dimensional representation, which is close under the same speakers and separate between different speakers. To improve the performance of speaker embedding, in Section 7.2, the supervised or discriminative manifold learning [247, 248] is presented to carry out the adversarial learning with threefold advances in speaker recognition. The first idea is to conduct deep learning to characterize the unknown and complicated relation within and between speakers for PLDA speaker recognition. The second thought is to build a latent variable representation [30] for an underlying model structure. The uncertainty of neural network is represented and compensated in a stochastic variant of error back-propagation algorithm by using minibatches of training utterances. The third advanced method is to carry out adversarial learning and manifold learning where an encoder is trained to generate random samples as the prior variables for discriminator. We present the way of merging the adversarial autoencoder [122] in PLDA model for speaker recognition. This solution is implemented as an *adversarial manifold learning* consisting of three components, which are encoder, decoder, and discriminator. The encoder is trained to identify latent representation corresponding to an i-vector, and the decoder is learned to transform the latent variable to reconstruct input i-vector. This reconstruction is used for PLDA scoring [249]. Also, the discriminator is trained to achieve the worst binary classification for judgment between real samples using the PLDA prior and fake data using the generator. We present the approach to this minimax learning problem according to the stochastic optimization using i-vectors in minibatches.

In Section 7.3, we present the approach to data augmentation that can increase the size of training utterances as well as balance the number of training utterances across various speakers. The augmentation of synthesized i-vectors in low-dimensional space is performed. In particular, an *adversarial augmentation learning* is developed to build a neural generative model for PLDA-based speaker recognition. As a result, the diversity of training utterances is increased so as to enhance the robustness in the trained speaker model. Minimax optimization is fulfilled to carry out a two-player game where the generator is optimized to estimate the artificial i-vectors by using random samples from standard Gaussian. These synthesized samples are hardly recognizable from true i-vectors based on the estimated discriminator. The neural networks of generator and discriminator are trained as a distribution model. We improve the system robustness in the presence of changing conditions of i-vectors due to various speakers. Furthermore, the auxiliary classifier GAN [250] is implemented as an extension of conditional GAN [251], which is merged with a class label so that the class conditional likelihood of true i-vectors and synthesized i-vectors are maximized. The speaker-embedded GAN is trained with a fusion of structural information in latent variables. For the sake of speaker recognition, we further minimize the cosine similarity between artificial i-vectors and true i-vectors in the observation level as well in the feature level. A dedicated GAN is formulated and trained in accordance with a multi-objective function where the Gaussian assumption in Gaussian PLDA is met and the reconstruction due to PLDA scoring is assured.

7.1 Variational Manifold PLDA

This section addresses a new variant of PLDA for speaker recognition where the manifold learning is incorporated into the construction of PLDA. A subspace PLDA that carries out low-dimensional speaker recognition is presented. In what follows, we first introduce the subspace method based on the stochastic neighbor embedding.

7.1.1 Stochastic Neighbor Embedding

In [245], an unsupervised learning method called stochastic neighbor embedding (SNE) was proposed to learn the nonlinear manifold in high-dimensional feature space. Assume that $\mathcal{X} = \{\mathbf{x}_n\}$ contains samples in a high-dimensional space. SNE aims to estimate a low-dimensional representations $\mathcal{Z} = \{\mathbf{z}_n\}$ in the feature space, where $\mathbf{z}_n \in \mathbb{R}^d$ preserves the pairwise similarity to $\mathbf{x}_n \in \mathbb{R}^D$ such that $d < D$. The joint probability p_{nm} of two samples $\mathbf{x}_n$ and $\mathbf{x}_m$ can be obtained from a Gaussian distribution

$$p_{nm} = \frac{\exp\left(-\|\mathbf{x}_n - \mathbf{x}_m\|^2\right)}{\sum_s \sum_{t \neq s} \exp\left(-\|\mathbf{x}_s - \mathbf{x}_t\|^2\right)}. \tag{7.1}$$

The probability corresponding to p_{nm} in the low-dimensional representation can also be modeled by a Gaussian using the pairwise similarity among $\{\mathbf{z}_n\}$:

$$q_{nm} = \frac{\exp\left(-\|\mathbf{z}_n - \mathbf{z}_m\|^2\right)}{\sum_s \sum_{t \neq s} \exp\left(-\|\mathbf{z}_s - \mathbf{z}_t\|^2\right)}. \tag{7.2}$$

It is meaningful that p_{nn} and q_{nn} are set to zero in Eqs. 7.1 and 7.2. In [252], a symmetric SNE was realized by minimizing the KL-divergence between two sets of probability densities: $P_n = \{p_{nm}\}_{m=1}^N$ and $Q_n = \{q_{nm}\}_{m=1}^N$, i.e.,

$$\min_{\mathcal{Z}} \sum_n \mathcal{D}_{\text{KL}}(P_n \| Q_n) = \min_{\mathcal{Z}} \sum_n \sum_m p_{nm} \log\left(\frac{p_{nm}}{q_{nm}}\right). \tag{7.3}$$

Nonlinear and nonparametric transformation can be used to preserve the neighbor embeddings of all samples in the original and embedded spaces.

Maaten and Hinton [4] proposed the t-distributed SNE (t-SNE) in which the joint distribution of two samples, $\mathbf{z}_n$ and $\mathbf{z}_m$, in low-dimensional space is modeled by a Student's t-distribution:

$$q_{nm} = \frac{\left(1 + \|\mathbf{z}_n - \mathbf{z}_m\|^2/\nu\right)^{-\frac{\nu+1}{2}}}{\sum_s \sum_{t \neq s} \left(1 + \|\mathbf{z}_s - \mathbf{z}_t\|^2/\nu\right)^{-\frac{\nu+1}{2}}}. \tag{7.4}$$

In Eq. 7.4, ν means the degree of freedom. t-SNE can alleviate the crowding problem in SNE. Figure 7.1 illustrates the concept of distribution construction based on neighbor embeddings in the original data space and a new data space. The front-end processing that uses t-SNE to reduce the dimensionality of i-vectors is performed for subspace speaker recognition using PLDA scoring function [247]. In [249], the PLDA subspace model was constructed by using the variational manifold learning. The stochastic gradient variational Bayesian [114] procedure was implemented to build the latent variable model with neighbor embeddings in PLDA subspace model. The performance of

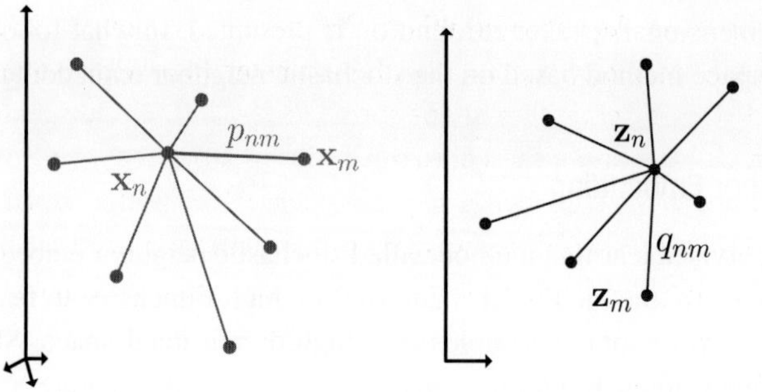

Figure 7.1 Illustration for joint distributions p_{nm} and q_{nm} calculated by neighboring samples centered at $\mathbf{x}_n$ and $\mathbf{z}_n$ in original data space $\mathbf{x}$ (left) and embedded data space $\mathbf{z}$ (right), respectively.

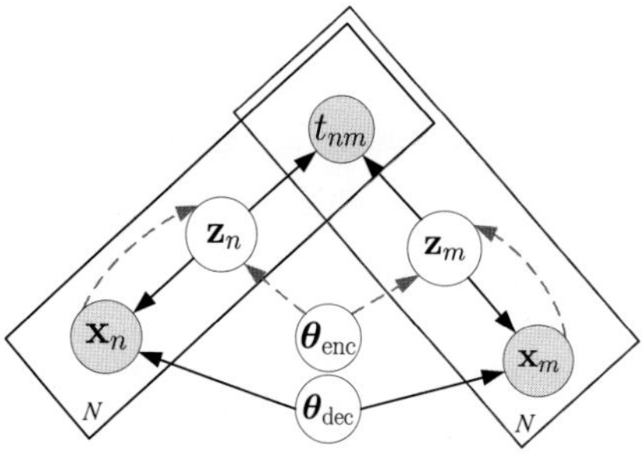

Figure 7.2 Graphical illustration for supervised manifold learning with an encoder (dash line) and a decoder (solid line) given by parameters $\boldsymbol{\theta}_{\text{enc}}$ and $\boldsymbol{\theta}_{\text{dec}}$, respectively. t_{nm} denotes the target value that indicates whether $\mathbf{x}_n$ and $\mathbf{x}_m$ belong to the same class. [Adapted from *Variational Manifold Learning for Speaker Recognition (Figure 2), J.T. Chien and C.W. Hsu, Proceedings International Conference on Acoustics, Speech, and Signal Processing, pp. 4935–4939, 2017, with permission of IEEE*]

speaker recognition based on PLDA model was elevated due to the uncertainty modeling of latent variables.

7.1.2 Variational Manifold Learning

In PLDA-based speaker recognition, the i-vector $\mathbf{x}_n$ in a speaker session is modeled by a latent variable representation based on a latent random vector $\mathbf{z}_n$. We present a neural network variant of a PLDA model where the variational inference is performed. In particular, this section addresses an approach to PLDA subspace representation, which is called the variational manifold PLDA (denoted by VM-PLDA) [249]. This approach aims to hold the property of neighbor embedding for low-dimensional latent representation $\mathbf{z}_n$ which is extracted from the original i-vector $\mathbf{x}_n$. Figure 7.2 illustrates a graphical representation of VM-PLDA. The underlying concept of this model is to implement a variational autoencoder [114] for manifold learning in supervised mode. Encoder and decoder are allocated. As shown in this figure, dash lines show the encoder based on the variational posterior $q(\mathbf{z}_n|\mathbf{x}_n, \boldsymbol{\theta}_{\text{enc}})$ with parameter $\boldsymbol{\theta}_{\text{enc}}$ while solid lines show the decoder based on the generative distribution $p(\mathbf{x}_n|\mathbf{z}_n, \boldsymbol{\theta}_{\text{dec}})$ with parameter $\boldsymbol{\theta}_{\text{dec}}$. Similar to Section 2.3, we apply the approximate or variational inference to estimate the variational parameter $\boldsymbol{\theta}_{\text{enc}}$ in VB-E step and then calculate the model parameters $\boldsymbol{\theta}_{\text{dec}}$ in VB-M step. In the encoder, a Gaussian distribution is used to represent the variational distribution given by the mean vector $\boldsymbol{\mu}(\mathbf{x}_n)$ and the diagonal covariance matrix $\sigma^2(\mathbf{x}_n)\mathbf{I}$

$$q(\mathbf{z}_n|\mathbf{x}_n, \boldsymbol{\theta}_{\text{enc}}) = \mathcal{N}(\mathbf{z}_n|\boldsymbol{\mu}(\mathbf{x}_n), \sigma^2(\mathbf{x}_n)\mathbf{I}), \tag{7.5}$$

where i-vector $\mathbf{x}_n$ is used as the input to a fully connected neural network with multiple layers of weights $\boldsymbol{\theta}_{\text{enc}} = \mathbf{W}_{\text{enc}}$ for Gaussian parameters. In the decoder, the likelihood of PLDA model using i-vector $\mathbf{x}_n$ is used for data reconstruction based on the PLDA parameters $\boldsymbol{\theta}_{\text{dec}} = \{\mathbf{m}, \mathbf{V}, \boldsymbol{\Sigma}\}$

$$p(\mathbf{x}_n|\mathbf{z}_n, \boldsymbol{\theta}_{\text{dec}}) = \mathcal{N}(\mathbf{x}_n|\mathbf{m} + \mathbf{V}\mathbf{z}_n, \boldsymbol{\Sigma}). \tag{7.6}$$

More specifically, the neighbor embedding is measured and adopted as a learning objective to conduct supervised manifold learning. This learning strategy focuses on estimating low-dimensional representation of two samples, $\mathbf{z}_n$ and $\mathbf{z}_m$, which sufficiently reflect the neighbor relation of the same samples, $\mathbf{x}_n$ and $\mathbf{x}_m$, in original i-vector space. The corresponding class targets t_n and t_m are merged in supervised manifold learning. For PLDA speaker recognition, the i-vectors within a specific speaker are represented by the shared latent variable. Under this assumption, the joint probability of two samples from the same speaker $t_n = t_m$ is defined by $p_{nn} = 0$ and $p_{nm} = 1$ while the probability is defined by $p_{nm} = 0$ for the case of different speaker $t_n \neq t_m$. We therefore use these joint probabilities $P = \{p_{nm}\}$ to describe the target probabilities for the associated low-dimensional representations $\mathbf{z}_n$ and $\mathbf{z}_m$. Correspondingly, the Bernoulli target value $t_{nm} \triangleq p_{nm}$ is seen as an observation sample in supervised training. In addition, t distribution shown in Eq. 7.4 is here adopted to characterize the neighbor embedding distribution q_{nm} in low-dimensional subspace. Using VM-PLDA model, the variational parameters $\boldsymbol{\theta}_{\text{enc}}$ in VB-E step and the model parameters $\boldsymbol{\theta}_{\text{dec}}$ in VB-M step are estimated by maximizing the variational lower bound to fulfill the hybrid learning objective for both manifold learning and PLDA reconstruction. Gaussian distribution $\mathcal{N}(\mathbf{z}_n|\boldsymbol{\mu}(\mathbf{x}_n), \mathbf{C}(\mathbf{x}_n))$ is used as a prior for latent variable $\mathbf{z}_n$ which is incorporated to estimate PLDA parameters $\boldsymbol{\theta}_{\text{dec}} = \{\mathbf{m}, \mathbf{V}, \boldsymbol{\Sigma}\}$.

The learning objective of VM-PLDA is constructed as a joint log-likelihood of observation data consisting of i-vectors and class targets $\log p(\mathcal{X}, \mathcal{T})$ where $\mathcal{X} = \{\mathbf{x}_n\}$ and $\mathcal{T} = \{t_{nm}\}$. Supervised learning is performed. Following the fundamental in variational inference, the variational lower bound is derived on the basis of a marginal likelihood with respect to the latent variables of different data pairs $\mathcal{Z} = \{\mathbf{z}_n, \mathbf{z}_m\}$, i.e.,

$$
\begin{aligned}
\log p(\mathcal{X},\mathcal{T}) &= \log \int \prod_m \prod_n p(t_{nm}|\mathbf{z}_n,\mathbf{z}_m)p(\mathbf{x}_n|\mathbf{z}_n)p(\mathbf{z}_n)d\mathcal{Z} \\
&\geq \sum_m \Bigg[\underbrace{\sum_n \mathbb{E}_{q_n}\mathbb{E}_{q_m}[\log p(t_{nm}|\mathbf{z}_n,\mathbf{z}_m)]}_{\text{manifold likelihood}} \\
&\quad + \underbrace{\mathbb{E}_{q_n}[\log p(\mathbf{x}_n|\mathbf{z}_n,\boldsymbol{\theta}_{\text{dec}})]}_{\text{PLDA likelihood}} \\
&\quad - \underbrace{\mathcal{D}_{\text{KL}}(q(\mathbf{z}_n|\mathbf{x}_n,\boldsymbol{\theta}_{\text{enc}})\|p(\mathbf{z}_n))}_{\text{Gaussian regularization } (\mathcal{L}_{\text{gau}})} \Bigg] \\
&\triangleq \mathcal{L}(\boldsymbol{\theta}_{\text{enc}},\boldsymbol{\theta}_{\text{dec}}),
\end{aligned}
\tag{7.7}
$$

where $\mathbb{E}_{q_n}$ means taking expectation with respect to $\mathbf{z}_n|\mathbf{x}_n \sim q(\mathbf{z}_n|\mathbf{x}_n,\boldsymbol{\theta}_{\text{enc}})$. $\log p(\mathcal{X},\mathcal{T})$ can be maximized by optimizing the variational lower bound or equivalently the evidence lower bound (ELBO) $\mathcal{L}(\boldsymbol{\theta}_{\text{enc}},\boldsymbol{\theta}_{\text{dec}})$. It is interesting that VM-PLDA fulfills a multi-objective programming for subspace learning and PLDA scoring. The optimal subspace with the most likely neighbor embedding as indicated in the first term of Eq. 7.7 and the smallest PLDA reconstruction error as indicated in the second term of Eq. 7.7 is constructed. Meaningfully, the remaining term $\mathcal{L}_{\text{gau}}$ is seen as a regularization in joint training that encourages the variational distribution $q(\mathbf{z}_n|\mathbf{x}_n,\boldsymbol{\theta}_{\text{enc}})$ to be close to a prior with standard Gaussian

$$p(\mathbf{z}_n) = \mathcal{N}(\mathbf{0},\mathbf{I}). \tag{7.8}$$

This regularization is imposed to reflect the Gaussian property in PLDA. The raining procedure of VM-PLDA is completed. In general, VM-PLDA is implemented to pursue neighbor embedding and PLDA reconstruction in a stochastic latent variable model where the single-player optimization is realized to estimate the variational neural network for speaker recognition. We argue that the latent variables $\{\mathbf{z}_n,\mathbf{z}_m\}$ may *not* be so discriminative or competitive. Single-player optimization may be improved to the two-player optimization that here means the adversarial learning based on the minimax optimization. Discriminative latent variables are focused. In what follows, we address the neural adversarial learning for speaker recognition.

7.2 Adversarial Manifold PLDA

This section addresses an adversarial manifold PLDA where the adversarial manifold learning is presented for the construction of PLDA-based speaker recognition [119]. The auxiliary classifier GAN is introduced for this implementation. The background of GAN has been addressed in Section 4.6.

7.2.1 Auxiliary Classifier GAN

For practical consideration, it is possible to improve the generation of the synthesized i-vectors $\widehat{\mathbf{x}}$ in GAN by incorporating auxiliary information such as class labels or speaker

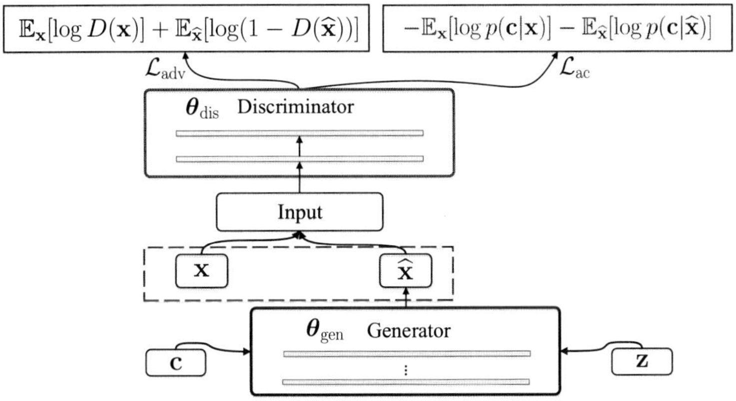

Figure 7.3 Procedure for calculation of adversarial loss $\mathcal{L}_{\text{adv}}$ and auxiliary classification loss $\mathcal{L}_{\text{ac}}$ in auxiliary classifier GAN. [Reprinted from *Adversarial Learning and Augmentation for Speaker Recognition (Figure 2), by J.T. Chien and K.T. Peng, Proceedings Odyssey 2018 The Speaker and Language Recognition Workshop, pp. 342–348, with permission of ISCA*]

identities $\mathbf{c}$ into the the generator and discriminator so that class conditional samples [251] or speaker conditional i-vectors can be produced. For example, the auxiliary classifier GAN (denoted by AC-GAN) [250] makes use of such an auxiliary information to train the discriminator and generator. Specifically, the discriminator was implemented and treated as an auxiliary neural network as decoder to output the class labels $\mathbf{c}$ given training data $\mathbf{x}$ or synthesize data $\widehat{\mathbf{x}}$. As shown in Figure 7.3, the generator receives a class label $\mathbf{c}$ and a noise sample $\mathbf{z}$ as input and synthesizes a vector $\widehat{\mathbf{x}}$ as output:

$$\widehat{\mathbf{x}} = G(\mathbf{z}, \mathbf{c}). \tag{7.9}$$

We therefore incorporate the auxiliary classification (AC) loss as an additional loss which consists of two terms. One is the class conditional distribution of class $\mathbf{c}$ measured by the real i-vector $\mathbf{x}$ while the other is the same distribution but measured by the synthesized i-vector $\widehat{\mathbf{x}}$, namely,

$$\mathcal{L}_{\text{ac}} = -\mathbb{E}_{\mathbf{x}\sim p(\mathbf{x})}[\log p(\mathbf{c}|\mathbf{x})] - \mathbb{E}_{\widehat{\mathbf{x}}\sim p_{\text{gen}}(\mathbf{x})}[\log p(\mathbf{c}|\widehat{\mathbf{x}})]. \tag{7.10}$$

There are two posterior labels in the outputs of discriminator. One is the posterior for the label of true or fake and the other is the posterior for the label of classes. AC-GAN deals with a minimax optimization for estimation of discriminator and generator. Both discriminator and generator are estimated by minimizing the AC loss in Eq. 7.10. At the same time, the adversarial loss $\mathcal{L}_{\text{adv}}$ (as defined in Eq. 4.68) is included in minimax optimization for AC-GAN. The discriminator parameters $\boldsymbol{\theta}_{\text{dis}}$ are optimized by maximizing $\mathcal{L}_{\text{adv}}$, and the generator parameters $\boldsymbol{\theta}_{\text{gen}}$ are optimized by minimizing $\mathcal{L}_{\text{adv}}$. The objectives $\mathcal{L}_{\text{adv}}$ and $\mathcal{L}_{\text{ac}}$ are both included in multi-objective learning of discriminator $\boldsymbol{\theta}_{\text{dis}}$ as well as generator $\boldsymbol{\theta}_{\text{gen}}$. The AC loss $\mathcal{L}_{\text{ac}}$ is introduced to learn for the class information $\mathbf{c}$ while the adversarial loss $\mathcal{L}_{\text{adv}}$ is considered to implement an adversarial generative network for speaker recognition.

7.2.2 Adversarial Manifold Learning

As we know, speaker recognition system using a PLDA-based latent variable model is considerably affected by the learned latent space $\mathcal{Z}$. This space is constructed and controlled by the encoder and decoder with parameters $\boldsymbol{\theta}_{\text{enc}}$ and $\boldsymbol{\theta}_{\text{dec}}$, respectively. The adversarial generative network (GAN) is implemented to enhance the discrimination in latent variable space based on PLDA and develop the adversarial manifold PLDA (denoted by AM-PLDA) for speaker recognition. An additional discriminator is merged as displayed in Figure 7.4. The discriminator is characterized by a fully connected neural network given by the parameter $\boldsymbol{\theta}_{\text{dis}} = \mathbf{W}_{\text{dis}}$. The posterior output of discriminator is denoted by $D(\mathbf{z}_n, \boldsymbol{\theta}_{\text{dis}})$ with the input of the latent code $\mathbf{z}_n$.

In implementation of AM-PLDA, a multi-objective learning, driven by minimax optimization, is run to jointly estimate the discriminator parameters $\boldsymbol{\theta}_{\text{dis}} = \mathbf{W}_{\text{dis}}$ and the generator parameters $\{\boldsymbol{\theta}_{\text{enc}} = \mathbf{W}_{\text{enc}}, \boldsymbol{\theta}_{\text{dec}} = \{\mathbf{m}, \mathbf{V}, \boldsymbol{\Sigma}\}\}$ in a form of

$$\min_{\{\boldsymbol{\theta}_{\text{enc}},\boldsymbol{\theta}_{\text{dec}}\}} \max_{\boldsymbol{\theta}_{\text{dis}}} \Bigg\{ \sum_n \Bigg[\underbrace{- \sum_m \mathbb{E}_{q_n}\mathbb{E}_{q_m}[\log p(t_{nm}|\mathbf{z}_n, \mathbf{z}_m)]}_{\mathcal{L}_{\text{m}}} \underbrace{- \mathbb{E}_{q_n}[\log p(\mathbf{x}_n|\mathbf{z}_n, \boldsymbol{\theta}_{\text{dec}})]}_{\mathcal{L}_{\text{rec}}} \underbrace{+ \mathbb{E}_{p_n}[\log D(\mathbf{z}_n, \boldsymbol{\theta}_{\text{dis}})] + \mathbb{E}_{q_n}[\log(1 - D(\mathbf{z}_n, \boldsymbol{\theta}_{\text{dis}}))]}_{\mathcal{L}_{\text{adv}}} \Bigg] \Bigg\}. \tag{7.11}$$

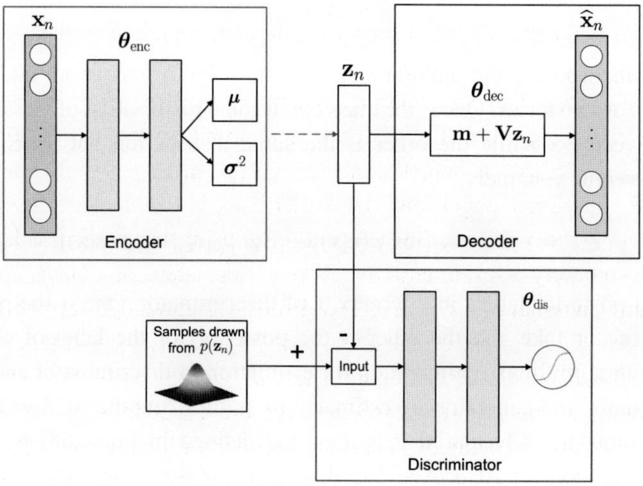

Figure 7.4 Adversarial manifold PLDA consisting of an encoder, a decoder, and a discriminator. A dashed line means sampling. Manifold loss is calculated from the neighbors of latent variables $\{\mathbf{z}_n, \mathbf{z}_m\}$. [Reprinted from *Adversarial Manifold Learning for Speaker Recognition (Figure 3), by J.T. Chien and K.T. Peng, Proceedings ASRU, 2017, with permission of IEEE*]

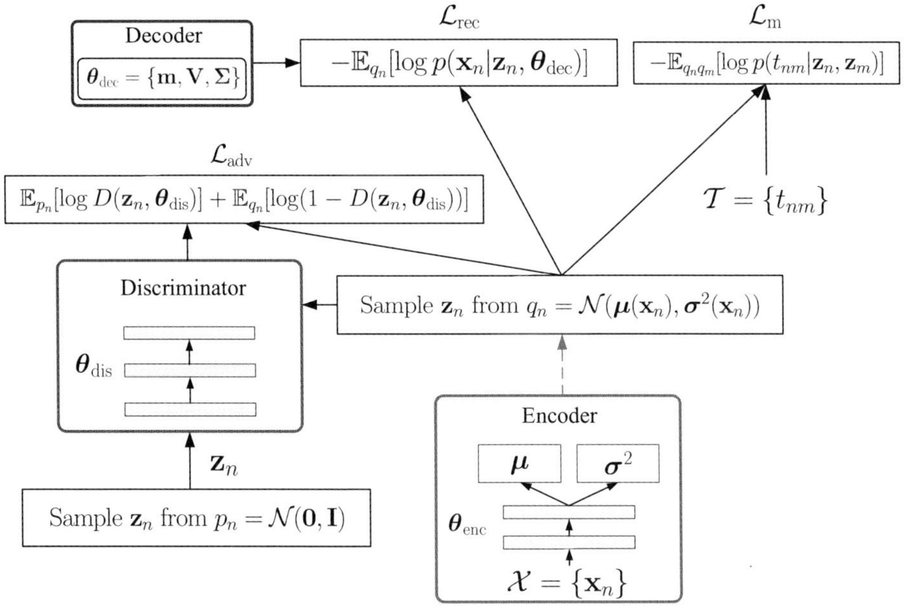

Figure 7.5 Procedure for calculation of manifold loss, PLDA reconstruction loss, and adversarial loss for adversarial manifold PLDA, which are displayed in different rectangles. [Adapted from *Adversarial Manifold Learning for Speaker Recognition (Figure 4), by J.T. Chien and K.T. Peng, Proceedings ASRU, 2017, with permission of IEEE*]

There are three loss functions in multi-objective learning that are the manifold loss $\mathcal{L}_{\text{m}}$, the PLDA reconstruction loss $\mathcal{L}_{\text{rec}}$, and the adversarial loss $\mathcal{L}_{\text{adv}}$. Adversarial loss is seen as the negative cross entropy function, which is calculated by using the class output posterior of discriminator $D(\mathbf{z}_n, \boldsymbol{\theta}_{\text{dis}})$ with the true distribution $p(\mathbf{z}_n) = \mathcal{N}(\mathbf{0}, \mathbf{I})$, and the class output posterior $1 - D(\mathbf{z}_n, \boldsymbol{\theta}_{\text{dis}})$ with the fake or variational distribution $q(\mathbf{z}_n|\mathbf{x}_n, \boldsymbol{\theta}_{\text{enc}})$. The system architecture of calculating different objectives $\{\mathcal{L}_{\text{m}}, \mathcal{L}_{\text{rec}}, \mathcal{L}_{\text{adv}}\}$ is depicted in Figure 7.5. The adversarial manifold PLDA is developed for speaker recognition. This AM-PLDA is seen as an extension of VM-PLDA by replacing the regularization term in Eq. 7.7 using the adversarial loss. The loss is calculated by a binary discriminator that judges if the latent variable $\mathbf{z}_n$ comes from a prior of standard Gaussian $p(\mathbf{z}_n) = \mathcal{N}(\mathbf{0}, \mathbf{I})$ or from a variational posterior $q(\mathbf{z}_n|\mathbf{x}_n, \boldsymbol{\theta}_{\text{enc}})$ using an encoder. In minimax optimization using Eq. 7.11, we maximize the hybrid objective over the discriminator with parameters $\boldsymbol{\theta}_{\text{dis}}$ (included in the third and fourth terms) and minimize the hybrid objective over the encoder with parameters $\boldsymbol{\theta}_{\text{enc}}$ (included in the terms involving $\mathbb{E}_{q_n}$) and the decoder with parameters $\boldsymbol{\theta}_{\text{dec}}$ (included in the second term). Importantly, PLDA scoring is treated as a model regularization that is incorporated to mitigate the mode collapse problem in GAN model [253]. The parameters of encoder $\boldsymbol{\theta}_{\text{enc}}$, discriminator $\boldsymbol{\theta}_{\text{dis}}$, and decoder $\boldsymbol{\theta}_{\text{dec}}$ are jointly trained via a two-player game. The optimal parameters are obtained to achieve the *worst* performance in classification of latent code $\mathbf{z}_n$ either sampled from $q(\mathbf{z}_n|\mathbf{x}_n, \boldsymbol{\theta}_{\text{enc}})$ or given by standard Gaussian prior $p(\mathbf{z}_n)$. The adversarial learning is performed with multi-objectives including manifold loss, cross-entropy loss, and reconstruction loss. We realize a two-player game theory for speaker recognition where the multi-objectives are maximized to estimate the discriminator and minimized to estimate the generator including an encoder and a decoder.

Therefore, the expectation of the negative log likelihood with Bernoulli distribution is calculated with respect to $\mathcal{Z}$ by using the samples $\mathbf{z}_{n,l}$ and $\mathbf{z}_{m,s}$ from a pair of neighbors, $\mathbf{z}_n$ and $\mathbf{z}_m$, which are obtained from the encoder distributions $q(\mathbf{z}_n|\mathbf{x}_n, \boldsymbol{\theta}_{\text{enc}})$ and $q(\mathbf{z}_m|\mathbf{x}_m, \boldsymbol{\theta}_{\text{enc}})$, respectively. The manifold loss $\mathcal{L}_m$ in Eq. 7.11 is implemented by

$$\begin{aligned}\mathcal{L}_m = \sum_n \sum_m \sum_l \sum_s \Bigg[& t_{nm} \frac{\nu+1}{2} \log\left(1 + \|\mathbf{z}_{n,l} - \mathbf{z}_{m,s}\|^2/\nu\right) \\ & - (1 - t_{nm}) \log\left(1 - \left(1 + \|\mathbf{z}_{n,l} - \mathbf{z}_{m,s}\|^2/\nu\right)^{-\frac{\nu+1}{2}}\right)\Bigg].\end{aligned} \tag{7.12}$$

This AM-PLDA is seen as a neural network solution to the supervised t-SNE [254] where the adversarial learning is merged. In addition, PLDA reconstruction error $\mathcal{L}_{\text{rec}}$ in the second term of Eq. 7.11 is seen as an expectation of negative log-likelihood, which is calculated by using the samples $\{\mathbf{z}_{n,l}\}_{l=1}^{L}$ of latent variable $\mathbf{z}_n$:

$$\begin{aligned}\mathcal{L}_{\text{rec}} = \frac{1}{2} \sum_{n=1}^{N} \sum_{l=1}^{L} \Bigg[& \log |2\pi \boldsymbol{\Sigma}| \\ & + (\mathbf{x}_n - \mathbf{m} - \mathbf{V}\mathbf{z}_{n,l})^\top \boldsymbol{\Sigma}^{-1} (\mathbf{x}_n - \mathbf{m} - \mathbf{V}\mathbf{z}_{n,l})\Bigg].\end{aligned} \tag{7.13}$$

where the decoder parameters $\boldsymbol{\theta}_{\text{dec}} = \{\mathbf{m}, \mathbf{V}, \boldsymbol{\Sigma}\}$ are applied.

There are twofold differences in comparison between VM-PLDA and AM-PLDA. First, VM-PLDA carries out a single-player optimization with a regularization term based on the KL divergence that is minimized to regularize the variational posterior $q(\mathbf{z}|\mathbf{x}, \boldsymbol{\theta}_{\text{enc}})$ to be close to a standard Gaussian prior $p(\mathbf{z})$, as seen in Eq. 7.7. Differently, as explained in Eqs. 4.67 and 4.74, AM-PLDA implements a two-player optimization with a maximum adversarial loss that is achieved to find an optimized discriminator where the Jensen–Shannon (JS) divergence $p_{\text{gen}}(\mathbf{x}) \to p(\mathbf{x})$ is minimum. The aspect of discriminator was ignored in latent variable representation in the implementation of VM-PLDA. In addition, different from VM-PLDA, AM-PLDA is viewed as an extension of adversarial autoencoder [122] for PLDA speaker recognition where the latent variable of i-vector is encoded and then used to decode for PLDA scoring. Learning representation is boosted by an adversarial training procedure.

This chapter not only addresses the adversarial learning for PLDA subspace modeling based on the adversarial manifold learning, but also presents the adversarial learning for data augmentation. The adversarial augmentation learning is developed for PLDA speaker verification in the condition that the collected i-vectors are imbalanced among various speakers.

7.3 Adversarial Augmentation PLDA

This section describes a specialized generative adversarial network (GAN) for data augmentation [120]. The i-vectors generated by the network can be augmented to the original training set to compensate the issues of imbalanced and insufficient data in PLDA-based speaker verification. The neural network is trained to ensure that the generated i-vectors reflect speaker behaviors or properties that are matching with those of the real i-vectors. We will address how practical constraints are imposed to perform the constrained optimization, so as to improve model capability for PLDA speaker recognition. An adversarial augmentation PLDA (denoted by AA-PLDA) is developed and implemented to generate new i-vectors for data augmentation. Similar to AM-PLDA, this AA-PLDA also considers speaker identity to build a kind of *speaker-dependent* i-vector generation model where the auxiliary classifier GAN (AC-GAN), as mentioned in Section 7.2.1, is incorporated in i-vector augmentation. Namely, the noise sample $\mathbf{z}$ and speaker label $\mathbf{c}$ are both merged in generator to find synthesized i-vector, i.e.,

$$\widehat{\mathbf{x}} = G(\mathbf{z}, \mathbf{c}). \tag{7.14}$$

Hereafter, the sample index n or m is neglected for notation simplicity.

In implementation of AA-PLDA for speaker recognition, the adversarial augmentation learning is performed by jointly optimizing an adversarial loss $\mathcal{L}_{\text{adv}}$ as given in Eq. 4.68 as well as an auxiliary classification (AC) loss $\mathcal{L}_{\text{ac}}$ as given in Eq. 7.10. AC loss generally expresses the expectation of the negative logarithm of class conditional likelihoods $-\mathbb{E}_{\mathbf{x}\sim p(\mathbf{x})}[\log p(\mathbf{c}|\mathbf{x})]$ and $-\mathbb{E}_{\widehat{\mathbf{x}}\sim p_{\text{gen}}(\mathbf{x})}[\log p(\mathbf{c}|\widehat{\mathbf{x}})]$ measured by the original i-vector and the synthesized i-vector, respectively. Nevertheless, it is insufficient to directly realize AC-GAN for data augmentation in PLDA speaker recognition. Some physical properties can be considered as constraints to compensate such an insufficiency. We therefore incorporate a couple of constraints to characterize practical meanings to generate new and informative i-vectors for PLDA speaker recognition. The cosine generative adversarial network and the PLDA generative adversarial network are accordingly developed as the specialized generative models for synthesis of i-vectors in PLDA speaker recognition.

7.3.1 Cosine Generative Adversarial Network

As we know, cosine distance is usually used as the scoring measure or similarity measure for i-vectors between target speaker and test speaker. This distance measure is an alternative to PLDA scoring based on the likelihood ratio. A Cosine measure generally works well as the scoring measure. A meaningful idea for implementing GAN for speaker recognition is to shape up the synthesized i-vector $\widehat{\mathbf{x}}$ based on the cosine distance between the generated i-vector and real i-vector $\mathbf{x}$. We introduce a regularization loss that is formed by the negative cosine distance between synthesized i-vector $\widehat{\mathbf{x}}$ and original i-vector $\mathbf{x}$

$$\mathcal{L}_{\text{cosx}} = -\mathbb{E}_{\mathbf{x}\sim p(\mathbf{x}),\widehat{\mathbf{x}}\sim p_{\text{gen}}(\mathbf{x})}\left[D_{\cos}(\mathbf{x},\widehat{\mathbf{x}})|G\right]. \tag{7.15}$$

This measure is affected by the generator with function $G(\mathbf{z},\mathbf{c})$, which is characterized by parameter $\boldsymbol{\theta}_{\text{gen}}$. We treat this measure as the regularization term and minimize it to pursue *distribution matching*.

In real implementation, the evaluation of similarity between original i-vector and synthesized i-vector may be degraded since i-vectors are sometimes deteriorated by channel factors and other variabilities. The representation capability for speaker identities may be insufficient. One reasonable idea is to extend the evaluation of cosine similarity between original i-vector and synthesized i-vector by measuring in i-vector raw space, which is denoted by $\mathcal{X}$ rather than in i-vector feature space, which is denoted by $\mathcal{Y}$. As a result, the cosine loss as regularization loss is calculated by using the hidden feature representations $\{\mathbf{y},\widehat{\mathbf{y}}\}$ to replace the i-vector representations $\{\mathbf{x},\widehat{\mathbf{x}}\}$:

$$\mathcal{L}_{\text{cosy}} = -\mathbb{E}_{\mathbf{x}\sim p(\mathbf{x}),\widehat{\mathbf{x}}\sim p_{\text{gen}}(\mathbf{x})}\left[D_{\cos}(\mathbf{y},\widehat{\mathbf{y}})|G,D\right]. \tag{7.16}$$

Notably, this loss function is related to both generator G and discriminator D given by the parameters $\boldsymbol{\theta}_{\text{gen}}$ and $\boldsymbol{\theta}_{\text{dis}}$, respectively. In the implementation, the discriminator D is modeled by a fully connected neural network and used as a binary classification for input vector that is either original i-vector $\mathbf{x}$ or synthesized i-vector $\widehat{\mathbf{x}}$. It is meaningful to extract the hidden units in the last layer of discriminator and use them as the training samples $\{\mathbf{y},\widehat{\mathbf{y}}\}$ to calculate the cosine loss $\mathcal{L}_{\text{cosy}}$ in feature space. We therefore investigate the cosine distance in feature domain $\mathcal{Y}$. Correspondingly, we implement the adversarial augmentation PLDA (AA-PLDA) by using the so-called cosine GANs, Cosx-GAN, and Cosy-GAN based on the cosine losses Cosx and Cosy, which are measured in the spaces $\mathcal{X}$ and $\mathcal{Y}$, respectively. As a result, there are two generative models that are estimated according to individual multi-objectives for solving minimax optimization problems as shown below

$$\text{Cosx-GAN:}\ \mathcal{L}_{\text{adv}} + \mathcal{L}_{\text{ac}} + \mathcal{L}_{\text{cosx}} \tag{7.17}$$

$$\text{Cosy-GAN:}\ \mathcal{L}_{\text{adv}} + \mathcal{L}_{\text{ac}} + \mathcal{L}_{\text{cosy}}. \tag{7.18}$$

Using AA-PLDA, we can carry out two variants of GAN, Cosx-GAN, and Cosy-GAN, where the generator and discriminator are trained in accordance with four loss functions including $\mathcal{L}_{\text{adv}}$, $\mathcal{L}_{\text{ac}}$, $\mathcal{L}_{\text{cosx}}$, and $\mathcal{L}_{\text{cosy}}$. The calculation of these loss functions is shown in Figure 7.6. Notably, the calculation of gradient of cosine distance with respect to the synthesized i-vector $\widehat{\mathbf{x}}$ is required for SGD parameter updating in Cosx-GAN and Cosy-GAN. The approximation to this gradient can be derived by

$$\frac{\partial}{\partial\widehat{\mathbf{x}}}D_{\cos}(\mathbf{x},\widehat{\mathbf{x}}) \approx \frac{\mathbf{x}}{|\mathbf{x}|\,|\widehat{\mathbf{x}}|} - D_{\cos}(\mathbf{x},\widehat{\mathbf{x}})\frac{\widehat{\mathbf{x}}}{|\widehat{\mathbf{x}}|^2} \tag{7.19}$$

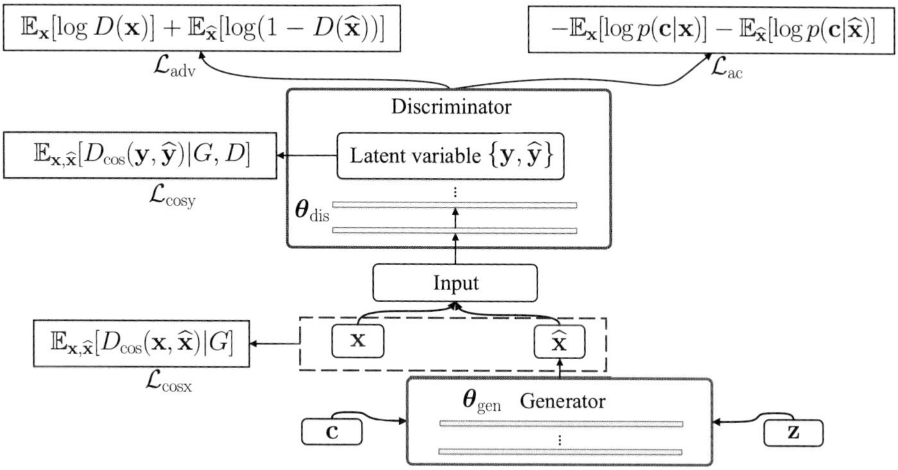

Figure 7.6 The loss functions in the minimax optimization procedure in the cosine GAN, where $\mathcal{L}_{\text{cosx}}$ and $\mathcal{L}_{\text{cosy}}$ denote the cosine loss functions in i-vector space $\mathcal{X}$ and feature space $\mathcal{Y}$, respectively. [Reprinted from *Adversarial Learning and Augmentation for Speaker Recognition (Figure 3), by J.T. Chien and K.T. Peng, Proceedings Odyssey 2018 The Speaker and Language Recognition Workshop, pp. 342–348, with permission of ISCA*]

because

$$
\begin{aligned}
D_{\cos}(\mathbf{x}, \widehat{\mathbf{x}} + d\widehat{\mathbf{x}}) &= \frac{\mathbf{x} \cdot \widehat{\mathbf{x}} + \mathbf{x} \cdot d\widehat{\mathbf{x}}}{|\mathbf{x}|\, |\widehat{\mathbf{x}} + d\widehat{\mathbf{x}}|} \\
&\approx \frac{\mathbf{x} \cdot \widehat{\mathbf{x}} + \mathbf{x} \cdot d\widehat{\mathbf{x}}}{|\mathbf{x}| \left(1 + \frac{\widehat{\mathbf{x}}}{|\widehat{\mathbf{x}}|^2} \cdot d\widehat{\mathbf{x}}\right) |\widehat{\mathbf{x}}|} \\
&\approx \frac{\mathbf{x} \cdot \widehat{\mathbf{x}} + \mathbf{x} \cdot d\widehat{\mathbf{x}}}{|\mathbf{x}|\, |\widehat{\mathbf{x}}|} \left(1 - \frac{\widehat{\mathbf{x}}}{|\widehat{\mathbf{x}}|^2} \cdot d\widehat{\mathbf{x}}\right) \\
&\approx \frac{\mathbf{x} \cdot \widehat{\mathbf{x}}}{|\mathbf{x}|\, |\widehat{\mathbf{x}}|} + \left(\frac{\mathbf{x}}{|\mathbf{x}|\, |\widehat{\mathbf{x}}|} - \frac{\mathbf{x} \cdot \widehat{\mathbf{x}}}{|\mathbf{x}|\, |\widehat{\mathbf{x}}|} \frac{\widehat{\mathbf{x}}}{|\widehat{\mathbf{x}}|^2}\right) \cdot d\widehat{\mathbf{x}} \\
&= D_{\cos}(\mathbf{x}, \widehat{\mathbf{x}}) + \left(\frac{\mathbf{x}}{|\mathbf{x}|\, |\widehat{\mathbf{x}}|} - D_{\cos}(\mathbf{x}, \widehat{\mathbf{x}}) \frac{\widehat{\mathbf{x}}}{|\widehat{\mathbf{x}}|^2}\right) \cdot d\widehat{\mathbf{x}}.
\end{aligned}
\tag{7.20}
$$

As illustrated in Algorithms 7 and 8, the learning procedures based on Cosx-GAN and Cosy-GAN are implemented as two realizations of AC-GAN where the model regularization based on the cosine similarity using original i-vector representation $\mathcal{X}$ and feature space $\mathcal{Y}$ is considered, respectively. The feedforward neural network parameters in generator $\boldsymbol{\theta}_{\text{gen}}$ and discriminator $\boldsymbol{\theta}_{\text{dis}}$ are accordingly estimated. The parameter updating is performed by using SGD algorithm with momentum where minibatches of i-vectors and noise samples are used. The empirical setting is to run three updating steps (i.e., $k = 3$) for discriminator, and then run one updating step for generator at each training iteration. In implementation of Cosx-GAN, the cosine loss in i-vector space $\mathcal{L}_{\text{cosx}}$ is considered in updating the generator without updating the discriminator. We do perform the minimax optimization with respect to discriminator and generator. It is

Algorithm 7 Adversarial augmentation learning where Cosx-GAN is implemented

Initialize the parameters $\{\boldsymbol{\theta}_{\text{gen}}, \boldsymbol{\theta}_{\text{dis}}\}$
For number of training iterations
  For k steps
    sample a minibatch of i-vectors $\{\mathbf{x}_n\}_{n=1}^N$ from $p(\mathbf{x})$
    sample a minibatch of noise examples $\{\mathbf{z}_n\}_{i=n}^N$ from $p(\mathbf{z})$
    update the discriminator $\boldsymbol{\theta}_{\text{dis}}$ by descending the stochastic gradient
      $\frac{1}{N}\sum_{n=1}^N \nabla_{\boldsymbol{\theta}_{\text{dis}}}(-\mathcal{L}_{\text{adv}} + \mathcal{L}_{\text{ac}})$
  End For
  sample a minibatch of i-vectors $\{\mathbf{x}_n\}_{n=1}^N$ from $p(\mathbf{x})$
  sample a minibatch of noise examples $\{\mathbf{z}_n\}_{n=1}^N$ from $p(\mathbf{z})$
  update the generator $\boldsymbol{\theta}_{\text{gen}}$ by descending the stochastic gradient
    $\frac{1}{N}\sum_{n=1}^N \nabla_{\boldsymbol{\theta}_{\text{gen}}}(\mathcal{L}_{\text{adv}} + \mathcal{L}_{\text{ac}} + \mathcal{L}_{\text{cosx}})$
End For

Algorithm 8 Adversarial augmentation learning where Cosy-GAN is implemented

Initialize the parameters $\{\boldsymbol{\theta}_{\text{gen}}, \boldsymbol{\theta}_{\text{dis}}\}$
For number of training iterations
  For k steps
    sample a minibatch of i-vectors $\{\mathbf{x}_n\}_{n=1}^N$ from $p(\mathbf{x})$
    sample a minibatch of noise examples $\{\mathbf{z}_n\}_{i=n}^N$ from $p(\mathbf{z})$
    update the discriminator $\boldsymbol{\theta}_{\text{dis}}$ by descending the stochastic gradient
      $\frac{1}{N}\sum_{i=1}^N \nabla_{\boldsymbol{\theta}_{\text{dis}}}(-\mathcal{L}_{\text{adv}} + \mathcal{L}_{\text{ac}} + \mathcal{L}_{\text{cosy}})$
  End For
  sample a minibatch of i-vectors $\{\mathbf{x}_n\}_{n=1}^N$ from $p(\mathbf{x})$
  sample a minibatch of noise examples $\{\mathbf{z}_n\}_{n=1}^N$ from $p(\mathbf{z})$
  update the generator $\boldsymbol{\theta}_{\text{gen}}$ by descending the stochastic gradient
    $\frac{1}{N}\sum_{i=1}^N \nabla_{\boldsymbol{\theta}_{\text{gen}}}(\mathcal{L}_{\text{adv}} + \mathcal{L}_{\text{ac}} + \mathcal{L}_{\text{cosy}})$
End For

because the gradients of adversarial loss function $\mathcal{L}_{\text{adv}}$ in updating discriminator and generator have different signs. On the other hand, Cosy-GAN is implemented by using the cosine loss function $\mathcal{L}_{\text{cosy}}$, which depends both on the parameters of discriminator and generator.

7.3.2 PLDA Generative Adversarial Network

In this subsection, we further explore how to incorporate PLDA constraint in adversarial augmentation learning for speaker recognition. Here, the property of PLDA is sufficiently reflected or represented as the constraint or regularization in GAN training. The variational autoencoder (VAE) [237, 249] is considered to move forward an advanced solution where different regularization terms are derived and forced to artificially synthesize i-vectors with good-quality PLDA speaker recognition is performed by using real i-vectors as well as synthesized i-vectors. The resulting PLDA GAN is different from the cosine GAN mentioned in Section 7.3.1. Cosine GAN uses a fully connected neural network to extract *deterministic* latent features $\mathbf{y}$ to calculate the cosine loss function $\mathcal{L}_{\text{cosy}}$. An extension of combining cosine GAN and PLDA scoring is to carry out the so-called PLDA-Cos-GAN where the stochastic or variational neural network is introduced with the regularization terms from cosine similarity and some others. The variational distribution is used as the encoder to find the stochastic latent variable of $\mathbf{x}$ or $\widehat{\mathbf{x}}$ based on a Gaussian distribution driven by mean vector $\boldsymbol{\mu}(\mathbf{x}) = \{\mu_d(\mathbf{x})\}$ and diagonal covariance matrix $\boldsymbol{\sigma}^2(\mathbf{x})\mathbf{I} = \text{diag}\{\sigma_d^2(\mathbf{x})\}$ using feedforward neural network parameters $\boldsymbol{\theta}_{\text{enc}}$, i.e.,

$$\mathbf{y} \sim q(\mathbf{y}|\mathbf{x}) = \mathcal{N}(\boldsymbol{\mu}(\mathbf{x}), \boldsymbol{\sigma}^2(\mathbf{x})\mathbf{I}). \tag{7.21}$$

The adversarial augmentation learning is developed through variational learning where a decoder with PLDA parameters $\boldsymbol{\theta}_{\text{dec}} = \{\mathbf{m}, \mathbf{V}, \boldsymbol{\Sigma}\}$. Such a decoder was similarly adopted in the variational manifold learning as mentioned in Section 7.1.2 and the adversarial manifold learning as addressed in Section 7.2.2. According to the variational inference, we minimize the negative evidence lower bound (ELBO) based on negative log-likelihood using a collection of i-vectors $\mathbf{x} = \{\mathbf{x}_n\}$ [30, 249]:

$$\begin{aligned}
-\log p(\mathbf{x}) &= -\int q(\mathbf{y}|\mathbf{x}) \log(p(\mathbf{x}))d\mathbf{y} \\
&= -\int q(\mathbf{y}|\mathbf{x}) \log\left(\frac{p(\mathbf{y},\mathbf{x})}{q(\mathbf{y}|\mathbf{x})}\right) d\mathbf{y} - \int q(\mathbf{y}|\mathbf{x}) \log\left(\frac{q(\mathbf{y}|\mathbf{x})}{p(\mathbf{y}|\mathbf{x})}\right) d\mathbf{y} \\
&\leq -\int q(\mathbf{y}|\mathbf{x}) \log(p(\mathbf{x}|\mathbf{y}))d\mathbf{y} - \int q(\mathbf{y}|\mathbf{x}) \log\left(\frac{p(\mathbf{y})}{q(\mathbf{y}|\mathbf{x})}\right) d\mathbf{y} \\
&= \underbrace{-\mathbb{E}_{q(\mathbf{y}|\mathbf{x})}[\log p(\mathbf{x}|\mathbf{y})]}_{\mathcal{L}_{\text{rec}}} + \underbrace{\mathcal{D}_{\text{KL}}(q(\mathbf{y}|\mathbf{x})\|p(\mathbf{y}))}_{\mathcal{L}_{\text{gau}}}.
\end{aligned} \tag{7.22}$$

We can see that the PLDA reconstruction loss $\mathcal{L}_{\text{rec}}$ and the Gaussian regularization $\mathcal{L}_{\text{gau}}$ are simultaneously minimized. In this learning objective, the conditional likelihood function

$$p(\mathbf{x}|\mathbf{y}) = \mathcal{N}(\mathbf{m} + \mathbf{V}\mathbf{y}, \boldsymbol{\Sigma}) \tag{7.23}$$

and the standard Gaussian prior $p(\mathbf{y}) = \mathcal{N}(\mathbf{0}, \mathbf{I})$ are used. To implement Eq. (7.22), we calculate the PLDA scoring or PLDA loss $\mathcal{L}_{\text{rec}}$ by using Eq. (7.13) and also calculate the Gaussian regularization term $\mathcal{L}_{\text{gau}}$, which is derived as

$$\mathcal{D}_{\text{KL}}(\mathcal{N}(\boldsymbol{\mu}, \boldsymbol{\sigma}^2)\|\mathcal{N}(\mathbf{0}, \mathbf{I})) = \frac{1}{2} \sum_d \left[\mu_d^2 + \sigma_d^2 + \log(\sigma_d^2) - 1\right]. \tag{7.24}$$

We therefore develop a so-called PLDA-GAN, which is a special realization of adversarial autoencoder [122] for speaker recognition based on the PLDA scoring.

Using PLDA-Cos-GAN, there are five loss terms derived in adversarial augmentation learning for speaker recognition. The original i-vector $\mathbf{x}$ and the synthesized i-vector $\widehat{\mathbf{x}}$ from the *generator* are merged in calculation of these five loss functions. The

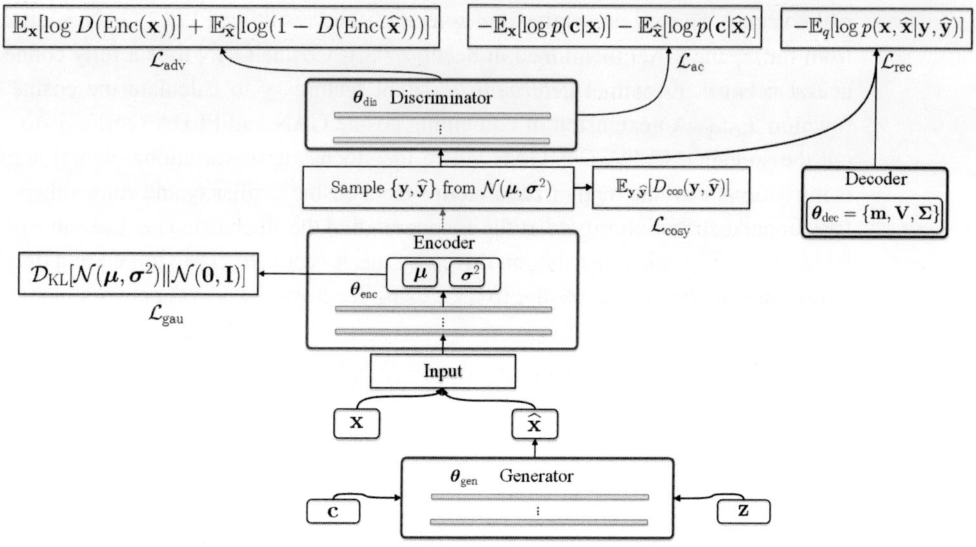

Figure 7.7 The loss functions in the minimax optimization procedure in the PLDA-Cos-GAN, where $\mathcal{L}_{\text{gau}}$ and $\mathcal{L}_{\text{rec}}$ denote the loss functions for Gaussian regularization and PLDA reconstruction, respectively. [Reprinted from *Adversarial Learning and Augmentation for Speaker Recognition (Figure 4), by J.T. Chien and K.T. Peng, Proceedings Odyssey 2018 The Speaker and Language Recognition Workshop, pp. 342–348, with permission of ISCA*]

training of *encoder*, *discriminator*, and *decoder* in PLDA-Cos-GAN is based on the loss functions $\{\mathcal{L}_{\text{gau}}, \mathcal{L}_{\text{cosy}}\}$, $\{\mathcal{L}_{\text{adv}}, \mathcal{L}_{\text{ac}}\}$, and $\mathcal{L}_{\text{rec}}$, respectively, as illustrated in Figure 7.7. When calculating the objectives $\mathcal{L}_{\text{adv}}$, $\mathcal{L}_{\text{ac}}$, $\mathcal{L}_{\text{cosy}}$, and $\mathcal{L}_{\text{rec}}$, the samples of variational Gaussian distribution $q(\mathbf{y}|\mathbf{x})$ are used. The key difference between Cosy-GAN and PLDA-Cos-GAN is the calculation of cosine loss $\mathcal{L}_{\text{cosy}}$. Cosy-GAN calculates this term from last layer of discriminator while PLDA-Cos-GAN calculates this loss by using encoder. In addition, the hidden variable $\mathbf{y}$ is random in PLDA-Cos-GAN while the latent code $\mathbf{y}$ is deterministic in Cosy-GAN. Using PLDA-Cos-GAN, the parameters of decoder $\boldsymbol{\theta}_{\text{dec}}$ based on PLDA model and the parameters of generator, encoder, and discriminator using the fully connected neural network $\{\boldsymbol{\theta}_{\text{gen}}, \boldsymbol{\theta}_{\text{enc}}, \boldsymbol{\theta}_{\text{dis}}\}$ are jointly estimated according to minimax optimization. The training procedure for different PLDA-Cos-GAN parameters is formulated and implemented in Algorithm 9. The updating of individual parameters is performed by taking derivatives for those related loss terms. $\mathbf{z}$ and $\mathbf{y}$ are stochastic and used as the latent codes for generator and decoder, respectively. In Algorithm 9, we introduce regularization parameter λ to adjust the tradeoff between different loss functions. This hyperparameter can be chosen by using a validation set in accordance with some selection criterion.

Algorithm 9 Adversarial augmentation learning for PLDA-Cos-GAN

Initialize the parameters $\{\boldsymbol{\theta}_{\text{gen}}, \boldsymbol{\theta}_{\text{enc}}, \boldsymbol{\theta}_{\text{dec}}, \boldsymbol{\theta}_{\text{dis}}\}$
For number of training iterations
  For k steps
    sample a minibatch of i-vectors $\{\mathbf{x}_n\}_{n=1}^N$ from $p(\mathbf{x})$
    sample a minibatch of noise examples $\{\mathbf{z}_n, \mathbf{y}_n\}_{n=1}^N$ from $\{p(\mathbf{z}), p(\mathbf{y})\}$
    update the discriminator $\boldsymbol{\theta}_{\text{dis}}$ by descending the stochastic gradient
      $\frac{1}{N}\sum_{n=1}^N \nabla_{\boldsymbol{\theta}_{\text{dis}}} \lambda(-\mathcal{L}_{\text{adv}} + \mathcal{L}_{\text{ac}})$
    update the encoder $\boldsymbol{\theta}_{\text{enc}}$ by descending the stochastic gradient
      $\frac{1}{N}\sum_{n=1}^N \nabla_{\boldsymbol{\theta}_{\text{enc}}} (\lambda(\mathcal{L}_{\text{adv}} + \mathcal{L}_{\text{ac}} - \mathcal{L}_{\text{cosy}}) + \mathcal{L}_{\text{rec}} + \mathcal{L}_{\text{gau}})$
    update the decoder $\boldsymbol{\theta}_{\text{dec}}$ by descending the stochastic gradient
      $\frac{1}{N}\sum_{n=1}^N \nabla_{\boldsymbol{\theta}_{\text{dec}}} (\mathcal{L}_{\text{rec}})$
  End For
  sample a minibatch of i-vectors $\{\mathbf{x}_n\}_{n=1}^N$ from $p(\mathbf{x})$
  sample a minibatch of noise examples $\{\mathbf{z}_n, \mathbf{y}_n\}_{n=1}^N$ from $\{p(\mathbf{z}), p(\mathbf{y})\}$
  update the generator $\boldsymbol{\theta}_{\text{gen}}$ by descending the stochastic gradient
    $\frac{1}{N}\sum_{n=1}^N \nabla_{\boldsymbol{\theta}_{\text{gen}}} (\lambda(\mathcal{L}_{\text{adv}} + \mathcal{L}_{\text{ac}} - \mathcal{L}_{\text{cosy}}) + \mathcal{L}_{\text{rec}})$
End For

7.4 Concluding Remarks

We have presented two deep adversarial learning approaches to PLDA speaker recognition. The generative adversarial networks were implemented in different model architectures to cope with various problems for speaker verification. A two-player game was formulated and handled by a learning procedure where the PLDA parameters and fully connected neural network parameters were optimized to build the decoder, generator, encoder, and discriminator. An adversarially learned latent representation for PLDA reconstruction or scoring was obtained by solving a minimax optimization with multiple objectives. For an integrated work, we might carry out a hybrid adversarial manifold learning and adversarial augmentation learning that jointly optimized multi-objectives to meet neighbor embedding, classification accuracy, PLDA reconstruction, and adversarial training for speaker recognition. We considered various regularization terms including cosine distance, Gaussianity, and PLDA scoring to benefit the training procedure. A theoretical solution to deep machine learning was presented to carry out state-of-the-art speaker recognition system. We integrated different learning models in this chapter including stochastic neighbor embedding, variational autoencoder, generative adversarial network, and conditional adversarial neural network.

8 Future Direction

Deep learning has been extensively used in speaker recognition. It will continue to play an important role in the foreseeable future, particularly in feature learning, end-to-end systems, and domain-invariant speaker recognition.

8.1 Time-Domain Feature Learning

So far, state-of-the-art speaker recognition systems use handcrafted features inspired by the human auditory systems. MFCC, PLP, and log-mel filterbank magnitude coefficients are some of the common handcrafted features. These features can be efficiently extracted from waveforms using signal processing methods. In speech recognition, it has been shown that deep neural networks can be trained to extract filter-bank like features directly from speech waveforms using a long context window [255–258]. This type of feature learning approach has the advantage that the feature extractor and the classifier can be jointly training, resulting in end-to-end systems.

Because speech signals are nonstationary waveforms with short-time stationarity, it is natural to apply a one-D convolutional neural network to process long-term waveform segments ($\approx$ 300ms), with a hop size of 10ms [259]. The waveform of an utterance is partitioned into into long-term segments (310ms in [260]), each of which is further partitioned into a short time frame. Denote $\{x_n\}_{n=1}^{N}$ as N consecutive samples in a long-term segment of an utterance. Then, the short-term segment centered at time point t within the long-term segment is

$$\mathbf{x}_t = \left[x_{t-(K_1-1)/2}, \ldots, x_t, \ldots, x_{t+(K_1-1)/2}\right]^{\mathsf{T}},$$

where K_1 is the kernel width (50 samples in [260]). In the first convolutional layer, $\mathbf{x}_t$ is convolved with F_1 FIR filters to give a matrix of size $F_1 \times K_1$. Then, temporal max-pooling is independently applied to the output sequences of the FIR filters. Figure 8.1 shows the operation of convolution and temporal max-pooling. In the remaining convolutional layers, the kernel width is in the unit of the number of time frames (rectangular blocks in Figure 8.1). For example, in [258], $K_2 = K_3 = 7$ and the frame shift (hopped distance) is 1.

Comparing Figure 8.1 and Figure 5.5 reveals that the CNN architecture is similar to the x-vector extractor. By changing the classes in Figure 8.1 to speaker labels and adding

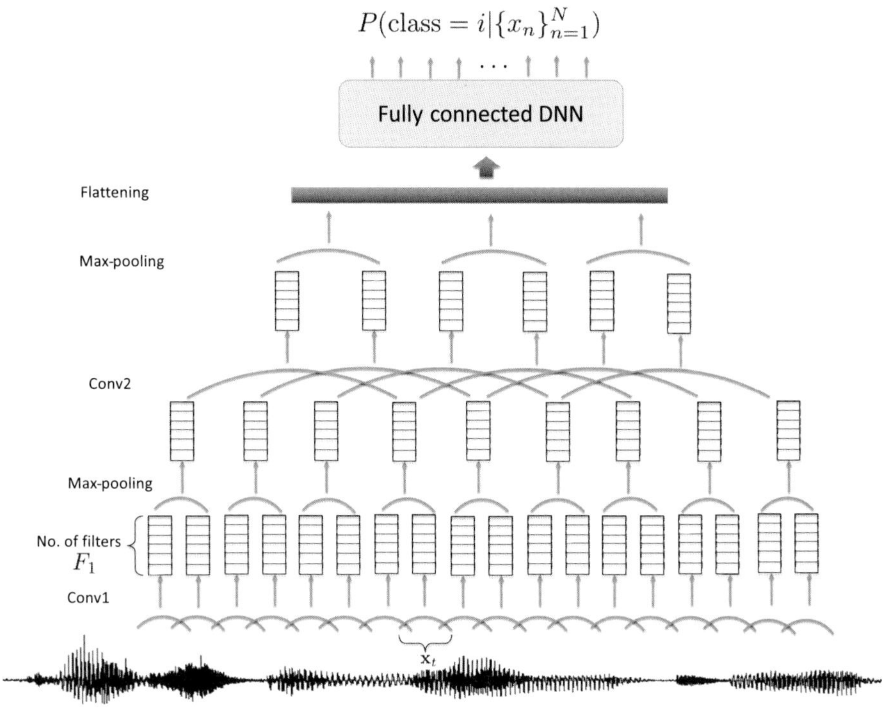

Figure 8.1 The convolution and max-pooling processes of a CNN for extracting features from speech waveforms. Nonlinearity is applied after max-pooling.

a statistical pooling layer (for pooling over time) and an embedded layer (for x-vector representation), it is possible to extract x-vectors from speech waveforms.

Another approach to extracting features from waveforms is to combine CNN, DNN, and LSTM [261]. Specifically, 25–35ms of time-domain waveform is fed to a few CNN layers that perform both time and frequency convolutions to reduce spectral variation. Then, the output of the last CNN layer is fed to a few LSTM layers. Finally, the output of the last LSTM layer is fed to a DNN. The process is repeated for every 10ms. The idea is based on the CLDNN in [262]. But unlike [262], the CLDNN in [261] receives raw waveforms as input. The advantage of using waveform as input is that the fine structure of the time-domain signals can be captured. It was shown that these networks have complementary modeling capabilities. Specifically, the CNN can reduce frequency variations, the LSTM can model long-term temporal variations, and the DNN can produce a higher-order representation that can be easily separated into distinct classes.

The first layer in the raw-waveform CLDNN [261] is a time-convolutional layer over the waveform. This layer can be considered as a finite impulse-response filterbank (such as a gammatone filterbank) followed by nonlinear operations. Therefore, the outputs of different convolutional nodes in the first layer correspond to different frequencies. By convolving M time points in a moving window with P convolutional filters with

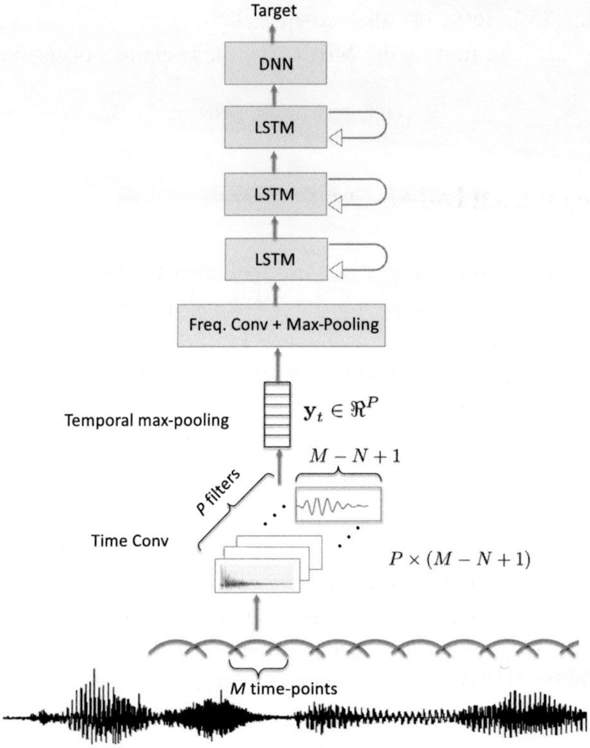

Figure 8.2 The architecture of CLDNN that accepts waveforms as input. The first time-convolutional layer has P filters (feature maps) of length N. For each frame, the first convolutional layer produces a vector $\mathbf{y}_t$, where t denotes the frame index. The elements in $\boldsymbol{y}_t$ are considered as frequency components, which are subject to frequency convolution and max-pooling. The resulting vectors are fed to the LSTM, followed by a DNN.

length N, the output of the first convolutional layer has dimension $P \times (M - N + 1)$ in time $\times$ frequency. These outputs are subjected to temporal pooling to produce $P \times 1$ outputs. Performing pooling in time has the effect of suppressing phrase information and making the feature representation in upper layers phase invariant. This convolutional and pooling process effectively convert time-domain signals into frame-level feature vectors of dimension P ($\mathbf{y}_t$ in Figure 8.2). The window is shifted by 10ms for each frame-level feature vector. Each $\mathbf{y}_t$ is subject to frequency convolution to reduce spectral variation.

In Figure 8.2, raw waveforms are processed by CNN followed by LSTM rather than directly processed by the LSTM. The advantage of the former is that each step in the LSTM corresponds to one frame (typically 10 ms), whereas in the latter, each time step corresponds to one time point (typically 0.125 ms). To model variations across 300ms, using the LSTM alone requires unrolling 2,400 time steps, which is not feasible in practice. On the other hand, using the CNN+LSTM to model the same time span, we only need to unroll 30 time steps.

To apply the architectures in Figures 8.1 and 8.2 for speaker recognition, we may extract embedded feature vectors from the outputs of the convolutional filters.

As demonstrated in [259, 260], the first convolutional layer acts like a filter bank. Therefore, it is reasonable to replace the MFCCs by the responses of the convolutional filter.

8.2 Speaker Embedding from End-to-End Systems

Recent advances in deep learning has led to the use of deep neural networks for speaker feature extraction. The main idea is to replace the commonly used i-vectors by creating an embedded layer in a deep neural network and train the network in an end-to-end manner. Speaker features are then extracted from the embedded layer.

In [263, 264], the authors proposed to use the Inception-Resnet-v1 architecture that is commonly used in the computer vision community for such purpose. Specifically, short segments of a spectrogram are presented to the input of an inception network, as shown in Figures 8.3 and 8.4. At the embedding layer, utterance-level features are obtained by averaging the activations across all of the short segments in the utterance. The network is optimized by minimizing the triplet loss, as shown in Figure 8.3, using a Euclidean distance metric. Because of the averaging operation, the feature dimension at the embedding layer is fixed regardless of the length of the utterance. A similar strategy has also been used in the VGGVox network that creates the Voxceleb2 dataset [265] and the use of residual blocks in [266, 267]. Another example is the small footprint speaker embedding in [268] in which knowledge distillation was used to reduce the performance gap between large and small models.

To create fix-length embedding in the inception network, [263, 264] spatial pyramid pooling (SPP) [269] was applied to a convolutional layer of the inception network. Instead of using a sliding window as in conventional CNNs, SPP has bin size propor-

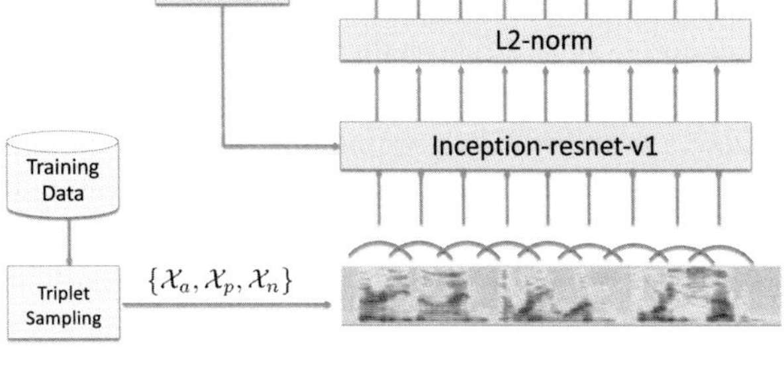

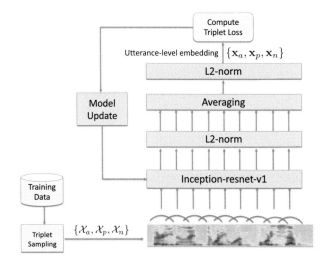

Figure 8.3 Training an inception-resnet-v1 to create an utterance-level embedded vector.

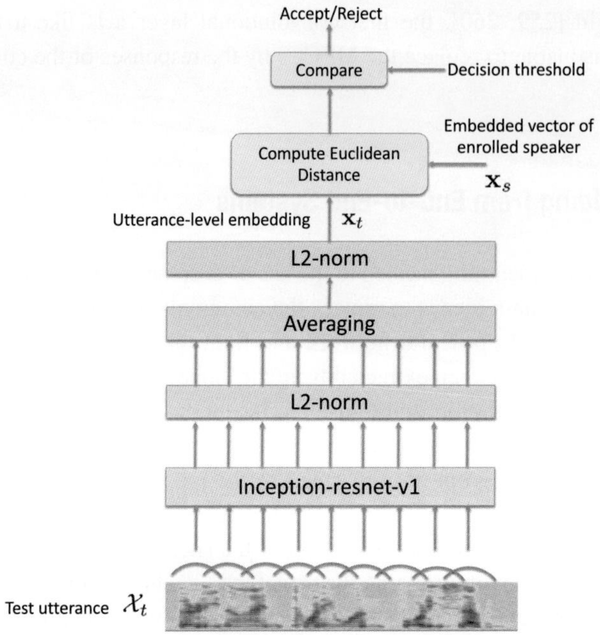

Figure 8.4 Given a test utterance, the inception network gives an utterance-level embedded vector, whose distance to the embedded vector of the enrolled speaker is compared with a decision threshold for decision making.

tional to the input feature size. As a result, the number of bins is fixed regardless of the length of the spectrogram. The embedded feature vectors can then be scored by either cosine-distance scoring or PLDA scoring.

8.3 VAE–GAN for Domain Adaptation

Adversarial learning [270] has been applied to create a domain-invariant space [266, 271–274], to reduce the language mismatch [275], channel mismatch [276], noise-type mismatch [277], acoustic condition mismatch [278], and to map unlabeled speech from one domain to another domain [279]. For instance, in [207], an encoder and a discriminator network are adversarially trained to produce bottleneck features that are robust against noise. Wang et al. [211] applied domain adversarial training (DAT) [272] to generate speaker discriminative and domain-invariant feature representations by incorporating a speaker classifier into an adversarial autoencoder, which outperforms traditional DA approaches on the 2013 domain adaptation challenge. In [279], cycle-GANs are used to map microphone speech in the SITW corpus to telephone speech and the generated telephone speech are mapped back to microphone speech. The inputs to the cycle-GANs are 11 frames of 40-dimensional filter bank features. The microphone-to-telephone features were passed to an x-vector extractor trained by using telephone speech.

Although adversarial learning based unsupervised domain adaptation [207, 211] has greatly boosted the performance of speaker verification systems under domain mismatch conditions, the adversarial training may lead to non-Gaussian latent vectors, which do not meet the Gaussianity requirement of the PLDA back end. One possible way to make the x-vectors more Gaussian is to apply Gaussian constraint on the embedding layer of the x-vector extractor during training [280]. The problem can also be overcome by using variational autoencoders [237]. One desirable property of a VAE is that its KL-divergence term can be considered as a regularizer that constrains the encoder to produce latent vectors that follow a desired distribution. This property can be leveraged to cause the encoder to produce Gasussian latent vectors, which will be amenable to PLDA modeling.

8.3.1 Variational Domain Adversarial Neural Network (VDANN)

By applying domain adversarial training (DAT), the domain adversarial neural networks (DANN) in [211, 272] can create a latent space from which domain-invariant speaker discriminative features can be obtained. It is possible to incorporate variational regularization into the DAT to make the embedded features follow a Gaussian distribution so that the Gaussian PLDA back end can be applied. The resulting network is referred to as variational domain adversarial neural network (VDANN) [281].

Denote $\mathcal{X} = \{\mathcal{X}^{(r)}\}_{r=1}^{R}$ as a training set containing samples from R domains, where $\mathcal{X}^{(r)} = \{\mathbf{x}_1^{(r)}, \ldots, \mathbf{x}_{N_r}^{(r)}\}$ contains N_r samples from domain r. Also, denote $\mathbf{y}$ and $\mathbf{d}$ as the speaker and domain labels in one-hot format, respectively. Variational autoencoders (VAE) were originally proposed to solve the variational approximate inference problem (see Section 4.5). A VAE is trained by maximizing the evidence lower bound (ELBO) [237]:

$$\begin{aligned} L_{\text{ELBO}}(\theta,\phi) = \sum_{r=1}^{R}\sum_{i=1}^{N_r}\Big\{ & -\mathcal{D}_{\text{KL}}\left(q_\phi(\mathbf{z}|\mathbf{x}_i^{(r)}) \,\|\, p_\theta(\mathbf{z})\right) \\ & + \mathbb{E}_{q_\phi(\mathbf{z}|\mathbf{x}_i^{(r)})}\left[\log p_\theta\left(\mathbf{x}_i^{(r)}|\mathbf{z}\right)\right]\Big\}, \end{aligned} \tag{8.1}$$

where $q_\phi(\mathbf{z}|\mathbf{x})$ is an approximate posterior to the intractable true posterior $p_\theta(\mathbf{z}|\mathbf{x})$ and ϕ and θ are parameters of the encoder and decoder, respectively. The network parameterized by ϕ represents a recognition model that encodes input samples into the latent space and the network parameterized by θ represents a generative model that transforms the latent representations back to the original data space.

A VAE has a desirable property in that the first term in Eq. 8.1 can constrain the variational posterior $q_\phi(\mathbf{z}|\mathbf{x})$ to be similar to the desired prior $p_\theta(\mathbf{z})$. As a result, if $p_\theta(\mathbf{z})$ is a multivariate Gaussian distribution, the encoder will make the latent vectors to follow a Gasussian distribution, which is amenable to PLDA modeling.

Suppose $p_\theta(\mathbf{z}) = \mathcal{N}(\mathbf{z}; \mathbf{0}, \mathbf{I})$. Also assume that the true posterior follows a Gaussian distribution with an approximate diagonal covariance matrix. Then, the approximate posterior becomes

$$\log q_\phi(\mathbf{z}|\mathbf{x}_i) = \log \mathcal{N}\left(\mathbf{z}; \boldsymbol{\mu}_i, \boldsymbol{\sigma}_i^2 \mathbf{I}\right), \tag{8.2}$$

where the mean $\boldsymbol{\mu}_i$ and standard derivation $\boldsymbol{\sigma}_i$ are outputs of the encoder given input $\mathbf{x}_i$, and they are parameterized by ϕ. Using the reparameterization trick (see Section 6.5.1), we obtain the lth latent sample $\mathbf{z}_{il} = \boldsymbol{\mu}_i + \boldsymbol{\sigma}_i \odot \boldsymbol{\epsilon}_l$, where $\boldsymbol{\epsilon}_l \sim \mathcal{N}(\mathbf{0}, \mathbf{I})$ and $\odot$ is the Hadamard product. Substitute these terms into Eq. 8.1, we have the Gaussian VAE loss [237]:

$$\mathcal{L}_{\text{VAE}}(\theta, \phi) \simeq -\sum_{r=1}^{R}\sum_{i=1}^{N_r}\left\{\frac{1}{2}\sum_{j=1}^{J}\left[1 + \log\left(\sigma_{ij}^{(r)}\right)^2 - \left(\mu_{ij}^{(r)}\right)^2 - \left(\sigma_{ij}^{(r)}\right)^2\right] + \frac{1}{L}\sum_{l=1}^{L}\log p_\theta\left(\mathbf{x}_i^{(r)}|\mathbf{z}_{il}\right)\right\}, \tag{8.3}$$

where J is the dimension of $\mathbf{z}$ and L denotes the number of latent samples. In practice, we set $L = 1$.

Figure 8.5 shows the structure of the VDANN. It comprises a speaker predictor C, a domain classifier D and a VAE, where the VAE comprises an encoder E and a decoder G. The parameters of C, D, E, and G are denoted as θ_c, θ_d, ϕ_e, and θ_g, respectively. Through adversarial training, a VDANN can form a domain-invariant space based on training data from multiple domains. More precisely, by applying adversarial training

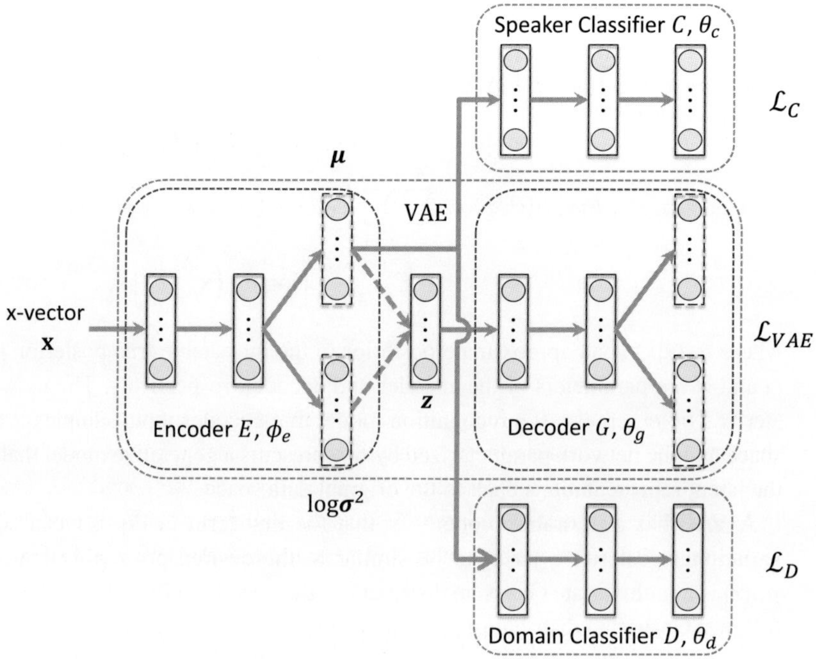

Figure 8.5 Structure of a VDANN. The dashed and solid arrows represent stochastic sampling and network connections, respectively.

on E while fixing θ_d and minimizing the cross-entropy loss of C with respect to ϕ_e will make E to produce a domain-invariant but speaker discriminative representation.

The VDANN is trained by minimizing the following loss function:

$$\mathcal{L}_{\text{VDANN}}(\theta_c, \theta_d, \phi_e, \theta_g) = \mathcal{L}_C(\theta_c, \phi_e) - \alpha \mathcal{L}_D(\theta_d, \phi_e) + \beta \mathcal{L}_{\text{VAE}}(\phi_e, \theta_g), \tag{8.4}$$

where

$$\mathcal{L}_C(\theta_c, \phi_e) = \sum_{r=1}^{R} \mathbb{E}_{p_{\text{data}}(\mathbf{x}^{(r)})} \left\{ - \sum_{k=1}^{K} y_k^{(r)} \log C\left(E\left(\mathbf{x}^{(r)}\right)\right)_k \right\}, \tag{8.5}$$

$$\mathcal{L}_D(\theta_d, \phi_e) = \sum_{r=1}^{R} \mathbb{E}_{p_{\text{data}}(\mathbf{x}^{(r)})} \left\{ - \log D\left(E\left(\mathbf{x}^{(r)}\right)\right)_r \right\}, \tag{8.6}$$

and $\mathcal{L}_{\text{VAE}}$ has a form similar to Eq. 8.3. The only exception is that the parameters of the encoder and decoder are changed to ϕ_e and θ_g, respectively. The subscript k in the categorical cross-entropy loss of the speaker classifier C in Eq. 8.5 represents the kth output of the classifier. α and β are hyperparameters controling the contribution of different losses that shape the features produced by E.

For each minibatch in the training process, the weights in D are optimized by minimizing the domain classification loss. Then, its weights are frozen when training the remaining part of the VDANN. To incorporate speaker information into E, the speaker cross-entropy loss as computed at the output of C is minimized. Meanwhile, the domain classification loss is maximized to ensure that E can create a domain-invariant space. The VAE loss is also minimized to make the embedded features to follow a Gaussian distribution. In summary, the training process of VDANN can be expressed by the following minimax operation:

$$\min_{\theta_c, \phi_e, \theta_g} \max_{\theta_d} \mathcal{L}_{\text{VDANN}}(\theta_c, \theta_d, \phi_e, \theta_g). \tag{8.7}$$

This minimax optimization process can be divided into two operations:

$$\hat{\theta}_d = \operatorname*{argmax}_{\theta_d} \mathcal{L}_{\text{VDANN}}(\hat{\theta}_c, \theta_d, \hat{\phi}_e, \hat{\theta}_g), \tag{8.8}$$

$$\left(\hat{\theta}_c, \hat{\phi}_e, \hat{\theta}_g\right) = \operatorname*{argmin}_{\theta_c, \phi_e, \theta_g} \mathcal{L}_{\text{VDANN}}(\theta_c, \hat{\theta}_d, \phi_e, \theta_g), \tag{8.9}$$

where symbols with a hat (e.g., $\hat{\theta}_c$) on the right-hand side of Eq. 8.8 and Eq. 8.9 mean that they are fixed during thc optimization process. After training, embedded features can be extracted from the last (linear) hidden layer of the encoder E (denoted by $\boldsymbol{\mu}$ in Figure 8.5). Because the approximate variational posterior is regularized to follow a Gaussian distribution, features extracted from the encoder will tend be Gaussian.

8.3.2 Relationship with Domain Adversarial Neural Network (DANN)

DANNs have been used for reducing the mismatch between the source and target domains, where data from the source domain have speaker labels while training data

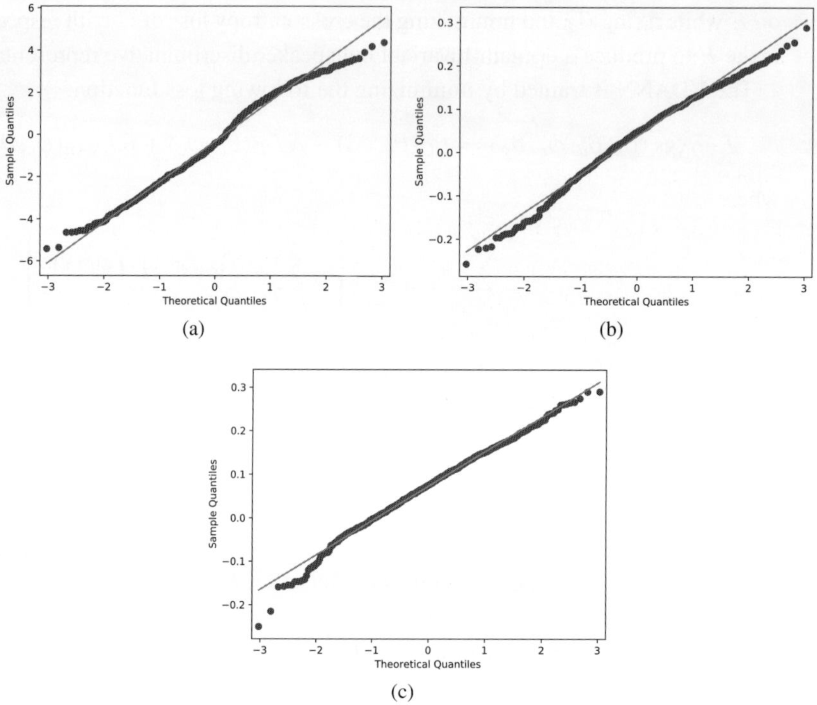

Figure 8.6 Quantile-quantile (Q–Q) plots of the eleventh component of (a) x-vectors, (b) DANN-transformed x-vectors, and (c) VDANN-transformed x-vectors. The vertical and horizontal axes correspond to the samples under test and the samples drawn from a standard normal distribution, respectively. The straight line represents the situation of perfectly Gaussian. The p-values above the graphs were obtained from Shapiro–Wilk tests in which $p > 0.05$ means failing to reject the null hypothesis that the test samples come from a Gaussian distribution.

from the target domain are unlabeled. In that case $R = 2$ in Eq. 8.5. For example, Wang et al. [211] proposed a DANN comprising a feature extractor E, a speaker predictor C and a domain classifier D. Note that the DANN in [211] is a special case of VDANN, because by setting $\beta = 0$ in Eq. 8.4, we obtain the loss function of DANNs:

$$\mathcal{L}_{\text{DANN}}(\theta_c, \theta_d, \phi_e) = \mathcal{L}_C(\theta_c, \phi_e) - \alpha \mathcal{L}_D(\theta_d, \phi_e), \tag{8.10}$$

where θ_c, θ_d, and ϕ_e are the parameters for C, D, E, respectively and α controls the trade-off between the two objectives. $\mathcal{L}_C$ and $\mathcal{L}_D$ are identical to Eq. 8.5 and Eq. 8.6, respectively. The parameters of DANN are optimized by:

$$\hat{\theta}_d = \underset{\theta_d}{\text{argmax}}\, \mathcal{L}_{\text{DANN}}\left(\hat{\theta}_c, \theta_d, \hat{\phi}_e\right), \tag{8.11}$$

$$\left(\hat{\theta}_c, \hat{\phi}_e\right) = \underset{\theta_c, \phi_e}{\text{argmin}}\, \mathcal{L}_{\text{DANN}}\left(\theta_c, \hat{\theta}_d, \phi_e\right). \tag{8.12}$$

Because there is no extra constraint on the distribution of the embedded features, adversarial training may lead to non-Gaussian embedded vectors, which is not desirable for the PLDA back end.

8.3.3 Gaussianality Analysis

Figure 8.6 shows the normal Q–Q plots of the three dimensions of x-vectors and the x-vectors transformed by a DANN and a VDANN. Obviously, the distribution of the x-vectors transformed by the VDANN is closer to a Gaussian distribution than the others. This suggests that the VAE loss can make the embedded vectors $\mathbf{z}$'s and the hidden layer outputs $\boldsymbol{\mu}$ to follow a Gaussian distribution. The p-values obtained from Shapiro–Wilk tests [282] also suggest that the distribution of VDANN-transformed vectors is the closest to the standard Gaussian.

Appendix: Exercises

This appendix comprises some of the exam questions that the authors used in their machine learning, human computer interaction, and speaker recognition courses in the last five years.

Q1 Denote a dataset as $\mathcal{X} \times \mathcal{L} = \{(\mathbf{x}_n, \ell_n); n = 1, \ldots, N\}$, where $\mathbf{x}_n \in \mathbb{R}^D$ and ℓ_n is the class label of $\mathbf{x}_n$. Assume that the dataset is divided into two classes such that the sets C_1 and C_2 comprise the vector indexes for which the vectors belong to Class 1 and Class 2, respectively. In Fisher discriminant analysis (FDA), $\mathbf{x}_n$ is projected onto a line to obtain a score $y_n = \mathbf{w}^{\mathsf{T}}\mathbf{x}_n$, where $\mathbf{w}$ is a weight vector defining the orientation of the line. Given that the objective function of FDA is

$$J(\mathbf{w}) = \frac{(\mu_1^y - \mu_2^y)^2}{(\sigma_1^y)^2 + (\sigma_2^y)^2},$$

where μ_k^y and $(\sigma_k^y)^2$ are the mean and variance of the FDA-projected scores for Class k, respectively. Also given is the mean of Class k:

$$\boldsymbol{\mu}_k = \frac{1}{N_k} \sum_{n \in C_k} \mathbf{x}_n,$$

where N_k is the number of samples in Class k.

(a) Show that the mean of the projected scores for Class k is

$$\mu_k^y = \mathbf{w}^{\mathsf{T}} \boldsymbol{\mu}_k.$$

(b) Show that the variance of the projected scores for Class k is

$$(\sigma_k^y)^2 = \frac{1}{N_k} \sum_{n \in C_k} \mathbf{w}^{\mathsf{T}}(\mathbf{x}_n - \boldsymbol{\mu}_k)(\mathbf{x}_n - \boldsymbol{\mu}_k)^{\mathsf{T}}\mathbf{w}.$$

(c) Show that the optimal projection vector $\mathbf{w}^*$ is given by

$$\mathbf{w}^* = \underset{\mathbf{w}}{\operatorname{argmax}} = \frac{\mathbf{w}^{\mathsf{T}}\mathbf{S}_B\mathbf{w}}{\mathbf{w}^{\mathsf{T}}\mathbf{S}_W\mathbf{w}},$$

where

$$\mathbf{S}_B = (\boldsymbol{\mu}_1 - \boldsymbol{\mu}_2)(\boldsymbol{\mu}_1 - \boldsymbol{\mu}_2)^{\mathsf{T}}$$

and

$$\mathbf{S}_W = \sum_{k=1}^{2} \frac{1}{N_k} \sum_{n \in C_k} (\mathbf{x}_n - \boldsymbol{\mu}_k)(\mathbf{x}_n - \boldsymbol{\mu}_k)^{\mathsf{T}}.$$

Q2 In factor analysis, an observed vector $\mathbf{x}$ can be expressed as

$$\mathbf{x} = \boldsymbol{\mu} + \mathbf{V}\mathbf{z} + \boldsymbol{\epsilon},$$

where $\boldsymbol{\mu}$ is the global mean of all possible $\mathbf{x}$'s, $\mathbf{V}$ is a low-rank matrix, $\mathbf{z}$ is the latent factor, and $\boldsymbol{\epsilon}$ is a residue term. Assume that the prior of $\mathbf{z}$ follows a standard Gaussian distribution $\mathcal{N}(\mathbf{0}, \mathbf{I})$ and that $\boldsymbol{\epsilon} \sim \mathcal{N}(\mathbf{0}, \boldsymbol{\Sigma})$. Show that the covariance matrix of $\mathbf{x}$'s is $\mathbf{V}\mathbf{V}^{\mathsf{T}} + \boldsymbol{\Sigma}$.

Q3 The kernel K-means algorithm aims to divide a set of training data $\mathcal{X} = \{\mathbf{x}_1, \ldots, \mathbf{x}_N\}$ into K disjoint sets $\{\mathcal{X}_1, \ldots, \mathcal{X}_K\}$ by minimizing the sum of squared error:

$$E_{\phi} = \sum_{k=1}^{K} \sum_{\mathbf{x} \in \mathcal{X}_k} \left\| \phi(\mathbf{x}) - \frac{1}{N_k} \sum_{\mathbf{z} \in \mathcal{X}_k} \phi(\mathbf{z}) \right\|^2, \tag{A.1}$$

where $\phi(\mathbf{x})$ is a function of $\mathbf{x}$. It can be shown that Eq. A.3 can be implemented by

$$E'_{\phi} = \sum_{k=1}^{K} \sum_{\mathbf{x} \in \mathcal{X}_k} \left[\frac{1}{N_k^2} \sum_{\mathbf{z} \in \mathcal{X}_k} \sum_{\mathbf{z}' \in \mathcal{X}_k} K(\mathbf{z}, \mathbf{z}') - \frac{2}{N_k} \sum_{\mathbf{z} \in \mathcal{X}_k} K(\mathbf{z}, \mathbf{x}) \right], \tag{A.2}$$

where $K(\mathbf{z}, \mathbf{z}') = \phi(\mathbf{z})^{\mathsf{T}} \phi(\mathbf{z}')$ is a nonlinear kernel.

(a) What is the purpose of the function $\phi(\mathbf{x})$?

(b) State an advantage of computing $\phi(\mathbf{z})^{\mathsf{T}} \phi(\mathbf{z}')$ using the nonlinear kernel $K(\mathbf{z}, \mathbf{z}')$.

(c) Give a function $\phi(\mathbf{x})$ so that $K(\mathbf{x}, \mathbf{y}) = \mathbf{x}^{\mathsf{T}}\mathbf{y}$.

Q4 Figure 3.8 shows the decision plane (solid line) and the two hyperplanes (dashed lines) that maximize the margin of separation d between the two classes with training samples represented by filled circles (•) and hollow circles (○), respectively. Denote $\mathbf{x}_i \in \Re^D$ and $y_i \in \{-1, +1\}$ as the ith training sample and its label, respectively, where $i = 1, \ldots, N$. Then, the training set is denoted as $\mathcal{T} = \{(\mathbf{x}_1, y_1), \ldots, (\mathbf{x}_N, y_N)\}$.

(a) Show that

$$d = \frac{2}{\|\mathbf{w}\|},$$

where $\|\mathbf{w}\|$ is the norm of the vector $\mathbf{w}$.

(b) Given the training set $\mathcal{T}$, the optimal solution of $\mathbf{w}$ can be found by maximizing d subject to the following constraints:

$$\mathbf{x}_i \cdot \mathbf{w} + b \geq +1 \text{ for } i \in \{1, \ldots, N\} \text{ where } y_i = +1$$
$$\mathbf{x}_i \cdot \mathbf{w} + b \leq -1 \text{ for } i \in \{1, \ldots, N\} \text{ where } y_i = -1.$$

(c) Explain why the solution of $\mathbf{w}$ can be obtained by solving the following constrained optimization problem:

$$\begin{array}{ll} \min & \frac{1}{2}\|\mathbf{w}\|^2 \\ \text{subject to} & y_i(\mathbf{x}_i \cdot \mathbf{w} + b) \geq 1 \quad \forall i = 1, \ldots, N. \end{array}$$

(d) Given that the Lagrangian function of this problem is

$$L(\mathbf{w}, b, \{\alpha_i\}) = \frac{1}{2}\|\mathbf{w}\|^2 - \sum_{i=1}^{N} \alpha_i [y_i(\mathbf{x}_i \cdot \mathbf{w} + b) - 1],$$

where α_i's are Lagrange multipliers. State the constraints for minimizing $L(\mathbf{w}, b, \{\alpha_i\})$.

(e) Show that the Wolfe dual of this constrained optimization problem is

$$\max_{\alpha} \sum_{i=1}^{N} \alpha_i - \frac{1}{2} \sum_{i=1}^{N} \sum_{j=1}^{N} \alpha_i \alpha_j y_i y_j (\mathbf{x}_i \cdot \mathbf{x}_j)$$

$$\text{subject to} \sum_{i=1}^{N} \alpha_i y_i = 0 \quad \text{and} \quad \alpha_i \geq 0, i = 1, \ldots, N.$$

Q5 The K-means algorithm aims to divide a set of training data $\mathcal{X} = \{\mathbf{x}_1, \ldots, \mathbf{x}_N\}$ into K disjoint sets $\{\mathcal{X}_1, \ldots, \mathcal{X}_K\}$ such that

$$\mathcal{X} = \bigcup_{k=1}^{K} \mathcal{X}_k \quad \text{and} \quad \mathcal{X}_i \cap \mathcal{X}_j = \emptyset \; \forall i \neq j.$$

This is achieved by minimizing the sum of the squared error:

$$E = \sum_{k=1}^{K} \sum_{\mathbf{x} \in \mathcal{X}_k} \|\mathbf{x} - \boldsymbol{\mu}_k\|^2.$$

(a) Show that E is minimal when $\boldsymbol{\mu}_k = \frac{1}{N_k} \sum_{\mathbf{x} \in \mathcal{X}_k} \mathbf{x}$, where N_k is the number of samples in $\mathcal{X}_k$.

(b) Explain why the K-means algorithm is not suitable for clustering the samples in Figure A.1.

(c) One possible approach to clustering the samples in Figure A.1 is to map $\mathbf{x}$'s to a high-dimensional feature space using a nonlinear function $\phi(\mathbf{x})$, followed by applying the K-means algorithm to cluster the mapped samples in the feature space. Specifically, we aim to find the K disjoint sets $\{\mathcal{X}_1, \ldots, \mathcal{X}_K\}$ that minimize the following objective function:

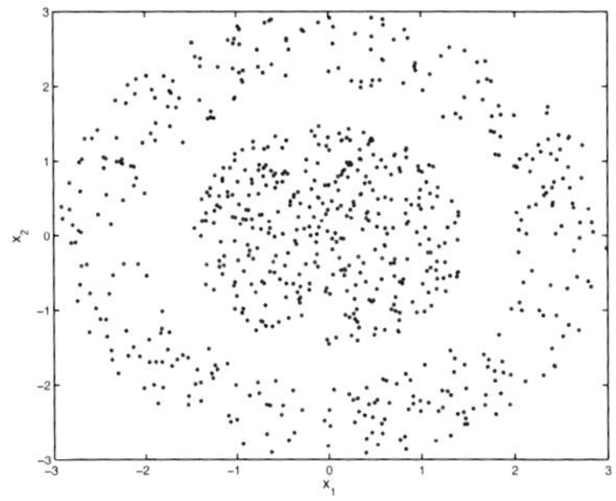

Figure A.1 Samples with a doughnut-shape pattern.

$$E_{\phi} = \sum_{k=1}^{K} \sum_{\mathbf{x} \in \mathcal{X}_k} \left\| \phi(\mathbf{x}) - \frac{1}{N_k} \sum_{\mathbf{z} \in \mathcal{X}_k} \phi(\mathbf{z}) \right\|^2 . \tag{A.3}$$

Show that minimizing Eq. A.3 is equivalent to minimizing the following objective function

$$E'_{\phi} = \sum_{k=1}^{K} \sum_{\mathbf{x} \in \mathcal{X}_k} \left[\frac{1}{N_k^2} \sum_{\mathbf{z} \in \mathcal{X}_k} \sum_{\mathbf{z}' \in \mathcal{X}_k} \phi(\mathbf{z})^{\mathsf{T}} \phi(\mathbf{z}') - \frac{2}{N_k} \sum_{\mathbf{z} \in \mathcal{X}_k} \phi(\mathbf{z})^{\mathsf{T}} \phi(\mathbf{x}) \right] . \tag{A.4}$$

(d) When the dimension of $\phi(\mathbf{x})$ is very high, Eq. A.4 cannot be implemented. Suggest a way to solve this problem and write the objective function.

Q6 Given a set of training vectors $\mathcal{X} = \{\mathbf{x}_1, \ldots, \mathbf{x}_N\}$ in D-dimensional space, principle component analysis (PCA) aims to find a projection matrix $\mathbf{U}$ that projects the vectors in $\mathcal{X}$ from D-dimensional space to M-dimensional space, where $M \leq D$. The projection matrix can be obtained by solving the following equation:

$$\mathbf{X}\mathbf{X}^{\mathsf{T}}\mathbf{U} = \mathbf{U}\mathbf{\Lambda}_M, \tag{A.5}$$

where $\mathbf{\Lambda}_M$ is an $M \times M$ diagonal matrix and $\mathbf{X}$ is a $D \times N$ centered data matrix whose nth column is given by $(\mathbf{x}_n - \frac{1}{N}\sum_{i=1}^{N} \mathbf{x}_i)$.

(a) How do the values of the diagonal elements of $\mathbf{\Lambda}_M$ related to the variances of the training vectors?

(b) If D is very large (say 100,000), solving Eq. A.5 is computationally demanding. Suggest a method to find $\mathbf{U}$ when $M < N \ll D$.

Q7 In principal component analysis (PCA), given a training set $\mathcal{X} = \{\mathbf{x}_1, \ldots, \mathbf{x}_N\}$ comprising vectors in D-dimensional space, the M-dimensional PCA-projected vectors $\{\mathbf{h}_i\}_{i=1}^{N}$ can be considered as latent variables that can approximately reconstruct the training data:

$$\mathbf{x}_i \approx \mathbf{\Phi}\mathbf{h}_i + \boldsymbol{\mu}, \quad i = 1, \ldots, N,$$

where $\mathbf{\Phi}$ is a $D \times M$ projection matrix and μ is the global mean. To determine $\mathbf{\Phi}$ from training data, we minimize the following cost function:

$$\hat{\mathbf{\Phi}}, \{\hat{\mathbf{h}}_i\}_{i=1}^N = \underset{\mathbf{\Phi}, \{\mathbf{h}_i\}_{i=1}^N}{\operatorname{argmin}} \left\{ \sum_{i=1}^N \left[\mathbf{x}_i - \mu - \mathbf{\Phi}\mathbf{h}_i\right]^\mathsf{T} \left[\mathbf{x}_i - \mu - \mathbf{\Phi}\mathbf{h}_i\right] \right\}.$$

(a) Show that $\hat{\mathbf{\Phi}}$ comprises the eigenvectors of

$$\sum_{i=1}^N (\mathbf{x}_i - \mu)(\mathbf{x}_i - \mu)^\mathsf{T}$$

Hints:

$$\frac{\partial}{\partial \mathbf{X}} \mathrm{tr}\{\mathbf{X}\mathbf{B}\mathbf{X}^\mathsf{T}\} = \mathbf{X}\mathbf{B}^\mathsf{T} + \mathbf{X}\mathbf{B} \quad \text{and} \quad \frac{\partial \mathbf{a}^\mathsf{T}\mathbf{X}^\mathsf{T}\mathbf{b}}{\partial \mathbf{X}} = \mathbf{b}\mathbf{a}^\mathsf{T}.$$

(b) If the dimension of $\mathbf{x}_i$ is 100,000, explain why computing $\hat{\mathbf{\Phi}}$ using the solution in Q7(a) is very expensive.

Q8 Assume that you are given a dataset $\mathcal{X} \times \mathcal{L} = \{(\mathbf{x}_n, \ell_n); n = 1, \ldots, N\}$, where $\mathbf{x}_n \in \mathbb{R}^D$ and ℓ_n is the class label of $\mathbf{x}_n$. Also assume that the dataset is divided into K classes such the set C_k comprises the vector indexes for which the vectors belong to the kth class. Linear discriminant analysis (LDA) can project $\mathbf{x}_n$ onto an M-dimensional space to give vector $\mathbf{y}_n$:

$$\mathbf{y}_n = \mathbf{W}^\mathsf{T}(\mathbf{x}_n - \mu), \quad \mathbf{y}_n \in \mathbb{R}^M$$

where $\mathbf{W} = [\mathbf{w}_1 \ \cdots \ \mathbf{w}_M]$ is a $D \times M$ projection matrix comprising M eigenvectors with the largest eigenvalues and $\mu = \frac{1}{N}\sum_{n=1}^N \mathbf{x}_n$ is the global mean of the vectors in $\mathcal{X}$.

(a) What is the maximum value of M? Briefly explain your answer.

(b) The projection matrix $\mathbf{W}$ can be obtained by maximizing the following objective function:

$$J(\mathbf{W}) = \mathrm{Tr}\left\{ \left(\mathbf{W}^\mathsf{T}\mathbf{S}_B\mathbf{W}\right) \left(\mathbf{W}^\mathsf{T}\mathbf{S}_W\mathbf{W}\right)^{-1} \right\},$$

where $\mathbf{S}_B$ and $\mathbf{S}_W$ are the between-class and the within-class scatter matrices, respectively, and Tr stands for matrix trace. Given that the total scatter matrix $\mathbf{S}_T$ is the sum of $\mathbf{S}_B$ and $\mathbf{S}_W$, i.e.,

$$\mathbf{S}_T = \sum_{k=1}^K \sum_{n \in C_k} (\mathbf{x}_n - \mu)(\mathbf{x}_n - \mu)^\mathsf{T} = \mathbf{S}_B + \mathbf{S}_W,$$

show that

$$\mathbf{S}_B = \sum_{k=1}^K N_k(\mu_k - \mu)(\mu_k - \mu)^\mathsf{T} \text{ and } \mathbf{S}_W = \sum_{k=1}^K \sum_{n \in C_k} (\mathbf{x}_n - \mu_k)(\mathbf{x}_n - \mu_k)^\mathsf{T},$$

where N_k is the number of samples in class k, i.e., $N_k = |C_k|$.

(c) The maximization of $J(\mathbf{W})$ with respect to $\mathbf{W}$ can be achieved by the following constrained optimization:

$$\begin{aligned} \max_{\mathbf{W}} \quad & \mathrm{Tr}\{\mathbf{W}^{\mathsf{T}}\mathbf{S}_B\mathbf{W}\} \\ \text{subject to} \quad & \mathbf{W}^{\mathsf{T}}\mathbf{S}_W\mathbf{W} = \mathbf{I} \end{aligned}$$

where $\mathbf{I}$ is an $M \times M$ identity matrix. Show that $\mathbf{W}$ comprises the M eigenvectors of $\mathbf{S}_W^{-1}\mathbf{S}_B$. *Hints*: The derivative of matrix trace is

$$\frac{\partial}{\partial \mathbf{X}}\mathrm{tr}\{\mathbf{X}^{\mathsf{T}}\mathbf{B}\mathbf{X}\mathbf{C}\} = \mathbf{B}\mathbf{X}\mathbf{C} + \mathbf{B}^{\mathsf{T}}\mathbf{X}\mathbf{C}^{\mathsf{T}}.$$

Q9 The output of a support vector machine (SVM) is given by

$$f(\mathbf{x}) = \sum_{i \in \mathcal{S}} a_i K(\mathbf{x}, \mathbf{x}_i) + b,$$

where $\mathbf{x}$ is an input vector on $\mathbb{R}^D$, $\mathcal{S}$ comprises the indexes to a set of support vectors, b is a bias term, a_i's are SVM parameters and $K(\cdot,\cdot)$ is a kernel. For a second-degree polynomial kernel, $K(\cdot,\cdot)$ is given by

$$K(\mathbf{x}, \mathbf{y}) = \left(1 + \mathbf{x}^{\mathsf{T}}\mathbf{y}\right)^2,$$

where $\mathbf{x}$ and $\mathbf{y}$ are any input vectors on the D-dimensional input space.

(a) Assume that $D = 2$, show that the kernel function maps the input vectors to a six-dimensional space, where the decision boundary becomes linear.

(b) Explain why $f(\mathbf{x})$ is a nonlinear function of $\mathbf{x}$.

Q10 Denote $\mathcal{X} = \{\mathbf{x}_1, \ldots, \mathbf{x}_N\}$ as a set of R-dimensional vectors. In factor analysis, $\mathbf{x}_i$'s are assumed to follow a linear model:

$$\mathbf{x}_i = \mathbf{m} + \mathbf{V}\mathbf{z}_i + \boldsymbol{\epsilon}_i \qquad i = 1, \ldots, N$$

where $\mathbf{m}$ is the global mean of vectors in $\mathcal{X}$, $\mathbf{V}$ is a low-rank $R \times D$ matrix, $\mathbf{z}_i$ is a D-dimensional latent factor with prior density $\mathcal{N}(\mathbf{z}|\mathbf{0}, \mathbf{I})$, and $\boldsymbol{\epsilon}_i$ is the residual noise following a Gaussian density with zero mean and covariance matrix $\boldsymbol{\Sigma}$. Based on the Bayes theorem, the posterior density of the latent factor $\mathbf{z}_i$ is given by

$$\begin{aligned} p(\mathbf{z}_i|\mathbf{x}_i) &\propto p(\mathbf{x}_i|\mathbf{z}_i)p(\mathbf{z}_i) \\ &= \mathcal{N}(\mathbf{x}_i|\mathbf{m} + \mathbf{V}\mathbf{z}_i, \boldsymbol{\Sigma})\mathcal{N}(\mathbf{z}|\mathbf{0}, \mathbf{I}), \end{aligned}$$

where $\mathcal{N}(\mathbf{z}|\boldsymbol{\mu}_z, \boldsymbol{\Sigma}_z)$ represents a Gaussian distribution with mean $\boldsymbol{\mu}_z$ and covariance matrix $\boldsymbol{\Sigma}_z$. Show that the posterior mean and posterior moment of $\mathbf{z}_i$ is given by

$$\begin{aligned} \langle \mathbf{z}_i|\mathbf{x}_i \rangle &= \mathbf{L}^{-1}\mathbf{V}^{\mathsf{T}}\boldsymbol{\Sigma}^{-1}(\mathbf{x}_i - \mathbf{m}) \\ \langle \mathbf{z}_i\mathbf{z}_i^{\mathsf{T}}|\mathbf{x}_i \rangle &= \mathbf{L}^{-1} + \langle \mathbf{z}_i|\mathcal{X} \rangle \langle \mathbf{z}_i^{\mathsf{T}}|\mathbf{x}_i \rangle, \end{aligned}$$

respectively, where $\mathbf{L}^{-1} = (\mathbf{I} + \mathbf{V}^{\mathsf{T}}\boldsymbol{\Sigma}^{-1}\mathbf{V})^{-1}$ is the posterior covariance matrix of $\mathbf{z}_i$.

Hints: The Gaussian distribution of random vectors $\mathbf{z}$'s can be expressed as

$$\mathcal{N}(\mathbf{z}|\boldsymbol{\mu}_z, \mathbf{C}_z) \propto \exp\left\{-\frac{1}{2}(\mathbf{z}-\boldsymbol{\mu}_z)^{\mathsf{T}}\mathbf{C}_z^{-1}(\mathbf{z}-\boldsymbol{\mu}_z)\right\}$$
$$\propto \exp\left\{\mathbf{z}^{\mathsf{T}}\mathbf{C}_z^{-1}\boldsymbol{\mu}_z - \frac{1}{2}\mathbf{z}^{\mathsf{T}}\mathbf{C}_z^{-1}\mathbf{z}\right\},$$

where $\boldsymbol{\mu}_z$ and $\boldsymbol{\Sigma}_z$ are the mean vector and covariance matrix, respectively.

Q11 The scoring functions of GMM–UBM and GMM–SVM speaker verification systems are closely related. Denote the acoustic vectors extracted from a test utterance as $O^{(t)} = \{\mathbf{o}_1, \ldots, \mathbf{o}_T\}$, where T is the number of frames in the utterance. Also denote $\Lambda^{(s)}$ and Λ^{ubm} as the GMM of client-speaker s and the UBM, respectively. The GMM–UBM score and the GMM–SVM score are respectively given by

$$S_{\text{GMM–UBM}}(O^{(t)}|\Lambda^{(s)}, \Lambda^{\text{ubm}}) = \log p(O^{(t)}|\Lambda^{(s)}) - \log p(O^{(t)}|\Lambda^{\text{ubm}})$$

and

$$S_{\text{GMM–SVM}}(O^{(t)}|\text{SVM}_s) = \alpha_0^{(s)} K\left(\text{vec}\,\boldsymbol{\mu}^{(s)}, \text{vec}\,\boldsymbol{\mu}^{(t)}\right) - \sum_{i \in \mathcal{S}_{\text{bkg}}} \alpha_i^{(s)} K\left(\text{vec}\,\boldsymbol{\mu}^{(i)}, \text{vec}\,\boldsymbol{\mu}^{(t)}\right) + b^{(s)},$$

where SVM_s is the SVM of speaker s, $\mathcal{S}_{\text{bkg}}$ comprises the support vector indexes of the background speakers, $\text{vec}\,\boldsymbol{\mu}^{(s)}$ and $\text{vec}\,\boldsymbol{\mu}^{(t)}$ are the GMM-supervector of speaker s and the test utterance, respectively, $\alpha_j^{(s)}$'s are the Lagrange multipliers, and $b^{(s)}$ is the bias term of the SVM.

(a) Based on the GMM–SVM scoring function, determine the lower-bound on the number of enrollment utterances from speaker s. Briefly justify your answer. You may assume that each enrollment utterance gives one GMM-supervector.

(b) In practical GMM–SVM systems, the kernel function $K(\cdot,\cdot)$ is always linear. Explain why it is the case. Also explain why it is inappropriate to use nonlinear kernels such as the RBF kernel.

(c) Discuss the advantages of GMM–SVM systems over the GMM–UBM systems.

Q12 In i-vector/PLDA speaker verification, the i-vectors are modeled by a factor analysis model:

$$\mathbf{x} = \mathbf{m} + \mathbf{V}\mathbf{z} + \boldsymbol{\epsilon} \tag{A.6}$$

where $\mathbf{x}$ is an i-vector, $\mathbf{m}$ is the global mean of all i-vectors, $\mathbf{V}$ represents the speaker subspace, $\mathbf{z}$ is a latent factor, and $\boldsymbol{\epsilon}$ is a residual term that follows a Gaussian distribution $\mathcal{N}(\boldsymbol{\epsilon}|\mathbf{0}, \boldsymbol{\Sigma})$.

(a) Explain why it is necessary to model i-vectors by Eq. A.6.

(b) Assume that the prior of $\mathbf{z}$ follows a standard Gaussian, i.e., $\mathbf{z} \sim \mathcal{N}(\mathbf{z}|\mathbf{0}, \mathbf{I})$, where $\mathbf{I}$ is an identity matrix. Use Eq. A.6 to explain why i-vectors follow a Gaussian distribution with mean $\mathbf{m}$ and covariance matrix $\mathbf{V}\mathbf{V}^{\mathsf{T}} + \boldsymbol{\Sigma}$, i.e.,

$$\mathbf{x} \sim \mathcal{N}(\mathbf{x}|\mathbf{m}, \mathbf{V}\mathbf{V}^\mathsf{T} + \boldsymbol{\Sigma})$$

Hints: Take the expectation of $\mathbf{x}$ and $(\mathbf{x} - \boldsymbol{\mu})(\mathbf{x} - \boldsymbol{\mu})^\mathsf{T}$ in Eq. A.6, i.e., $\mathbb{E}\{\mathbf{x}\}$ and $\mathbb{E}\{(\mathbf{x} - \boldsymbol{\mu})(\mathbf{x} - \boldsymbol{\mu})^\mathsf{T}\}$.

Q13 Assume that the acoustic vectors of an utterance are given by $O = \{\mathbf{o}_1, \ldots, \mathbf{o}_T\}$, where T is the number of frames in the utterance. Denote the parameters of a Gaussian mixture model (GMM) as $\boldsymbol{\Lambda} = \{\pi_j, \boldsymbol{\mu}_j, \boldsymbol{\Sigma}_j\}_{j=1}^M$, where π_j, $\boldsymbol{\mu}_j$, and $\boldsymbol{\Sigma}_j$ are the mixture coefficient, mean vector, and covariance matrix of the jth Gaussian, respectively. To use maximum-likelihood linear regression (MLLR) for adapting the mean vectors of the GMM, we may estimate the transformation parameters $(\hat{\mathbf{A}}, \hat{\mathbf{b}})$ as follows:

$$\begin{aligned}(\hat{\mathbf{A}}, \hat{\mathbf{b}}) &= \arg\max_{\mathbf{A},\mathbf{b}} \sum_{t=1}^{T} \log p(\mathbf{o}_t|\mathbf{A}, \mathbf{b}, \boldsymbol{\Lambda}) \\ &= \arg\max_{\mathbf{A},\mathbf{b}} \left\{ \sum_{t=1}^{T} \log \sum_{j=1}^{M} \pi_j \mathcal{N}(\mathbf{o}_t|\mathbf{A}\boldsymbol{\mu}_j + \mathbf{b}, \boldsymbol{\Sigma}_j) \right\},\end{aligned}$$

where $\mathcal{N}(\mathbf{x}|\boldsymbol{\mu}, \boldsymbol{\Sigma})$ represents a Gaussian distribution with mean $\boldsymbol{\mu}$ and covariance matrix $\boldsymbol{\Sigma}$.

(a) If the dimension of $\mathbf{o}_t$ is 39, what is the dimension of $\hat{\mathbf{A}}$?

(b) Express the adapted mean vector $\hat{\boldsymbol{\mu}}_j$ in terms of $\hat{\mathbf{A}}$, $\hat{\mathbf{b}}$, and $\boldsymbol{\mu}_j$.

(c) If we only want to adapt the mean vectors that are very close to the acoustic vectors in O, should we use MLLR for the adaptation? Briefly explain your answer.

Q14 In i-vector based speaker verification, the GMM-supervector $\boldsymbol{\mu}$ representing an utterance is assumed to follow a factor analysis model:

$$\boldsymbol{\mu} = \boldsymbol{\mu}^{(b)} + \mathbf{T}\mathbf{w} \tag{A.7}$$

where $\boldsymbol{\mu}^{(b)}$ is a GMM-supervector corresponding to a universal background model (UBM), $\mathbf{T}$ is a low-rank total variability matrix, and $\mathbf{w}$ is a low-dimensional latent factor.

(a) Discuss the purpose of matrix $\mathbf{T}$. Why is it important?

(b) To extract the i-vector from an utterance, we need to align the acoustic vectors of the utterance against the UBM and then compute the posterior mean of $\mathbf{w}$. Assuming that the acoustic vectors are of dimension 60 and that the UBM comprises 1024 Gaussians, what are the dimensions of $\boldsymbol{\mu}$ and $\mathbf{T}$ when the dimension of $\mathbf{w}$ is 500?

(c) Given the acoustic vectors $O = \{\mathbf{o}_1, \ldots, \mathbf{o}_T\}$ of an utterance, its i-vector $\mathbf{x}$ is the posterior mean of $\mathbf{w}$ in Eq. A.7, i.e., $\mathbf{x} = \mathbb{E}\{\mathbf{w}|O\}$. State one advantage of using i-vectors for speaker recognition. What is the advantage of using i-vectors rather than the GMM-supervectors ($\boldsymbol{\mu}$ in Eq. A.7) for speaker recognition?

Q15 In i-vector/PLDA speaker verification, the i-vectors are further modeled by another factor analysis model:

$$\mathbf{x} = \mathbf{m} + \mathbf{V}\mathbf{z} + \boldsymbol{\epsilon} \tag{A.8}$$

where $\mathbf{x}$ is an i-vector, $\mathbf{m}$ is the global mean of all i-vectors, $\mathbf{V}$ represents the speaker subspace, $\mathbf{z}$ is a latent factor, and $\boldsymbol{\epsilon}$ is a residual term with covariance matrix $\boldsymbol{\Sigma}$.

(a) Explain why it is necessary to model i-vectors by A.8.

(b) Denote $\mathbf{x}_s$ and $\mathbf{x}_t$ as the i-vectors of target-speaker s and test speaker t, respectively. Assume that the target-speaker and the test speaker are the same person. Show that the joint likelihood of $\mathbf{x}_s$ and $\mathbf{x}_t$ is given by

$$p(\mathbf{x}_s, \mathbf{x}_t|\text{Same speaker}) = \mathcal{N}\left(\begin{bmatrix}\mathbf{x}_s\\ \mathbf{x}_t\end{bmatrix}\middle|\begin{bmatrix}\mathbf{m}\\ \mathbf{m}\end{bmatrix}, \begin{bmatrix}\mathbf{V}\mathbf{V}^{\mathsf{T}} + \boldsymbol{\Sigma} & \mathbf{V}\mathbf{V}^{\mathsf{T}}\\ \mathbf{V}\mathbf{V}^{\mathsf{T}} & \mathbf{V}\mathbf{V}^{\mathsf{T}} + \boldsymbol{\Sigma}\end{bmatrix}\right)$$

where $\mathcal{N}(\mathbf{x}|\boldsymbol{\mu}_x, \boldsymbol{\Sigma}_x)$ denotes a Gaussian density function with mean vector $\boldsymbol{\mu}_x$ and covariance matrix $\boldsymbol{\Sigma}_x$. *Hint*: The convolution of Gaussians is also a Gaussian, i.e.,

$$\begin{aligned}\int p(\mathbf{x}|\mathbf{z})p(\mathbf{z})d\mathbf{z} &= \int \mathcal{N}(\mathbf{x}|\mathbf{m} + \mathbf{V}\mathbf{z}, \boldsymbol{\Sigma})\mathcal{N}(\mathbf{z}|\mathbf{0}, \mathbf{I})d\mathbf{z}\\ &= \mathcal{N}(\mathbf{x}|\mathbf{m}, \mathbf{V}\mathbf{V}^{\mathsf{T}} + \boldsymbol{\Sigma}).\end{aligned}$$

Q16 In i-vector based speaker verification, the dimension of the GMM-supervector $\boldsymbol{\mu}_s$ corresponding to Speaker s is reduced by the following factor analysis model:

$$\boldsymbol{\mu}_s = \boldsymbol{\mu} + \mathbf{T}\mathbf{w}_s,$$

where $\mathbf{T}$ is a low-rank matrix called the total variability matrix, $\mathbf{w}_s$ is a speaker factor and $\boldsymbol{\mu}$ is a mean supervector formed by stacking the mean vectors of a universal background model.

(a) Explain why this factor analysis model can reduce the dimension of $\boldsymbol{\mu}_s$.

(b) Why does the total variability matrix define both the speaker and session (channel) variability?

(c) The i-vector $\mathbf{x}_s$ of Speaker s is the posterior mean of $\mathbf{w}_s$, which can be obtained from $\mathbf{T}$ and the acoustic vectors derived from his/her utterance. The i-vector $\mathbf{x}_c$ of a claimant is obtained in the same manner. During a verification session, we are given the i-vectors of Speaker s and Claimant c. A naive way is to accept Claimant c if the cosine-distance score is larger than a decision threshold η, i.e.,

$$S_{\text{cosine}}(\mathbf{x}_s, \mathbf{x}_c) = \frac{\mathbf{x}_s^{\mathsf{T}}\mathbf{x}_c}{\|\mathbf{x}_s\|\|\mathbf{x}_c\|} > \eta,$$

where ^{T} and $\|\cdot\|$ denote vector transpose and vector norm, respectively. Explain why this naive approach is undesirable for speaker verification. Suggest a method to preprocess the i-vectors to remedy this undesirable situation.

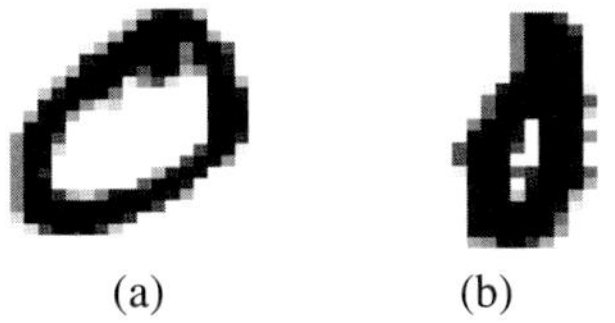

Figure A.2 Two of the digit '0' in the MNIST dataset.

Q17 A training set comprising 10,000 images was used to train a handwritten digit recognizer based on support vector machines (SVMs). Each of the digits ('0' to '9') has 1,000 images of size 28×28 pixels. The recognizer comprises 10 one-vs-rest SVMs, each responsible for classifying one digit against the remaining digits. Figure A.2 shows two of the Digit '0'; they are part of the training set for training the SVM responsible for classifying Digit '0' against Digits '1'–'9'.

(a) Which of the images [(a) or (b)] in Figure A.2 corresponds to a support vector? Briefly explain your answer.

(b) Assume that the SVMs in the recognizer use a radial basis function (RBF) as their kernel, i.e.,

$$K(\mathbf{x}, \mathbf{x}_i) = \exp\left\{-\frac{\|\mathbf{x} - \mathbf{x}_i\|^2}{2\sigma^2}\right\},$$

where σ is the kernel parameter and $\mathbf{x}_i$'s are support vectors. What will be the theoretical maximum number of support vectors in each SVM? Suggest the value of σ that will cause the SVMs to have the maximum number of support vectors.

(c) If the number of training images reduces to one per digit and linear SVMs are used in the recognizer, what will be the minimum number of support vectors for each SVM?

Q18 In GMM-UBM and GMM-SVM speaker verification, given a sequence of acoustic vectors $\mathcal{X}^{(s)}$ from a client speaker s, the maximum a posteriori (MAP) adaptation is used for adapting the universal background model (UBM) to create the speaker-dependent Gaussian mixture model (GMM). Typically, only the mean vectors of the UBM are adapted:

$$\boldsymbol{\mu}_j^{(s)} = \alpha_j E_j(\mathcal{X}^{(s)}) + (1 - \alpha_j)\boldsymbol{\mu}_j^{\text{ubm}}, j = 1, \ldots, M,$$

where M is the number of mixture components in the UBM, $E_j(\mathcal{X}^{(s)})$ is the sufficient statistics depending on $\mathcal{X}^{(s)}$, and $\boldsymbol{\mu}_j^{\text{ubm}}$ and $\boldsymbol{\mu}_j^{(s)}$ are the jth mean vector of the UBM and the adapted GMM, respectively.

(a) Discuss the value of α_j when the enrollment utterance is very long and when the enrollment utterance is very short.

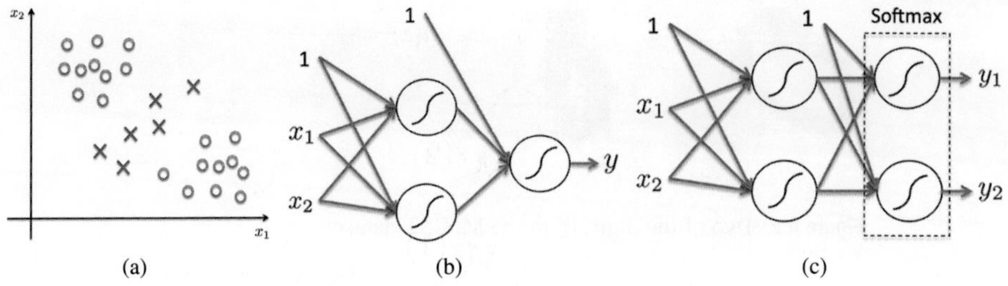

Figure A.3 (a) A two-class problem. (b) A neural network with one output node. (c) A neural network with two output nodes.

(b) In GMM-SVM speaker verification, we stack the mean vectors $\mu_j^{(s)}$ for $j = 1, \ldots, M$ to construct a speaker-dependent supervector vector:

$$\mu^{(s)} = \left[\left(\mu_1^{(s)}\right)^{\mathsf{T}} \cdots \left(\mu_M^{(s)}\right)^{\mathsf{T}} \right]^{\mathsf{T}}.$$

Why is it important to use MAP instead of directly applying the EM algorithm to compute $\mu_j^{(s)}$'s when constructing vec $\mu^{(s)}$?

Q19 Figure A.3(a) shows a binary classification problem.

(a) Explain why a perceptron (a network with only one neuron) will fail to solve this classification problem.

(b) Explain why the network in Figure A.3(b) can solve this problem perfectly as long as the activation function in the hidden layer is nonlinear.

Q20 The problem in Figure A.3(a) can also be solved by the network shown in Figure A.3(c). The network in Figure A.3(c) is trained by minimizing the multi-class cross-entropy loss function:

$$E_{\text{mce}} = -\sum_{\mathbf{x} \in \mathcal{X}} \sum_{k=1}^{2} t_k \log y_k, \quad k = 1,2$$

where $t_k \in \{0,1\}$ are the target outputs for the training sample $\mathbf{x} = [x_1 \;\; x_2]^{\mathsf{T}}$ in the input space and $\mathcal{X}$ is a minibatch. To use this cross-entropy function, the output nodes should use the softmax function, i.e.,

$$y_k = \frac{\exp(a_k)}{\sum_{j=1}^{2} \exp(a_j)},$$

where a_k is the activation of the kth node in the output layer.

(a) Show that $0 \leq y_k \leq 1$.

(b) Show that E_{mce} can be reduced to the binary cross-entropy:

$$E_{\text{bce}} = \sum_{\mathbf{x} \in \mathcal{X}} [-t_k \log y_k - (1 - t_k) \log(1 - y_k)], \quad k = 1 \text{ or } 2$$

Discuss the implication of this result.

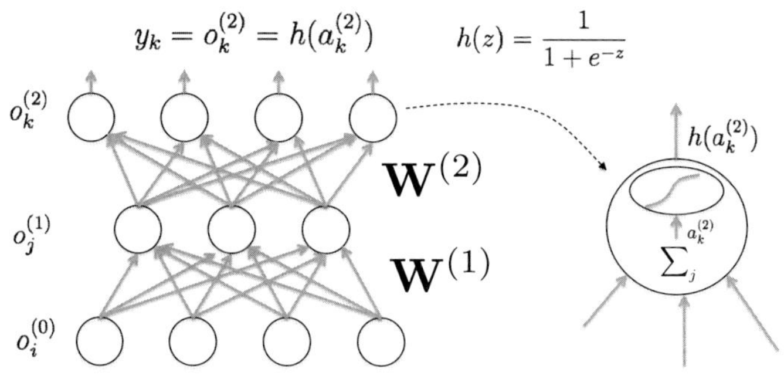

Figure A.4 A neural network with one hidden layer.

Q21 A deep neural networks has one input layer, $L - 1$ hidden layers, and one output layers with K outputs.

(a) If the network is to be used for classification, suggest an appropriate activation function for the output nodes. Express the suggested function in terms of the linear weighted sum $a_k^{(L)}$ at output node k, for $k = 1, \ldots, K$.

(b) For the network to be useful, each neuron in the hidden layers should have a nonlinear activation function such as the sigmoid function or the ReLU. Explain why the network will not be very useful if a linear activation function is used for all neurons, including the output neurons.

Q22 Figure A.4 shows a neural network with one hidden layer. Given a training vector $\mathbf{x} \in \mathbb{R}^D$, the instantaneous squared error between the actual outputs y_k's and the target outputs t_k's of the network is given by

$$E = \frac{1}{2}\sum_{k=1}^{K}(y_k - t_k)^2 = \frac{1}{2}\sum_{k=1}^{K}\left(o_k^{(2)} - t_k\right)^2, \tag{A.9}$$

where

$$o_k^{(2)} = h(a_k^{(2)}) \tag{A.10}$$

is the output of the kth output node of the network. In Eq. A.10,

$$h(a_k^{(2)}) = \frac{1}{1 + e^{-a_k^{(2)}}} \quad \text{and} \quad a_k^{(2)} = \sum_j o_j^{(1)} w_{kj}^{(2)},$$

where $o_j^{(1)}$ is the jth hidden node's output.

(a) Based on Figure A.4, determine the value of K in Eq. A.9.

(b) Show that the gradient of E with respect to the weight connecting the jth hidden node and the kth output node is given by

$$\frac{\partial E}{\partial w_{kj}^{(2)}} = \left(o_k^{(2)} - t_k\right) o_k^{(2)} \left(1 - o_k^{(2)}\right) o_j^{(1)}.$$

(c) Write the equation for updating the weight $w_{kj}^{(2)}$ based on stochastic gradient descent.

Q23 The stochastic gradient-descent algorithm of a single-output deep neural network (DNN) aims to minimize the following instantaneous cross-entropy error:

$$E_{\mathrm{ce}}(\mathbf{W}) = -t \log y - (1 - t)\log(1 - y),$$

where $\mathbf{W}$ comprises the weights of the network, y is the actual output subject to an input vector $\mathbf{x}$, and t is the target output. Assume that the DNN has L layers, show that the error gradient with respect to the weights in the output layer is

$$\frac{\partial E_{\mathrm{ce}}}{\partial w_j^{(L)}} = (y - t)o_j^{(L-1)},$$

where $w_j^{(L)}$ is the weight connecting the jth hidden neuron in the $(L - 1)$-th layer to the output node (at layer L) and $o_j^{(L-1)}$ is the output of this hidden neuron.

References

[1] C. M. Bishop, *Pattern Recognition and Machine Learning*. New York: Springer, 2006.

[2] Z. L. Tan and M. W. Mak, "Bottleneck features from SNR-adaptive denoising deep classifier for speaker identification," in *Proceedings of Asia-Pacific Signal and Information Processing Association Annual Summit and Conference (APSIPA ASC)*, 2015.

[3] Z. Tan, M. Mak, B. K. Mak, and Y. Zhu, "Denoised senone i-vectors for robust speaker verification," *IEEE/ACM Transactions on Audio, Speech, and Language Processing*, vol. 26, no. 4, pp. 820–830, Apr. 2018.

[4] L. v. d. Maaten and G. Hinton, "Visualizing data using t-SNE," *Journal of Machine Learning Research*, vol. 9, pp. 2579–2605, Nov. 2008.

[5] M. H. Moattar and M. M. Homayounpour, "A review on speaker diarization systems and approaches," *Speech Communication*, vol. 54, no. 10, pp. 1065–1103, 2012.

[6] S. B. Davis and P. Mermelstein, "Comparison of parametric representations for monosyllabic word recognition in continuously spoken sentences," *IEEE Transactions on Acoustics, Speech, and Signal Processing*, vol. 28, no. 4, pp. 357–366, Aug. 1980.

[7] D. A. Reynolds, T. F. Quatieri, and R. B. Dunn, "Speaker verification using adapted Gaussian mixture models," *Digital Signal Processing*, vol. 10, no. 1–3, pp. 19–41, Jan. 2000.

[8] A. P. Dempster, N. M. Laird, and D. B. Rubin, "Maximum likelihood from incomplete data via the EM algorithm," *Journal of the Royal Statistical Society: Series B (Methodological)*, vol. 39, no. 1, pp. 1–38, 1977.

[9] J. Pelecanos and S. Sridharan, "Feature warping for robust speaker verification," in *Proceedings of Speaker and Language Recognition Workshop (Odyssey)*, 2001, pp. 213–218.

[10] M. W. Mak, K. K. Yiu, and S. Y. Kung, "Probabilistic feature-based transformation for speaker verification over telephone networks," *Neurocomputing: Special Issue on Neural Networks for Speech and Audio Processing*, vol. 71, pp. 137–146, 2007.

[11] R. Teunen, B. Shahshahani, and L. Heck, "A model-based transformational approach to robust speaker recognition," in *Proc of International Conference on Spoken Language Processing (ICSLP)*, vol. 2, 2000, pp. 495–498.

[12] K. K. Yiu, M. W. Mak, and S. Y. Kung, "Environment adaptation for robust speaker verification by cascading maximum likelihood linear regression and reinforced learning," *Computer Speech and Language*, vol. 21, pp. 231–246, 2007.

[13] R. Auckenthaler, M. Carey, and H. Lloyd-Thomas, "Score normalization for text-independent speaker verification systems," *Digital Signal Processing*, vol. 10, no. 1–3, pp. 42–54, Jan. 2000.

[14] W. M. Campbell, D. E. Sturim, and D. A. Reynolds, "Support vector machines using GMM supervectors for speaker verification," *IEEE Signal Processing Letters*, vol. 13, no. 5, pp. 308–311, May 2006.

[15] P. Kenny, G. Boulianne, P. Ouellet, and P. Dumouchel, "Joint factor analysis versus eigenchannels in speaker recognition," *IEEE Transactions on Audio, Speech, and Language Processing*, vol. 15, no. 4, pp. 1435–1447, May 2007.

[16] N. Dehak, P. Kenny, R. Dehak, P. Dumouchel, and P. Ouellet, "Front-end factor analysis for speaker verification," *IEEE Transactions on Audio, Speech, and Language Processing*, vol. 19, no. 4, pp. 788–798, May 2011.

[17] S. Prince and J. Elder, "Probabilistic linear discriminant analysis for inferences about identity," in *Proceedings of IEEE International Conference on Computer Vision (ICCV)*, 2007, pp. 1–8.

[18] A. Martin, G. Doddington, T. Kamm, M. Ordowski, and M. Przybocki, "The DET curve in assessment of detection task performance," in *Proceedings of European Conference on Speech Communication and Technology (EUROSPEECH)*, 1997, pp. 1895–1898.

[19] D. Leeuwen and N. Brümmer, "The distribution of calibrated likelihood-ratios in speaker recognition," in *Proceedings of Annual Conference of International Speech Communication Association (INTERSPEECH)*, 2013, pp. 1619–1623.

[20] K. Hornik, M. Stinchcombe, and H. White, "Multilayer feedforward networks are universal approximators," *Neural Networks*, vol. 2, pp. 359–366, 1989.

[21] S. Kullback and R. A. Leibler, "On information and sufficiency," *Annals of Mathematical Statistics*, vol. 22, no. 1, pp. 79–86, 1951.

[22] M. Jordan, Z. Ghahramani, T. Jaakkola, and L. Saul, "An introduction to variational methods for graphical models," *Machine Learning*, vol. 37, no. 2, pp. 183–233, 1999.

[23] H. Attias, "Inferring parameters and structure of latent variable models by variational Bayes," in *Proceedings of Conference on Uncertainty in Artificial Intelligence (UAI)*, 1999, pp. 21–30.

[24] R. M. Neal, "Probabilistic inference using Markov chain Monte Carlo methods," Department of Computer Science, University of Toronto, Tech. Rep., 1993.

[25] J. S. Liu, *Monte Carlo Strategies in Scientific Computing*. New York, NY: Springer, 2008.

[26] C. Andrieu, N. De Freitas, A. Doucet, and M. I. Jordan, "An introduction to MCMC for machine learning," *Machine Learning*, vol. 50, no. 1-2, pp. 5–43, 2003.

[27] S. Geman and D. Geman, "Stochastic relaxation, Gibbs distributions, and the Bayesian restoration of images," *IEEE Transactions on Pattern Analysis and Machine Intelligence*, vol. 6, no. 1, pp. 721–741, 1984.

[28] W. K. Hastings, "Monte Carlo sampling methods using Markov chains and their applications," *Biometrika*, vol. 57, pp. 97–109, 1970.

[29] Y. W. Teh, M. I. Jordan, M. J. Beal, and D. M. Blei, "Hierarchical Dirichlet processes," *Journal of American Statistical Association*, vol. 101, no. 476, pp. 1566–1581, 2006.

[30] S. Watanabe and J.-T. Chien, *Bayesian Speech and Language Processing*. Cambridge, UK: Cambridge University Press, 2015.

[31] D. J. MacKay, "Bayesian interpolation," *Neural computation*, vol. 4, no. 3, pp. 415–447, 1992.

[32] S. Y. Kung, M. W. Mak, and S. H. Lin, *Biometric Authentication: A Machine Learning Approach*. Englewood Cliffs, NJ: Prentice Hall, 2005.

[33] V. N. Vapnik, *The Nature of Statistical Learning Theory*. New York: Springer-Verlag, 1995.

[34] S. P. Boyd and L. Vandenberghe, *Convex Optimization*. New York: Cambridge University Press, 2004.

[35] M. Mak and W. Rao, "Utterance partitioning with acoustic vector resampling for GMM-SVM speaker verification," *Speech Communication*, vol. 53, no. 1, pp. 119–130, Jan. 2011.

[36] G. Wu and E. Y. Chang, "KBA: Kernel boundary alignment considering imbalanced data distribution," *IEEE Transactions on Knowledge and Data Engineering*, vol. 17, no. 6, pp. 786–795, 2005.

[37] Y. Tang, Y. Q. Zhang, N. V. Chawla, and S. Krasser, "SVMs modeling for highly imbalanced classification," *IEEE Transactions on System, Man, and Cybernetics, Part B*, vol. 39, no. 1, pp. 281–288, Feb. 2009.

[38] M. W. Mak and W. Rao, "Acoustic vector resampling for GMMSVM-based speaker verification," in *Proceedings of Annual Conference of International Speech Communication Association (INTERSPEECH)*, 2010, pp. 1449–1452.

[39] W. Rao and M. W. Mak, "Addressing the data-imbalance problem in kernel-based speaker verification via utterance partitioning and speaker comparison," in *Interspeech*, 2011, pp. 2717–2720.

[40] A. Solomonoff, C. Quillen, and W. M. Campbell, "Channel compensation for SVM speaker recognition," in *Proceedings of Speaker and Language Recognition Workshop (Odyssey)*, 2004, pp. 57–62.

[41] A. Solomonoff, W. M. Campbell, and I. Boardman, "Advances in channel compensation for SVM speaker recognition," in *Proceedings of International Conference on Acoustics, Speech, and Signal Processing (ICASSP)*, 2005, pp. 629–632.

[42] W. M. Campbell, D. E. Sturim, D. A. Reynolds, and A. Solomonoff, "SVM based speaker verification using a GMM supervector kernel and NAP variability compensation," in *Proceedings of International Conference on Acoustics, Speech, and Signal Processing (ICASSP)*, vol. 1, 2006, pp. 97–100.

[43] E. Kokiopoulou, J. Chen, and Y. Saad, "Trace optimization and eigenproblems in dimension reduction methods," *Numerical Linear Algebra with Applications*, vol. 18, no. 3, pp. 565–602, 2011.

[44] P. Bromiley, "Products and convolutions of Gaussian probability density functions," *Tina-Vision Memo*, vol. 3, no. 4, 2003.

[45] P. Kenny, G. Boulianne, and P. Dumouchel, "Eigenvoice modeling with sparse training data," *IEEE Transactions on Speech and Audio Processing*, vol. 13, no. 3, pp. 345–354, 2005.

[46] S. M. Kay, *Fundamentals of Statistical Signal Processing*. Englewood Cliffs, NJ: Prentice-Hall, 1993.

[47] M. W. Mak and J. T. Chien, "PLDA and mixture of PLDA formulations," Supplementary Materials for "Mixture of PLDA for Noise Robust I-Vector Speaker Verification," *IEEE/ACM Trans. on Audio Speech and Language Processing*, vol. 24, No. 1, pp. 130–142, Jan. 2016. [Online]. Available: http://bioinfo.eie.polyu.edu.hk/mPLDA/SuppMaterials.pdf

[48] P. Rajan, A. Afanasyev, V. Hautamäki, and T. Kinnunen, "From single to multiple enrollment i-vectors: Practical PLDA scoring variants for speaker verification," *Digital Signal Processing*, vol. 31, pp. 93–101, 2014.

[49] L. Chen, K. A. Lee, B. Ma, W. Guo, H. Li, and L. R. Dai, "Minimum divergence estimation of speaker prior in multi-session PLDA scoring," in *Proceedings of International Conference on Acoustics, Speech, and Signal Processing (ICASSP)*, 2014, pp. 4007–4011.

[50] S. Cumani, O. Plchot, and P. Laface, "On the use of i-vector posterior distributions in probabilistic linear discriminant analysis," *IEEE/ACM Transactions on Audio, Speech, and Language Processing*, vol. 22, no. 4, pp. 846–857, 2014.

[51] L. Burget, O. Plchot, S. Cumani, O. Glembek, P. Matejka, and N. Briimmer, "Discriminatively trained probabilistic linear discriminant analysis for speaker verification," in

Acoustics, Speech, and Signal Processing (ICASSP), 2011 IEEE International Conference on, 2011, pp. 4832–4835.

[52] V. Vasilakakis, P. Laface, and S. Cumani, "Pairwise discriminative speaker verification in the I-vector space," *IEEE Transactions on Audio, Speech, and Language Processing*, vol. 21, no. 6, pp. 1217–1227, 2013.

[53] J. Rohdin, S. Biswas, and K. Shinoda, "Constrained discriminative plda training for speaker verification," in *Proceedings IEEE International Conference on Acoustics, Speech, and Signal Processing (ICASSP)*. IEEE, 2014, pp. 1670–1674.

[54] N. Li and M. W. Mak, "SNR-invariant PLDA modeling for robust speaker verification," in *Proceedings of Annual Conference of International Speech Communication Association (INTERSPEECH)*, 2015.

[55] N. Li and M. W. Mak, "SNR-invariant PLDA modeling in nonparametric subspace for robust speaker verification," *IEEE/ACM Trans. on Audio Speech and Language Processing*, vol. 23, no. 10, pp. 1648–1659, 2015.

[56] S. O. Sadjadi, J. Pelecanos, and W. Zhu, "Nearest neighbor discriminant analysis for robust speaker recognition," in *Proceedings of Annual Conference of International Speech Communication Association (INTERSPEECH)*, 2014, pp. 1860–1864.

[57] L. He, X. Chen, C. Xu, and J. Liu, "Multi-objective optimization training of plda for speaker verification," in *2019 IEEE International Conference on Acoustics, Speech, and Signal Processing (ICASSP)*. IEEE, 2019, pp. 6026–6030.

[58] O. Ghahabi and J. Hernando, "Deep belief networks for i-vector based speaker recognition," in *Proceedings of International Conference on Acoustics, Speech, and Signal Processing (ICASSP)*, 2014, pp. 1700–1704.

[59] T. Stafylakis, P. Kenny, M. Senoussaoui, and P. Dumouchel, "Preliminary investigation of Boltzmann machine classifiers for speaker recognition," in *Proceedings of Speaker and Language Recognition Workshop (Odyssey)*, 2012.

[60] O. Ghahabi and J. Hernando, "I-vector modeling with deep belief networks for multi-session speaker recognition," in *Proceedings of Speaker and Language Recognition Workshop (Odyssey)*, 2014, pp. 305–310.

[61] P. Kenny, "Bayesian speaker verification with heavy-tailed priors," in *Proceedings of Speaker and Language Recognition Workshop (Odyssey)*, 2010.

[62] N. Brummer, A. Silnova, L. Burget, and T. Stafylakis, "Gaussian meta-embeddings for efficient scoring of a heavy-tailed PLDA model," in *Proceedings of Speaker and Language Recognition Workshop (Odyssey)*, 2018, pp. 349–356.

[63] K. B. Petersen and M. S. Pedersen, "The matrix cookbook," Oct 2008. [Online]. Available: www2.imm.dtu.dk/pubdb/p.php?3274

[64] W. D. Penny, "KL-Divergences of Normal, Gamma, Direchlet and Wishart densities," Department of Cognitive Neurology, University College London, Tech. Rep., 2001.

[65] J. Soch and C. Allefeld, "Kullback-Leibler divergence for the normal-Gamma distribution," *arXiv preprint arXiv:1611.01437*, 2016.

[66] D. Garcia-Romero and C. Espy-Wilson, "Analysis of i-vector length normalization in speaker recognition systems," in *Proceedings of Annual Conference of International Speech Communication Association (INTERSPEECH)*, 2011, pp. 249–252.

[67] A. Silnova, N. Brummer, D. Garcia-Romero, D. Snyder, and L. Burget, "Fast variational Bayes for heavy-tailed PLDA applied to i-vectors and x-vectors," *arXiv preprint arXiv:1803.09153*, 2018.

[68] S. Shum, N. Dehak, E. Chuangsuwanich, D. Reynolds, and J. Glass, "Exploiting intra-conversation variability for speaker diarization," in *Proceedings of Annual Conference of International Speech Communication Association (INTERSPEECH)*, 2011, pp. 945–948.

[69] E. Khoury and M. Garland, "I-vectors for speech activity detection," in *Proceedings of Speaker and Language Recognition Workshop (Odyssey)*, 2016, pp. 334–339.

[70] N. Dehak, P. A. Torres-Carrasquillo, D. Reynolds, and R. Dehak, "Language recognition via i-vectors and dimensionality reduction," in *Proceedings of Annual Conference of International Speech Communication Association (INTERSPEECH)*, 2011, pp. 857–860.

[71] S. S. Xu, M.-W. Mak, and C.-C. Cheung, "Patient-specific heartbeat classification based on i-vector adapted deep neural networks," in *Proceedings of IEEE International Conference on Bioinformatics and Biomedicine*, 2018.

[72] P. Kenny, "A small footprint i-vector extractor," in *Proceedings of Speaker and Language Recognition Workshop (Odyssey)*, 2012.

[73] J. Luttinen and A. Ilin, "Transformations in variational bayesian factor analysis to speed up learning," *Neurocomputing*, vol. 73, no. 7–9, pp. 1093–1102, 2010.

[74] A. Hatch, S. Kajarekar, and A. Stolcke, "Within-class covariance normalization for SVM-based speaker recognition," in *Proceedings of International Conference on Spoken Language Processing (ICSLP)*, 2006, pp. 1471–1474.

[75] K. Fukunaga, *Introduction to Statistical Pattern Recognition*. Boston, MA: Academic Press, 1990.

[76] Z. Li, D. Lin, and X. Tang, "Nonparametric discriminant analysis for face recognition," *IEEE Transactions on Pattern Analysis and Machine Intelligence*, vol. 31, no. 4, pp. 755–761, 2009.

[77] F. Bahmaninezhad and J. H. Hansen, "I-vector/PLDA speaker recognition using support vectors with discriminant analysis," in *Proceedings of International Conference on Acoustics, Speech, and Signal Processing (ICASSP)*. IEEE, 2017, pp. 5410–5414.

[78] M. W. Mak and H. B. Yu, "A study of voice activity detection techniques for NIST speaker recognition evaluations," *Computer, Speech and Language*, vol. 28, no. 1, pp. 295–313, Jan. 2014.

[79] W. Rao and M. W. Mak, "Boosting the performance of i-vector based speaker verification via utterance partitioning," *IEEE Trans. on Audio, Speech, and Language Processing*, vol. 21, no. 5, pp. 1012–1022, May 2013.

[80] P. Kenny, T. Stafylakis, P. Ouellet, M. J. Alam, and P. Dumouchel, "PLDA for speaker verification with utterances of arbitrary duration," in *Proceedings of International Conference on Acoustics, Speech, and Signal Processing (ICASSP)*, 2013, pp. 7649–7653.

[81] W. Rao, M. W. Mak, and K. A. Lee, "Normalization of total variability matrix for i-vector/PLDA speaker verification," in *Proceedings of International Conference on Acoustics, Speech, and Signal Processing (ICASSP)*, 2015, pp. 4180–4184.

[82] W. W. Lin and M. W. Mak, "Fast scoring for PLDA with uncertainty propagation," in *Proceedings of Speaker and Language Recognition Workshop (Odyssey)*, 2016, pp. 31–38.

[83] W. W. Lin, M. W. Mak, and J. T. Chien, "Fast scoring for PLDA with uncertainty propagation via i-vector grouping," *Computer Speech & Language*, vol. 45, pp. 503–515, 2017.

[84] Y. Lei, N. Scheffer, L. Ferrer, and M. McLaren, "A novel scheme for speaker recognition using a phonetically-aware deep neural network," in *Proceedings of International Conference on Acoustics, Speech, and Signal Processing (ICASSP)*, 2014.

[85] L. Ferrer, Y. Lei, M. McLaren, and N. Scheffer, "Study of senone-based deep neural network approaches for spoken language recognition," *IEEE/ACM Transactions on Audio, Speech, and Language Processing*, vol. 24, no. 1, pp. 105–116, 2016.

[86] P. Kenny, "Joint factor analysis of speaker and session variability: Theory and algorithms," CRIM, Montreal, Tech. Rep. CRIM-06/08-13, 2005.

[87] P. Kenny, P. Ouellet, N. Dehak, V. Gupta, and P. Dumouchel, "A study of inter-speaker variability in speaker verification," *IEEE Transactions on Audio, Speech, and Language Processing*, vol. 16, no. 5, pp. 980–988, 2008.

[88] O. Glembek, L. Burget, N. Dehak, N. Brummer, and P. Kenny, "Comparison of scoring methods used in speaker recognition with joint factor analysis," in *Proceedings of International Conference on Acoustics, Speech, and Signal Processing (ICASSP)*, 2009, pp. 4057–4060.

[89] G. E. Hinton and R. R. Salakhutdinov, "Reducing the dimensionality of data with neural networks," *Science*, vol. 313, no. 5786, pp. 504–507, 2006.

[90] G. E. Hinton, *A Practical Guide to Training Restricted Boltzmann Machines*. Berlin Heidelberg: Springer, 2012, pp. 599–619.

[91] J. J. Hopfield, "Neural networks and physical systems with emergent collective computational abilities," *Proceedings of the National Academy of Sciences of the United States of America*, vol. 79, no. 8, pp. 2554–2558, 1982.

[92] G. E. Hinton, "Training products of experts by minimizing contrastive divergence," *Neural Computation*, vol. 14, no. 8, pp. 1771–1800, 2002.

[93] M. A. Carreira-Perpinan and G. E. Hinton, "On contrastive divergence learning," in *Proceedings of International Workshop on Artificial Intelligence and Statistics (AISTATS)*, 2005, pp. 33–40.

[94] G. E. Hinton, S. Osindero, and Y.-W. Teh, "A fast learning algorithm for deep belief nets," *Neural Computation*, vol. 18, no. 7, pp. 1527–1554, 2006.

[95] N. Li, M. W. Mak, and J. T. Chien, "DNN-driven mixture of PLDA for robust speaker verification," *IEEE/ACM Transactions on Audio, Speech, and Language Processing*, no. 6, pp. 1371–1383, 2017.

[96] Y. LeCun, L. Bottou, Y. Bengio, and P. Haffner, "Gradient-based learning applied to document recognition," *Proceedings of the IEEE*, pp. 2278–2324, 1998.

[97] D. Yu, G. Hinton, N. Morgan, J.-T. Chien, and S. Sagayama, "Introduction to the special section on deep learning for speech and language processing," *IEEE Transactions on Audio, Speech, and Language Processing*, vol. 20, no. 1, pp. 4–6, 2012.

[98] G. Hinton, L. Deng, D. Yu, G. Dahl, A. Mohamed, N. Jaitly, A. Senior, V. Vanhoucke, P. Nguyen, T. Sainath, and B. Kingsbury, "Deep neural networks for acoustic modeling in speech recognition: Four research groups share their views," *IEEE Signal Processing Magazine*, vol. 29, no. 6, pp. 82–97, 2012.

[99] G. Saon and J.-T. Chien, "Large-vocabulary continuous speech recognition systems: A look at some recent advances," *IEEE Signal Processing Magazine*, vol. 29, no. 6, pp. 18–33, 2012.

[100] J.-T. Chien and Y.-C. Ku, "Bayesian recurrent neural network for language modeling," *IEEE Transactions on Neural Networks and Learning Systems*, vol. 27, no. 2, pp. 361–374, 2016.

[101] M. D. Zeiler, G. W. Taylor, and R. Fergus, "Adaptive deconvolutional networks for mid and high level feature learning," in *Proceedings of IEEE International Conference on Computer Vision (ICCV)*, 2011, pp. 2018–2025.

[102] J. Xie, L. Xu, and E. Chen, "Image denoising and inpainting with deep neural networks," in *Advances in Neural Information Processing Systems 25*, F. Pereira, C. J. C. Burges, L. Bottou, and K. Q. Weinberger, Eds., 2012, pp. 341–349.

[103] R. Salakhutdinov and H. Larochelle, "Efficient learning of deep Boltzmann machines," in *Proceedings of International Conference on Artificial Intelligence and Statistics (AISTATS)*, 2010, pp. 693–700.

[104] P. Vincent, H. Larochelle, I. Lajoie, Y. Bengio, and P.-A. Manzagol, "Stacked denoising autoencoders: Learning useful representations in a deep network with a local denoising criterion," *Journal of Machine Learning Research*, vol. 11, pp. 3371–3408, 2010.

[105] M. Schuster and K. K. Paliwal, "Bidirectional recurrent neural networks," *IEEE Transactions on Signal Processing*, vol. 45, no. 11, pp. 2673–2681, 1997.

[106] D. E. Rumelhart, G. E. Hinton, and R. J. Williams, "Learning internal representation by backpropagating errors," *Nature*, vol. 323, pp. 533–536, 1986.

[107] I. Goodfellow, Y. Bengio, and A. Courville, *Deep Learning*. Cambridge, MA: MIT Press, 2016.

[108] Y. Bengio, P. Lamblin, D. Popovici, and H. Larochelle, "Greedy layer-wise training of deep networks," in *Advances in Neural Information Processing Systems 19*, B. Schölkopf, J. C. Platt, and T. Hoffman, Eds. Cambridge, MA: MIT Press, 2007, pp. 153–160.

[109] G. E. Hinton and R. R. Salakhutdinov, "Reducing the dimensionality of data with neural networks," *Science*, vol. 313, pp. 504–507, 2006.

[110] G. E. Hinton, S. Osindero, and Y.-W. Teh, "A fast learning algorithm for deep belief nets," *Neural Computation*, vol. 18, no. 7, pp. 1527–1554, 2006.

[111] R. Salakhutdinov and G. E. Hinton, "Deep Boltzmann machines," in *Proceedings of International Conference on Artificial Intelligence and Statistics (AISTATS)*, 2009, p. 3.

[112] P. Vincent, H. Larochelle, Y. Bengio, and P.-A. Manzagol, "Extracting and composing robust features with denoising autoencoders," in *Proceedings of International Conference on Machine Learning (ICML)*, 2008, pp. 1096–1103.

[113] A. Hyvärinen, "Estimation of non-normalized statistical models by score matching," *Journal of Machine Learning Research*, vol. 6, pp. 695–709, 2005.

[114] D. P. Kingma and M. Welling, "Auto-encoding variational Bayes," in *Proceedings of International Conference on Learning Representation (ICLR)*, 2014.

[115] J.-T. Chien and K.-T. Kuo, "Variational recurrent neural networks for speech separation," in *Proceedings of Annual Conference of International Speech Communication Association (INTERSPEECH)*, 2017, pp. 1193–1197.

[116] J.-T. Chien and C.-W. Hsu, "Variational manifold learning for speaker recognition," in *Proceedings of International Conference on Acoustics, Speech, and Signal Processing (ICASSP)*, 2017, pp. 4935–4939.

[117] D. J. Rezende, S. Mohamed, and D. Wierstra, "Stochastic backpropagation and approximate inference in deep generative models," in *Proceedings of International Conference on Machine Learning (ICML)*, 2014, pp. 1278–1286.

[118] I. Goodfellow, J. Pouget-Abadie, M. Mirza, B. Xu, D. Warde-Farley, S. Ozair, A. Courville, and Y. Bengio, "Generative adversarial nets," in *Advances in Neural Information Processing Systems (NIPS)*, 2014, pp. 2672–2680.

[119] J.-T. Chien and K.-T. Peng, "Adversarial manifold learning for speaker recognition," in *Prof. of IEEE Automatic Speech Recognition and Understanding Workshop (ASRU)*, 2017, pp. 599–605.

[120] J.-T. Chien and K.-T. Peng, "Adversarial learning and augmentation for speaker recognition," in *Proceedings of Speaker and Language Recognition Workshop (Odyssey)*, 2018, pp. 342–348.

[121] Y. Bengio, E. Laufer, G. Alain, and J. Yosinski, "Deep generative stochastic networks trainable by backprop," in *Proceedings of International Conference on Machine Learning (ICML)*, 2014, pp. 226–234.

[122] A. Makhzani, J. Shlens, N. Jaitly, and I. Goodfellow, "Adversarial autoencoders," *arXiv preprint arXiv:1511.05644*, 2015.

[123] A. B. L. Larsen, S. K. Sønderby, and O. Winther, "Autoencoding beyond pixels using a learned similarity metric," in *Proceedings of International Conference on Machine Learning (ICML)*, no. 1558–1566, 2015.

[124] S. J. Pan and Q. Yang, "A survey on transfer learning," *IEEE Transactions on Knowledge and Data Engineering*, vol. 22, no. 10, pp. 1345–1359, 2009.

[125] A. Evgeniou and M. Pontil, "Multi-task feature learning," *Advances in Neural Information Processing Systems (NIPS)*, vol. 19, p. 41, 2007.

[126] R. K. Ando and T. Zhang, "A framework for learning predictive structures from multiple tasks and unlabeled data," *Journal of Machine Learning Research*, vol. 6, pp. 1817–1853, 2005.

[127] A. Argyriou, M. Pontil, Y. Ying, and C. A. Micchelli, "A spectral regularization framework for multi-task structure learning," in *Advances in Neural Information Processing Systems (NIPS)*, 2007, pp. 25–32.

[128] W. Lin, M. Mak, and J. Chien, "Multisource i-vectors domain adaptation using maximum mean discrepancy based autoencoders," *IEEE/ACM Transactions on Audio, Speech, and Language Processing*, vol. 26, no. 12, pp. 2412–2422, Dec 2018.

[129] W. W. Lin, M. W. Mak, L. X. Li, and J. T. Chien, "Reducing domain mismatch by maximum mean discrepancy based autoencoders," in *Proceedings of Speaker and Language Recognition Workshop (Odyssey)*, 2018, pp. 162–167.

[130] M. Sugiyama, S. Nakajima, H. Kashima, P. V. Buenau, and M. Kawanabe, "Direct importance estimation with model selection and its application to covariate shift adaptation," in *Advances in Neural Information Processing Systems (NIPS)*, 2008, pp. 1433–1440.

[131] S. Bickel, M. Brückner, and T. Scheffer, "Discriminative learning under covariate shift," *Journal of Machine Learning Research*, vol. 10, pp. 2137–2155, 2009.

[132] J. Blitzer, R. McDonald, and F. Pereira, "Domain adaptation with structural correspondence learning," in *Proceedings of Conference on Empirical Methods in Natural Language Processing (EMNLP)*, 2006, pp. 120–128.

[133] P. von Bünau, F. C. Meinecke, F. C. Király, and K.-R. Müller, "Finding stationary subspaces in multivariate time series," *Physical Review Letters*, vol. 103, no. 21, p. 214101, 2009.

[134] S. J. Pan, J. T. Kwok, and Q. Yang, "Transfer learning via dimensionality reduction," in *Proceedings of AAAI Conference on Artificial Intelligence*, vol. 8, 2008, pp. 677–682.

[135] A. Gretton, K. M. Borgwardt, M. Rasch, B. Schölkopf, and A. J. Smola, "A kernel method for the two-sample-problem," in *Advances in Neural Information Processing Systems (NIPS)*, 2007, pp. 513–520.

[136] K. M. Borgwardt, A. Gretton, M. J. Rasch, H.-P. Kriegel, B. Schölkopf, and A. J. Smola, "Integrating structured biological data by kernel maximum mean discrepancy," *Bioinformatics*, vol. 22, no. 14, pp. e49–e57, 2006.

[137] A. Ahmed, K. Yu, W. Xu, Y. Gong, and E. Xing, "Training hierarchical feed-forward visual recognition models using transfer learning from pseudo-tasks," in *Proceedings of European Conference on Computer Vision (ECCV)*, 2008, pp. 69–82.

[138] S. Ji, W. Xu, M. Yang, and K. Yu, "3D convolutional neural networks for human action recognition," *IEEE Transactions on Pattern Analysis and Machine Intelligence*, vol. 35, no. 1, pp. 221–231, 2013.

[139] G. Hinton, L. Deng, D. Yu, G. Dahl, A. Mohamed, N. Jaitly, A. Senior, V. Vanhoucke, P. Nguyen, T. Sainath, and B. Kingsbury, "Deep neural networks for acoustic modeling in speech recognition: The shared views of four research groups," *IEEE Signal Processing Magazine*, vol. 29, pp. 82–97, 2012.

[140] G. E. Dahl, D. Yu, L. Deng, and A. Acero, "Context-dependent pre-trained deep neural networks for large-vocabulary speech recognition," *IEEE Transactions on Audio, Speech, and Language Processing*, vol. 20, no. 1, pp. 30–42, 2012.

[141] L. Deng, "A tutorial survey of architectures, algorithms, and applications for deep learning," *APSIPA Transactions on Signal and Information Processing*, vol. 3, p. e2, 2014.

[142] A. R. Mohamed, G. E. Dahl, and G. Hinton, "Acoustic modeling using deep belief networks," *IEEE Transactions on Audio, Speech, and Language Processing*, vol. 20, no. 1, pp. 14–22, 2012.

[143] Y. Z. Işik, H. Erdogan, and R. Sarikaya, "S-vector: A discriminative representation derived from i-vector for speaker verification," in *Proceedings of European Signal Processing Conference (EUSIPCO)*, 2015, pp. 2097–2101.

[144] S. Novoselov, T. Pekhovsky, O. Kudashev, V. S. Mendelev, and A. Prudnikov, "Non-linear PLDA for i-vector speaker verification," in *Proceedings of Annual Conference of the International Speech Communication Association (INTERSPEECH)*, 2015.

[145] T. Pekhovsky, S. Novoselov, A. Sholohov, and O. Kudashev, "On autoencoders in the i-vector space for speaker recognition," in *Proceedings of Speaker and Language Recognition Workshop (Odyssey)*, 2016, pp. 217–224.

[146] S. Mahto, H. Yamamoto, and T. Koshinaka, "I-vector transformation using a novel discriminative denoising autoencoder for noise-robust speaker recognition," in *Proceedings of Annual Conference of International Speech Communication Association (INTERSPEECH)*, 2017, pp. 3722–3726.

[147] Y. Tian, M. Cai, L. He, and J. Liu, "Investigation of bottleneck features and multilingual deep neural networks for speaker verification," in *Proceedings of Annual Conference of International Speech Communication Association (INTERSPEECH)*, 2015, pp. 1151–1155.

[148] Z. L. Tan, Y. K. Zhu, M. W. Mak, and B. Mak, "Senone i-vectors for robust speaker verification," in *Proceedings of International Symposium on Chinese Spoken Language Processing (ISCSLP)*, Tianjin, China, October 2016.

[149] S. Yaman, J. Pelecanos, and R. Sarikaya, "Bottleneck features for speaker recognition," in *Proceedings of Speaker and Language Recognition Workshop (Odyssey)*, vol. 12, 2012, pp. 105–108.

[150] E. Variani, X. Lei, E. McDermott, I. Lopez M., and J. Gonzalez-Dominguez, "Deep neural networks for small footprint text-dependent speaker verification," in *Proceedings of International Conference on Acoustics, Speech, and Signal Processing (ICASSP)*, 2014, pp. 4052–4056.

[151] T. Yamada, L. B. Wang, and A. Kai, "Improvement of distant-talking speaker identification using bottleneck features of DNN," in *Proceedings of Annual Conference of International Speech Communication Association (INTERSPEECH)*, 2013, pp. 3661–3664.

[152] P. Kenny, V. Gupta, T. Stafylakis, P. Ouellet, and J. Alam, "Deep neural networks for extracting Baum-Welch statistics for speaker recognition," in *Proceedings of Speaker and Language Recognition Workshop (Odyssey)*, 2014, pp. 293–298.

[153] D. Garcia-Romero and A. McCree, "Insights into deep neural networks for speaker recognition," in *Proceedings of Annual Conference of International Speech Communication Association (INTERSPEECH)*, 2015, pp. 1141–1145.

[154] M. McLaren, Y. Lei, and L. Ferrer, "Advances in deep neural network approaches to speaker recognition," in *Proceedings of International Conference on Acoustics, Speech, and Signal Processing (ICASSP)*, 2015, pp. 4814–4818.

[155] D. Snyder, D. Garcia-Romero, D. Povey, and S. Khudanpur, "Deep neural network embeddings for text-independent speaker verification," in *Proceedings of Annual Conference of International Speech Communication Association (INTERSPEECH)*, 2017, pp. 999–1003.

[156] D. Snyder, D. Garcia-Romero, G. Sell, D. Povey, and S. Khudanpur, "X-vectors: Robust DNN embeddings for speaker recognition," in *Proceedings of International Conference on Acoustics, Speech, and Signal Processing (ICASSP)*, 2018, pp. 5329–5333.

[157] Y. Tang, G. Ding, J. Huang, X. He, and B. Zhou, "Deep speaker embedding learning with multi-level pooling for text-independent speaker verification," in *Proceedings IEEE International Conference on Acoustics, Speech, and Signal Processing (ICASSP)*, 2019, pp. 6116–6120.

[158] C.-P. Chen, S.-Y. Zhang, C.-T. Yeh, J.-C. Wang, T. Wang, and C.-L. Huang, "Speaker characterization using tdnn-lstm based speaker embedding," in *Proceedings IEEE International Conference on Acoustics, Speech, and Signal Processing (ICASSP)*. IEEE, 2019, pp. 6211–6215.

[159] W. Zhu and J. Pelecanos, "A bayesian attention neural network layer for speaker recognition," in *Proceedings IEEE International Conference on Acoustics, Speech, and Signal Processing (ICASSP)*. IEEE, 2019, pp. 6241–6245.

[160] Y. Zhu, T. Ko, D. Snyder, B. Mak, and D. Povey, "Self-attentive speaker embeddings for text-independent speaker verification," in *Proceedings Interspeech*, vol. 2018, 2018, pp. 3573–3577.

[161] N. Brummer, L. Burget, P. Garcia, O. Plchot, J. Rohdin, D. Romero, D. Snyder, T. Stafylakis, A. Swart, and J. Villalba, "Meta-embeddings: A probabilistic generalization of embeddings in machine learning," in *JHU HLTCOE 2017 SCALE Workshop*, 2017.

[162] N. Li and M. W. Mak, "SNR-invariant PLDA modeling for robust speaker verification," in *Proceedings of Annual Conference of International Speech Communication Association (INTERSPEECH)*, 2015, pp. 2317–2321.

[163] N. Li, M. W. Mak, W. W. Lin, and J. T. Chien, "Discriminative subspace modeling of SNR and duration variabilities for robust speaker verification," *Computer Speech & Language*, vol. 45, pp. 83–103, 2017.

[164] S. Prince and J. Elder, "Probabilistic linear discriminant analysis for inferences about identity," in *Proceedings of IEEE International Conference on Computer Vision (ICCV)*, 2007, pp. 1–8.

[165] S. J. Prince, *Computer Vision: Models, Learning, and Inference*. New York: Cambridge University Press, 2012.

[166] A. Sizov, K. A. Lee, and T. Kinnunen, "Unifying probabilistic linear discriminant analysis variants in biometric authentication," in *Structural, Syntactic, and Statistical Pattern Recognition*. Berlin, Heidelberg: Springer, 2014, pp. 464–475.

[167] T. Hasan, R. Saeidi, J. H. L. Hansen, and D. A. van Leeuwen, "Duration mismatch compensation for I-vector based speaker recognition system," in *Proceedings of International Conference on Acoustics, Speech, and Signal Processing (ICASSP)*, 2013, pp. 7663–7667.

[168] A. Kanagasundaram, D. Dean, S. Sridharan, J. Gonzalez-Dominguez, J. Gonzalez-Rodriguez, and D. Ramos, "Improving short utterance i-vector speaker verification using utterance variance modelling and compensation techniques," *Speech Communication*, vol. 59, pp. 69–82, 2014.

[169] K. H. Norwich, *Information, Sensation, and Perception*. San Diego: Academic Press, 1993.

[170] P. Billingsley, *Probability and Measure*. New York: John Wiley & Sons, 2008.

[171] M. W. Mak, X. M. Pang, and J. T. Chien, "Mixture of PLDA for noise robust i-vector speaker verification," *IEEE/ACM Trans. on Audio Speech and Language Processing*, vol. 24, no. 1, pp. 132–142, 2016.

[172] M. W. Mak, "SNR-dependent mixture of PLDA for noise robust speaker verification," in *Proceedings of Annual Conference of International Speech Communication Association (INTERSPEECH)*, 2014, pp. 1855–1859.

[173] X. M. Pang and M. W. Mak, "Noise robust speaker verification via the fusion of SNR-independent and SNR-dependent PLDA," *International Journal of Speech Technology*, vol. 18, no. 4, 2015.

[174] M. E. Tipping and C. M. Bishop, "Mixtures of probabilistic principal component analyzers," *Neural Computation*, vol. 11, no. 2, pp. 443–482, 1999.

[175] T. Pekhovsky and A. Sizov, "Comparison between supervised and unsupervised learning of probabilistic linear discriminant analysis mixture models for speaker verification," *Pattern Recognition Letters*, vol. 34, no. 11, pp. 1307–1313, 2013.

[176] N. Li, M. W. Mak, and J. T. Chien, "Deep neural network driven mixture of PLDA for robust i-vector speaker verification," in *Proceedings of IEEE Workshop on Spoken Language Technology (SLT)*, San Diego, CA, 2016, pp. 186–191.

[177] S. Cumani and P. Laface, "Large-scale training of pairwise support vector machines for speaker recognition," *IEEE/ACM Transactions on Audio, Speech, and Language Processing*, vol. 22, no. 11, pp. 1590–1600, 2014.

[178] D. Snyder, P. Ghahremani, D. Povey, D. Garcia-Romero, Y. Carmiel, and S. Khudanpur, "Deep neural network-based speaker embeddings for end-to-end speaker verification," in *Proceedings of IEEE Spoken Language Technology Workshop (SLT)*, 2016, pp. 165–170.

[179] M. I. Mandasari, R. Saeidi, M. McLaren, and D. A. van Leeuwen, "Quality measure functions for calibration of speaker recognition systems in various duration conditions," *IEEE Transactions on Audio, Speech, and Language Processing*, vol. 21, no. 11, pp. 2425–2438, Nov. 2013.

[180] M. I. Mandasari, R. Saeidi, and D. A. van Leeuwen, "Quality measures based calibration with duration and noise dependency for speaker recognition," *Speech Communication*, vol. 72, pp. 126–137, 2015.

[181] A. O. J. Villalba, A. Miguel and E. Lleida, "Bayesian networks to model the variability of speaker verification scores in adverse environments," *IEEE/ACM Transactions on Audio, Speech, and Language Processing*, vol. 24, no. 12, pp. 2327–2340, 2016.

[182] A. Nautsch, R. Saeidi, C. Rathgeb, and C. Busch, "Robustness of quality-based score calibration of speaker recognition systems with respect to low-SNR and short-duration conditions," in *Proceedings of Speaker and Language Recognition Workshop (Odyssey)*, 2016, pp. 358–365.

[183] L. Ferrer, L. Burget, O. Plchot, and N. Scheffer, "A unified approach for audio characterization and its application to speaker recognition," in *Proceedings of Speaker and Language Recognition Workshop (Odyssey)*, 2012, pp. 317–323.

[184] Q. Hong, L. Li, M. Li, L. Huang, L. Wan, and J. Zhang, "Modified-prior PLDA and score calibration for duration mismatch compensation in speaker recognition system," in *Proceedings of Annual Conference of International Speech Communication Association (INTERSPEECH)*, 2015.

[185] A. Shulipa, S. Novoselov, and Y. Matveev, "Scores calibration in speaker recognition systems," in *Proceedings of International Conference on Speech and Computer*, 2016, pp. 596–603.

[186] N. Brümmer, A. Swart, and D. van Leeuwen, "A comparison of linear and non-linear calibrations for speaker recognition," in *Proceedings of Speaker and Language Recognition Workshop (Odyssey)*, 2014, pp. 14–18.

[187] N. Brümmer and G. Doddington, "Likelihood-ratio calibration using prior-weighted proper scoring rules," in *Proceedings of Annual Conference of International Speech Communication Association (INTERSPEECH)*, 2013, pp. 1976–1980.

[188] N. Brümmer and D. Garcia-Romero, "Generative modelling for unsupervised score calibration," in *Proceedings of IEEE International Conference on Acoustics, Speech, and Signal Processing (ICASSP)*, 2014, pp. 1680–1684.

[189] R. Caruana, "Multitask learning: A knowledge-based source of inductive bias," *Machine Learning*, vol. 28, pp. 41–75, 1997.

[190] D. Chen and B. Mak, "Multitask learning of deep neural networks for low-resource speech recognition," *IEEE/ACM Transactions on Audio, Speech, and Language Processing*, vol. 23, no. 7, pp. 1172–1183, 2015.

[191] Q. Yao and M. W. Mak, "SNR-invariant multitask deep neural networks for robust speaker verification," *IEEE Signal Processing Letters*, vol. 25, no. 11, pp. 1670–1674, Nov. 2018.

[192] D. Garcia-Romero and A. McCree, "Supervised domain adaptation for i-vector based speaker recognition," in *Proceedings of International Conference on Acoustics, Speech, and Signal Processing (ICASSP)*, 2014, pp. 4047–4051.

[193] J. Villalba and E. Lleida, "Bayesian adaptation of PLDA based speaker recognition to domains with scarce development data," in *Proceedings of Speaker and Language Recognition Workshop (Odyssey)*, Singapore, 2012.

[194] J. Villalba and E. Lleida, "Unsupervised adaptation of PLDA by using variational Bayes methods," in *Proceedings of International Conference on Acoustics, Speech, and Signal Processing (ICASSP)*, 2014, pp. 744–748.

[195] B. J. Borgström, E. Singer, D. Reynolds, and O. Sadjadi, "Improving the effectiveness of speaker verification domain adaptation with inadequate in-domain data," in *Proceedings of Annual Conference of International Speech Communication Association (INTERSPEECH)*, 2017, pp. 1557–1561.

[196] S. Shum, D. A. Reynolds, D. Garcia-Romero, and A. McCree, "Unsupervised clustering approaches for domain adaptation in speaker recognition systems," in *Proceedings of Speaker and Language Recognition Workshop (Odyssey)*, 2014, pp. 266–272.

[197] D. Garcia-Romero, X. Zhang, A. McCree, and D. Povey, "Improving speaker recognition performance in the domain adaptation challenge using deep neural networks," in *Proceedings of IEEE Spoken Language Technology Workshop (SLT)*. IEEE, 2014, pp. 378–383.

[198] Q. Q. Wang and T. Koshinaka, "Unsupervised discriminative training of PLDA for domain adaptation in speaker verification," in *Proceedings of Annual Conference of International Speech Communication Association (INTERSPEECH)*, 2017, pp. 3727–3731.

[199] S. Shon, S. Mun, W. Kim, and H. Ko, "Autoencoder based domain adaptation for speaker recognition under insufficient channel information," in *Proceedings of Annual Conference of International Speech Communication Association (INTERSPEECH)*, 2017, pp. 1014–1018.

[200] H. Aronowitz, "Inter dataset variability compensation for speaker recognition," in *Proceedings of International Conference on Acoustics, Speech, and Signal Processing (ICASSP)*, 2014, pp. 4002–4006.

[201] H. Aronowitz, "Compensating inter-dataset variability in PLDA hyper-parameters for robust speaker recognition," in *Proceedings of Speaker and Language Recognition Workshop (Odyssey)*, 2014, pp. 282–286.

[202] H. Rahman, A. Kanagasundaram, D. Dean, and S. Sridharan, "Dataset-invariant covariance normalization for out-domain PLDA speaker verification," in *Proceedings of Annual Conference of International Speech Communication Association (INTERSPEECH)*, 2015, pp. 1017–1021.

[203] A. Kanagasundaram, D. Dean, and S. Sridharan, "Improving out-domain PLDA speaker verification using unsupervised inter-dataset variability compensation approach," in *Proceedings of International Conference on Acoustics, Speech, and Signal Processing (ICASSP)*, 2015, pp. 4654–4658.

[204] O. Glembek, J. Ma, P. Matejka, B. Zhang, O. Plchot, L. Burget, and S. Matsoukas, "Domain adaptation via within-class covariance correction in i-vector based speaker recognition systems," in *Proceedings of International Conference on Acoustics, Speech, and Signal Processing (ICASSP)*, 2014, pp. 4032–4036.

[205] E. Singer and D. A. Reynolds, "Domain mismatch compensation for speaker recognition using a library of whiteners," *IEEE Signal Processing Letters*, vol. 22, no. 11, pp. 2000–2003, 2015.

[206] F. Bahmaninezhad and J. H. L. Hansen, "Compensation for domain mismatch in text-independent speaker recognition," in *Proceedings of Annual Conference of International Speech Communication Association (INTERSPEECH)*, 2018, pp. 1071–1075.

[207] H. Yu, Z. H. Tan, Z. Y. Ma, and J. Guo, "Adversarial network bottleneck features for noise robust speaker verification," *arXiv preprint arXiv:1706.03397*, 2017.

[208] D. Michelsanti and Z. H. Tan, "Conditional generative adversarial networks for speech enhancement and noise-robust speaker verification," *arXiv preprint arXiv:1709.01703*, 2017.

[209] J. C. Zhang, N. Inoue, and K. Shinoda, "I-vector transformation using conditional generative adversarial networks for short utterance speaker verification," in *Proceedings Interspeech*, 2018, pp. 3613–3617.

[210] Z. Meng, J. Y. Li, Z. Chen, Y. Zhao, V. Mazalov, Y. F. Gong, and B. H. Juang, "Speaker-invariant training via adversarial learning," *arXiv preprint arXiv:1804.00732*, 2018.

[211] Q. Wang, W. Rao, S. Sun, L. Xie, E. S. Chng, and H. Z. Li, "Unsupervised domain adaptation via domain adversarial training for speaker recognition," in *Proceedings of International Conference on Acoustics, Speech, and Signal Processing (ICASSP)*, 2018, pp. 4889–4893.

[212] I. Viñals, A. Ortega, J. Villalba, A. Miguel, and E. Lleida, "Domain adaptation of PLDA models in broadcast diarization by means of unsupervised speaker clustering," in

Proceedings of Annual Conference of International Speech Communication Association (INTERSPEECH), 2017, pp. 2829–2833.

[213] J. Y. Li, M. L. Seltzer, X. Wang, R. Zhao, and Y. F. Gong, "Large-scale domain adaptation via teacher-student learning," in *Proceedings of Annual Conference of International Speech Communication Association (INTERSPEECH)*, 2017, pp. 2386–2390.

[214] H. Aronowitz, "Inter dataset variability modeling for speaker recognition," in *Proceedings International Conference on Acoustics, Speech, and Signal Processing (ICASSP)*, 2017, pp. 5400–5404.

[215] M. McLaren and D. Van Leeuwen, "Source-normalized LDA for robust speaker recognition using i-vectors from multiple speech sources," *IEEE Transactions on Audio, Speech, and Language Processing*, vol. 20, no. 3, pp. 755–766, 2012.

[216] M. H. Rahman, I. Himawan, D. Dean, C. Fookes, and S. Sridharan, "Domain-invariant i-vector feature extraction for PLDA speaker verification," in *Proceedings of Speaker and Language Recognition Workshop (Odyssey)*, 2018, pp. 155–161.

[217] S. E. Shepstone, K. A. Lee, H. Li, Z.-H. Tan, and S. H. Jensen, "Total variability modeling using source-specific priors," *IEEE/ACM Transactions on Audio, Speech, and Language Processing*, vol. 24, no. 3, pp. 504–517, 2016.

[218] M. J. Alam, G. Bhattacharya, and P. Kenny, "Speaker verification in mismatched conditions with frustratingly easy domain adaptation," in *Proceedings of Speaker and Language Recognition Workshop (Odyssey)*, 2018, pp. 176–180.

[219] B. Sun, J. Feng, and K. Saenko, "Return of frustratingly easy domain adaptation," in *Proceedings of AAAI Conference on Artificial Intelligence*, vol. 6, no. 7, 2016.

[220] J. Alam, P. Kenny, G. Bhattacharya, and M. Kockmann, "Speaker verification under adverse conditions using i-vector adaptation and neural networks," in *Proceedings of Annual Conference of International Speech Communication Association (INTERSPEECH)*, 2017, pp. 3732–3736.

[221] C. Szegedy, S. Ioffe, V. Vanhoucke, and A. A. Alemi, "Inception-v4, inception-resnet and the impact of residual connections on learning," in *Proceedings of AAAI Conference on Artificial Intelligence*, 2017.

[222] G. Bhattacharya, J. Alam, P. Kenn, and V. Gupta, "Modelling speaker and channel variability using deep neural networks for robust speaker verification," in *Proceedings of IEEE Spoken Language Technology Workshop (SLT)*, 2016, pp. 192–198.

[223] *Domain Adaptation Challenge*, John Hopkins University, 2013.

[224] A. Storkey, "When training and test sets are different: Characterizing learning transfer," in *Dataset Shift in Machine Learning*, J. Quinonero-Candela, M. Sugiyama, A. Schwaighofer, and N. Lawrence, Eds. Cambridge, MA: MIT Press, 2009, pp. 3–28.

[225] H. Shimodaira, "Improving predictive inference under covariate shift by weighting the log-likelihood function," *Journal of Statistical Planning and Inference*, vol. 90, no. 2, pp. 227–244, 2000.

[226] S. B. David, T. Lu, T. Luu, and D. Pál, "Impossibility theorems for domain adaptation," in *Proceedings International Conference on Artificial Intelligence and Statistics (AISTATS)*, 2010, pp. 129–136.

[227] Y. Mansour, M. Mohri, and A. Rostamizadeh, "Domain adaptation: Learning bounds and algorithms," *arXiv preprint arXiv:0902.3430*, 2009.

[228] P. Germain, A. Habrard, F. Laviolette, and E. Morvant, "A PAC-Bayesian approach for domain adaptation with specialization to linear classifiers," in *Proceedings International Conference on Machine Learning (ICML)*, 2013, pp. 738–746.

[229] H.-Y. Chen and J.-T. Chien, "Deep semi-supervised learning for domain adaptation," in *IEEE International Workshop on Machine Learning for Signal Processing (MLSP)*, 2015, pp. 1–6.

[230] A. Gretton, K. M. Borgwardt, M. Rasch, B. Schölkopf, and A. J. Smola, "A kernel method for the two-sample-problem," in *Advances in Neural Information Processing Systems (NIPS)*, 2007, pp. 513–520.

[231] Y. Li, K. Swersky, and R. Zemel, "Generative moment matching networks," in *Proceedings International Conference on Machine Learning (ICML)*, 2015, pp. 1718–1727.

[232] M. Long, Y. Cao, J. Wang, and M. Jordan, "Learning transferable features with deep adaptation networks," in *Proceedings International Conference on Machine Learning (ICML)*, 2015, pp. 97–105.

[233] A. Smola, A. Gretton, L. Song, and B. Schölkopf, "A Hilbert space embedding for distributions," in *International Conference on Algorithmic Learning Theory*. Berlin, Heidelberg: Springer, 2007, pp. 13–31.

[234] P. Vincent, H. Larochelle, I. Lajoie, Y. Bengio, and P. Manzagol, "Stacked denoising autoencoders: Learning useful representations in a deep network with a local denoising criterion," *Journal of Machine Learning Research*, vol. 11, pp. 3371–3408, 2010.

[235] F. Schroff, D. Kalenichenko, and J. Philbin, "Facenet: A unified embedding for face recognition and clustering," in *Proceedings of IEEE Conference on Computer Vision and Pattern Recognition (CVPR)*, 2015, pp. 815–823.

[236] Y. Wen, K. Zhang, Z. Li, and Y. Qiao, "A discriminative feature learning approach for deep face recognition," in *Proceedings of European Conference on Computer Vision (ECCV)*, 2016, pp. 499–515.

[237] D. P. Kingma and M. Welling, "Auto-encoding variational Bayes," in *Proceedings of International Conference on Learning Representations (ICLR)*, 2014.

[238] D. P. Kingma, S. Mohamed, D. J. Rezende, and M. Welling, "Semi-supervised learning with deep generative models," in *Advances in Neural Information Processing Systems (NIPS)*, 2014, pp. 3581–3589.

[239] D. J. Rezende, S. Mohamed, and D. Wierstra, "Stochastic backpropagation and approximate inference in deep generative models," in *Proceedings of International Conference on Machine Learning (ICML)*, 2014.

[240] C. Doersch, "Tutorial on variational autoencoders," *arXiv preprint arXiv:1606.05908*, 2016.

[241] E. Wilson, "Backpropagation learning for systems with discrete-valued functions," in *Proceedings of the World Congress on Neural Networks*, vol. 3, 1994, pp. 332–339.

[242] X. Glorot and Y. Bengio, "Understanding the difficulty of training deep feedforward neural networks," in *Proceedings of International Conference on Artificial Intelligence and Statistics (AISTATS)*, vol. 9, 2010, pp. 249–256.

[243] D. Kingma and J. Ba, "Adam: A method for stochastic optimization," in *Proceedings of International Conference on Learning Representations (ICLR)*, San Diego, CA, 2015.

[244] W. Rao and M. W. Mak, "Alleviating the small sample-size problem in i-vector based speaker verification," in *Proceedings of International Symposium on Chinese Spoken Language Processing (ISCSLP)*, 2012, pp. 335–339.

[245] G. E. Hinton and S. T. Roweis, "Stochastic neighbor embedding," in *Advances in Neural Information Processing Systems (NIPS)*, S. Becker, S. Thrun, and K. Obermayer, Eds., Baltimore, MD: MIT Press, 2003, pp. 857–864.

[246] H.-H. Tseng, I. E. Naqa, and J.-T. Chien, "Power-law stochastic neighbor embedding," in *Proceedings of International Conference on Acoustics, Speech, and Signal Processing (ICASSP)*, 2017, pp. 2347–2351.

[247] J.-T. Chien and C.-H. Chen, "Deep discriminative manifold learning," in *Proceedings of IEEE International Conference on Acoustics, Speech, and Signal Processing (ICASSP)*, 2016, pp. 2672–2676.

[248] K. Chen and A. Salman, "Learning speaker-specific characteristics with a deep neural architecture," *IEEE Transactions on Neural Networks*, vol. 22, no. 11, pp. 1744–1756, 2011.

[249] J.-T. Chien and C.-W. Hsu, "Variational manifold learning for speaker recognition," in *Proceedings of IEEE International Conference on Acoustics, Speech, and Signal Processing (ICASSP)*, 2017, pp. 4935–4939.

[250] A. Odena, C. Olah, and J. Shlens, "Conditional image synthesis with auxiliary classifier GANs," *arXiv preprint arXiv:1610.09585*, 2016.

[251] M. Mirza and S. Osindero, "Conditional generative adversarial nets," *arXiv preprint arXiv:1411.1784*, 2014.

[252] J. Cook, I. Sutskever, A. Mnih, and G. E. Hinton, "Visualizing similarity data with a mixture of maps," in *Proceedings of International Conference on Artificial Intelligence and Statistics (AISTATS)*, 2007, pp. 67–74.

[253] T. Che, Y. Li, A. P. Jacob, Y. Bengio, and W. Li, "Mode regularized generative adversarial networks," *arXiv preprint arXiv:1612.02136*, 2016.

[254] M. R. Min, L. Maaten, Z. Yuan, A. J. Bonner, and Z. Zhang, "Deep supervised *t*-distributed embedding," in *Proceedings of International Conference on Machine Learning (ICML)*, 2010, pp. 791–798.

[255] D. Palaz, R. Collobert, and M. M. Doss, "Estimating phoneme class conditional probabilities from raw speech signal using convolutional neural networks," in *Proceedings of Annual Conference of International Speech Communication Association (INTERSPEECH)*, 2013, pp. 1766–1770.

[256] N. Jaitly and G. Hinton, "Learning a better representation of speech soundwaves using restricted Boltzmann machines," in *Proceedings of International Conference on Acoustics, Speech, and Signal Processing (ICASSP)*, 2011, pp. 5884–5887.

[257] Z. Tüske, P. Golik, R. Schlüter, and H. Ney, "Acoustic modeling with deep neural networks using raw time signal for LVCSR," in *Proceedings of Annual Conference of International Speech Communication Association (INTERSPEECH)*, 2014.

[258] D. Palaz, M. Magimai-Doss, and R. Collobert, "End-to-end acoustic modeling using convolutional neural networks for HMM-based automatic speech recognition," *Speech Communication*, 2019.

[259] Y. Hoshen, R. J. Weiss, and K. W. Wilson, "Speech acoustic modeling from raw multi-channel waveforms," in *Proceedings of International Conference on Acoustics, Speech, and Signal Processing (ICASSP)*, 2015, pp. 4624–4628.

[260] D. Palaz, M. Magimai-Doss, and R. Collobert, "Analysis of CNN-based speech recognition system using raw speech as input," in *Proceedings of Annual Conference of International Speech Communication Association (INTERSPEECH)*, 2015, pp. 11–15.

[261] T. N. Sainath, R. J. Weiss, A. Senior, K. W. Wilson, and O. Vinyals, "Learning the speech front-end with raw waveform CLDNNs," in *Proceedings of Annual Conference of International Speech Communication Association (INTERSPEECH)*, 2015.

[262] T. N. Sainath, O. Vinyals, A. Senior, and H. Sak, "Convolutional, long short-term memory, fully connected deep neural networks," in *Proceedings of International Conference on Acoustics, Speech, and Signal Processing (ICASSP)*, 2015, pp. 4580–4584.

[263] C. Zhang, K. Koishida, and J. H. Hansen, "Text-independent speaker verification based on triplet convolutional neural network embeddings," *IEEE/ACM Transactions on Audio, Speech, and Language Processing*, vol. 26, no. 9, pp. 1633–1644, 2018.

[264] C. Zhang and K. Koishida, "End-to-end text-independent speaker verification with triplet loss on short utterances," in *Proceedings of Annual Conference of International Speech Communication Association (INTERSPEECH)*, 2017, pp. 1487–1491.

[265] J. S. Chung, A. Nagrani, and A. Zisserman, "Voxceleb2: Deep speaker recognition," in *Proceedings Interspeech*, 2018, pp. 1086–1090.

[266] G. Bhattacharya, J. Alam, and P. Kenny, "Adapting end-to-end neural speaker verification to new languages and recording conditions with adversarial training," in *Proceedings IEEE International Conference on Acoustics, Speech, and Signal Processing (ICASSP)*. IEEE, 2019, pp. 6041–6045.

[267] Y.-Q. Yu, L. Fan, and W.-J. Li, "Ensemble additive margin softmax for speaker verification," in *Proceedings IEEE International Conference on Acoustics, Speech, and Signal Processing (ICASSP)*. IEEE, 2019, pp. 6046–6050.

[268] S. Wang, Y. Yang, T. Wang, Y. Qian, and K. Yu, "Knowledge distillation for small foot-print deep speaker embedding," in *Proceedings IEEE International Conference on Acoustics, Speech, and Signal Processing (ICASSP)*. IEEE, 2019, pp. 6021–6025.

[269] K. He, X. Zhang, S. Ren, and J. Sun, "Spatial pyramid pooling in deep convolutional networks for visual recognition," *IEEE Transactions on Pattern Analysis and Machine Intelligence*, vol. 37, no. 9, pp. 1904–1916, 2015.

[270] A. Kurakin, I. Goodfellow, and S. Bengio, "Adversarial machine learning at scale," *arXiv preprint arXiv:1611.01236*, 2016.

[271] A. Makhzani, J. Shlens, N. Jaitly, and I. J. Goodfellow, "Adversarial autoencoders," *CoRR*, vol. abs/1511.05644, 2015. [Online]. Available: http://arxiv.org/abs/1511.05644

[272] Y. Ganin, E. Ustinova, H. Ajakan, P. Germain, H. Larochelle, F. Laviolette, M. Marchand, and V. Lempitsky, "Domain-adversarial training of neural networks," *Journal of Machine Learning Research*, vol. 17, no. 59, pp. 1–35, 2016.

[273] J. C. Tsai and J. T. Chien, "Adversarial domain separation and adaptation," in *Proceedings IEEE MLSP*, Tokyo, 2017.

[274] G. Bhattacharya, J. Monteiro, J. Alam, and P. Kenny, "Generative adversarial speaker embedding networks for domain robust end-to-end speaker verification," in *Proceedings IEEE International Conference on Acoustics, Speech, and Signal Processing (ICASSP)*. IEEE, 2019, pp. 6226–6230.

[275] J. Rohdin, T. Stafylakis, A. Silnova, H. Zeinali, L. Burget, and O. Plchot, "Speaker verification using end-to-end adversarial language adaptation," in *Proceedings IEEE International Conference on Acoustics, Speech, and Signal Processing (ICASSP)*. IEEE, 2019, pp. 6006–6010.

[276] X. Fang, L. Zou, J. Li, L. Sun, and Z.-H. Ling, "Channel adversarial training for cross-channel text-independent speaker recognition," in *Proceedings IEEE International Conference on Acoustics, Speech, and Signal Processing (ICASSP)*, 2019, pp. 6221–6225.

[277] J. Zhou, T. Jiang, L. Li, Q. Hong, Z. Wang, and B. Xia, "Training multi-task adversarial network for extracting noise-robust speaker embedding," in *Proceedings IEEE Interna-*

tional Conference on Acoustics, Speech, and Signal Processing (ICASSP). IEEE, 2019, pp. 6196–6200.

[278] Z. Meng, Y. Zhao, J. Li, and Y. Gong, "Adversarial speaker verification," in *Proceedings IEEE International Conference on Acoustics, Speech, and Signal Processing (ICASSP)*. IEEE, 2019, pp. 6216–6220.

[279] P. S. Nidadavolu, J. Villalba, and N. Dehak, "Cycle-gans for domain adaptation of acoustic features for speaker recognition," in *Proceedings IEEE International Conference on Acoustics, Speech, and Signal Processing (ICASSP)*. IEEE, 2019, pp. 6206–6210.

[280] L. Li, Z. Tang, Y. Shi, and D. Wang, "Gaussian-constrained training for speaker verification," in *Proceedings IEEE International Conference on Acoustics, Speech, and Signal Processing (ICASSP)*. IEEE, 2019, pp. 6036–6040.

[281] Y. Tu, M.-W. Mak and J.-T. Chien, "Variational domain adversarial learning for speaker verification," in *Proceedings of Annual Conference of International Speech Communication Association (INTERSPEECH)*, 2019, pp. 4315–4319.

[282] S. S. Shapiro and M. B. Wilk, "An analysis of variance test for normality (complete samples)," *Biometrika*, vol. 52, no. 3/4, pp. 591–611, 1965.

Index